PRINCIPLES OF CMOS VLSI DESIGN

This book is in the Addison-Wesley VLSI Systems Series.

Lynn Conway and Charles Seitz, *Consulting Editors*

The VLSI Systems Series

Circuits, Interconnections, and Packaging for VLSI by H. B. Bakoglu

Analog VLSI and Neural Systems by Carver Mead

The CMOS3 Cell Library edited by Dennis Heinbuch

Computer Aids for VLSI Design by Steven Rubin

The Design and Analysis of VLSI Circuits by Lance Glasser and Daniel Dobberpuhl

Principles of CMOS VLSI Design by Neil H. E. Weste and Kamran Eshraghian

Also from Addison-Wesley:
An Introduction to VLSI Systems by Carver Mead and Lynn Conway

PRINCIPLES OF CMOS VLSI DESIGN
A Systems Perspective

Second Edition

Neil H. E. Weste
TLW, Inc.

Kamran Eshraghian
University of Adelaide

ADDISON-WESLEY PUBLISHING COMPANY
Reading, Massachusetts • Menlo Park, California • New York
Don Mills, Ontario • Wokingham, England • Amsterdam • Bonn
Sydney • Singapore • Tokyo • Madrid • San Juan • Milan • Paris

Sponsoring Editor: Peter S. Gordon
Production Supervisor: Peggy McMahon
Marketing Manager: Bob Donegan
Manufacturing Supervisor: Roy Logan
Cover Designer: Eileen Hoff
Composition Services: Mike Wile
Technical Art Supervisor: Joseph K. Vetere
Technical Art Consultant: Loretta Bailey
Technical Art Coordinator: Alena B. Konecny

Library of Congress Cataloging-in-Publication Data
Weste, Neil H. E.
    Principles of CMOS VLSI design : a systems perspective / Neil
Weste, Kamran Eshraghian -- 2nd ed.
        p. cm.
    Includes bibliographical references and index.
    ISBN 0-201-53376-6
    1. Intergrated circuits--Very large scale integration--design and
construction   2. Metal oxide semiconductors, Complementary.
I. Eshraghian, Kamran.   . II. Title.
TK7874.W46 1992
621.3'95--dc20   92-16564

Cover Photo: Dick Morton
Cover Art: Neil Weste
Photo Credit: Plates 5, 12, and 13, Melgar Photography, Inc., Santa Clara, CA

AT&T

To Avril, Melissa, Tammy and Nicky
and Shohreh, Michelle, Kylie, Natasha and Jason

ABOUT THE AUTHORS

Neil Weste is President of TLW, Inc., a VLSI engineering company in Burlington, Massachusetts. Before cofounding TLW, he was Director of VLSI Systems at Symbolics, Inc., where he led the team that developed the Ivory Lisp microprocessor and the NS design system. Prior to joining Symbolics, Inc., Weste spent six years at AT&T Bell Labs in Holmdel, New Jersey. He worked one year at the Microelectronics Center of North Carolina, with teaching duties at Duke University and the University of North Carolina, Chapel Hill. Weste received his B. Sc., B.E., and Ph.D. from the University of Adelaide, South Australia.

Kamran Eshraghian is an Associate Professor in Electrical Engineering and Director of the Center for Gallium Arsenide VLSI Technology at The University of Adelaide, South Australia. His research interests include very high performance circuits, systems and architectures with applications in digital signal processing. Eshraghian received his B.Tech., B.E., and Ph.D. from the University of Adelaide. Prior to teaching, Eshraghian was with Philips Ltd. as an IC designer.

PREFACE

In the eight years since this book was first published, CMOS technology has steadily moved to occupy a central position in modern electronic system design. Whether digital systems are high speed, high density, low power, or low cost, CMOS technology finds ubiquitous use in the majority of leading-edge commercial applications. CMOS processes have shrunk, and more automated design tools have become commonplace, leading to far more complex chips operating at much higher speeds than a decade ago. While the basic theory of CMOS design remains unchanged, the emphasis and approach to design have changed. With smaller processes and higher speeds comes an increased emphasis on clocking and power distribution, while with complex chip designs and short time-to-market constraints, less emphasis is now placed on die size and the physical details of chip design. The requirement for higher-quality CMOS chips has also increased the need for good approaches to testing.

This edition was updated with these changes in mind. All chapters have undergone extensive revision, and a new chapter on testing replaces one on symbolic layout. Sections on emerging technologies such as BiCMOS, logic synthesis, and parallel scan testing have been added. The overall emphasis has been to include as much as possible of the engineering (and to some extent, the economic) side of CMOS-system design. The artwork has been completely redone and many new figures have been added. All figures were captured on a CMOS VLSI design system. Thus, where possible, diagrams were checked via simulation or net comparison. The tendency has been to include figures where possible ("a picture is worth a thousand words") to trigger the reader's thinking.

As a text, this book provides students with the necessary background to complete CMOS designs and assess which particular design style to use on a given design, from Field Programmable Gate Arrays to full custom design. For the practicing designer, the book provides an extensive source of reference material that covers contemporary CMOS logic, circuit, design, and processing technology.

In common with the first edition, the text is divided into three main sections. The first deals with basic CMOS logic and circuit design and CMOS processing technology. This includes design issues such as speed, power dissipation, and clocking and subsystem design. The second section deals with design approaches and testing. The final section describes three examples of CMOS module/chip designs to provide working examples of the material presented in the first two parts of the book.

In the eighties, designers struggled with tools, circuit techniques, and technology to build CMOS digital systems that could frequently be mastered by one person. The design issues, for example, related to whether a simulation for a circuit could be done and, if so, how accurately. Or perhaps the success of a project depended on a router or a design-rule checker that could deal with large databases. Today, the technology has moved to a point where, to a first order, the technology always works. Failures in design relate to incomplete specifications, inadequate testing, poor communication between designers in a team, or other issues that are somewhat removed from the detailed engineering that still has to take place. That engineering is supported by well-developed design tools. A significant task to be mastered in today's world (once the basics have been learnt) is to take a specification, turn it into a design, enter the design into a CAD system, test it, have it manufactured, and then be able to ship the product.

Increasingly, CMOS VLSI design is being seen as an ideal medium in which to teach the general digital (and analog) system design principles required in such a design process by introducing such issues as structured design and testing. Coupled with education-based Field Programmable Logic Array tools and prototyping kits, courses can be crafted around the basic principles of CMOS design, such as logic design and delay estimation, and coupled with more advanced topics such as simulation, timing analysis, placement and routing, and testing. With reprogrammable hardware, the concept-to-reality delay is reduced to minutes, and the education dynamics of almost-real-time feedback can only help in the education of tomorrow's system designers. The principles used in these laboratory systems are then applicable, with suitable modifications and information, to real-world products, whether such products employ gate-array, standard-cell, or full-custom CMOS design techniques.

Burlington, Mass.   N. H. E. W.

Technical Note: The text was revised using Microsoft Word 4/5 on an Apple Mac II (8Mb RAM, 1.2Gb disk) from a scanned OCR'ed version of the first-edition text. The figures were captured by the author using the TLW NS VLSI design software (developed at Symbolics) with custom Lisp code for specialized EPS output and for capturing SPICE simulation results. The NS design system was run under the Genera operating system on the Mac II, using a Symbolics MacIvory 2 board (2.6 Mwords physical memory, 400Mb of paging space), and a Symbolics XL 1200 Lisp machine. All design work (symbolic layout and schematic capture, net comparison, SPICE, timing and switch simulation, compaction, and timing analysis) dealt with in the book was completed on these machines. In fact, an interesting example of "the wheel of reincarnation" applies: the first edition of the book was used in part to create the Ivory Lisp microprocessor, while the processor was used in turn to create the second edition of the text.

ACKNOWLEDGMENTS

Bruce Edwards, Chris Terman, Jud Leonard, and Brian Ogilvie of TLW, Inc. supplied a number of the circuit designs used in this edition. They also provided comments and encouragement during the rewrite. Kurt Keutzer provided material on logic synthesis. Tom Knight, André DeHon, and Thomas Simon provided material on high-performance CMOS pads. Jim Cherry provided material on timing analyzers. Fred Rosenberger helped with metastability and provided corrections to the first edition. Don MacLennan provided feedback on the section on design economics. The author would like to acknowledge the support of Amihai Miron and that of Philips Laboratories for permission to include the ghost canceller system example in Chapter 9. Analog Devices provided the means to fabricate and test the A/D described in Chapter 9. Diane DeCastro was of assistance in the early stages of this rewrite with OCR support for the capture of the previous text. Bryan Ackland and Ismail Eldumiati of AT&T Bell Labs also provided assistance.

A number of reviewers were instrumental in determining the direction of this revision, and their detailed comments on the first drafts were much appreciated. In particular, the author would like to thank Don Trotter for sharing his course notes and his detailed comments on the first draft. Lex A. Akers, Jonathan Allen, Andreas Andreou, James H. Aylor, Prithvraj Banaerjee, Hans van den Biggelaar, Ray Chen, Willhelm Eggimann, Joseph Ku, Ronald K. Lomax, and John Uyemura also provided valuable feedback.

Finally, the author would like to thank his family—Avril, Melissa, Tammy, and Nicky—for their continued support and their work on the book through the long period it took to complete this revision.

Burlington, Mass.   N. H. E. W.

The author would like to thank the following people for their input on corrections to the first printing of the second edition: Professor Fred Rosenberger (and staff and students) of Washington University, St. Louis, MO; Professor H. Hwang of Mankato State University, Mankato, MN; Professor Douglas Fouts (and students) of the Naval Postgraduate School, Monterey, CA; Mr. Joseph Skudlarek of Mentor Graphics, Wilsonville, OR.

N.H.E.W.3/94

KEY TO SCHEMATICS USED IN THIS BOOK

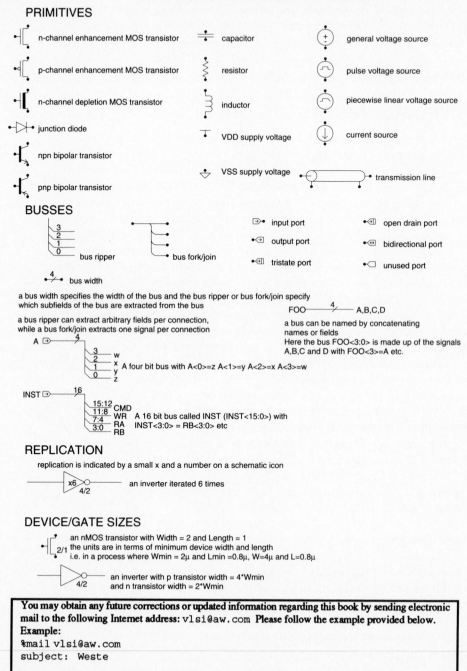

PRIMITIVES

n-channel enhancement MOS transistor

p-channel enhancement MOS transistor

n-channel depletion MOS transistor

junction diode

npn bipolar transistor

pnp bipolar transistor

capacitor

resistor

inductor

VDD supply voltage

VSS supply voltage

general voltage source

pulse voltage source

piecewise linear voltage source

current source

transmission line

BUSSES

3
2
1
0
bus ripper

bus fork/join

4
bus width

input port

output port

tristate port

open drain port

bidirectional port

unused port

a bus width specifies the width of the bus and the bus ripper or bus fork/join specify
which subfields of the bus are extracted from the bus

a bus ripper can extract arbitrary fields per connection,
while a bus fork/join extracts one signal per connection

FOO ——4—— A,B,C,D

a bus can be named by concatenating
names or fields
Here the bus FOO<3:0> is made up of the signals
A,B,C and D with FOO<3>=A etc.

A
4
3
2
1
0
w
x
y
z
A four bit bus with A<0>=z A<1>=y A<2>=x A<3>=w

INST
16
15:12 CMD
11:8 WR
7:4 RA
3:0 RB
A 16 bit bus called INST (INST<15:0>) with
INST<3:0> = RB<3:0> etc

REPLICATION

replication is indicated by a small x and a number on a schematic icon

x6
4/2
an inverter iterated 6 times

DEVICE/GATE SIZES

2/1
an nMOS transistor with Width = 2 and Length = 1
the units are in terms of minimum device width and length
i.e. in a process where Wmin = 2μ and Lmin =0.8μ, W=4μ and L=0.8μ

4/2
an inverter with p transistor width = 4*Wmin
and n transistor width = 2*Wmin

You may obtain any future corrections or updated information regarding this book by sending electronic
mail to the following Internet address: vlsi@aw.com Please follow the example provided below.
Example:
```
%mail vlsi@aw.com
subject: Weste


path return@address.edu <your email address>
send information
```

CONTENTS

3

CMOS PROCESSING TECHNOLOGY   **109**

4

CIRCUIT CHARACTERIZATION
AND PERFORMANCE ESTIMATION

5

CMOS CIRCUIT AND LOGIC DESIGN   261

PART **2**

SYSTEMS DESIGN AND DESIGN METHODS

6

CMOS DESIGN METHODS

7

CMOS TESTING

8

PART **3**
CMOS SYSTEM CASE STUDIES   625

9

CMOS SYSTEM DESIGN EXAMPLES   627

INTRODUCTION TO CMOS TECHNOLOGY

This part introduces the system designer to CMOS technology. Chapter 1 gives a brief overview of CMOS logic design and design representations. Chapter 2 deals with the theory of operation of MOS transistors, CMOS inverters, and BiCMOS inverters. Chapter 3 summarizes current CMOS processing technologies and introduces typical geometric design rules. Chapter 4 introduces techniques to estimate the performance (speed, power) of CMOS circuits. Chapter 5 covers in some depth the various alternatives available to the CMOS circuit designer. It also covers the important subjects of clocking and I/O design.

PART 1

INTRODUCTION TO CMOS CIRCUITS

1.1 A Brief History

Over the past decade, Complementary Metal Oxide Silicon (CMOS for short) technology has played an increasingly important role in the global integrated circuit industry. Not that CMOS technology is that new. In fact, the basic principle behind the MOS field-effect transistor was proposed by J. Lilienfeld as early as 1925, and a similar structure closely resembling a modern MOS transistor was proposed by O. Heil in 1935. Problems with materials foiled these early attempts. Experiments with early field-effect transistors led to the invention of the bipolar transistor. The success of the latter device led to a decline of interest in the MOS transistor. MOS devices remained an oddity until the invention of the silicon planar process around 1960. Although the first MOS calculator was introduced in 1965, material and quality-control problems dogged the expansion of the MOS device into a variety of commercial uses until about 1967.[1] Even then, single-polarity p-type transistors were favored until the emergence of the nMOS-silicon-gate technology in about 1971. The use of both polarity devices on the same substrate was invented by at least two people in the early 1960s. P. K. Weimer, of RCA, filed a patent (U.S. 3,191,061) on May 31st, 1962, issued on May 22, 1965, that featured the elements of modern CMOS flip-flops, demonstrating possible implementations in thin-film-transistor technology. Frank Wanlass, of Fairchild Semiconductor Research and Development, filed a patent on June 18th, 1963, (U.S. Patent 3,356,858), granted on December 5th, 1967,[2] that covered the CMOS concept and three circuits, the inverter, NOR gate, and NAND gate implemented as MOS devices. Wanlass had to build his own nMOS transistors because only pMOS devices were available. The initial circuits were developed using discrete MOS transistors and demonstrated what was for many years the hallmark of CMOS—low power dissipation. The first inverter dissipated nanowatts of power compared with milliwatts for pMOS or the then popular bipolar gates. The low-power attribute led CMOS to be initially used for very low power applications, such as watches. Since the processing technology required in the fabrication of CMOS circuits was more complex and the required silicon area was significantly larger than that for single polarity transistors, CMOS was applied sparingly to general system designs. As nMOS production processes became more complicated, the additional complexity of the basic CMOS process decreased in importance. Additionally, as the technology improved to support very large chip sizes, system designers were faced with power consumption problems. For this, and for other reasons that will become evident during the course of this book, CMOS technology has increased in level of importance to the point where it now clearly holds center stage as the dominant VLSI technology.

The purpose of this book is to provide designers of hardware or software systems with an understanding of CMOS technology, circuit design, layout, and system design sufficient to feel confident with the technology. The text deals with the technology from a digital systems level down to the layout level of detail, thereby providing a view of the technology for both the system level ASIC designer and the full custom designer.

1.2 Book Summary

This book is divided into three main sections. Chapters 1–5 provide a circuit view of the CMOS IC design. In the first chapter, a simplified view of CMOS technology will be taken and some basic forms of logic and memory will be introduced. The aim is to provide an unencumbered picture of the technology without delving into unnecessary detail. A small chip project is used to illustrate the steps in modern CMOS design. Chapter 2 deals at greater depth with the operation of the MOS transistor and the DC operation of the CMOS inverter and a few other basic circuits of interest. It also introduces the junction diode and bipolar transistor. A summary of CMOS processing technology is presented in Chapter 3. The basic processes in current use are described along with some interesting process enhancements. Some representative geometric design rules are also presented in this chapter. Chapter 4 treats the important subject of performance estimation and characterization of circuit operation. This covers circuit speed and power dissipation. A section summarizing some first-order scaling effects is also included. A summary of basic CMOS circuit forms is provided in Chapter 5. Various clocking schemes are discussed, with emphasis on good engineering practice.

The second section of this book comprises Chapters 6–8. These chapters present a *subsystem* view of CMOS design. Chapter 6 focuses on a range of current design methods, identifying where appropriate the issues peculiar to CMOS. Testing and test techniques are discussed in Chapter 7. Chapter 8 is a rather hefty chapter on subsystem design, using for illustration the circuits discussed in Chapter 5. A discussion of a variety of datapath operators opens the chapter. RAMs, ROMs, and the implementation of control logic are then covered.

The book's final section is contained in Chapter 9. It consists of several examples of CMOS VLSI designs that combine many of the design approaches covered in the preceding chapters, and demonstrate some of the practical tradeoffs in the design of actual chips.

The remainder of the current chapter provides a basic introduction to CMOS switches, logic gates, memory elements, and the various abstractions that are used to design integrated systems.

1.3   MOS Transistors

Silicon, a semiconductor, forms the basic starting material for a large class of integrated circuits. An MOS (Metal-Oxide-Silicon) structure is created by superimposing several layers of conducting, insulating, and transistor-forming materials to create a sandwich-like structure. These structures are created by a series of chemical processing steps involving oxidation of the silicon, diffusion of impurities into the silicon to give it certain conduction characteristics, and deposition and etching of aluminum on the silicon to provide interconnection in the same way that a printed wiring board is constructed. This construction process is carried out on a single crystal of silicon, which is available in the form of thin, flat circular wafers around 15cm in diameter. CMOS technology provides two types of transistors (also called *devices* in this text), an n-type transistor (nMOS) and a p-type transistor (pMOS). These are fabricated in silicon by using either *nega-*tively diffused (doped) silicon that is rich in electrons (negatively charged) or *p*ositively doped silicon that is rich in holes (the dual of electrons, and positively charged). After the fabrication steps, a typical MOS structure includes distinct layers called diffusion (silicon which has been doped), polysilicon (polycrystalline silicon used for interconnect), and aluminum, separated by insulating layers. Typical physical structures for the two types of MOS transistors are shown in Fig. 1.1. For the n-transistor, the structure consists of a section of p-type silicon (called the substrate) separating two areas of n-type silicon. This structure is constructed by using a chemical process that changes selected areas in the positive substrate into negative regions rich in electrons. The area separating the n regions is capped with a sandwich consisting of silicon dioxide (an insulator) and a conducting electrode (usually polycrystalline silicon–poly) called the *gate*. Similarly, for the p-transistor the structure consists of a section of n-type silicon separating two p-type areas. In common with the n-transistor, the p-transistor also has a gate electrode. For the purpose of introduction, we will assume that the transistors have two additional connections, designated the *source* and the *drain*, these being formed by the n (p in the case of a p-device)

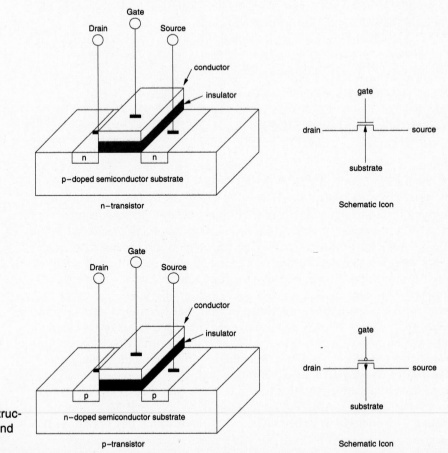

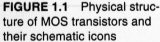

FIGURE 1.1 Physical structure of MOS transistors and their schematic icons

diffused regions. The gate is a control input—it affects the flow of electrical current between the source and the drain. In fact, the drain and source may be viewed as two switched terminals. They are physically equivalent; the name assignment depends on the direction of current flow. For now, we will regard them as interchangeable. The fourth terminal of an MOS transistor, the substrate will be ignored for this discussion.

1.4   MOS Transistor Switches

The gate controls the passage of current between the source and the drain. Simplifying this to the extreme allows the MOS transistors to be viewed as simple on/off switches. In the following discussion, we will assume that a '1' is a high voltage that is normally set to a value between 1.5 and 15 volts and called POWER (PWR) or V_{DD}. The symbol '0' will be assumed to be a low voltage that is normally set to zero volts and called GROUND (GND) or V_{SS}. The strength of the '1' and '0' signals can vary. The "strength" of a signal is measured by its ability to sink or source current. In general, the stronger a signal, the more current it can source or sink. By convention, current is sourced from POWER and GROUND sinks current. Where the terms *output* and *input* are used, an output will be a source of stronger '1's and '0's than an input. The power supplies (V_{DD} and V_{SS}) are the source of the strongest '1's and '0's.

The nMOS switch (N-SWITCH) is shown in Fig. 1.2(a). The conventional schematic icon representation is shown along with that for the switch notation. The gate has been labeled with the signal s, the drain a, and the source b. In an N-SWITCH, the switch is closed or 'ON' if the drain and the

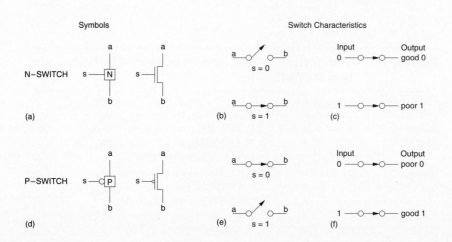

FIGURE 1.2 nMOS and pMOS switch symbols and characteristics

source are connected. This occurs when there is a '1' on the gate. The switch is open or 'OFF' if the drain and source are disconnected. A '0' on the gate ensures this condition. These conditions are summarized in Fig. 1.2(b). An N-SWITCH is almost a perfect switch when a '0' is to be passed from an output to an input (say a to b in Fig. 1.2b). However the N-SWITCH is an imperfect switch when passing a '1.' In doing this, the '1' voltage level is reduced a little (this is explained in Section 2.5). These cases are shown in Fig. 1.2(c). The pMOS switch (P-SWITCH) is shown in Fig. 1.2(d). It has different properties from the N-SWITCH. The P-SWITCH is closed or 'ON' when there is a '0' on the gate. The switch is open or 'OFF' when there is a '1' on the gate. Figure 1.2(e) depicts these conditions. Notice that the pMOS and nMOS switches are ON and OFF for complementary values of the gate signal. We denote this difference for a P-SWITCH by including the inversion bubble in the schematic icon notation. A P-SWITCH is almost perfect for passing '1' signals but imperfect when passing '0' signals. This is illustrated in Fig 1.2(f).

The output logic levels of an N-SWITCH or a P-SWITCH are summarized in Table 1.1.

By combining an N-SWITCH and a P-SWITCH in parallel (Fig. 1.3a), we obtain a switch in which '0's and '1's are passed in an acceptable fashion (Fig. 1.3b). We term this a complementary switch, or C-SWITCH. In a circuit where only a '0' or a '1' has to be passed, the appropriate subswitch (n or p) may be deleted, reverting to a P-SWITCH or an N-SWITCH. Note that a double-rail logic is implied for the complementary switch (the control input and its complement are routed to all switches where necessary. The control signal is applied to the n-transistor and the complement to the p-transistor). The complementary switch is also called a transmission gate or pass gate (complementary). Commonly used schematic icons for the transmission gate are shown in Fig. 1.3(c).

TABLE 1.1   The Output Logic Levels of N-SWITCHES and P-SWITCHES

LEVEL	SYMBOL	SWITCH CONDITION
Strong 1	**1**	P-SWITCH gate = 0, source = V_{DD}
Weak 1	1	N-SWITCH gate = 1, source = V_{DD} or P-SWITCH connected to V_{DD}
Strong 0	**0**	N-SWITCH gate = 1, source = V_{SS}
Weak 0	0	P-SWITCH gate = 0, source = V_{SS} or N-SWITCH connected to V_{SS}
High impedance	Z	N-SWITCH gate = 0 or P-SWITCH gate = 1

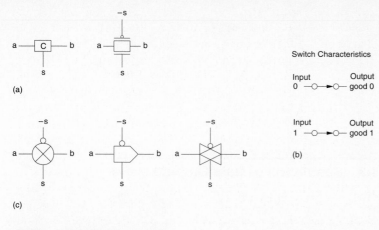

Switch Characteristics

(a)

(b)

(c)

FIGURE 1.3 A complementary CMOS switch

1.5   CMOS Logic

1.5.1   The Inverter

Table 1.2 outlines the truth table required to implement a logical inverter.

If we examine this table, we find that when there is a '0' on the input, there is a '1' at the output. This suggests a P-SWITCH connected from a '1' source (V_{DD}) to the output, as shown in Fig 1.4(a). When there is a '1' on the input, a '0' has to be connected to the output. This suggests the addition of an N-SWITCH between the output and a '0' source (V_{SS}). The completed circuit is shown in Fig 1.4(b). Note that as the lower switch only has to pass a '0' (the V_{SS} source of '0's is stronger than the output of the inverter), only an N-SWITCH is needed. By similar reasoning, the upper switch, which only has to pass a '1,' needs only a P-SWITCH. The transistor schematic and the schematic icon forms for this are shown in Fig 1.4(c). In general, a fully complementary CMOS gate always has an N-SWITCH (pull-down) array to connect the output to '0' (V_{SS}) and a P-SWITCH (pull-up) array to connect the output to '1' (V_{DD}).

When we join a P-SWITCH to an N-SWITCH to form a logic-gate output, both will attempt to exert a logic level at the output. For a structure consisting of a pull-down connected to '0' and a pull-up connected to '1' with

TABLE 1.2   Inverter Truth Table

INPUT	OUTPUT
0	1
1	0

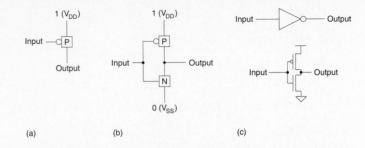

FIGURE 1.4   A CMOS inverter

(a)   (b)   (c)

TABLE 1.3   Resolution of Gate Output Levels

PULL-DOWN OUTPUT	PULL-UP OUTPUT	COMBINED OUTPUT
0	Z	**0**
Z	1	1
Z	Z	Z
0	1	Crowbarred

independent control of the inputs, the possible levels at the output of the pull-up and pull-down are shown in Table 1.3.

From this table it may be seen that the output of a CMOS logic gate can be in four states. The **1** and **0** levels have been encountered with the inverter, where either the pull-up or pull-down are in a high-impedance state and the other structure is turned on. When both pull-up and pull-down are in a high-impedance state, the Z-output state results. This is of importance in multiplexers, storage elements, and bus drivers. The *crowbarred* level exists when both pull-up and pull-down are simultaneously turned on. This causes an "indeterminate" logic level, and also causes static power to be dissipated. It is usually an unwanted condition in any CMOS digital circuit.

1.5.2   Combinational Logic

If two N-SWITCHES are placed in series, the composite switch constructed by this action is closed (or ON) if both switches are closed (or ON) (see Fig 1.5a). This yields an 'AND' function. The corresponding structure for P-SWITCHES is shown in Fig. 1.5(b). The composite switch is closed if both inputs are set to '0'.

When two N-SWITCHES are placed in parallel (Fig. 1.5c), the composite switch is closed if either switch is closed (if either input is a '1'). Thus an 'OR' function is created. The switch shown in Fig. 1.5(d) is composed of two P-SWITCHES placed in parallel. In contrast to the previous case, if either input is a '0' the switch is closed.

By using combinations of these constructions, CMOS combinational gates may be constructed.

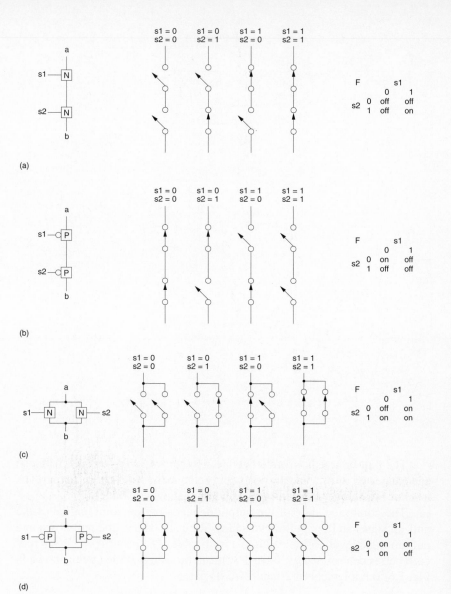

(a)

(b)

(c)

(d)

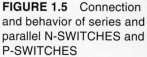

FIGURE 1.5 Connection and behavior of series and parallel N-SWITCHES and P-SWITCHES

1.5.3 The NAND Gate

Figure 1.6 outlines the construction of a 2-input NAND gate using the constructions introduced in Fig. 1.5(a) and Fig. 1.5(d). The pull-down tree is a series pair of N-SWITCHES with one end connected to V_{SS} and the other end connected to the output. The output level of this structure, given the logic levels on the control inputs, is shown in Table 1.4.

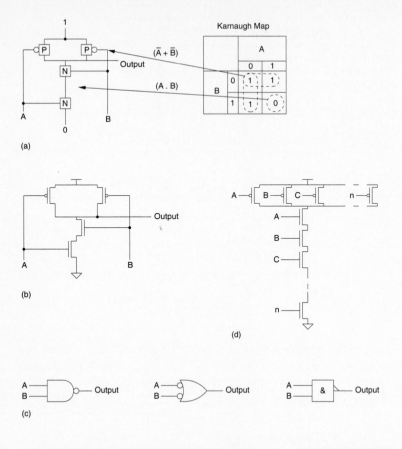

FIGURE 1.6   A CMOS NAND gate

The pull-up tree is a parallel connection pair of P-SWITCHES with one end connected to V_{DD} and the other connected to the NAND gate output. The level of the output of the combined switch is shown in Table 1.5.

The combined state of the output depends on the combination of the pull-up states and the pull-down states. Table 1.6 shows the pull-down and pull-up logic levels combined into a truth table. The resulting logic level of each cell is determined by Table 1.3. It can be seen that the circuit shown in Fig. 1.6(a) implements a 2-input NAND gate.

TABLE 1.4   Nand Gate Pull-down Truth Table

SWITCH A CONTROL INPUT	SWITCH B CONTROL INPUT	OUTPUT
0	0	Z
0	1	Z
1	0	Z
1	1	**0**

TABLE 1.5 Nand Gate Pull-up Truth Table

SWITCH A CONTROL INPUT	SWITCH B CONTROL INPUT	OUTPUT
0	0	**1**
0	1	**1**
1	0	**1**
1	1	Z

TABLE 1.6 2-input CMOS NAND Gate Truth Table

OUTPUT		A INPUT 0	A INPUT 1
B INPUT	0	**1** Z	**1** Z
	1	**1** Z	Z **0**

The circuit and logic schematics for the 2-input NAND gate are shown in Fig. 1.6(b) and Fig. 1.6(c). Note that larger input NAND gates are constructed by placing one N-SWITCH in series on the n side and one P-SWITCH in parallel for each additional input to the gate (Fig. 1.6d).

1.5.4 The NOR Gate

For the NOR gate we will start with the conventional Karnaugh map shown in Table 1.7.

Here the '0's and '1's have been grouped together. The '0' term (pull-down to '0') dictates an OR structure $(A + B)$. Grouping the '1's together results in a structure that requires $\overline{A}.\overline{B}$. This is realized by the series p AND structure. The complemented signals are obtained automatically through the operation of the p-device. The p-structure is the logical dual of the

TABLE 1.7 Karnaugh Map for 2-input NOR Gate

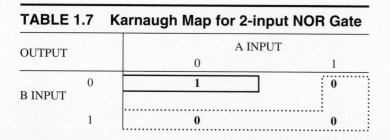

OUTPUT		A INPUT 0	A INPUT 1
B INPUT	0	**1**	**0**
	1	**0**	**0**

TABLE 1.8   2-input CMOS NOR Gate Truth Table

OUTPUT		A INPUT	
		0	1
B INPUT	0	**1**(p) Z(n)	Z(p) **0**(n)
	1	Z(p) **0**(n)	Z(p) **0**(n)

n-structure. This property is used in most complementary CMOS logic gates (but not necessarily in dynamic gates or static gates that dissipate static power). The truth table and switch levels are shown in Table 1.8. By inspection, one may see that this implements the NOR function.

The resulting 2-input NOR gate schematic is shown in Fig. 1.7(a). It is composed from sections introduced in Fig. 1.5(b) and Fig. 1.5(c), according

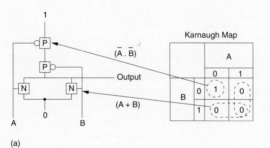

(a)

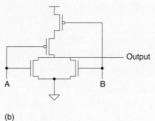

(b)

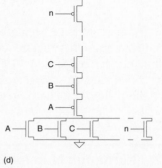

(d)

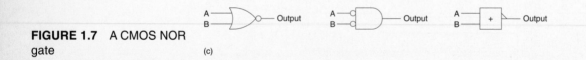

(c)

FIGURE 1.7   A CMOS NOR gate

to the Karnaugh map in Table 1.7. Note that the N- and P-SWITCH combinations are the dual or complement of the combination for the NAND gate. In contrast to the NAND gate, extra inputs are accommodated in the NOR structure by adding N-SWITCHES in parallel and P-SWITCHES in series with the corresponding switch structures (Fig. 1.7d).

Some further points may be noted from this example. First, note that for all inputs there is always a path from '1' or '0' (V_{DD} or V_{SS} supplies) to the output and that the full supply voltages appear at the output. The latter feature leads to a *fully restored* logic family. This simplifies the circuit design considerably. In comparison with other forms of logic, where the pull-up and pull-down switch transistors have to be ratioed in some manner, the transistors in the CMOS gate do not have to be ratioed for the gate to function correctly. Second, there is never a path from the '1' to the '0' supplies for any combination of inputs (in contrast to single channel MOS, GaAs, or bipolar technologies). As we will learn in subsequent chapters, this is the basis for the low static power dissipation in CMOS.

1.5.5 Compound Gates

A compound gate is formed by using a combination of series- and parallel-switch structures. For example, the derivation of the switch connection diagram for the function $F = \overline{((A.B) + (C.D))}$ is shown in Fig. 1.8. The decomposition of this function and generation of the diagram may be approached as follows. For the n-side, take the uninverted expression $((A.B) + (C.D))$. The AND expressions $(A.B)$ and $(C.D)$ may be implemented by series connections of switches, as shown in Fig. 1.8(a). Now, taking these as subswitches and ORing the result requires the parallel connection of these two structures. This is shown in Fig. 1.8(b). For the p-side we invert the expression used for the n-expansion, yielding $((A + B) . (C + D))$. This suggests two OR structures, which are subsequently connected in series. This progression is evident in Fig. 1.8(c) and Fig. 1.8(d). The final step requires connecting one end of the p-structure to '1' (V_{DD}) and the other to the output. One side of the n-structure is connected to '0' (V_{SS}) and the other to the output in common with the p-structure. This yields the final connection diagram (Fig. 1.8e). The schematic icon is shown in Fig. 1.8(f), which shows that this gate may be used in a 2-input multiplexer. If $C = \overline{B}$, then $F = \overline{A}$ if B is true, while $F = \overline{D}$ if B is false.

The Karnaugh map for a second function $F = \overline{((A + B + C). D)}$ is shown in Fig. 1.9(a). The subfunction $(A + B + C)$ is implemented as three parallel N-SWITCHES. This structure is then placed in series with an N-SWITCH with D on the input. The p-function is $(\overline{D} + \overline{A}.\overline{B}.\overline{C})$ (Fig. 1.9b). This requires three P-SWITCHES in series connected in turn in parallel with a P-SWITCH with D on the input. The completed gate is shown in Fig. 1.9(c). In general, CMOS gates may be implemented by analyzing the relevant

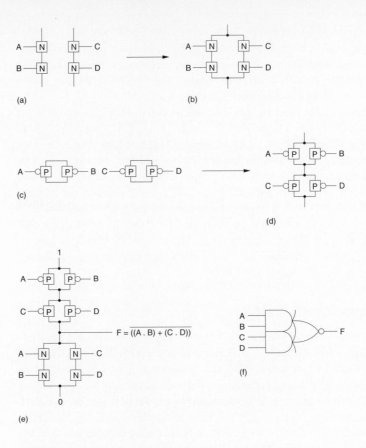

FIGURE 1.8 CMOS compound gate for function $F = \overline{((A.B) + (C.D))}$

Karnaugh map for both n- and p-logic structures and subsequently generating the required series and parallel combinations of transistors.

Often the function required might require the output of the gate to be inverted or one or more of the inputs to be inverted. For instance, if you required a 4-input AND gate, you could implement this with a 4-input NAND

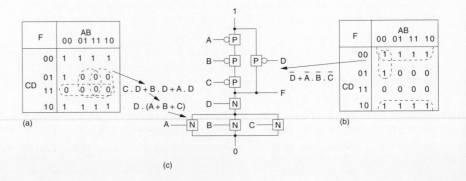

FIGURE 1.9 CMOS compound gate for function $F = \overline{((A + B + C). D)}$

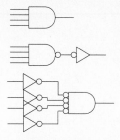

FIGURE 1.10 Various implementations of a CMOS 4-input AND gate

gate and an inverter, or by DeMorgan's theorem, one 4-input NOR gate and four input inverters. Figure 1.10 shows these options. Obviously, in isolation, the former is the most compact implementation. In a larger logic system one may optimize the gates depending on the speed and density required.

Exercises

1. Design CMOS logic gates for the following functions:
   a. $Z = \overline{A.B.C.D}$
   b. $Z = \overline{A + B + C + D}$
   c. $Z = \overline{((A.B.C) + D)}$
   d. $Z = \overline{(((A.B) + C).D)}$
   e. $Z = \overline{(A.B) + C.(A + B)}$

2. Use a combination of CMOS gates to generate the following functions:
   a. $Z = A$ (buffer)
   b. $Z = A.\overline{B} + \overline{A}.B$ (XOR)
   c. $Z = A.B + \overline{A} . \overline{B}$ (XNOR)
   d. $Z = A.\overline{B}.\overline{C} + \overline{A}.\overline{B}.C + \overline{A}.\overline{C}.B + A.B.C$

   (SUM function in binary adder)

3. Design the following logic functions:
   a. A 2:4 decoder defined by
      $Z0 = \overline{A0} . \overline{A1}$
      $Z1 = A0. \overline{A1}$
      $Z2 = \overline{A0} .A1$
      $Z3 = A0.A1$
   b. A 3:2 priority encoder defined by
      $Z0 = \overline{A0} .(A1 + \overline{A2})$
      $Z1 = \overline{A0} . \overline{A1}$

1.5.6 Multiplexers

Complementary switches may be used to select between a number of inputs, thus forming a multiplexer function. Figure 1.11(a) shows a connection dia-

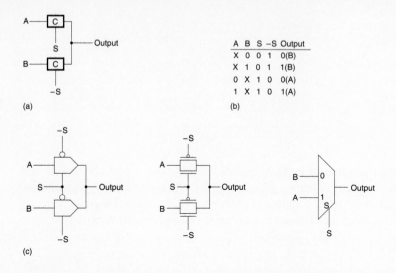

A B S –S Output
X 0 0 1   0(B)
X 1 0 1   1(B)
0 X 1 0   0(A)
1 X 1 0   1(A)

(b)

(a)

(c)

FIGURE 1.11   A 2-input CMOS multiplexer

TABLE 1.9   2-input Multiplexer Karnaugh Map

OUTPUT		S($\bar{S}$) INPUTS	
		0(1)	1(0)
	00	0	0
AB INPUTS	01	1	0
	11	1	1
	10	0	1

gram for a 2-input multiplexer. As the switches have to pass '0's and '1's equally well, complementary switches with n- and p-transistors are used. The Karnaugh map for the structure in Fig. 1.11(a) is shown in Table 1.9. It can be seen that this implements the function

$$\text{Output} = A.S + B.\bar{S}$$

The multiplexer connection in terms of this symbol and transistor symbols is shown in Fig. 1.11(c).

Multiplexers are key components in CMOS memory elements and data manipulation structures.

Exercises

1. Design a 2-input multiplexer (defined by Table 1.9) that uses CMOS logic gates in place of CMOS switches.

2. Design a 4:1 multiplexer

    a. using a combination of CMOS switches and logic gates.

    b. using only CMOS logic gates.

Assess the efficiency of each implementation by counting the total number of switches used in each implementation. Which is more efficient? Why?

1.5.7 Memory—Latches and Registers

We have now constructed enough CMOS structures to enable a memory element to be constructed. A structure called a D latch using one 2-input multiplexer and two inverters is shown in Fig. 1.12(a). It consists of a data input, D, a clock input, CLK, and outputs Q and $-Q$. When $CLK = $ '1', Q is set to D and $-Q$ is set to $-D$ (the logical NOT of D) (Fig. 1.12b). (Note: A number of ways are used to indicate the logical NOT (or inverse) of a signal. The form $\overline{D}$ is often used in texts. However, CAD systems due to the use of an ASCII character set commonly use $-D$, DN or $D.L.$) When CLK is switched to '0', a feedback path around the inverter pair is established (Fig. 1.12c). This causes the current state of Q to be stored. While $CLK = $ '0' the input D is ignored. This is known as a *level-sensitive latch*. That is, the state of the output is dependent on the level of the clock signal. The latch shown is a positive level-sensitive latch. By reversing the control connections to the multiplexer, a negative level-sensitive latch may be constructed.

By combining two level-sensitive latches, one positive sensitive and one negative sensitive, one may construct an *edge-triggered register* as shown in

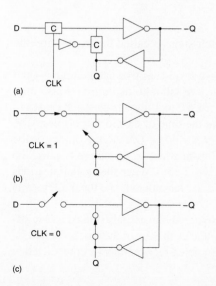

(a)

(b)

(c)

FIGURE 1.12 A CMOS positive-level-sensitive D latch

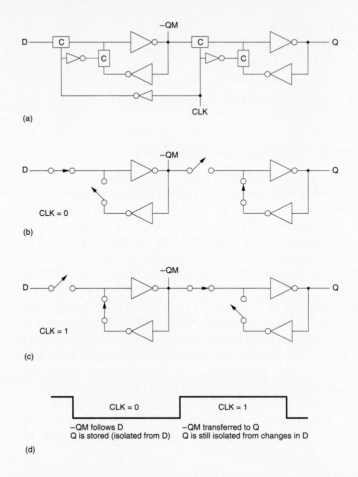

(a)

(b)

(c)

FIGURE 1.13 A CMOS positive edge-triggered D register

(d)

Fig. 1.13(a). By convention, the first latch-stage is called the master and the second is called the slave.

While *CLK* is low, the master negative level-sensitive latch output (*–QM*) follows the *D* input while the slave positive latch holds the previous value (Fig. 1.13b). When the clock transitions from 0 to 1, the master latch ceases to sample the input and stores the *D* value at the time of the clock transition. The slave latch opens, passing the stored master value (*–QM*) to the output of the slave latch (*Q*). The *D* input is prevented from affecting the output because the master is disconnected from the *D* input (Fig. 1.13c). When the clock transitions from 1 to 0, the slave latch locks in the master latch output and the master starts sampling the input again.

Thus this device is a positive edge-triggered register (also called a *D* register or *D* flip-flop) because it samples the input at an edge of the clock. By reversing the latch polarities, a negative edge-triggered register may be constructed.

Apart from RAM and ROM, these structures form the basis of most CMOS storage elements.

Exercises

1. Design a positive level-sensitive D latch in which the Q output, by a signal *RESET,* may be reset to '0' independently of the state of the *CLK* signal (i.e., *RESET*=1 $\rightarrow$ *Q*=0). This is the basis for an asynchronously resettable latch (asynchronous because it resets independent of the state of the clock).

2. Design a positive edge-triggered D register that can be asynchronously set (i.e., *SET*=1 $\rightarrow$ *Q*=1 irrespective of the *CLK* state).

3. Design a positive edge-triggered D register in which the Q output may be reset, synchronous with the clock input.

1.6   Circuit and System Representations

In the previous section we developed the basic functions required in any digital system. Any complex digital system may be eventually broken down into component gates and memory elements by successively subdividing the system in a hierarchical manner. This subdivision may be done manually or may be mechanized. Highly automated techniques now exist for taking very high level descriptions of system behavior and converting the descriptions into a form that eventually may be used to specify how a chip is manufactured. To do this, a specific set of abstractions have been developed to describe integrated electronic systems. These are well captured by the diagram shown in Fig. 1.14.[3] In this figure three distinct design domains are represented by three radial lines. These domains are the

- behavioral,
- structural, and
- physical domains.

The behavioral domain specifies what a particular system does. The structural domain specifies how entities are connected together to effect the prescribed behavior. Finally, the physical domain specifies how to actually build a structure that has the required connectivity to implement the prescribed behavior.

Each design domain may be specified at a variety of levels of abstraction. Concentric circles around the center indicate the various levels of

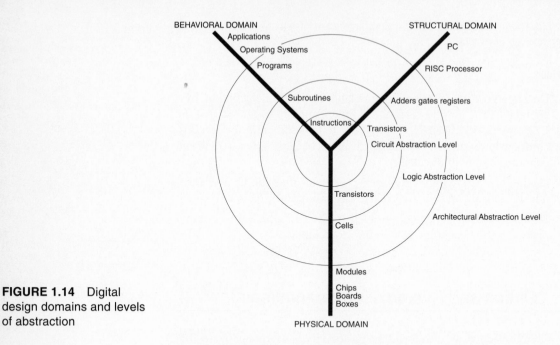

FIGURE 1.14 Digital design domains and levels of abstraction

abstraction that are common in electronic design. From highest to lowest they might include

- architectural,
- algorithmic,
- module or functional block,
- logical,
- switch, and
- circuit levels.

Generally, a design is expressed in terms of the three design domains, while the levels of abstraction that are used vary depending on design style and circuit complexity.

1.6.1 Behavioral Representation

A behavioral representation describes how a particular design should respond to a given set of inputs. Behavior may be specified by Boolean equations, tables of input and output values, or algorithms written in standard high level computer languages or special Hardware Description Languages (HDLs). The latter include VHDL,[4] Verilog®,[5] and ELLA.[6]

Within the behavioral domain there are many levels of abstraction. As one descends through these levels, more information about a particular implementation is evident. For instance, one might start with an algorithm describing a system and progress to a description of the specific hardware registers and the communication between them that is required to implement the original algorithm. At lower levels of abstraction the Boolean equations to implement the algorithm would be specified. The aim of most modern design systems is to convert a specification at as high a level as possible into a system design in minimum time and with maximum likelihood that the system will perform as desired.

Because addition is so pervasive in digital processing, we will use this as an example throughout this section. Imagine an algorithm with the following code segment (this is the heart of the Bresenham algorithm[7] that is the basis for line drawing in most raster scan displays):

```
if (d < 0) d = d+a;
else d = d+b;
```

It is clear that an adder of some precision is required for both the additions and the evaluation of the conditional. The precision of the adder would normally be the precision of the machine on which the code is implemented (unless you are running Lisp, in which case you get numbers of virtually infinite precision). A n-bit adder is constructed by cascading n 1-bit adders. A 1-bit adder has two operand inputs, A and B; a carry input, C; a carry output, CO; and a sum output, S. The truth table for an adder is shown in Table 1.10.

The Boolean equations that implement this function are as follows:

$$S = A.\bar{B}.\bar{C} + \bar{A}.\bar{B}.C + \bar{A}.\bar{C}.B + A.B.C$$
$$CO = A.B + A.C + B.C$$

Implemented in the Verilog® language, at the algorithm level the description for the carry function (CO) might look like this:

```
module carry (co, a, b, c) ;
      output co;
      input a,b,c;
      assign
            co = (a&b)|(a&c)|(b&c);
endmodule
```

TABLE 1.10 Truth Table for 1-bit Binary Adder

A	B	C	CO	S
0	0	0	0	0
0	0	1	0	1
0	1	0	0	1
0	1	1	1	0
1	0	0	0	1
1	0	1	1	0
1	1	0	1	0
1	1	1	1	1

As a primitive, the Boolean behavioral specification for the carry gate might look like this:

```
primitive carry (co, a, b, c) ;
      output co;
      input a,b,c;
      table
      // a b c   co
         1 1 ? : 1 ;
         1 ? 1 : 1 ;
         ? 1 1 : 1 ;
         ? 0 0 : 0 ;
         0 ? 0 : 0 ;
         0 0 ? : 0 ;
      endtable
endprimitive
```

In this description a primitive `carry` is defined with a behavior defined by a table. The first line of the table, 1 1 ? : 1 ;, defines that if a = 1 and b = 1 and c = 1, 0 or X (? means "Don't Care"), the output co = 1. Both of these descriptions are technology independent behavioral specifications at the logical level. No notion of how to implement the function is implied, nor is any speed performance implied.

The speed of a gate is part of the timing behavior of a gate, so it is often necessary to have a means of specifying the rise and fall times of a gate. For the example given above, the functional specification would be augmented with timing information as shown below:

```
module carry (co, a, b, c);
      output co;
      input a,b,c;
      wire #10 co = (a&b)|(a&c)|(b&c);
endmodule
```

This specifies that the co signal changes 10 time units after a or b or c change. More detailed timing may be specified when necessary.

While standard computer languages such as C, Pascal, FORTRAN, and Lisp tend to be much more supported in software development terms, HDLs support specific hardware concepts such as concurrency, time, word size, and bit vectors in a convenient manner. Modern HDLs tend to be increasingly favored for describing VLSI system designs.

1.6.2   Structural Representation

A structural specification specifies how components are interconnected to perform a certain function (or achieve a designated behavior). In general, this description is a list of modules and their interconnections. Whereas in the behavioral domain one could move through a hierarchy of algorithm, register-level transfer, and Boolean equation, at the structural level the levels of abstraction include the module level, the gate level, the switch level, and the circuit level. In each successive level more detail is revealed about the implementation.

In the case of the adder required by the behavioral specification, the cascading of 1-bit adders to form a 4-bit adder needs to be specified. In the Verilog® HDL at the module or functional block level, this further level of detail might be specified for a 4-bit adder as follows:[8]

```
module add4 (s,c4,ci,a,b) ;
     input [3:0]a,b;
     input ci;
     output [3:0]s;
     output c4;
     wire [2:0] co;
          add a0 (co[0],s[0],a[0],b[0],ci);
          add a1 (co[1],s[1],a[1],b[1],co[0]);
          add a2 (co[2],s[2],a[2],b[2],co[1]);
          add a3 (c4,s[3],a[3],b[3],co[2]);
endmodule
```

The first line declares a module called add4. The inputs and outputs are defined in the next four lines. The inputs comprise a carry input, ci and two 4-bit input operands a and b. The outputs are a carry out, c4; and a 4-bit sum, s. A 3-bit internal signal co, is then specified. Following this, four calls to a module called add are made. The add module is specified as follows:

```
module add (co,s,a,b,c) ;
     input a,b,c;
     output s,co;
          sum s1 (s,a,b,c);
          carry c1 (co,a,b,c);
endmodule
```

This specifies that a 1-bit adder (add) is comprised of two modules, a module to compute the sum and a module to compute the carry. In the case of the carry gate the module may be specified at the logic level as follows:

```
module carry (co,a,b,c) ;
     input a,b,c;
     output co;
     wire x,y,z;
          and g1 (x,a,b);
          and g2 (y,a,c);
          and g3 (z,b,c);
          or g4 (co,x,y,z);
endmodule
```

This is a technology-independent structural description, because generic gates have been used and the actual gate implementations have not been specified. In terms of CMOS switches the carry gate might be implemented as follows (Note: In this implementation the pull-up tree is not the exact complement of the pull-down tree. Prove that the pull-up structure shown performs the same function as that constructed by complementing the pull-down tree. This form is used as it reduces the physical layout size.):

```
module carry (co,a,b,c) ;
        input a,b,c;
        output co;
        wire i1,i2,i3,i4,cn;
            nmos n1 (i1,vss,a);
            nmos n2 (i1,vss,b);
            nmos n3 (cn,i1,c);
            nmos n4 (i2,vss,b);
            nmos n5 (cn,i2,a);
            pmos p1 (i3,vdd,b);
            pmos p2 (cn,i3,a);
            pmos p3 (cn,i4,c);
            pmos p4 (i4,vdd,b);
            pmos p5 (i4,vdd,a);
            pmos p6 (co,vdd,cn);
            nmos n6 (co,vss,cn);
endmodule
```

Following the input declaration is a list of transistors with their type and connections in the form given below:

Transistor-type	Name	Output (drain)	Data (source)	Control (gate)
nmos	n1	i1	vss	a

Thus the first statement describes an n-transistor, n1, with drain = i1, gate = a, source = vss.

A graphical view of the adder hierarchy is shown in Fig. 1.15(a) along with the icons that might be used to represent each level of the hierarchy. The carry gate schematic is shown in Fig. 1.15(b) for both logic and switch implementations. Compared with the behavioral description, the structural description for the carry gate details the internal nodes and connections between the primitive gates or switch elements required to actually implement the gate. In the higher level descriptions these connections are irrelevant. In essence, as we ascend the hierarchy a kind of "information hiding" takes place.

However, with the description generated so far, we still do not have the information required to assess the timing behavior of the gate. We need to specify the size of the transistors and the stray capacitance. Because the Verilog® language was designed as a switch-level and gate-level language, it is

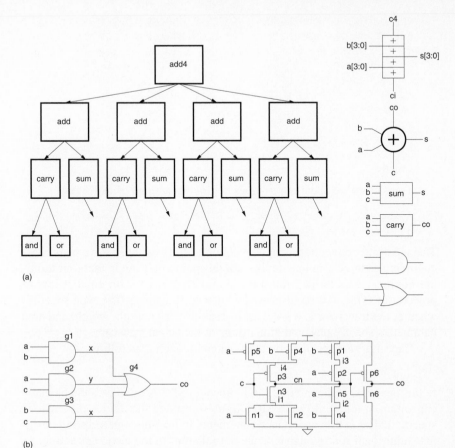

FIGURE 1.15 Structural decomposition of a CMOS adder

not that suitable for structural descriptions at this level of detail. At this point we turn to another common structural language used by the circuit simulator SPICE.[9,10] The specification of the CARRY subcircuit at the circuit level might be represented as follows:

```
.SUBCKT CARRY VDD VSS A B C CO
MN1 I1 A VSS VSS NFET W=8U L=1U AD=8P AS=8P
MN2 I1 B VSS VSS NFET W=8U L=1U AD=8P AS=8P
MN3 CN C I1 VSS NFET W=8U L=1U AD=8P AS=8P
MN4 I2 B VSS NFET W=8U L=1U AD=8P AS=8P
MN5 CN A I2 NFET W=8U L=1U AD=8P AS=8P
MP1 I3 B VDD VDD PFET W=8U L=1U AD=8P AS=8P
MP2 CN A I3 VDD PFET W=8U L=1U AD=8P AS=8P
MP3 CN C I4 VDD PFET W=8U L=1U AD=8P AS=8P
MP4 I4 B VDD VDD PFET W=8U L=1U AD=8P AS=8P
MP5 I4 A VDD VDD PFET W=8U L=1U AD=8P AS=8P
```

```
MP6 CO CN VDD VDD PFET W=16U L=1U AD=16P AS=16P
MN6 CO CN VSS VSS NFET W=16U L=1U AD=16P AS=16P
C1 I1 VSS 50fF
C2 I2 VSS 50fF
C3 I3 VSS 50fF
C4 I4 VSS 50fF
CA A VSS 100fF
CB B VSS 100fF
CC C VSS 100fF
CCO CO VSS 150fF
.ENDS
```

Transistors are specified by lines beginning with an M as follows:

```
Mname drain gate source substrate type W=width L=length
AD=drain area  AS=source area
```

The substrate connection is new. Although MOS switches have been masquerading as three terminal devices till this point, they are in fact four terminal devices; a technical detail that is taken care of with correct layout procedures. This will be discussed further in Chapter 2. The type specifies whether the transistor is a p-device or n-device. The width, length, and area parameters specify physical dimensions of the actual transistors.

Capacitors are specified by lines beginning with C as follows:

```
Cname node-1 node-2 value
```

In this description the internal MOS model in SPICE calculates the parasitic capacitances inherent in the MOS transistor using the device dimensions specified. The extra capacitance statements in the above description designate additional routing capacitance not inherent to the device structure. At the circuit level of structural specification, all connections are specified that are necessary to fully characterize the carry gate in terms of speed, power, and connectivity. In some design systems, SPICE simulations are run with the structural detail shown above. Rise and fall times are measured, and these might be communicated back to the Verilog® logic level in the form of delays. More advanced design systems use a single structural description in which all structural and behavioral information is merged with the use of well-defined data structures.

1.6.3   Physical Representation

The physical specification for a circuit is used to define how a particular part has to be constructed to yield a specific structure and hence behavior. In an IC process, the lowest level of physical specification is the photo-mask information required by the various processing steps in the fabrication process (see Chapter 3).

Similar to the behavioral and structural domains, various levels of abstraction may be defined for the physical representation of a chip. At the module level, the physical layout for the 4-bit adder may be defined by a rectangle or polygon that specifies the outer boundary of all the geometry for the adder, a set of calls to submodules, and a collection of ports. Each port corresponds to an I/O connection in the structural description of the adder. The position, layer, name, and width are specified for each port. For instance, the following is a partial representation of the physical description of the 4-bit adder in an imaginary physical description language.

```
module add4;
      input a[3;0],b[3:0];
      input ci;
      output s[3:0];
      output c4;
      boundary [0,0,100,400];
      port a[0] aluminum width=1 origin=[0,25];
      port b[0] aluminum width=1 origin=[0,75];
      port ci polysilicon width=1 origin=[50,0];
      port s[0] aluminum width=1 origin=[100,50];
          .
      add a0 origin = [0,0];
      add a1 origin = [0,100];
          .
endmodule
```

Here the ports are denoted by the keyword **port**. Calls to a submodule add (a 1-bit adder) are shown. At the lowest level, the physical description makes calls to transistors, wires, and contacts. These in turn specify sized rectangles on the various layers used by the CMOS process. At this stage, we will not dwell on these details but leave discussion of them for Chapter 3. At this time we can think of the physical representation of a CMOS gate as a boundary rectangle with ports. Each port has a position, connection layer, width, and name. This information might be used by an automatic routing program to interconnect this module with others in a design. This level of physical information may be represented symbolically, as shown in Fig. 1.16, for the add4 and add modules. The add module is constructed in a way that allows the interconnection of the carry signal between adder bits via vertical abutment. The inputs enter on the left, and the sum outputs are available on the right.

Exercises

1. Develop a behavioral model for the sum gate of the adder.

2. Design a set of CMOS gates to implement the sum function.

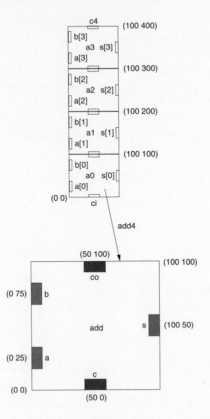

FIGURE 1.16 An abstract view of the physical representation of a CMOS adder

3. Complete the gate-level and switch-level hierarchy for the sum gate in Verilog® or another HDL with which you are familiar.

4. Derive a circuit level model in SPICE or other circuit level HDL to describe the sum gates.

1.7 An Example

CMOS VLSI design today can be highly automated and will become more automated in the future. While very high performance, low power, or low manufacture price generally mean that the CMOS designer has to get close to the technology, medium speed CMOS chips of relatively high complexity may be specified at very high levels. To appreciate how this is done, a simple example will be followed through from behavioral specification to physical design.

1.7.1   Specification

The specification for the function to be implemented is a chip to generate a triangular waveform. The generator is to have 4-bit precision. Thus the output will ramp from 0 to 15 then to 0 and so on. The period of the ramp will be dependent on a clock input, and the period of the triangle waveform will be $32T_{CLK}$. Additionally, the chip will have a reset input and a 4-bit triangle waveform output. The chip will have one V_{DD} and one V_{SS} supply pad. Thus a total of 8 pads are required. The circuit is to fit in an 8-bit mini-DIP package. It has to run from a supply of 5 volts, operate with a clock frequency of up to 1 MHz ($F_{triangle} = \frac{1}{32}$ MHz) and dissipate less than 10 milliwatts. A system-level diagram of the chip in use is shown in Fig. 1.17.

1.7.2   Behavioral Description

One might start out by writing the function as a *C* subroutine. The following code is representative.

```
main (){
triangle ()   }
triangle ()
{       int j = 1;
        int i = wave = 0;
        while (1){
            if (wave == 15) j = -1;
            else if (wave == 0) j = 1;
            wave = wave + j;
            printf(stdout,"i=%d wave=%d\n", i++, wave); }
```

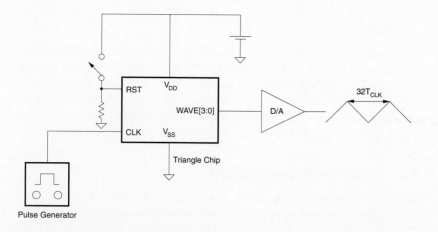

FIGURE 1.17 A system level diagram of the triangle generator chip

Here wave is the current value of the waveform. When executed, this routine would run until aborted.

The behavior may also be coded in a Hardware Description Language (HDL). The following behavioral Verilog® code specifies the triangle generator. (Verilog® is representative of any HDL.)

```
module triangle (wave);
output [0:3]wave;
reg clock;
reg [0:3] acc;
initial begin
    acc = 0;
end
// clock waveform
always
    begin
        #100 clock =0;
        #100 clock =1;
    end
// triangle functionality
always @(posedge, clock)
    begin
        if (wave == 15)
            begin
                inc = -1;
            end
        else if (wave == 0)
            begin
                inc = 1;
            end
        acc = acc + inc;
        wave = acc;
    end
endmodule
```

A number of differences are evident. First, the bit width of the input and output signals has been specified. In addition, the notion of time and a clock have been added. Delays could also be added to the Verilog® description.

With the behavioral description in hand, one would normally run functional simulations to verify the behavior and the compliance with the specification. Once this is satisfactory one moves to the structural domain.

1.7.3   Structural Description

Conversion between the behavioral and structural domains might be done automatically or manually. Automatic programs exist to take HDLs and convert these to structural descriptions (synthesizers—see Chapter 6). In essence, these programs examine the behavior and extract the signal flow graph that will result in the desired behavior. Following this, the logic and registers required by the signal flow graph are synthesized.

In the above program, it may be deduced that an incrementer/decrementer, a 0 detect, a 1s detect, and a register are required. We will assume that the cell library that is available has only an input pad, an output pad, a register, an adder, and some combinational gates. As incrementers and comparators may be implemented with adders, we will assume that the behavioral to structural converter (which might be human!) makes the correct decision and generates the following structural description. The first module represents the complete chip. It has 8 I/O pads and a module called triangle_gen. A clock input is added because it was internally specified in the behavioral description. A reset input is also added because it was implied by the initial conditions in the behavioral description.

```
module chip (wave,clk,rst);
input clk,rst;
output [3:0] wave;
wire [3:0] output;
wire chip_clk, chip_rst;
    input_pad i1 (chip_clk, clk);
    input_pad i2 (chip_rst,rst);
    triangle_gen tr (output, chip_clk, chip_rst);
    output_pad o1 (wave[0], output[0]);
    output_pad o2 (wave[1], output[1]);
    output_pad o3 (wave[2], output[2]);
    output_pad o4 (wave[3], output[3]);
endmodule
```

Next, the structure of the triangle_gen module is specified. It has a module called inc_dec and some state logic to control the increment signal, inc. When inc = 1 the signal output is incremented, while when inc = 0, output is decremented. A 4-input AND gate detects when the output = 15 and a 4-input NOR gate detects output = 0. At either of these two endpoint conditions, a register is loaded with the complement of the current value of inc.

```
module triangle_gen (output,clk,rst);
output [3:0]output;
input clk, rst;
wire inc;
    inc_dec id1 (output,inc,clk,rst);
    and a1 (s1,output[0],output[1],output[2],output[3]);
    nor n1 (s2,output[0],output[1],output[2],output[3]);
    or  o1 (s3,s1,s2);
    xor x1 (s4,s3,inc);
    dreg d1 (inc,s4,clk,rst);
endmodule
```

Module `inc_dec` is then specified. It has an inverter to generate the complement of the `inc` signal four instances of a module called `inc_dec_bit` that are connected to form a 4-bit incrementer or decrementer.

```
module inc_dec (output,inc,clk,rst);
output [3:0] output;
input inc,cin,clk,rst;
wire [3:0]co;
    not inv1 (-inc,inc);
    inc_dec_bit id1 (output[0],co[0],inc,_inc,clk,rst);
    inc_dec_bit id2 (output[1],co[1],co[0],_inc,clk,rst);
    inc_dec_bit id3 (output[2],co[2],co[1],_inc,clk,rst);
    inc_dec_bit id4 (output[3],co[3],co[2],_inc,clk,rst);
endmodule
```

Finally, module `inc_dec_bit` is defined.

```
module inc_dec_bit (sum,co,ci,a,clk,rst);
output sum,co;
input ci,a,clk,rst;
    adder a1 (sum,co,a,q,ci);
    dreg r1 (q,sum,clk,rst);
endmodule
```

It consists of a resettable *D* register (`dreg`) and an adder. At this point the expansion of the hierarchy stops because all of the primitive modules are available as library cells. A conventional (or alternate hierarchy) schematic diagram for the triangle generator is shown in Fig. 1.18. The hierarchy diagram for the structural description of the chip is shown in Fig. 1.19. The cells

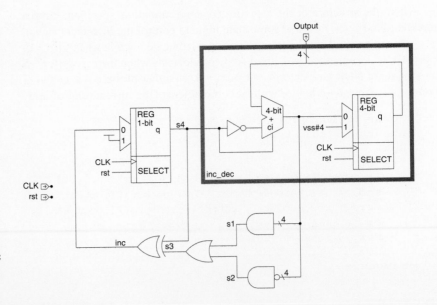

FIGURE 1.18 A schematic diagram for the triangle generator module

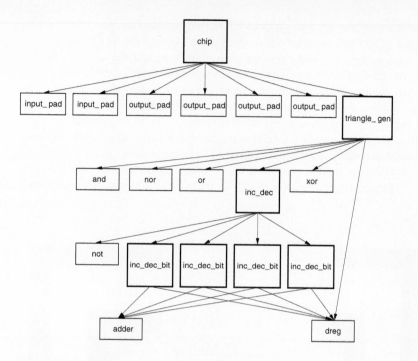

FIGURE 1.19   A hierarchy diagram of the triangle generator chip

at the bottom of the hierarchy tree are called *leaf cells*. These are indicated by thin boxes. There is a corresponding physical description for every cell that is a leaf cell in the structural hierarchy.

Once the structural description is complete, simulations would be run on it to verify compliance with the behavioral specification. Even if the structural description was generated automatically, we now have a logic description into which timing information may be inserted, so resimulation verifies the behavioral at the gate level. One might run a set of other structural tools to check the fault coverage (Chapter 7) of test programs or the estimated size of the chip.

1.7.4   Physical Description

The structural description provides a list of leaf cells and their interconnectivity. The generic leaf cells specified by the structural description are mapped to specific library cells. For instance, a 2-input NOR gate might be mapped to one vendor's cell called NR2 for normal speed and power or NR2H for high speed requirements. This information may be represented in a "net-list." For instance, for the signal s3 in module `tr` denoted by `tr.s3`, an internal signal in the `triangle_gen` module might be represented as follows:

```
tr.s3 o1(OR2.Z) x1(XOR.A)
```

This describes the net `tr.s3`, which has two connections, one to the Z connection of an OR2 gate and one to the A connection of the XOR gate.

This information and a specification of the type and placement of the I/O pads on the chip boundary may be used to construct a physical layout for the circuit. Again this might be done manually or automatically. These days, most "standard cell" layouts are completed using automatic placement and routing algorithms. Placement involves finding the most suitable arrangement in the 2D plane for the cells in the design. Routing then solves the nonplanar interconnection problem created by the placement.

A symbolic representation of the resulting chip layout is shown in Fig. 1.20. The description that specifies this diagram is a hierarchical geometric

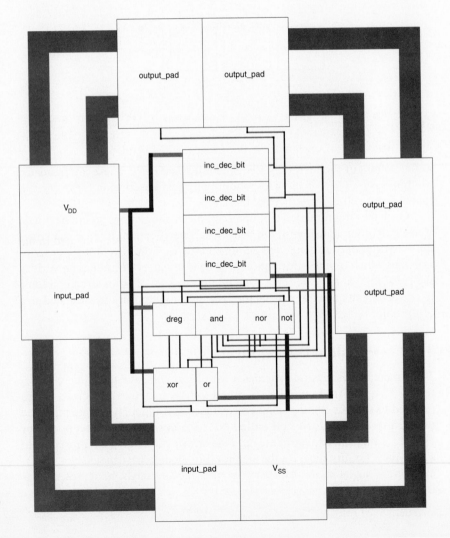

FIGURE 1.20 The triangle generator chip layout

language that at the lowest levels specifies polygons or rectangles that are the interface to the manufacturing process.

Once this description is available, the structural connectivity (that is, the connections between modules) may be reestablished from the geometry by special software. In addition, the actual transistor sizes and physical capacitances may be calculated to a high degree of accuracy. This information may be mapped back to the structural description to place more accurate delays on gates and nets. This *back annotation* is crucial to obtaining accurate performance estimations. The simulations may be run again to confirm the behavior at the required speed and to estimate power dissipation.

Exercise

1. Redesign the triangle waveform generator so that it has a maximum output amplitude of 8 bits and the ability to set the output amplitude by an additional signal called `level`. The redesign includes rewriting

   • the C (or other language) functional model and
   • the Verilog® (or other) structural description.

   The Verilog® structural description detailed above took a "bit-slice" approach to describing the module `inc_dec`. It might be more appropriate here to use a "functional block" hierarchy that consists of 8-bit registers, adders, etc. as in Fig. 1.18.

1.7.5 Summary

To a large extent, most CMOS IC design involves the steps illustrated in the preceding sections with steps either being completed manually or automatically or by a combination of both, depending on the complexity and nature of the chip design. The majority of chip design is and must be highly automated to improve productivity. The overall design flow is summarized in Fig. 1.21. Once a behavior is defined, it is verified against the specification for compliance. The logic corresponding to that behavior is then designed or synthesized. The structural description represented by that logic is compared functionally with the behavioral description to ensure that the logic still does the right thing. The structural description also extends to the transistor level. Finally, a layout may be designed or generated for the particular structural description. The layout provides a path to the manufacturing process. It is verified against the structural description by extracting a structural description from the physical layout and comparing this with the original structural description. Timing and power may be checked at this level. Thus the complete behavioral to physical translation is in theory guaranteed to produce a working silicon. The rest of this book concerns itself with ensuring that the CMOS system designer can make this recipe succeed and contribute new and interesting products to the world.

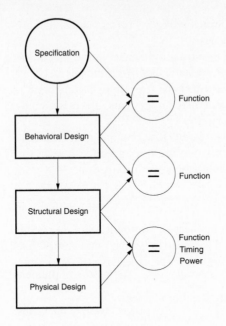

FIGURE 1.21   The design flow for a CMOS chip

1.8   CMOS Scorecard

CMOS technology is one option in a range of technology options available to the electronic system designer. Other options include silicon bipolar technology, Gallium Arsenide (GaAs) technology, and Josephson junction technology. Of the generally available technologies, GaAs technologies often demonstrate the fastest raw gate speed (that is, the speed of an individual gate). Bipolar technologies are not far behind, and advanced CMOS technologies are close behind bipolar. CMOS technologies in general show the highest densities and lowest power per gate. CMOS technology is adequate for analog circuits, but better performing bipolar circuits may be constructed. Straight-forward CMOS technologies are the cheapest to manufacture for high-density digital circuits with moderate analog requirements. Design costs are the cheapest for CMOS technologies due to the large investment already made in design tools and cell libraries. A combination of CMOS and bipolar—called BiCMOS—is emerging as a popular technology, especially for mixed signal chips. For an overwhelming percentage of today's system electronics, CMOS will be the technology of choice. You should be cautioned, though, that it is not the only choice, and you should be aware of the advantages and disadvantages of other technologies when making system-level design decisions.

For CMOS, a brief summary of the main attributes are provided below:

- Fully restored Logic Levels; i.e., output settles at V_{DD} or V_{SS}.
- Transition Times—Rise and fall times are of the same order.
- Memories are implemented both densely and with low power dissipation.
- Transmission Gates pass both logic levels well, allowing use of efficient, widely used logic structures such as multiplexers, latches, and registers.
- Power Dissipation—Almost zero static power dissipation for fully complementary circuits. Power is dissipated during logic transitions.
- Precharging Characteristics—Both n-type and p-type devices are available for precharging a bus to V_{DD} and V_{SS}. Nodes can be charged fully to V_{DD} or alternatively to V_{SS} in a short time.
- Power Supply—Voltage required to switch a gate is a fixed percentage of V_{DD}. Variable range is 1.5 to 15 volts.
- Packing Density—Requires $2n$ devices for n inputs for complementary static gates. Less for dynamic gates or ratioed logic forms.
- Layout—CMOS encourages regular and easily automated layout styles.

At the system level, the reason for CMOS dominance is probably that it is a forgiving technology. Complementary gates are almost guaranteed to function correctly, and if the speed requirements of the application are sufficiently separated from the capability of the technology, timing issues can be simplified. The density of processes and the automated CAD available have reached a point where the majority of systems may be implemented in a highly automated fashion. However, leading-edge products continue to push the technology in terms of cost, density, speed, and power.

1.9   Summary

This chapter introduced a simple switch model for a MOS transistor and developed logic that uses p-transistors and n-transistors available in CMOS processes. This led to a basic discussion of the various levels of representation of circuits and methods of composing these representations. The remainder of this book will expand on the material introduced in this chapter.

1.10   References

1. R. S. C. Cobbold, *Theory and Application of Field Transistors,* New York: Wiley Interscience, 1970.

2. Michael J. Riezenman, "Wanlass's CMOS Circuit," *IEEE Spectrum,* May 1991, p. 44.

3. Daniel D. Gajski, *Silicon Compilation,* Reading, Mass.: Addison-Wesley, 1988.

4. *VHDL Reference Manual,* IEEE Standard 1076, IEEE, Washington, D.C.

5. *Verilog Hardware Description Language Reference Manual—Draft Release 0.1,* Open Verilog International, Sunnyvale, Calif., July 1991.

6. *The ELLA User Manual,* Edition 2.0, Bath, U.K.: Praxis Systems, 1986.

7. J. E. Bresenham, "Algorithm for Computer Control of Digital Plotters," *IBM Systems Journal,* 4:1, 1965, pp. 25–30.

8. *Verilog Hardware Description Language Reference Manual—Draft Release 0.1, op. cit.*

9. L. W. Nagel, "SPICE2: A Computer Program to Simulate Semiconductor Circuits," Memo ERL-M520, University of California, Berkeley, Calif., May 9, 1975.

10. L. W. Nagel, "ADVICE for Circuit Simulation," IEEE International Symposium on Circuits and Systems, Houston, Tex., April 1980.

MOS TRANSISTOR THEORY

2

2.1   Introduction

In Chapter 1 the MOS transistor was introduced in terms of its operation as an ideal switch. In this chapter we will examine the characteristics of MOS transistors in more detail to lay the foundation for predicting the performance of the switches, which is less than ideal. Figure 2.1 shows some of the symbols that are commonly used for MOS transistors. The symbols in Fig. 2.1(a) will be used where it is necessary only to indicate the switch logic required to build a function. If the substrate connection needs to be shown, the symbols in Fig. 2.1(b) will be used. Figure 2.1(c) shows an example of the many symbols that may be encountered in the literature.

This chapter will concentrate on the static or DC operation of MOS transistors. This is the first design goal that must be satisfied to ensure that logic gates operate as logic gates. All circuits are analog in nature and the digital abstraction only remains an abstraction as long as certain design goals are met. Design for timing constraints is covered in Chapter 4.

An MOS transistor is termed a majority-carrier device, in which the current in a conducting channel between the source and the drain is modulated by a voltage applied to the gate. In an n-type MOS transistor (i.e., nMOS), the majority characters are electrons. A positive voltage applied on the gate with respect to the substrate enhances the number of electrons in the channel

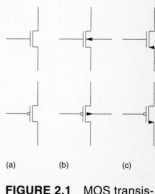

(a)   (b)   (c)

FIGURE 2.1   MOS transistor symbols

(the region immediately under the gate) and hence increases the conductivity of the channel. For gate voltages less than a threshold value denoted by V_t, the channel is cut off, thus causing a very low drain-to-source current. The operation of a p-type transistor (i.e., pMOS) is analogous to the nMOS transistor, with the exception that the majority carriers are holes and the voltages are negative with respect to the substrate.

The first parameter of interest that characterizes the switching behavior of an MOS device is the threshold voltage, V_t. This is defined as the voltage at which an MOS device begins to conduct ("turn on"). We can graph the relative conduction against the difference in gate-to-source voltage in terms of the source-to-drain current (I_{ds}) and the gate-to-source voltage (V_{gs}). These graphs for a fixed drain-source voltage, V_{ds}, are shown in Fig. 2.2. It is possible to make n-devices that conduct when the gate voltage is equal to the source voltage, while others require a positive difference between gate and source voltages to bring about conduction (negative for p-devices). Those devices that are normally cut off (i.e., nonconducting) with zero gate bias (gate voltage–source voltage) are further classed as enhancement-mode devices, whereas those devices that conduct with zero gate bias are called depletion-mode devices. The n-channel transistors and p-channel transistors are the duals of each other; that is, the voltage polarities required for correct operation are the opposite. The threshold voltages for n-channel and p-channel devices are denoted by V_{tn} and V_{tp}, respectively.

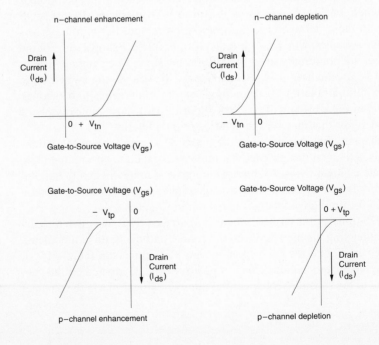

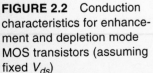

FIGURE 2.2 Conduction characteristics for enhancement and depletion mode MOS transistors (assuming fixed V_{ds})

In CMOS technologies both n-channel and p-channel transistors are fabricated on the same chip. Furthermore, most CMOS integrated circuits, at present, use transistors of the enhancement type.

2.1.1 nMOS Enhancement Transistor

The structure for an n-channel enhancement–type transistor, shown in Fig. 2.3, consists of a moderately doped p-type silicon substrate into which two heavily doped n^+ regions, the *source* and the *drain*, are diffused. Between these two regions there is a narrow region of p-type substrate called the *channel*, which is covered by a thin insulating layer of silicon dioxide (SiO_2) called *gate oxide*. Over this oxide layer is a polycrystalline silicon (polysilicon) electrode, referred to as the *gate*. Polycrystalline silicon is silicon that is not composed of a single crystal. Since the oxide layer is an insulator, the DC current from the gate to channel is essentially zero. Because of the inherent symmetry of the structure, there is no physical distinction between the drain and source regions. Since SiO_2 has relatively low loss and high dielectric strength, the application of high gate fields is feasible.

In operation, a positive voltage is applied between the source and the drain (V_{ds}). With zero gate bias ($V_{gs} = 0$), no current flows from source to drain because they are effectively insulated from each other by the two reversed biased *pn* junctions shown in Fig. 2.3 (indicated by the diode symbols). However, a voltage applied to the gate, which is positive with respect to the source and the substrate, produces an electric field E across the substrate, which attracts electrons toward the gate and repels holes. If the gate voltage is sufficiently large, the region under the gate changes from p-type to n-type (due to accumulation of attracted electrons) and provides a conduction path between the nsource and the drain. Under such a condition, the surface of the underlying p-type silicon is said to be *inverted*. The term *n-channel* is applied to the structure. This concept is further illustrated by Fig. 2.4(a), which shows the initial distribution of mobile positive holes in a p-type silicon substrate of an MOS structure for a voltage, V_{gs}, much less than a voltage, V_t, which is

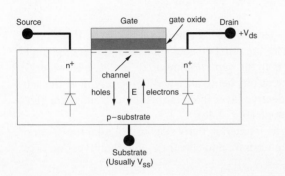

FIGURE 2.3 Physical structure of an nMOS transistor

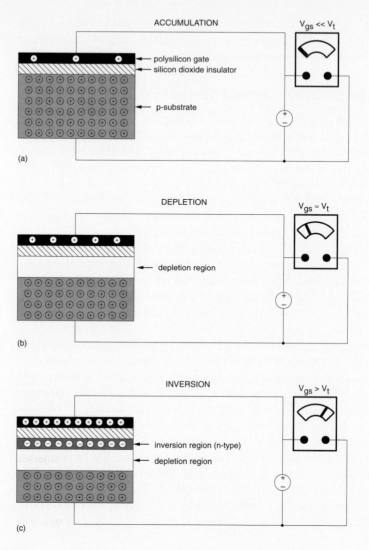

FIGURE 2.4 Accumulation, Depletion and Inversion modes in an MOS structure.

the threshold voltage. This is termed the *accumulation* mode. As V_{gs} is raised above V_t in potential, the holes are repelled causing a depletion region under the gate. Now the structure is in the *depletion* mode (Fig. 2.4b). Raising V_{gs} further above V_t results in electrons being attracted to the region of the substrate under the gate. A conductive layer of electrons in the p substrate gives rise to the name *inversion* mode (Fig. 2.4c).

The difference between a *pn* junction that exists in a bipolar transistor or diode (or between the source or drain and substrate) and the inversion layer substrate junction is that in the *pn* junction, the n-type conductivity is brought about by a metallurgical process; that is, the electrons are introduced into the semiconductor by the introduction of donor ions. In an inversion layer substrate junction, the n-type layer is induced by the electric field E applied to the gate. Thus, this junction, instead of being a metallurgical junction, is a *field-induced* junction.

Electrically, an MOS device therefore acts as a voltage-controlled switch that conducts initially when the gate-to-source voltage, V_{gs}, is equal to the threshold voltage, V_t. When a voltage V_{ds} is applied between source and drain, with $V_{gs} = V_t$, the horizontal and vertical components of the electrical field due to the source-drain voltage and gate-to-substrate voltage interact, causing conduction to occur along the channel. The horizontal component of the electric field associated with the drain-to-source voltage (i.e., $V_{ds} > 0$) is responsible for sweeping the electrons in the channel from the source toward the drain. As the voltage from drain to source is increased, the resistive drop along the channel begins to change the shape of the channel characteristic. This behavior is shown in Fig. 2.5. At the source end of the channel, the full gate voltage is effective in inverting the channel. However, at the drain end of the channel, only the difference between the gate and drain voltages is effective. When the effective gate voltage ($V_{gs} - V_t$) is greater than the drain voltage, the channel becomes deeper as V_{gs} is increased. This is termed the "linear," "resistive," "nonsaturated," or "unsaturated" region, where the channel current I_{ds} is a function of both gate and drain voltages. If $V_{ds} > V_{gs} - V_t$, then $V_{gd} < V_t$ (V_{gd} is the gate to drain voltage), and the channel becomes pinched off— the channel no longer reaches the drain. This is illustrated in Fig. 2.5(c). However, in this case, conduction is brought about by a drift mechanism of electrons under the influence of the positive drain voltage. As the electrons leave the channel, they are injected into the drain depletion region and are subsequently accelerated toward the drain. The voltage across the pinched-off channel tends to remain fixed at ($V_{gs} - V_t$). This condition is the "saturated" state in which the channel current is controlled by the gate voltage and is almost independent of the drain voltage. For fixed drain-to-source voltage and fixed gate voltage, the factors that influence the level of drain current, I_{ds}, flowing between source and drain (for a given substrate resistivity) are:

- the distance between source and drain
- the channel width
- the threshold voltage V_t
- the thickness of the gate-insulating oxide layer
- the dielectric constant of the gate insulator
- the carrier (electron or hole) mobility, μ.

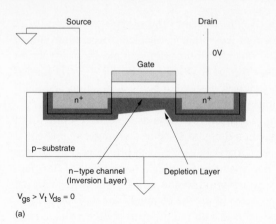

$V_{gs} > V_t \; V_{ds} = 0$

(a)

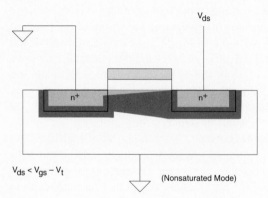

$V_{ds} < V_{gs} - V_t$ (Nonsaturated Mode)

(b)

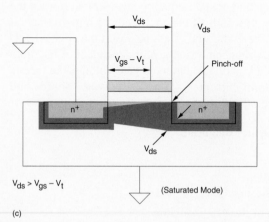

FIGURE 2.5   nMOS device behavior under the influence of different terminal voltages

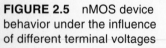

$V_{ds} > V_{gs} - V_t$ (Saturated Mode)

(c)

The normal conduction characteristics of an MOS transistor can be categorized as follows:

- "Cut-off" region: where the current flow is essentially zero (accumulation region).
- "Nonsaturated" region: weak inversion region where the drain current is dependent on the gate and the drain voltage (with respect to the substrate).
- "Saturated" region: channel is strongly inverted and the drain current flow is ideally independent of the drain-source voltage (strong inversion region).

An abnormal conduction condition called avalanche breakdown or punch-through can occur if very high voltages are applied to the drain. Under these circumstances, the gate has no control over the drain current.

2.1.2   pMOS Enhancement Transistor

So far, our discussions have been primarily directed toward nMOS; however, a reversal of n-type and p-type regions yields a p-channel MOS transistor. This is illustrated by Fig. 2.6. Application of a negative gate voltage (w.r.t. source) draws holes into the region below the gate, resulting in the channel changing from n-type to p-type. Thus, similar to nMOS, a conduction path is created between the source and the drain. In this instance, however, conduction results from the movement of holes (versus electrons) in the channel. A negative drain voltage sweeps holes from the source through the channel to the drain.

2.1.3   Threshold Voltage

The threshold voltage, V_t, for an MOS transistor can be defined as the voltage applied between the gate and the source of an MOS device below which the drain-to-source current I_{ds} effectively drops to zero. The word "effec-

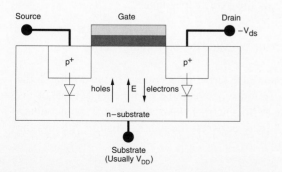

FIGURE 2.6 Physical structure of a pMOS transistor

tively" is used because the drain current never really is zero but drops to a very small value that may be deemed insignificant for the current application (i.e., fast digital CMOS circuits). In general, the threshold voltage is a function of a number of parameters including the following:

- Gate conductor material.
- Gate insulation material.
- Gate insulator thickness–channel doping.
- Impurities at the silicon-insulator interface.
- Voltage between the source and the substrate, V_{sb}.

In addition, the absolute value of the threshold voltage decreases with an increase in temperature. This variation is approximately −4 mV/°C for high substrate doping levels, and −2 mV/°C for low doping levels.[1]

2.1.3.1   Threshold Voltage Equations

Threshold voltage, V_t, may be expressed as

$$V_t = V_{t\text{-}mos} + V_{fb} \tag{2.1}$$

where $V_{t\text{-}mos}$ is the ideal threshold voltage of an ideal MOS capacitor and V_{fb} is what is termed the flat-band voltage. $V_{t\text{-}mos}$ is the threshold where there is no work function difference between the gate and substrate materials.

The MOS threshold voltage, $V_{t\text{-}mos}$, is calculated by considering the MOS capacitor structure that forms the gate of the MOS transistor (see for example[2] or[3]). The ideal threshold voltage may be expressed as

$$V_{t\text{-}mos} = 2\phi_b + \frac{Q_b}{C_{ox}} \tag{2.2}$$

where $\phi_b = \dfrac{kT}{q} ln\left(\dfrac{N_A}{N_i}\right)$, C_{ox} is the oxide capacitance

and $Q_b = \sqrt{2\varepsilon_{Si}qN_A2\phi_b}$ which is called the bulk charge term.

The symbol ϕ_b is the bulk potential, a term that accounts for the doping of the substrate. It represents the difference between the Fermi energy level of the doped semiconductor and the Fermi energy level of the intrinsic semiconductor. The intrinsic level is midway between the valence-band edge and the conduction-band edge of the semiconductor. In a p-type semiconductor the Fermi level is closer to the valence band, while in an n-type semiconductor it is closer to the conduction band. N_A is the density of carriers in the doped semiconductor substrate, and N_i is the carrier concentration in intrinsic (undoped) silicon. N_i is equal to 1.45×10^{10} cm^{-3} at 300°K. The lowercase k is Boltzmann's constant (1.380×10^{-23} J/°K). T is the temperature (°K) and q is the electronic charge (1.602×10^{-19} Coulomb). The expression kT/q equals .02586 Volts at 300°K. The term ε_{Si} is the permittivity of silicon (1.06×10^{-12} Farads/cm). The term C_{ox} is the gate-oxide capacitance, which is inversely proportional to the gate-oxide thickness (t_{ox}). The threshold voltage, $V_{t\text{-}mos}$, is positive for n-transistors and negative for p-transistors.

The flatband voltage, V_{fb}, is given by

$$V_{fb} = \phi_{ms} - \frac{Q_{fc}}{C_{ox}} \tag{2.3}$$

The term V_{fb} is the flat-band voltage. The term Q_{fc} represents the fixed charge due to surface states that arise due to imperfections in the silicon-oxide interface and doping. The term ϕ_{ms} is the work function difference between the gate material and the silicon substrate ($\phi_{gate} - \phi_{Si}$), which may be calculated for an n^+ gate over a p substrate (the normal way for an n transistor) as follows:[4]

$$\phi_{ms} = -(\frac{Eg}{2} + \phi_b) \approx -0.9V \ \ (N_A = 1\times10^{16} \ cm^{-3}) \tag{2.4a}$$

where

$$E_g = \text{is the band gap energy of silicon} \left(1.16 - .704 \times 10^{-3} \frac{T^2}{T + 1108} \right)^5$$

and T is the temperature (°K). For an n^+ poly gate on an n-substrate (a normal p-transistor)

$$\phi_{ms} = -(\frac{Eg}{2} - \phi_b) \approx -0.2V \ \ (N_A = 1\times10^{16} \ cm^{-3}) \tag{2.4b}$$

From these equations it may be seen that for a given gate and substrate material the threshold voltage may be varied by changing the doping concentration of the substrate (N_A), the oxide capacitance (C_{ox}), or the surface state charge (Q_{fc}). In addition, the temperature variation mentioned above may be seen.

It is often necessary to adjust the native (original) threshold voltage of an MOS device. Two common techniques used for the adjustment of the threshold voltage entail varying the doping concentration at the silicon-insulator interface through ion implantation (i.e., affecting Q_{fc}) or using different insulating material for the gate (i.e., affecting C_{ox}). The former approach introduces a small doped region at the oxide/substrate interface that adjusts the flat-band voltage by varying the Q_{fc} term in Eq. (2.3). In the latter approach for instance, a layer of silicon nitride (Si_3N_4) (relative permittivity of 7.5) is combined with a layer of silicon dioxide (relative permittivity of 3.9), resulting in an effective relative permittivity of about 6, which is substantially larger than the dielectric constant of SiO_2. Consequently, for the same thickness as an insulating layer consisting of only silicon dioxide, the dual dielectric process will be electrically equivalent to a thinner layer of SiO_2, leading to a higher C_{ox} value.

In order to prevent the surface of the silicon from inverting in the regions between transistors, the threshold voltage in these field regions is increased by heavily doped diffusions, by implants of the silicon surface, or by making the oxide layer very thick. MOS transistors are self-isolating as long as the surface of the silicon can be inverted under the gate, but not in the regions between devices by normal circuit voltages.

Example

1. Calculate the native threshold voltage for an n-transistor at 300°K for a process with a Si substrate with $N_A = 1.80 \times 10^{16}$, a SiO_2 gate oxide with thickness 200 Å. (Assume $\phi_{ms} = -0.9V$, $Q_{fc} = 0$.)

$$\phi_b = .02586 \, ln\left(\frac{1.8 \times 10^{16}}{1.45 \times 10^{10}}\right)$$

$$= .36 \text{ volts}$$

with

$$C_{ox} = \frac{\varepsilon_{ox}}{t_{ox}}$$

$$= \frac{3.9 \times 8.85 \times 10^{-14}}{0.2 \times 10^{-5}}$$

$$= 1.726 \times 10^{-7} \, Farads/cm^2$$

$$V_t = \phi_{ms} + \frac{\sqrt{2\varepsilon_{Si}qN_A2\phi_b}}{C_{ox}} + 2\phi_b$$

$$= -0.9 + .384 + .72$$

$$= 0.16 \text{ volts}$$

2.1.4 Body Effect

As we have seen so far, all devices comprising an MOS device are made on a common substrate. As a result, the substrate voltage of all devices is normally equal. (In some analog circuits this may not be true.) However, in arranging the devices to form gating functions it might be necessary to connect several devices in series as shown in Fig. 2.7 (for example, the NAND gate shown in Fig. 1.6). This may result in an increase in source-to-substrate voltage as we proceed vertically along the series chain ($V_{sb1} = 0$, $V_{sb2} \neq 0$).

Under normal conditions—that is, when $V_{gs} > V_t$—the depletion-layer width remains constant and charge carriers are pulled into the channel from the source. However, as the substrate bias V_{sb} ($V_{source} - V_{substrate}$) is increased, the width of the channel-substrate depletion layer also increases, resulting in an increase in the density of the trapped carriers in the depletion layer. For charge neutrality to hold, the channel charge must decrease. The resultant effect is that the substrate voltage, V_{sb}, adds to the channel-substrate junction potential. This increases the gate-channel voltage drop. The overall effect is an increase in the threshold voltage, V_t ($V_{t2} > V_{t1}$).

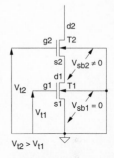

FIGURE 2.7
The effect of substrate bias on series-connected n-transistors

2.2 MOS Device Design Equations

2.2.1 Basic DC Equations

As stated previously, MOS transistors have three regions of operation:

- Cutoff or subthreshold region.
- Nonsaturation or linear region.
- Saturation region.

The ideal (first order, Shockley) equations[6,7,8] describing the behavior of an nMOS device in the three regions are:

The cutoff region:

$$I_{ds} = 0 \qquad V_{gs} \leq V_t \qquad \textbf{(2.5a)}$$

The nonsaturation, linear, or triode region:

$$I_{ds} = \beta \left[(V_{gs} - V_t) V_{ds} - \frac{V_{ds}^2}{2} \right] \qquad 0 < V_{ds} < V_{gs} - V_t \qquad \textbf{(2.5b)}$$

[Although this region is commonly called the linear region, I_{ds} varies linearly with V_{gs} and V_{ds} when the quadratic term $V_{ds}^2/2$ is very small (i.e., $V_{ds} \ll V_{gs} - V_t$).]

The saturation region:

$$I_{ds} = \beta \frac{(V_{gs} - V_t)^2}{2} \qquad 0 < V_{gs} - V_t < V_{ds} \qquad \textbf{(2.5c)}$$

where I_{ds} is the drain-to-source current, V_{gs} is the gate-to-source voltage, V_t is the device threshold, and β is the MOS transistor gain factor. The last factor is dependent on both the process parameters and the device geometry, and is given by

$$\beta = \frac{\mu \varepsilon}{t_{ox}} \left(\frac{W}{L} \right) \qquad \textbf{(2.6)}$$

where μ is the effective surface mobility of the carriers in the channel, ε is the permittivity of the gate insulator, t_{ox} is the thickness of the gate insulator, W is the width of the channel, and L is the length of the channel. The gain factor β thus consists of a process dependent factor $\mu \varepsilon / t_{ox}$, which contains all the process terms that account for such factors as doping density and gate-oxide thickness and a geometry dependent term (W / L), which depends on the actual layout dimensions of the device. The process dependent factor is sometimes written as μC_{ox}, where $C_{ox} = \varepsilon / t_{ox}$ is the gate oxide capacitance. The geometric terms in Eq. (2.6) are illustrated in Fig. 2.8 in relation to the physical MOS structure.

The voltage-current characteristics of the n- and p-transistors in the non-saturated and saturated regions are represented in Fig. 2.9 (with the SPICE circuit for obtaining these characteristics for an n-transistor). Note that we use the absolute value of the voltages concerned to plot the characteristics of the p- and n-transistors on the same axes. The boundary between the linear and saturation regions corresponds to the condition $|V_{ds}| = |V_{gs} - V_t|$ and appears as a dashed line in Fig. 2.9. The drain voltage at which the device

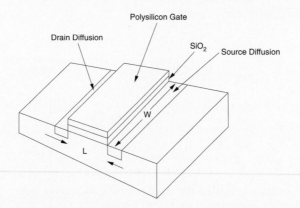

FIGURE 2.8 Geometric terms in the MOS device equation

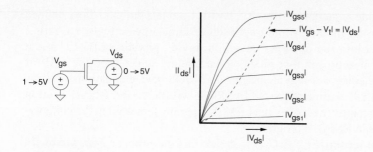

FIGURE 2.9   *VI* characteristics for n- and p-transistors

becomes saturated is called V_{dsat}, or the drain saturation voltage. In the above equations that is equal to $V_{gs} - V_t$.

Example

Typical values (for an n-device) for current ($\sim 1\mu$) processes are as follows:

$\mu_n = 500 \ cm^2 / V\text{-}sec$

$\varepsilon = 3.9\varepsilon_0 = 3.9 \times 8.85 \times 10^{-14} \ F/cm$ (permittivity of silicon dioxide, S_iO_2)

$t_{ox} = 200 \ \text{Å}$

Hence a typical n-device β would be

$$\frac{500 \times 3.9 \times 8.85 \times 10^{-14}}{.2 \times 10^{-5}} \frac{W}{L} = 88.5 \frac{W}{L} \mu A/V^2$$

On the other hand, p-devices have hole mobilities (μ_p) of about 180 $cm^2/V\text{-}sec$, yielding a β of

$$= 31.9 \frac{W}{L} \mu A/V^2$$

Thus the ratio of n-to-p gain factors in this example is about 2.8. This ratio varies from about 2 to 3 depending on the process.

2.2.2   Second Order Effects

Eq. (2.5) represents the simplest view of the MOS transistor DC voltage current equations. There have been many research papers published on more detailed and accurate models that have been created to fill a variety of requirements, such as accuracy, computational efficiency, and the conservation of charge. The circuit simulation program SPICE[9] and its commercial and proprietary derivations generally use a parameter called LEVEL to specify which model equations are used. LEVEL 1 models build on those defined in Eq. (2.5) and include some important second-order effects. LEVEL 2 models calculate the currents based on device physics. LEVEL 3 is a semiempirical approach that relies on parameters selected on the basis of matching the equations to real circuits. The MOS device equations in terms of the LEVEL 1 parameters used in SPICE will be covered here; Section 2.10, in this chapter, describes the LEVEL 3 parameters used in the commercially available HSPICE program.

First the term $\mu\varepsilon/t_{ox}$ (μC_{ox}) is defined as the *process gain factor*. In SPICE this is referred to as *KP*. Depending on the vintage of the process and the type of transistor, *KP* may vary from 10–100 $\mu A/V^2$. In addition, it is not unusual to expect a variation of 10%–20% in *KP* within a given process as a result of variations in starting materials and variation in SiO_2 growth.

2.2.2.1 Threshold Voltage–Body Effect

The threshold voltage V_t is not constant with respect to the voltage difference between the substrate and the source of the MOS transistor. This is known as the *substrate-bias effect* or *body effect*. The expression for the threshold voltage may be modified to incorporate V_{sb}, the difference between the source and the substrate.

$$V_t = V_{fb} + 2\phi_b + \frac{\sqrt{2\varepsilon_{Si}qN_A(2\phi_b + |V_{sb}|)}}{C_{ox}}$$

$$V_t = V_{t0} + \gamma\left[\sqrt{(2\phi_b + |V_{sb}|)} - \sqrt{2\phi_b}\right] \tag{2.7}$$

where V_{sb} is the substrate bias, V_{t0} is the threshold voltage for $V_{sb} = 0$ (Eq. 2.1), and γ is the constant that describes the substrate bias effect. The term ϕ_b is defined in Eq. 2.2.

Typical values for γ lie in the range of 0.4 to 1.2. It may be expressed as

$$\gamma = \frac{t_{ox}}{\varepsilon_{ox}}\sqrt{2q\varepsilon_{Si}N_A} = \frac{1}{C_{ox}}\sqrt{2q\varepsilon_{Si}N_A} \tag{2.8}$$

in which q is the charge on an electron, ε_{ox} is the dielectric constant of the silicon dioxide, ε_{Si} is the dielectric constant of the silicon substrate, and N_A is the doping concentration density of the substrate. The term γ is the SPICE parameter called GAMMA. V_{t0} is the parameter *VTO*, N_A is the parameter *NSUB*, and $\phi_s = 2\phi_b$ is *PHI*, the surface potential at the onset of strong inversion.

Example

For with $N_A = 3 \times 10^{16}$ cm^{-3}, $t_{ox} = 200$Å, $\varepsilon_{ox} = 3.9 \times 8.85 \times 10^{-14}$ F/cm, $\varepsilon_{Si} = 11.7 \times 8.85 \times 10^{-14}$ F/cm, and $q = 1.6 \times 10^{-19}$ Coulomb

$$\gamma = \frac{0.2 \times 10^{-5}}{3.9 \times 8.85 \times 10^{-14}} \sqrt{2 \times 1.6 \times 10^{-19} \times 11.7 \times 8.85 \times 10^{-14} \times 3 \times 10^{16}}$$

$$= .57$$

$$\phi_b = .02586 \; ln \left(\frac{3 \times 10^{16}}{1.5 \times 10^{10}} \right)$$

$$= .375$$

At a V_{sb} of 2.5 volts, and with

$$V_{t2.5} = V_{t0} + .57 \left[\sqrt{.75 + 2.5} - \sqrt{.75} \right]$$

$$= V_{t0} + .53$$

Thus the threshold shifts by approximately half a volt with the source at 2.5 volts for these process parameters.

As we shall learn in Chapter 3, the type of CMOS process can have a large impact on this parameter for both n- and p-transistors. The increase in threshold voltage leads to lower device currents, which in turn leads to slower circuits.

2.2.2.2   Subthreshold Region

The cutoff region described by Eq. (2.5a) is also referred to as the subthreshold region, where I_{ds} increases exponentially with V_{ds} and V_{gs}. Although the value of I_{ds} is very small ($I_{ds} \approx 0$), the finite value of I_{ds} may be used to advantage to construct very low power circuits[10] or it may adversely affect circuits such as dynamic-charge storage nodes. As an approximation, Level 1 SPICE models set the subthreshold current to 0. (See Section 2.11 for the SPICE Level 3 subthreshold equations.)

2.2.2.3   Channel-length Modulation

Simplified equations that describe the behavior of an MOS device assume that the carrier mobility is constant, and do not take into account the variations in channel length due to the changes in drain-to-source voltage, V_{ds}.

For long channel lengths, the influence of channel variation is of little consequence. However, as devices are scaled down, this variation should be taken into account.

When an MOS device is in saturation, the effective channel length actually is decreased such that

$$L_{eff} = L - L_{short} \tag{2.9}$$

where

$$L_{short} = \sqrt{2 \frac{\varepsilon_{Si}}{q N_A} (V_{ds} - (V_{gs} - V_t))}$$

The reduction in channel length increases the (W/L) ratio, thereby increasing β as the drain voltage increases. Thus rather than appearing as a constant current source with infinite output impedance, the MOS device has a finite output impedance. An approximation that takes this behavior into account[11] is represented by the following equation:

$$I_{ds} = \frac{k}{2} \frac{W}{L} (V_{gs} - V_t)^2 (1 + \lambda V_{ds}) \tag{2.10}$$

where k is the process gain factor $\mu \varepsilon / t_{ox}$ and λ is an empirical *channel-length modulation* factor having a value in the range $0.02 V^{-1}$ to $0.005 V^{-1}$. In the SPICE level 1 model λ is the parameter *LAMBDA*.

2.2.2.4 Mobility Variation

The mobility, μ, describes the ease with which carriers drift in the substrate material. It is defined by

$$\mu = \frac{average\ carrier\ drift\ velocity\ (V)}{Electric\ Field\ (E)} \tag{2.11}$$

If the velocity, V, is given in cm/sec, and the electric field, E, in V/cm, the mobility has the dimensions cm^2/V-sec. The mobility may vary in a number of ways. Primarily, mobility varies according to the type of charge carrier. Electrons (negative-charge carriers) in silicon have a much higher mobility than holes (positive-charge carriers), resulting in n-devices having higher current-producing capability than the corresponding p-devices. Mobility decreases with increasing doping-concentration and increasing temperature. The temperature variation becomes less pronounced as the doping density increases. In SPICE μ is specified by the parameter *UO*.

2.2.2.5   Fowler-Nordheim Tunneling

When the gate oxide is very thin, a current can flow from gate to source or drain by electron tunneling through the gate oxide. This current is proportional to the area of the gate of the transistor as follows:[12,13,14]

$$I_{FN} = C_1 WLE_{ox}^2 e^{\frac{-E_0}{E_{ox}}} \tag{2.12}$$

where $E_{ox} \approx \dfrac{V_{gs}}{t_{ox}}$ is the electric field across the gate oxide and

E_0 and C_1 are constants.

This effect limits the thickness of the gate oxide as processes are scaled. However, it is of great use in electrically alterable programmable logic devices.

2.2.2.6   Drain Punchthrough

When the drain is at a high enough voltage with respect to the source, the depletion region around the drain may extend to the source, thus causing current to flow irrespective of the gate voltage (i.e., even if it is zero). This is known as a punchthrough condition. Currently, this effect is used in I/O protection circuits to limit the voltages across internal circuit nodes, although it will impact design as devices are scaled down by requiring that internal circuit voltages be reduced to a point where the effect does not occur.

2.2.2.7   Impact Ionization—Hot Electrons

As the length of the gate of an MOS transistor is reduced, the electric field at the drain of a transistor in saturation increases (for a fixed drain voltage). For submicron gate lengths, the field can become so high that electrons are imparted with enough energy to become what is termed "hot." These hot electrons impact the drain, dislodging holes that are then swept toward the negatively charged substrate and appear as a substrate current. This effect is known as *impact ionization*. Moreover, the electrons can penetrate the gate oxide, causing a gate current. Eventually this can lead to degradation of the MOS device parameters (threshold voltage, subthreshold current, and transconductance), which in turn can lead to the failure of circuits.[15,16,17] While the substrate current may be used in a positive manner to estimate the severity of the hot-electron effect, it can lead to poor refresh times in dynamic memories, noise in mixed signal systems, and possibly latchup. Hot holes do not normally present a problem because of their lower mobility.

The presence of hot electrons has guided CMOS device engineering over the last few years. Chapter 3 shows some examples of the process steps that are used to provide long-lifetime submicron devices at 5 volts. Various circuit techniques that aim at reducing the voltage stress at the drains of n-transistors have also been proposed. Hot electrons will eventually push 3-volt and lower power supplies into prominence in CMOS design as the reduction in drain voltage markedly improves device lifetimes and reliability.

As an illustration of the relative magnitude of the substrate current, the following equation is representative[18] (for an $L = 0.8$ µ, $t_{ox} = 160$Å CMOS process):

$$I_{substrate} = I_{ds}C1 \ (V_{ds} - V_{dsat})^{C2} \tag{2.13}$$

where

$$C1 = 2.24 \times 10^{-5} - .1 \times 10^{-5} V_{ds}$$

$$C2 = 6.4$$

$$V_{dsat} = \frac{V_{tm}L_{eff}E_{sat}}{V_{tm} + L_{eff}E_{sat}}$$

with

$$V_{tm} = V_{gs} - V_{tn} - 0.13V_{bs} - 0.25V_{gs}$$

$$E_{sat} = 1.10 \times 10^{7} + 0.25 \times 10^{7} V_{gs}$$

L_{eff} is the effective channel length in meters.

2.2.3   MOS Models

In Section 2.2.2 we presented the ideal equations that describe the behavior of MOS transistors. While these incorporate some nonideal effects (channel-length modulation, threshold-voltage variation), they may not accurately model a specific device in a particular process. That is especially true for devices that have very small dimensions (gate lengths, gate widths, oxide thicknesses) as the modeling process becomes increasingly 3D in nature. Researchers have developed and refined a wide range of MOS models in an effort to predict more accurately the performance of MOS devices before they are fabricated for varying design scenarios. For instance, one might predict DC currents very accurately from raw process parameters, thus helping predict the behavior of an as yet untested device. However, because of the complexity of the model, it might not be appropriate for a fast-execution-time model that might be needed for digital simulation purposes. In that case, a model based on parameters measured from an actual process might be appropriate.

Depending on the particular circuit level simulator that may be available, a wide variety of MOS simulation models may be used. For instance in one commercial circuit simulator there are over 10 different MOS models.[19] Many semiconductor vendors expend a great deal of effort to model the devices they manufacture. Many times these efforts are aimed at internal circuit simulators and proprietary models. Most CMOS digital foundry operations have been standardized on the LEVEL 3 models in SPICE as the level of circuit modeling that is required for CMOS digital system design. Table 2.1 is a summary of the main SPICE DC parameters that are used in Levels 1, 2, and 3 with representative values for a 1μ n-well CMOS process.

SPICE Level 3 model parameters also include process parameters that are used to calculate *VTO*, *KP*, *GAMMA*, *PHI*, and *LAMBDA* if they are not specified. For instance, if *GAMMA* is not specified, *TOX* and *NSUB* may be used to calculate it. Section 2.11 has a full description of the SPICE LEVEL 3 parameters and their use.

Table 2.1 SPICE DC Parameters

Parameter	nMOS	pMOS	Units	Description
VTO	0.7	0.7	volt	Threshold voltage
KP	8×10^{-5}	2.5×10^{-5}	A/V^2	Transconductance coefficient
GAMMA	.4	.5	$V^{0.5}$	Bulk threshold parameter
PHI	.37	.36	volt	Surface potential at strong inversion
LAMBDA	.01	.01	$volt^{-1}$	Channel length modulation parameter
LD	0.1×10^{-6}	0.1×10^{-6}	meter	Lateral diffusion
TOX	2×10^{-8}	2×10^{-8}	meter	Oxide thickness
NSUB	2×10^{16}	4×10^{16}	$1/cm^3$	Substrate doping density

2.2.4 Small Signal AC Characteristics

The MOS transistor can be represented by the simplified ($V_{sb} = 0$) small-signal equivalent model shown in Fig. 2.10 when biased appropriately. Here the MOS transistor is modeled as a voltage-controlled current source (g_m), an output conductance (g_{ds}), and the interelectrode capacitances. These values may be used, for instance, to calculate voltage amplification factors (gain) or bandwidth characteristics when considered along with other circuit elements.

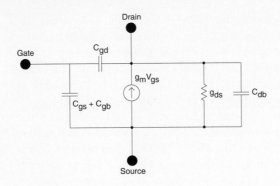

FIGURE 2.10 Small signal model for an MOS transistor

The output conductance (g_{ds}) in the linear region can be obtained by differentiating Eq. (2.5b) with respect to V_{ds}, which results in an output drain-source conductance of

$$g_{ds} = \beta [(V_{gs} - V_t) - 2V_{ds}]$$

$$= V_{ds} \overset{lim}{\to} 0 \approx \beta (V_{gs} - V_t) \tag{2.14}$$

Note that consistent with Eq. (2.5b), V_{ds} must be small compared to V_{gs} for the MOS device to be in a linear operating regime.

On rearrangement, the channel resistance R_c is approximated by

$$R_{c\,(linear)} = \frac{1}{\beta (V_{gs} - V_t)} \tag{2.15}$$

which indicates that it is controlled by the gate-to-source voltage. The relation defined by Eq. (2.15) is valid for gate to source voltages that maintain constant mobility in the channel. In contrast, in saturation [i.e., $V_{ds} \geq (V_{gs} - V_t)$], the MOS device behaves like a current source, the current being almost independent of V_{ds}. This may be verified from Eq. (2.5c) since

$$\frac{dI_{ds}}{dV_{ds}} = \frac{d\left[\frac{\beta}{2} (V_{gs} - V_t)^2 \right]}{dV_{ds}} = 0 \tag{2.16}$$

In practice, however, due to channel shortening (Eq. 2.9) and other effects, the drain-current characteristics have some slope. This slope defines the g_{ds} of the transistor. The output conductance can be decreased by lengthening the channel (i.e., L).

The transconductance g_m expresses the relationship between output current, I_{ds}, and the input voltage, V_{gs}, and is defined by

$$g_m = \frac{dI_{ds}}{dV_{gs}} \,|V_{ds} = \text{constant} \tag{2.17}$$

It is used to measure the gain of an MOS device. In the linear region g_m is given by

$$g_{m\,(linear)} = \beta V_{ds} \tag{2.18}$$

and in the saturation region by

$$g_{m\,(sat)} = \beta\,(V_{gs} - V_t)\,. \tag{2.19}$$

Since transconductance must have a positive value, the absolute value is used for voltages applied to p-type devices.

2.3   The Complementary CMOS Inverter–DC Characteristics

A complementary CMOS inverter is realized by the series connection of a p- and an n-device, as shown in Fig. 2.11. In order to derive the DC-transfer characteristics for the inverter (output voltage, V_{out}, as a function of the inverter, V_{in}), we start with Table 2.1, which outlines various regions of operation for the n- and p-transistors. In this table, V_{tn} is the threshold voltage of the n-channel device, and V_{tp} is the threshold voltage of the p-channel

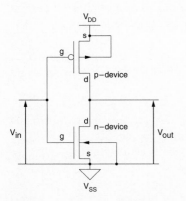

FIGURE 2.11   A CMOS inverter (with substrate connections)

TABLE 2.2   Relations Between Voltages for the Three Regions of Operation of a CMOS Inverter

	CUTOFF	NONSATURATED	SATURATED
p-device	$V_{gsp} > V_{tp}$ $V_{in} > V_{tp} + V_{DD}$	$V_{gsp} < V_{tp}$ $V_{in} < V_{tp} + V_{DD}$ $V_{dsp} > V_{gsp} - V_{tp}$ $V_{out} > V_{in} - V_{tp}$	$V_{gsp} < V_{tp}$ $V_{in} < V_{tp} + V_{DD}$ $V_{dsp} < V_{gsp} - V_{tp}$ $V_{out} < V_{in} - V_{tp}$
n-device	$V_{gsn} < V_{tn}$ $V_{in} < V_{tn}$	$V_{gsn} > V_{tn}$ $V_{in} > V_{tn}$ $V_{dsn} < V_{gs} - V_{tn}$ $V_{out} < V_{in} - V_{tn}$	$V_{gsn} > V_{tn}$ $V_{in} > V_{tn}$ $V_{dsn} > V_{gs} - V_{tn}$ $V_{out} > V_{in} - V_{tn}$

device. The objective is to find the variation in output voltage (V_{out}) for changes in the input voltage (V_{in}).

We begin with the graphical representation of the simple algebraic equations described by Eq. (2.5) for the two inverter transistors shown in Fig. 2.12(a).[20] The absolute value of the p-transistor drain current I_{ds} inverts this characteristic. This allows the *VI* characteristics for the p-device to be reflected about the *x*-axis (Fig. 2.12b). This step is followed by taking the absolute value of the p-device, V_{ds}, and superimposing the two characteristics yielding the resultant curves shown in Fig. 2.12(c). The input/output transfer curve may now be determined by the points of common V_{gs} intersection in Fig. 2.12(c). Thus, solving for $V_{inn} = V_{inp}$ and $I_{dsn} = I_{dsp}$ gives the desired transfer characteristics of a CMOS inverter as illustrated in Fig. 2.13. The switching point is typically designed to be 50 percent of the magnitude of the supply voltage: $\approx V_{DD}/2$. During transition, both transistors in the CMOS inverter are momentarily "ON," resulting in a short pulse of current drawn from the power supply. This is shown by the dotted line in Fig. 2.13.

The operation of the CMOS inverter can be divided into five regions (Fig. 2.13). The behavior of n- and p-devices in each of the regions may be found by using Table 2.2.

Region A.   This region is defined by $0 \leq V_{in} \leq V_{tn}$ in which the n-device is cut off ($I_{dsn} = 0$), and the p-device is in the linear region. Since $I_{dsn} = -I_{dsp}$, the drain-to-source current I_{dsp} for the p-device is also zero. But for $V_{dsp} = V_{out} - V_{DD}$, with $V_{dsp} = 0$, the output voltage is

$$V_{out} = V_{DD} \qquad\qquad (2.20)$$

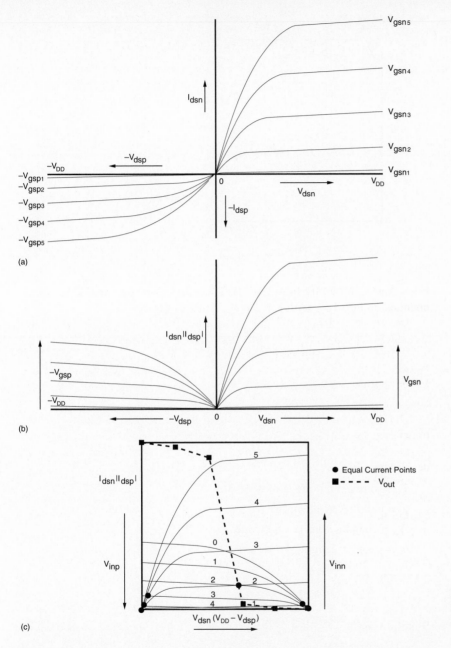

(a)

(b)

(c)

● Equal Current Points

■ - - - - V_{out}

FIGURE 2.12 Graphical derivation of CMOS inverter characteristic

Region B.   This region is characterized by $V_{tn} \leq V_{in} < V_{DD}/2$ in which the p-device is in its nonsaturated region ($V_{ds} \neq 0$) while the n-device is in saturation. The equivalent circuit for the inverter in this region can be represented by a resistor for the p-transistor and a current source for the n-

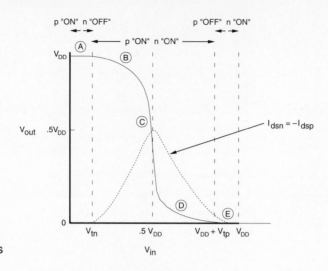

FIGURE 2.13 CMOS inverter DC transfer characteristic and operating regions

transistor as shown by Fig. 2.14(a). The saturation current I_{dsn} for the n-device is obtained by setting $V_{gs} = V_{in}$. This results in

$$I_{dsn} = \beta_n \frac{[V_{in} - V_{tn}]^2}{2} \qquad (2.21)$$

where

$$\beta_n = \frac{\mu_n \, \varepsilon}{t_{ox}} \left(\frac{W_n}{L_n} \right)$$

and

V_{tn} = threshold voltage of n-device

μ_n = mobility of electrons

W_n = channel width of n-device

L_n = channel length of n-device.

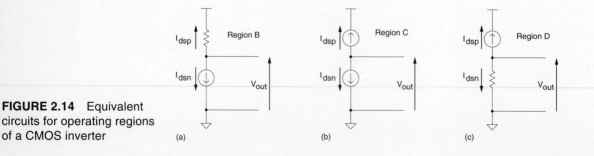

FIGURE 2.14 Equivalent circuits for operating regions of a CMOS inverter

The current for the p-device can be obtained by noting that

$$V_{gs} = (V_{in} - V_{DD})$$

and

$$V_{ds} = (V_{out} - V_{DD})$$

and therefore

$$I_{dsp} = -\beta_p \left[(V_{in} - V_{DD} - V_{tp}) (V_{out} - V_{DD}) - \frac{(V_{out} - V_{DD})^2}{2} \right] , \quad \textbf{(2.22)}$$

where

$$\beta_p = \frac{\mu_p \varepsilon}{t_{ox}} \left(\frac{W_p}{L_p} \right)$$

and

V_{tp} = threshold voltage of p-device
μ_p = mobility of holes
W_p = channel width of p-device
L_p = channel length of p-device.

Substituting

$$I_{dsp} = -I_{dsn}$$

the output voltage V_{out} can be expressed as

$$V_{out} = (V_{in} - V_{tp}) + \sqrt{(V_{in} - V_{tp})^2 - 2 (V_{in} - \frac{V_{DD}}{2} - V_{tp}) V_{DD} - \frac{\beta_n}{\beta_p} (V_{in} - V_{tn})^2}$$

$$\textbf{(2.23)}$$

Region C.   In this region both the n- and p-devices are in saturation. This is represented by the schematic in Fig. 2.14(b) which shows two current sources in series.

The saturation currents for the two devices are given by

$$I_{dsp} = -\frac{\beta_p}{2} (V_{in} - V_{DD} - V_{tp})^2$$

$$I_{dsn} = \frac{\beta_n}{2} (V_{in} - V_{tn})^2$$

with

$$I_{dsp} = -I_{dsn}.$$

This yields

$$V_{in} = \frac{V_{DD} + V_{tp} + V_{tn}\sqrt{\frac{\beta_n}{\beta_p}}}{1 + \sqrt{\frac{\beta_n}{\beta_p}}}. \qquad (2.24)$$

By setting

$$\beta_n = \beta_p \text{ and } V_{tn} = -V_{tp},$$

we obtain

$$V_{in} = \frac{V_{DD}}{2}, \qquad (2.25)$$

which implies that region C exists only for one value of V_{in}. The possible values of V_{out} in this region can be deduced as follows:

$$\text{n-channel:} \quad V_{in} - V_{out} < V_{tn}$$
$$V_{out} > V_{in} - V_{tn}$$
$$\text{p-channel:} \quad V_{in} - V_{out} > V_{tp}$$
$$V_{out} < V_{in} - V_{tp}.$$

Combining the two inequalities results in

$$V_{in} - V_{tn} < V_{out} < V_{in} - V_{tp}. \qquad (2.26)$$

This indicates that with $V_{in} = \frac{V_{DD}}{2}$, V_{out} varies within the range shown. Of course, we have assumed that an MOS device in saturation behaves like an ideal current source with drain-to-source current being independent of V_{ds}. In reality, as V_{ds} increases, I_{ds} also increases slightly; thus region C has a finite slope. The significant factor to be noted is that in region C we have two current sources in series, which is an "unstable" condition. Thus a small input voltage has a large effect at the output. This makes the output transition very steep, which contrasts with the equivalent nMOS inverter characteristic. (See Section 2.4.) The relation defined by Eq. (2.24) is particularly useful since it provides the basis for defining the gate threshold V_{inv}, which corresponds to the state where $V_{out} = V_{in}$. This region also defines the "gain" of the CMOS inverter when used as a small signal amplifier.

Region D.   This region is described by $V_{DD}/2 < V_{in} \leq V_{DD} + V_{tp}$. The p-device is in saturation while the n-device is operating in its nonsaturated region. This condition is represented by the equivalent circuit shown in Fig. 2.14(c). The two currents may be written as

$$I_{dsp} = -\frac{1}{2}\beta_p (V_{in} - V_{DD} - V_{tp})^2$$

and

$$I_{dsn} = \beta_n \left[(V_{in} - V_{tn}) V_{out} - \frac{V_{out}^2}{2} \right]$$

with

$$I_{dsp} = -I_{dsn} .$$

The output voltage becomes

$$V_{out} = (V_{in} - V_{tn}) - \sqrt{(V_{in} - V_{tn})^2 - \frac{\beta_p}{\beta_n}(V_{in} - V_{DD} - V_{tp})^2} \qquad \textbf{(2.27)}$$

Region E.   This region is defined by the input condition $V_{in} \geq V_{DD} - V_{tp}$, in which the p-device is cut off ($I_{dsp} = 0$), and the n-device is in the linear mode. Here, $V_{gsp} = V_{in} - V_{DD}$, which is more positive than V_{tp}. The output in this region is

$$V_{out} = 0. \qquad \textbf{(2.28)}$$

From the transfer curve of Fig. 2.13, it may be seen that the transition between the two states is very steep. This characteristic is very desirable because the noise immunity is maximized. This is covered in more detail in Section 2.3.2. For convenience, the characteristics associated with the five regions are summarized in Table 2.3.

TABLE 2.3   Summary of CMOS Inverter Operation

REGION	CONDITION	p-device	n-device	OUTPUT
A	$0 \leq V_{in} < V_{tn}$	nonsaturated	cutoff	$V_{out} = V_{DD}$
B	$V_{tn} \leq V_{in} < \dfrac{V_{DD}}{2}$	nonsaturated	saturated	Eq. (2.23)
C	$V_{in} = \dfrac{V_{DD}}{2}$	saturated	saturated	$V_{out} \neq f(V_{in})$
D	$\dfrac{V_{DD}}{2} < V_{in} \leq V_{DD} - |V_{tp}|$	saturated	nonsaturated	Eq. (2.27)
E	$V_{in} > V_{DD} - |V_{tp}|$	cutoff	nonsaturated	$V_{out} = V_{SS}$

2.3.1 β_n/β_p Ratio

In order to explore the variations of the transfer characteristic as a function of β_n/β_p, the transfer curve for several values of β_n/β_p are plotted in Fig. 2.15(a). Here, we note the gate-threshold voltage, V_{inv}, where $V_{in} = V_{out}$ is

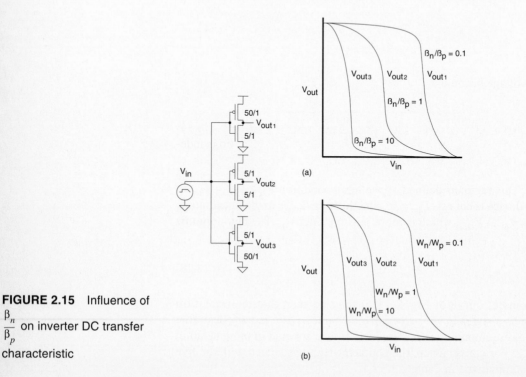

FIGURE 2.15   Influence of $\dfrac{\beta_n}{\beta_p}$ on inverter DC transfer characteristic

dependent on β_n/β_p. Thus, for a given process, if we want to change β_n/β_p, we need to change the channel dimensions, i.e., channel-length L and channel-width W. From Fig. 2.15(a) it can be seen that as the ratio β_n/β_p is decreased, the transition region shifts from left to right; however, the output voltage transition remains sharp (compare to the inverter responses in Figures 2.19, 2.21, 2.23, and 2.24). For the CMOS inverter a ratio of

$$\frac{\beta_n}{\beta_p} = 1 \qquad (2.29)$$

may be desirable since it allows a capacitive load to charge and discharge in equal times by providing equal current-source and -sink capabilities. This will be discussed further in Chapter 4. For interest, the inverter transfer curve is also plotted (Figure 2.15b) for W_n/W_p (the width of the n- and p-transistors). This shows a relative shift to the left compared with the β ratioed case because the p-device has inherently lower gain.

Temperature also has an effect on the transfer characteristic of an inverter.[21] As the temperature of an MOS device is increased, the effective carrier mobility, μ, decreases. This results in a decrease in β, which is related to temperature T by

$$\beta \alpha T^{-1.5} \qquad (2.30)$$

Therefore

$$I_{ds} \alpha T^{-1.5} \qquad (2.31)$$

Since the voltage transfer characteristics depend on the ratio β_n/β_p, and the mobility of both holes and electrons are similarly affected, this ratio is independent of temperature to a good approximation. Both V_{tn} and V_{tp} decrease slightly as temperature increases, and the extent of region A is reduced while the extent of region E increases. Thus the overall transfer characteristics of Fig. 2.15 shift to the left as temperature increases. Based on the figures given earlier, if the temperature rises by 50°C , the thresholds drop by 200mV each. This would cause a .2 V shift in the input threshold of the inverter (although due to the idealized model, less shift is seen in practice).

2.3.2   Noise Margin

Noise margin is a parameter closely related to the input-output voltage characteristics. This parameter allows us to determine the allowable noise voltage on the input of a gate so that the output will not be affected. The specification most commonly used to specify noise margin (or noise immunity) is in terms of two parameters—the *LOW* noise margin, NM_L, and the *HIGH* noise margin, NM_H. With reference to Fig. 2.16, NM_L is defined as the difference in magnitude between the maximum LOW output voltage of the driving gate and the maximum input LOW voltage recognized by the driven gate. Thus

$$NM_L = \left| V_{ILmax} - V_{OLmax} \right| . \tag{2.32}$$

The value of NM_H is the difference in magnitude between the minimum HIGH output voltage of the driving gate and the minimum input HIGH voltage recognized by the receiving gate. Thus

$$NM_H = \left| V_{OHmin} - V_{IHmin} \right| , \tag{2.33}$$

where

V_{IHmin} = minimum HIGH input voltage

V_{ILmax} = maximum LOW input voltage

V_{OHmin} = minimum HIGH output voltage

V_{OLmax} = maximum LOW output voltage.

These definitions are illustrated in Fig. 2.16.

Generally, it is desirable to have $V_{IH} = V_{IL}$ and for this to be a value that is midway in the "logic swing," V_{OL} to V_{OH}. This implies that the transfer characteristic should switch abruptly; that is, there should be high gain in the

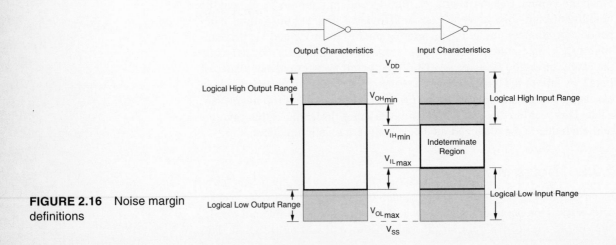

FIGURE 2.16   Noise margin definitions

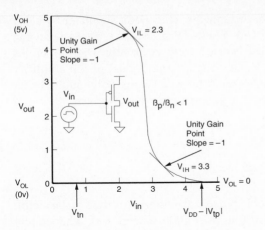

FIGURE 2.17 CMOS inverter noise margins

transition region. For the purpose of calculating noise margins, the transfer characteristic of a typical inverter and the definition of voltage levels V_{IL}, V_{OL}, V_{IH}, V_{OH} are shown in Fig. 2.17. To determine V_{IL}, we note that the inverter is in region B of operation, where the p-device is in its linear region while the n-device is in saturation. The V_{IL} is found by determining the unity gain point in the inverter transfer characteristic where the output transitions from V_{OH}. Similarly, V_{IH} is found by using the unity gain point at the V_{OL} end of the characteristic. For the inverter shown the NM_L is 2.3 volts while the NM_H is 1.7 volts.[22]

Note that if either NM_L or NM_H for a gate are reduced ($\approx 0.1\ V_{DD}$), then the gate may be susceptible to switching noise that may be present on the inputs. Apart from considering a single gate, one must consider the net effect of noise sources and noise margins on cascaded gates in assessing the overall noise immunity of a particular system. This is the reason to keep track of noise margins. Quite often noise margins are compromised to improve speed. Circuit examples later in this book will illustrate this trade-off.

2.3.3   The CMOS Inverter As an Amplifier

It should be noted that the CMOS inverter when used as a logic element is in reality an analog amplifier operated under saturating conditions. In region C in Fig. 2.14, the CMOS inverter acts as an inverting linear amplifier with a characteristic of

$$V_{out} = -AV_{in} \tag{2.34}$$

where A is the stage gain.

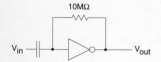

FIGURE 2.18 The CMOS inverter as an amplifier

This region may be further examined with a circuit simulator by using the circuit shown in Fig. 2.18, with a high-value resistor between input and output (10M Ω). The input is DC isolated using a capacitor. The gain of this amplifier is estimated by using the small-signal model of the amplifier shown in Fig. 2.10. This circuit is valid for small signals around the linear operating point of the amplifier. The gain is approximately given by

$$
\begin{aligned}
A &= g_{mtotal} R_{dseffective} \\
&= (g_{mn} + g_{mp})(r_{dsn} \parallel r_{dsp}) \\
&= g_m r_{ds} \text{ (if } g_{mn} = g_{mp} \text{ and } r_{dsn} = r_{sdp})
\end{aligned}
\tag{2.35}
$$

This gain is very dependent on the process and transistors used in the circuit but can be in the range from 100 to over 1000. The gain is enhanced by lengthening the transistors to improve the r_{ds} values. This improvement comes at the expense of speed and bandwidth of the amplifier.

2.4 Static Load MOS Inverters

Apart from the CMOS inverter, there are many other forms of MOS inverter that may be used to build logic gates. Figure 2.19(a) shows a generic nMOS inverter that uses either a resistive load or a constant current source. For the resistor case, if we superimpose the resistor-load line on the VI characteristics of the pull-down transistor (Fig. 2.19b), we can see that at a V_{gs} of 5 volts, the output is some small V_{ds} (V_{OL}) (Fig. 2.19c). When $V_{gs} = 0$ volts, V_{ds} rises to 5 volts. As the resistor is made larger, the V_{OL} decreases and the current flowing when the inverter is turned on decreases. Correspondingly, as the load resistor is decreased in value, the V_{OL} rises and the on current rises. Selection of the resistor value would seek a compromise between V_{OL}, the current drawn and the pull-up speed, which vary with the value of the load resistor.

The resistor- and current-source-load inverters shown in Fig. 2.19 are normally implemented using transistors in CMOS processes. In some memory processes, resistors are implemented using highly resistive undoped polysilicon. When transistors are used the inverter is called a saturated load inverter if the load transistor is operated in saturation as a constant current source. If the load transistor is biased for use as a resistor, then it is called an unsaturated load inverter.

In this section we will examine a number of static load inverters that one can implement in CMOS processes. Usually the reason for doing this is to reduce the number of transistors used for a gate to improve density and/or to lower dynamic power consumption.

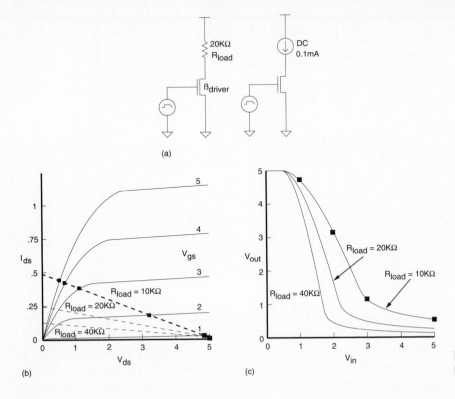

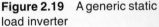

Figure 2.19   A generic static load inverter

2.4.1   The Pseudo-nMOS Inverter

Figure 2.20(a) shows an inverter that uses a p-device pull-up or load that has its gate permanently grounded. An n-device pull-down or driver is driven with the input signal. This is roughly equivalent to the use of a depletion load in nMOS technology (which preceded CMOS technology as a major systems technology) and is thus called "pseudo-nMOS." This circuit is used in a variety of CMOS logic circuits. Similar to the complementary inverter, a graphical solution to the transfer characteristic is shown in Fig. 2.20(b) for various sized p-devices for a particular CMOS process. This shows that the ratio of β_n/β_p affects the shape of the transfer characteristic and the V_{OL} of the inverter (shown in Fig. 2.20c). Figure 2.20(d) shows that when the driver is turned on, a constant DC current flows in the circuit. This is to be contrasted with the CMOS inverter in which no DC current flows when the input is either the terminal high or low state. The importance of whether DC current flows, and hence whether one can use the pseudo-nMOS inverter, depends on the application. CMOS watch circuits rely on the fact that when the circuit is not switching, no current is drawn from the small battery that powers

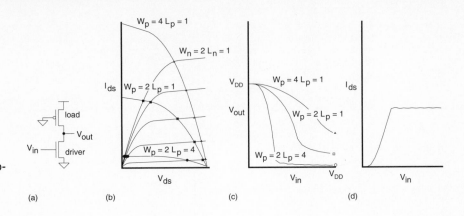

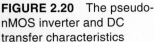

FIGURE 2.20 The pseudo-nMOS inverter and DC transfer characteristics

the watch. In this application, having circuits that consumed DC current would not be advisable. Similarly in circuits which required a power-down mode (as in palmtop or portable computers) one might not want such circuits. Finally, the fact that CMOS complementary circuits do not draw DC current has led some semiconductor manufacturers to have a gross test of CMOS chips that tests the DC current of a chip (IDDQ testing—see Chapter 7). If there is DC current, they assume there is some fault internally and have to do no more testing of that die. Notwithstanding these applications where pseudo-nMOS gates are not applicable, they do find wide application in high-speed circuits and circuits that require large fan-in NOR gates. Even in DC power critical applications, the pseudo-nMOS gate may be used by selectively grounding the gate of the p-device pull-up transistor. (*Note:* The output voltage of a pseudo-nMOS inverter with both driver and load transistors turned off will depend on the subthreshold characteristics of the transistors. This should be rigorously simulated if contemplated, or the output should be clamped to a known voltage.)

For the circuit shown in Fig. 2.20 the current in the n driver transistor is given by

$$I_{dsn} = \frac{\beta_n}{2}(V_{inv} - V_{tn})^2 \ (V_{out} > V_{in} - V_{tn}).$$

The p-device I_{ds} with $V_{gsp} = -V_{DD}$ is

$$I_{dsp} = \beta_p\left[(-V_{DD} - V_{tp})(V_{out} - V_{DD}) - \frac{(V_{out} - V_{DD})^2}{2}\right].$$

Equating the two currents we obtain

$$\frac{\beta_n}{2}(V_{in} - V_{tn})^2 = \beta_p \left[(-V_{DD} - V_{tp})(V_{out} - V_{DD}) - \frac{(V_{out} - V_{DD})^2}{2} \right].$$

Solving for V_{out},

$$V_{out} = -V_{tp} + \sqrt{(V_{DD} + V_{tp})^2 - C} \qquad \textbf{(2.36)}$$

where $C = k(V_{in} - V_{tn})^2$

and $k = \dfrac{\beta_n}{\beta_p}$

also

$$\frac{\beta_n}{\beta_p} = \frac{(V_{DD} + V_{tp})^2 - (V_{out} + V_{tp})^2}{(V_{in} - V_{tn})^2} \qquad \textbf{(2.37)}$$

Figure 2.21(a) shows two cascaded pseudo-nMOS inverters. For equal noise margins, the gate-threshold voltage V_{inv} might be set to approximately $0.5V_{DD}$. (Another criteria might set V_{inv} to be halfway between V_{IL} and V_{IH}.) At this operating point, the n-device (pull-down) is in saturation ($0 < V_{gsn} - V_{tn} < V_{dsn}$), and the p-device (pull-up) is in the linear mode of operation ($0 < V_{dsp} < V_{gsp} - V_{tp}$).

With $V_{inv} = 0.5V_{DD}$, $V_{tn} = |V_{tp}| = 0.2V_{DD}$, $V_{DD} = 5$ volts, the following result is obtained

$$\frac{\beta_n}{\beta_p} = 6$$

Recalling that the technology and geometry contributions to β, the ratio of widths of the n-device to the p-device might range between approximately 3/1 for $\mu_n/\mu_p = 2$ and 2/1 where $\mu_n/\mu_p = 3$. Figure 2.21(b) shows some typical transfer characteristics for varying β_n/β_p ratios. The noise margins are as follows:

β_n/β_p	V_{IL}	V_{IH}	V_{OL}	V_{OH}	NM_L	NM_H
2	3.4	4.5	1.4	5	2.0	0.5
4	1.8	3.3	0.6	5	1.2	1.7
6	1.4	2.8	0.35	5	1.05	2.2
8	1.1	2.4	0.24	5	0.86	2.6
100	0.5	1.1	0.00	5	0.5	3.9

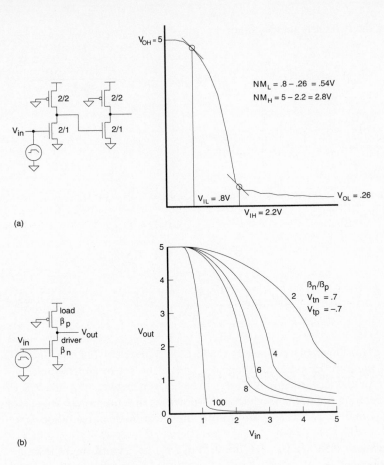

FIGURE 2.21 Cascaded pseudo-nMOS inverters

From this one can see that the low noise margin is considerably worse than the high noise margin. The overall noise margin of a pseudo-nMOS circuit can be enhanced considerably by following such a stage with a CMOS stage ($\beta_n/\beta_p = 1$). In this case for $\beta_n/\beta_p = 6$,

V_{IL}	V_{IH}	V_{OL}	V_{OH}	NM_L	NM_H
2.3	3.3	.35	5	1.95	1.7

This inverter finds widespread use in circuits where an "n-rich" circuit is required and the power dissipation can be tolerated. Typical uses include static ROMs and PLAs. Note that the circuit could use n-load devices and p-active pull-ups, if this were of advantage.

Rather than operate the p-transistor in the linear region it is possible to operate it as a constant current source (saturated load). Figure 2.22(a) shows an inverter with a p-transistor biased to be a constant current source ($V_{out} >$

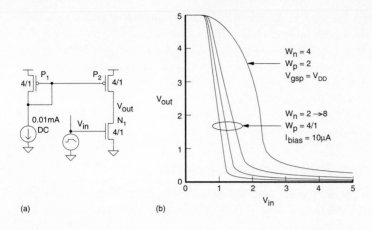

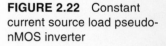

(a)   (b)

FIGURE 2.22   Constant current source load pseudo-nMOS inverter

$V_{gsp} - V_{tp}$). The constant current p load allows the inverter characteristics to be set to compensate for process changes (see also Fig. 5.27). Figure 2.22(b) shows transfer characteristics for a variety of n-transistor widths. (See also Section 5.4.3.)

2.4.2   Saturated Load Inverters

Figure 2.23(a) shows an inverter using an nMOS transistor load. This type of inverter was used in nMOS technologies prior to the availability of nMOS depletion loads and in pMOS technologies prior to the availability of nMOS technologies. It is included here for completeness. The high level is an n threshold down from V_{DD} (but remember that the threshold is modified by

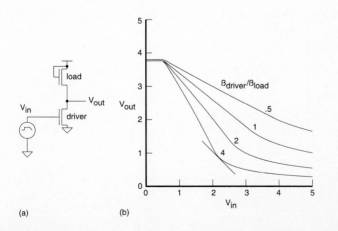

(a)   (b)

FIGURE 2.23   Saturated load inverter

the body effect because the source of the n-load transistor is above V_{SS}). Figure 2.23(b) shows the transfer characteristics for a variety of pull-up to pull-down ratios. For $k = 4$ $V_{OL} = .24$ volts, $V_{IH} = 2.2$ volts, $V_{OH} = 3.8$ volts and $V_{IL} = .56$ volts. Thus the low noise margin is .32 volts and the high noise margin is 1.6 volts for cascaded circuits. The small low noise margin would make this inverter nonoptimal as a conventional logic circuit. However, it might be used in isolated circumstances where p-transistors were not wanted (for instance, in some I/O structures).

2.4.3   More Saturated Load Inverters

A number of other "pseudo-nMOS" inverter configurations are possible. Figure 2.24(a) shows a p load with its gate connected to the output. The transfer characteristic is shown in Fig. 2.24(b) for a number of pull-up/pull-down ratios. The output rises to a p threshold down from V_{DD}. In addition as the output voltage approaches $V_{DD} - |V_{tp}|$, the V_{ds} across the pull-up is reduced, thus decreasing the current flowing in the pull-up, which has a detrimental effect on the pull-up speed. While $V_{out} > V_{in} - V_{tn}$ (i.e., for small V_{in} values), the driver transistor is in saturation

$$I_{dsdriver} = \frac{\beta_{driver}}{2}(V_{in} - V_{tn})^2.$$ **(2.38)**

Similarly the load device I_{ds} is permanently in the saturated or cutoff region

$$I_{dsload} = \frac{\beta_{load}}{2}(V_{out} - V_{DD} - V_{tp})^2.$$ **(2.39)**

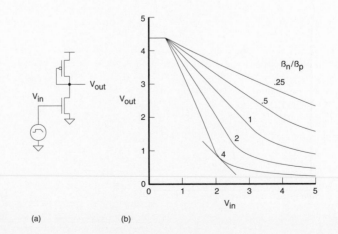

FIGURE 2.24   Saturated load inverter

(a)   (b)

Equating the two currents we obtain

$$\frac{\beta_{driver}}{2}(V_{in} - V_{tn})^2 = \frac{\beta_{load}}{2}(V_{out} - V_{DD} - V_{tp})^2.$$

Upon rearrangement,

$$V_{out} = V_{DD} + V_{tp} + \sqrt{k}\,(V_{in} - V_{tn}) \qquad \textbf{(2.40)}$$

where $k = \dfrac{\beta_{driver}}{\beta_{load}}$.

This effectively gives the V_{OH} value ($V_{in} = V_{tn}$). Similar calculations can yield the V_{OL}. From Fig. 2.24(b), for $k = 4$ $V_{OL} = .24$ volts, $V_{IH} = 2.1$ volts, $V_{OH} = 4.4$ volts, and $V_{IL} = .5$ volts. Thus the low noise margin is .26 volts and the high noise margin is 2.3 volts. The small low-noise-margin makes this inverter unsuitable for cascaded logic use, but it is of use in other circumstances and forms the basis for the differential pair inverter, which we will examine subsequently.

Finally, Fig. 2.25 shows an nMOS depletion load inverter. This inverter relies on the existence of a depletion nMOS transistor to form the load device. That is, the threshold of the depletion transistor is negative. While this is relatively rare in CMOS processes, this inverter formed the basis for the generation of MOS technology that ushered in the VLSI era. By connecting the gate of the load to the output, a constant current load is formed. Unlike the inverter shown in Fig. 2.24, which uses a p-device as a constant current load, the output of this inverter can rise to a full V_{DD} level.

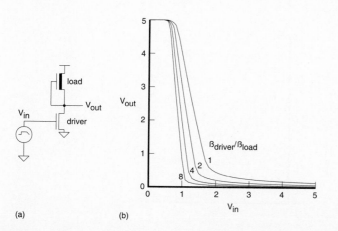

(a)

(b)

FIGURE 2.25 Depletion load inverter

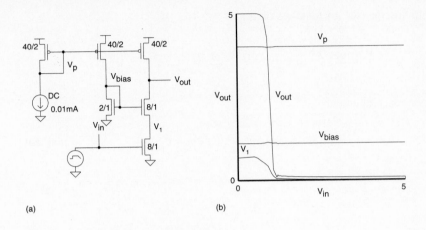

FIGURE 2.26 Cascode inverter

(a)

(b)

2.4.4 The Cascode Inverter

The cascode inverter is shown in Fig. 2.26. It resembles a pseudo-nMOS inverter but with an n-transistor connected in series with the pull-down n-transistor. If the gate of the series transistor is held at a constant voltage, V_{bias}, the drain of the driver transistor (V_1) will be held to an n threshold below V_{bias}. The output node, V_{out}, swings from V_{DD} to V_{SS}. The series transistor acts as a "common gate" amplifier and in effect isolates the V_1 node from the V_{out} node and keeps the signal swing on V_1 between V_{SS} and $V_{bias} - V_{tn}$. This feature will be used in a logic family discussed in Chapter 5.

2.4.5 TTL Interface Inverter

One final CMOS inverter is shown in Fig. 2.27.[23] This is of use in interfacing to TTL logic systems. The series-p load basically feeds a conventional

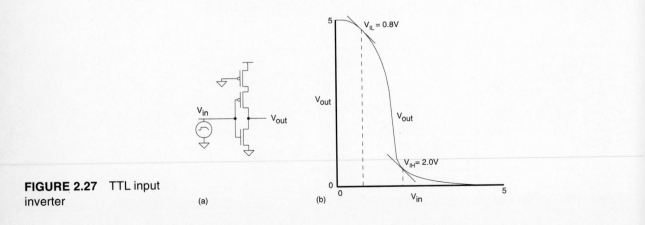

FIGURE 2.27 TTL input inverter

(a)

(b)

CMOS inverter with a reduced V_{DD} supply. This changes the input threshold to suit a TTL output. ($V_{IL} = 0.8V$ $V_{IH} = 2.0V$).

2.5   The Differential Inverter

All of the inverters that we have examined thus far have been singled-ended; that is, they have a single input signal and produce a single output signal. An inverter that uses two differential inputs and produces two differential outputs is shown in Fig. 2.28(a). Two n-transistors have their sources commoned and fed by a constant current source that is in turn connected to ground. The drains of each n-transistor are connected to resistor loads that are connected to the supply voltage.

If the input voltages V_{left} and V_{right} are set to the same voltage $V_{quiescent}$, then each transistor has a V_{gs} of $V_{quiescent} - V_N$, where V_N is the voltage across the constant current source. Thus the I_{ds} for each transistor is equal and the output voltages V_{out1} and V_{out2} are equal. If the voltages V_{left} and V_{right} are increased equally, then V_N rises to maintain the constant current through the current source. The output voltages, V_{out1} and V_{out2}, will stay at the same value. Applying this common signal to both inputs therefore results in no gain (ideally); this gain is referred to as the Common Mode Gain. If V_{left} is increased by δV, and V_{right} is decreased by δV, then the current in N_1 will increase by δI and the current in N_2 will decrease by δI. V_{out1} will decrease by δIR and V_{out2} will increase by δIR. Thus the differential gain from V_{left} to V_{out1} is

$$A_{diff} = -\frac{2\delta IR}{2\delta V} = -\frac{\delta IR}{\delta V}. \qquad (2.41)$$

The term $\delta I / \delta V$ may be recognized as the g_m of the driver transistor. Thus the gain is

$$A_{diff} = -g_m R. \qquad (2.42)$$

This is called the Differential Gain because it resulted from applying a differential signal to the inputs. In practical circuits, ideal constant current sources are hard to find so the Common Mode Gain and Differential Gains vary from the ideal. The Common Mode Rejection Ratio ($CMRR$) is defined as

$$CMRR = \frac{Differential\ Gain}{Common\ Mode\ Gain}. \qquad (2.43)$$

The value of the load resistor, R_{load}, is a tradeoff between gain (large R) and bandwidth (low R). Also, the value of the current source, I_{source}, represents a balance between power dissipation (low I, small power dissipation) and bandwidth (high I, low R, high bandwidth). As R_{load} is decreased for a given I_{source}, the minimum voltage at the output increases ($V_{outmin} = V_{DD} - I_{source}R_{load}$). As R_{load} is increased, V_{outmin} increases usually until a point at which the current source ceases to act as such or some other bias condition prevents the amplifier from operating as such. The size of the driver transistor affects the gain. The larger the transistor the higher the gain, but the larger are the associated parasitic capacitances.

For instance, in the circuit shown a tail current of 100μA is chosen. The quiescent conditions required are as follows:

$$V_{left} = V_{right} = 2.5 \text{ volts}$$
$$V_{out1} = V_{out2} = 3.5 \text{ volts}$$

$$\text{Thus} \quad I_{source}R_{load} = V_{DD} - 3.5$$
$$= 1.5$$

$$R_{load} = \frac{1.5}{50\mu A}$$
$$= 30K\Omega$$

Figure 2.28(b) shows the I/O characteristic for the circuit shown in Fig. 2.28(a) for a number of transistor widths. As the transistor width is increased, the gain increases. In addition, as the transistor width is increased,

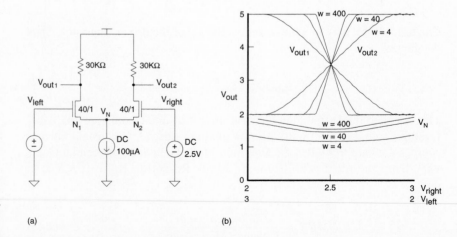

FIGURE 2.28 Basic differential amplifier

(a)   (b)

the V_N voltage rises as the required V_{gs} to establish the tail current decreases. At the quiescent point the driver transistors are in saturation, and for instance the β for the process is .124mA/V^2 and V_{tn} = .7 volts. Hence,

$$g_m = \beta(V_{gs} - V_t)$$
$$= .124 \times 20 \times (2.5 - 1.5 - .7)$$
$$= .74mS \text{ (milliSiemens)}$$
$$A = g_m R_{load}$$
$$= .74 \times 10^{-3} \times 30 \times 10^3$$
$$= 22.3.$$

From the characteristics in Fig. 2.28(b)

$$A = 22.2,$$

which shows good correspondence.

In Fig. 2.28 we used an ideal current source for the differential pair. An MOS transistor may be used to provide a very good constant current source provided certain operating conditions are met. From the DC operating equations, we know that when a transistor is in the saturation region, the drain current to a first approximation is independent of drain-source voltage. We can improve the characteristics of the MOS constant current source by lengthening the device beyond the minimum dimensions allowed. This reduces the effect of channel-length modulation.

A CMOS differential pair with an nMOS current source and pMOS load resistors is shown in Fig. 2.29. A voltage V_{bias} sets the current in the current source. The constant current source will act as such provided that $V_N > V_{bias} - V_{tn}$. To keep V_{bias} low while providing a reasonable current requires the current source to have a large β.

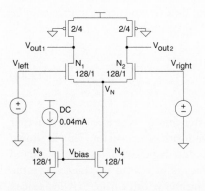

FIGURE 2.29 CMOS differential amplifier

In the circuit, V_{bias} is set by what is termed a current mirror. If a current is forced in N_3, then an identical current will flow in transistor N_4. The reason for this is as follows. With the drain connected to the gate, N_3 is in saturation. Forcing a current I_{s3} in N_3 yields a V_{gs3} of

$$V_{gs3} = \sqrt{\frac{2I_{s3}}{\beta}} + V_t \;.$$

Now, because N_4 has a $V_{gs} = V_{gs1}$,

$$I_{s4} = \frac{\beta}{2}(V_{gs} - V_t)^2 = I_{s3}.$$

One may cascade current mirrors to provide a variety of current tracking arrangements. If a current multiplication is required, this may be achieved by appropriate ratioing of the current mirror transistors.

Figure 2.30(a) shows a differential amplifier that employs an active current-mirror load structure rather than resistive p-transistors. This structure forms the basis for many RAM sense amplifiers. In this application, the current source is often connected as an unsaturated device. In these circumstances, one has to ensure that the DC conditions are such that the amplifier operates correctly. The active p loads have to be able to source the total current developed by the current source n-transistor. A starting point is to make $\beta_{N_3} = \beta_{P_1} = \beta_{P_2}$. Figure 2.30(b) shows the amplifier characteristic for varying load device sizes. If the p-devices are too small, then when $V_{left} = V_{DD}$, the high value at V_{out} will be lower than possible because P_1 will not be able to source all of the current from N_3. If P_1 and P_2 are made larger with respect to N_3, the low value of the amplifier increases, the gain of the amplifier decreases, and the transition region moves to the left as shown in Fig. 2.30(b). The gain is then determined by the g_m of N_1 and the output conductance of P_2 and N_2. Figure 2.30(c) and Fig. 2.30(d) show the I/O characteristics for the amplifier and the currents that flow in the current source and the two load devices. The small signal gain is given by[24]

$$A = \frac{g_{mn}}{g_o} \tag{2.44}$$

where g_{mn} is the g_m of the driver transistor and g_o is the combined output conductance of the p current load and the n-driver transistor. This is shown in Fig. 2.30(e) for various values of load- and driver-device sizes for a fixed current source. As the length of the devices is increased (r_{ds} increases), the gain of the amplifier increases. Increasing the width of the driver devices

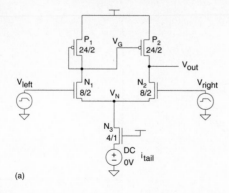

(a)

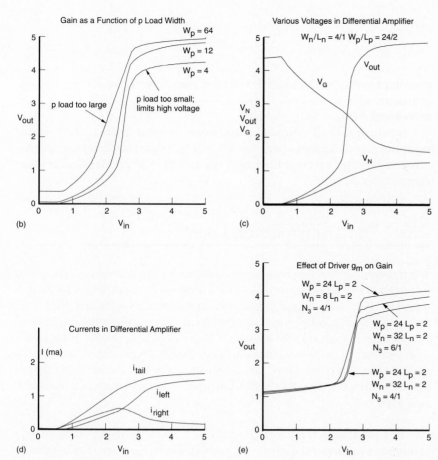

(b)

(c)

(d)

(e)

FIGURE 2.30   Active load CMOS differential amplifier

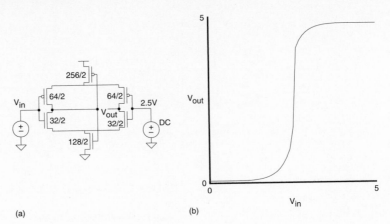

FIGURE 2.31 Self-biased
CMOS differential amplifier   (a)   (b)

does not have as marked an effect on the gain as the $g_m = \beta(V_{gs} - V_t)$. For
instance if the β of the driver transistors is quadrupled, then the $(V_{gs} - V_t)$ is
halved and the g_m is only doubled.

A further CMOS differential amplifier is shown in Fig. 2.31.[25] It has
twice the gain of the amplifier shown in Fig. 2.30 and has the advantage that
it is self-biasing. This amplifier is of use in TTL-CMOS input buffers and
comparators.

2.6   The Transmission Gate

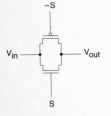

FIGURE 2.32
Transistor connec-
tion for CMOS
transmission gate

The transistor connection for a complementary switch or transmission gate is
reviewed in Fig. 2.32. It consists of an n-channel transistor and a p-channel
transistor with separate gate connections and common source and drain connections. The control signal is applied to the gate of the n-device, and its
complement is applied to the gate of the p-device. The operation of the transmission gate can be best explained by considering the characteristics of both
the n-device and p-device as pass transistors individually. We will address
this by treating the charging and discharging of a capacitor via a transmission gate.

nMOS Pass Transistor.   Referring to Fig. 2.33(a), the load capacitor C_{load}
is initially discharged (i.e., $V_{out} = V_{SS}$). With $S = 0$ (V_{SS}) (i.e., $V_{gs} = 0$ volts),
$I_{ds} = 0$, then $V_{out} = V_{SS}$ irrespective of the state of the input V_{in}. When $S = 1$
(V_{DD}), and $V_{in} = 1$, the pass transistor begins to conduct and charges the load
capacitor toward V_{DD}, i.e., initially $V_{gs} = V_{DD}$. Since initially V_{in} is at a

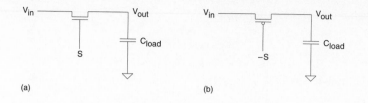

(a)   (b)

FIGURE 2.33 nMOS and pMOS transistor operation in transmission gate

higher potential than V_{out}, the current flows through the device from left to right. As the output voltage approaches $V_{DD} - V_{tn}$, the n-device begins to turn off. Load capacitor, C_{load}, will remain charged when S is changed back to 0. Therefore the output voltage V_{out} remains at $V_{DD} - V_{tn(V_{dd})}$. $V_{tn(V_{dd})}$ is the n-transistor body affected threshold with the source at $V_{DD} - V_{tn(V_{dd})}$. This implies that the transmission of logic one is degraded as it passes through the gate. With $V_{in} = 0$, $S = 1$, and $V_{out} = V_{DD} - V_{tn(V_{dd})}$, the pass transistor begins to conduct and discharge the load capacitor toward V_{SS}, i.e., $V_{gs} = V_{DD}$. Since initially V_{in} is at a lower potential than V_{out} the current flows through the device from right to left. As the output voltage approaches V_{SS}, the n-device current diminishes. Because V_{out} falls to V_{SS}, the transmission of a logic zero is not degraded.

pMOS Pass Transistor.   Once again a similar approach can be taken in analyzing the operation of a pMOS pass transistor as shown in Fig. 2.33(b). With $-S = 1$ ($S = 0$), $V_{in} = V_{DD}$, and $V_{out} = V_{SS}$, the load capacitor C_{load} remains uncharged. When $-S = 0$ ($S = 1$), current begins to flow and charges the load capacitor toward V_{DD}. However, when $V_{in} = V_{SS}$ and $V_{out} = V_{DD}$, the load capacitor discharges through the p-device until $V_{out} = V_{tp(V_{ss})}$, at which point the transistor ceases conducting. Thus transmission of a logic zero is somewhat degraded through the p-device.

The resultant behavior of the n-device and p-device are shown in Table 2.4. By combining the two characteristics we can construct a transmission gate that can transmit both a logic one and a logic zero without degradation. As can be deduced from the discussion so far, the operation of the transmission gate requires both the true and the complement version of the control signal.

TABLE 2.4   Transmission Gate Characteristics

DEVICE	TRANSMISSION OF '1'	TRANSMISSION OF '0'
n	poor	good
p	good	poor

The overall behavior can be expressed as:

$$S = 0\,(-S = 1);\quad\left(\begin{array}{l} \text{n-device} = \text{off} \\ \text{p-device} = \text{off} \\ V_{in} = V_{SS},\ V_{out} = Z \\ V_{in} = V_{DD},\ V_{out} = Z \end{array}\right. \tag{2.45}$$

where Z refers to a high impedance state and

$$S = 1\,(-S = 0);\quad\left(\begin{array}{l} \text{n-device} = \text{on} \\ \text{p-device} = \text{on} \\ V_{in} = V_{SS},\ V_{out} = V_{SS}. \\ V_{in} = V_{DD},\ V_{out} = V_{DD} \end{array}\right. \tag{2.46}$$

The transmission gate is a fundamental and ubiquitous component in MOS logic. It finds use as a multiplexing element, a logic structure, a latch element, and an analog switch. The transmission gate acts as a voltage controlled resistor connecting the input and the output.

Figure 2.34(a) shows a typical circuit configuration for a transmission gate in which the output is connected to a capacitor and the input to an inverter. The control input is shown turning the transmission gate on. That is, the gate of the n-channel transmission gate switch is changing from $0 \rightarrow 1$ and the gate of the p-channel is changing from $1 \rightarrow 0$. First consider the case where the control input changes rapidly, the inverter input is low (V_{SS}), the

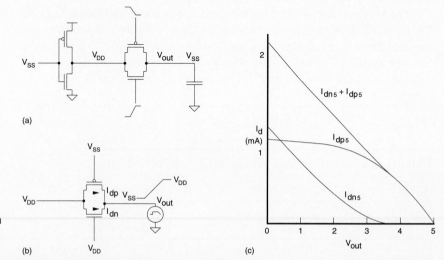

FIGURE 2.34 Transmission gate output characteristic for control input changing

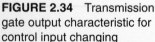

inverter output is high (V_{DD}), and the capacitor on the transmission gate output is discharged (V_{SS}). The currents that flow in this situation may be modeled by the circuit shown in Fig. 2.34(b) in which the input is held at V_{DD} and the output is ramped from V_{SS} to V_{DD}, while the currents in the pass transistors are monitored (in SPICE by using zero-volt voltage sources). In reality, the capacitor charge would be exponential, but a linear ramp serves to show what happens to the pass transistor currents. As V_{out} rises, the p-transistor current follows a constant V_{gs} of –5 volts (Fig. 2.34c). That is, it starts out in saturation and transitions to the nonsaturated case when $|V_{gsp} - V_{tp}| < |V_{dsp}|$. The n-transistor is always in the saturated region as $V_{dsn} = V_{gsn}$ and $V_{gsn} - V_{tn} < V_{dsn}$. When V_{out} reaches a V_{tn} below V_{DD}, the n-transistor turns off. Thus there are three regions of operation:

> Region A. n saturated, p saturated ($V_{out} < |V_{tp}|$)
> Region B. n saturated, p nonsaturated ($|V_{tp}| < V_{out} < V_{DD} - V_{tn}$)
> Region C. n off, p nonsaturated ($V_{DD} - V_{tn} < V_{out}$)

In region A, we can approximate the p-current as a constant current while the n-current varies quadradically with V_{out}. Hence the total current is roughly linear with V_{in}. In region B both currents yield a sum that varies almost linearly with V_{out}. Finally in region C the p-current varies linearly with V_{out}. Thus the transmission gate acts as a resistor, with contributions to its resistance from both n- and p-transistors. This can be seen in Fig. 2.34(c) ($I_{dn5} + I_{dp5}$). Similar simulations may be carried out for $V_{in} = V_{SS}$ and $V_{out} = V_{DD} \rightarrow V_{SS}$.

Another operation mode that the transmission gate encounters in lightly loaded circuits is where the output closely follows the input, such as shown in Fig. 2.35(a). Figure 2.35(b) shows a model of this while Fig. 2.35(c) shows the SPICE circuit used to model this condition including current monitoring voltage sources. Figure 2.35(d) shows the n- and p-pass transistor currents for $V_{out} - V_{in} = -0.1$ volts. It can be seen that again there are three regions of operation:

> Region A. n nonsaturated, p off
> Region B. n nonsaturated, p nonsaturated
> Region C. n off, p nonsaturated

The total current decreases in magnitude as V_{in} increases until $V_{in} = |V_{tp(body-affected)}|$. Here the p-transistor turns on and in this case slows the decrease of current. When $V_{in} > V_{DD} - V_{tn(body-effected)}$, the current starts to increase in magnitude as the p current continues to increase while the n transistor is off. In this simulation the p and n gains were matched. For the region $|V_{tp}| < V_{in} < V_{DD} - V_{tn}$, the transmission gate will have a roughly constant resistance. The effect of having only one polarity transistor in the transmission gate is also seen. If only an n-transistor is used, the output will rise to an n threshold below V_{DD} as current stops flowing at this point. Similarly, with a single p-transistor, the output would fall to a p threshold above V_{SS}, as

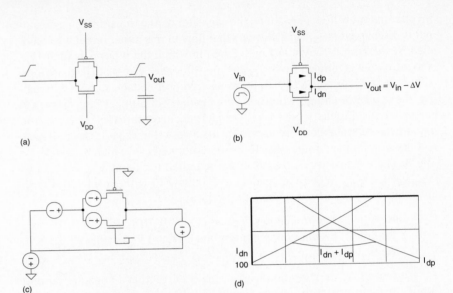

(a)

(b)

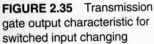

FIGURE 2.35 Transmission gate output characteristic for switched input changing

(c)

(d)

current stops flowing in the p-transistor at this point. Note also that as either the p or n current approaches zero, the speed of any circuit would be prejudiced. If the surrounding circuitry can deal with these imperfect high and low values, then single polarity transmission gates may be used. Figure 2.36 shows a plot of the transmission gate "on" resistance for the test circuit shown in Fig. 2.35(c).

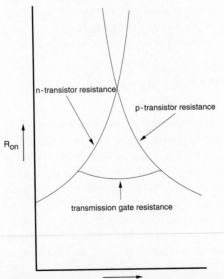

FIGURE 2.36 Resistance of a transmission gate for conditions in Figure 2.35

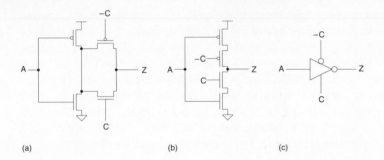

FIGURE 2.37 Tristate inverter

2.7 The Tristate Inverter

By cascading a transmission gate with an inverter the tristate inverter shown in Fig. 2.37(a) is constructed. When $C = 0$ and $-C = 1$, the output of the inverter is in a tristate condition (the Z output is not driven by the A input). When $C = 1$ and $-C = 0$, the output Z is equal to the complement of A. The connection between the n- and p-driver transistors may be omitted (Fig. 2.37b) and the operation remains substantially the same (except for a small speed difference). Figure 2.37(c) shows the schematic icon that represents the tristate inverter. For the same size n- and p-devices, this inverter is approximately half the speed of the inverter shown in Fig. 2.11. This inverter will be discussed in more detail in Chapter 5, because it forms the basis for various types of clocked logic, latches, bus drivers, multiplexers, and I/O structures.

2.8 Bipolar Devices

Thus far we have treated the MOS transistor in isolation as the device of interest. However, there are other semiconductor devices that are fabricated either parasitically or deliberately in a CMOS process. In particular, the junction diode and the bipolar transistor will be examined. The former is of use primarily in digital circuits as a protection device in I/O structures. The latter may be constructed to improve the speed of CMOS in BiCMOS processes. Of concern to all CMOS designers, however, are the parasitic bipolar transistors constructed as a by-product of building the basic nMOS/ pMOS structures in CMOS. These can lead to a circuit debilitating condition known as latchup. This will be covered in detail in Chapter 3.

2.8.1 Diodes

The diode is the most basic of semiconductor devices and is created when a metal and a semiconductor or two semiconductors form a junction When two diffusions of opposite polarity form a junction, a junction diode is formed. When a metal and semiconductor merge either an ohmic contact is made or a Schottky diode is created. In most CMOS processes only ohmic contacts are formed where metal contacts diffusions.

For instance, in an nMOS (or pMOS) transistor, the source and drain terminals form np (or pn) junction diodes to the substrate (or well). The schematic symbol for a junction diode is shown in Fig. 2.38(a). The two terminals are designated the anode and cathode. The *VI* characteristics of a diode are shown in Fig. 2.38(b). The current in a diode is given by[26]

$$I = A_d I_s \left(e^{\frac{qV}{kmt}} - 1 \right) \tag{2.47}$$

where

A_d = area of the diode

I_s = the saturation current/unit area

q = electronic charge

k = Boltzmann's constant

t = Temperature

m = a constant between 1 and 2 to account for various nonlinearities

($m \sim 2$ for pn junction diodes and $m \sim 1.2$ for Schottky diodes).

There are a number of characteristics of interest. When a positive voltage is applied to the cathode with respect to the anode, electrons are attracted to the supply and holes are repelled, leading to a "reverse-biased" condition

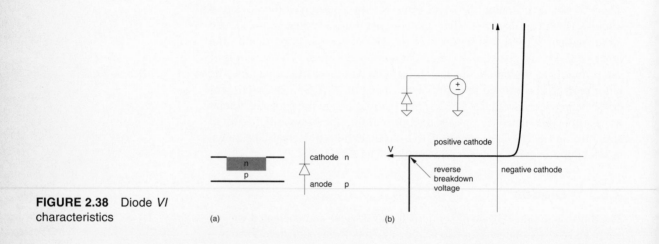

FIGURE 2.38 Diode *VI* characteristics

in which a very small reverse current flows. This results in a depletion region similar to that in the MOS transistor when it is in the depletion regime before inversion. In the above equation the exponential term is reduced in importance and the current is approximated by ($A_d = 1$)

$$I_{reverse} = -I_s \; (\sim 1 \times 10^{-15}A) \tag{2.48}$$

This condition applies until the voltage exceeds the reverse breakdown voltage of the junction, at which point the current increases rapidly due to avalanche multiplication. This occurs when electrons accelerated by the high field across the junction impact silicon atoms, thereby producing electron-hole pairs. When a negative voltage is applied to the cathode, the diode becomes forward biased. The current is approximated by ($A_d = 1$)

$$I_{forward} = I_s e^{\frac{qV}{kmt}} \tag{2.49}$$

As Fig. 2.38(b) shows, the current rapidly increases when the cathode-anode voltage is less than –0.6 volts. The x axis is reflected.

2.8.2 Bipolar Transistors

By building an NPN diffusion sandwich, as shown in Fig. 2.39(a), an NPN bipolar transistor may be constructed. Similarly a PNP transistor may be constructed by sandwiching an n diffusion between two p diffusions. The terminals of a bipolar transistor are called the collector, base, and emitter. The behavior of a transistor may be modeled (and is in the SPICE simulation program) by the structure shown in Fig. 2.39(b) for an NPN transistor. If V_{BE}, the base-emitter voltage, is set at around .7 volts and V_{CE} the collector-

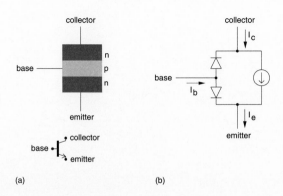

(a)   (b)

FIGURE 2.39 Structure and model of an NPN bipolar transistor

the collector base diode is reverse biased. By using the Ebers-Moll model,[27] the collector current may be calculated as

$$I_C = I_s \left(e^{\frac{qV_{BE}}{mkt}} - 1 \right) \left(1 + \frac{V_{CE}}{V_A} \right). \qquad (2.50)$$

While the emitter current is given by

$$I_E = I_C \left(1 + \frac{1}{\beta \left(1 + \frac{V_{CE}}{V_A} \right)} \right) \qquad (2.51)$$

where $kT/q = .026$ (at $300°K$)

V_{CE} = the collector-emitter voltage

V_{BE} = the base-emitter voltage

m = a constant between 1 and 2

V_A = the Early voltage (an approximation to allow for nonideal phenomena that result in finite output conductance)

β = forward current gain

I_S = the junction saturation current.

The forward current gain, β, (not to be confused with MOS β's) typically ranges from 20–500.

The *VI* characteristics of a typical NPN transistor are shown in Fig. 2.40.

The basic design equations for use with digital bipolar circuits are described in association with the inverter shown in Fig. 2.41. Here, the collector of an NPN transistor is connected to a positive supply via resistor R_c. The base is connected via resistor R_b to an input voltage V_{in}. The base current I_b is given by

$$I_b = \frac{V_{in} - V_{be}}{R_b} \qquad (2.52)$$

where V_{be} = the base emitter voltage (~0.7 volts)

and V_{in} = the input voltage.

The collector current is given by

$$I_c = \beta I_b$$

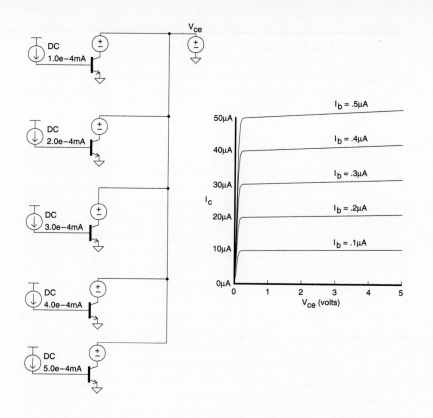

FIGURE 2.40 NPN transistor *VI* characteristics

and hence the collector voltage is given by

$$V_{out} = V_{DD} - I_c R_c$$

$$V_{out} = V_{DD} - \beta \frac{V_{in} - V_{be}}{R_b} R_c.$$

The gain, A, is given by

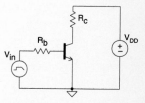

FIGURE 2.41 Inverter using an NPN transistor

$$\frac{dV_{out}}{dV_{in}} = \frac{\beta R_c}{R_b}. \qquad (2.53)$$

An n-well CMOS process inherently has a PNP transistor that is created between the substrate (collector), well (base), and source/drain diffusions (emitter). This PNP transistor is not that useful except for application as a current reference. This transistor is a vertical PNP because the transistor is formed by the vertical stacking of junctions.

Extra processing steps must be added to CMOS processes to build more useful NPN transistors. These steps result in what is termed a BiCMOS process (for Bipolar and CMOS). Similar to the case with p- and n-channel transistors in CMOS, NPN bipolar transistors have much higher gain and better high-frequency response than PNP transistors. Thus BiCMOS processes concentrate on adding a high-performance NPN transistor.

2.8.3   BiCMOS Inverters

The availability of an NPN transistor can markedly improve the output drive capability of a conventional CMOS inverter due to the high current gain of the NPN transistor.[28,29] Figure 2.42 shows one version of a BiCMOS inverter. When the input is low, P_1 is turned on and supplies base current to NPN_1 and sets the base voltage to V_{DD}. N_3 is turned on and clamps the base of NPN_2 to V_{SS}. Thus NPN_2 is off and the output rises to a V_{be} below V_{DD}. When the input is high, the base of NPN_1 is clamped to V_{SS} by N_1 and N_2 supplies the base current for NPN_2. The output falls to a small voltage above V_{SS}. This voltage is called V_{CEsat}, the collector emitter voltage with the transistor in saturation. This is due to the finite "on resistance" of the transistor and may be reduced by increasing I_b. It is normally in the range of $0.1 \rightarrow 0.3$ volts. Thus this inverter has an output swing between $V_{DD} - V_{be}$ and $V_{SS} + V_{CEsat}$. The V_{be} drop causes DC dissipation in any following CMOS gates, a problem which is not improved as the supply voltage is reduced.

A second BiCMOS inverter is shown in Fig. 2.43. Transistors P_2 and N_2 are used as resistors to bias the NPN transistors. When the input is low, P_1 feeds base current to NPN_1 and P_2 serves to pull the output high. The value of P_2 is a compromise to achieve high speed pull-up without bypass of the base current to NPN_1. When the input is high, N_2 feeds the base of NPN_2

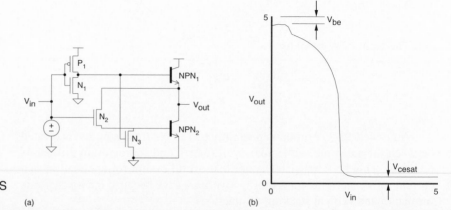

FIGURE 2.42   Basic BiCMOS inverter

(a)   (b)

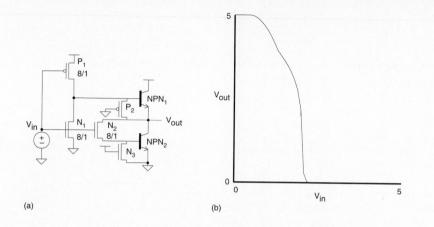

(a)

(b)

FIGURE 2.43 BiCMOS inverter using MOS transistors as resistors

while N_3 serves to pull the output to V_{SS}. The primary advantage of this implementation is that the output falls to V_{SS} and rises to V_{DD}.

A third BiCMOS inverter is shown in Fig. 2.44. In this inverter a feedback inverter is added to control the "resistor" transistors. When the input is low and the output is high, the feedback inverter places a zero on P_2 and N_2, thereby pulling the output high. When the input is high, the output becomes low and the feedback inverter places a high on N_2 and P_2, which pulls the output low.

A final BiCMOS inverter is shown in Fig. 2.45[30] which only uses a pull-up NPN transistor. When the input is low, P_1, P_2, NPN_1, and the feedback inverter combine to pull the output high. When the input is high, N_3 pulls the output low. This inverter is of particular use for 3.3 Volt supply circuits. The technique of using an nMOS transistor as the sole pull-down element can be used for the BiCMOS inverters shown in Figs. 2.42 and 2.43. Section 5.4.2 has an extended reference list of research on BiCMOS.

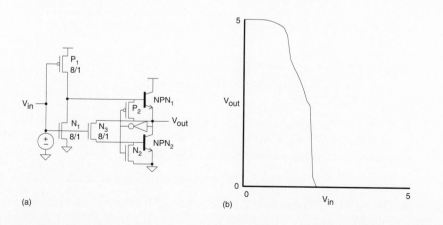

(a)

(b)

FIGURE 2.44 BiCMOS inverter with feedback inverter.

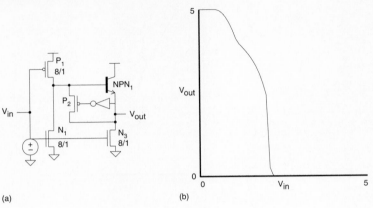

FIGURE 2.45 BiCMOS
inverter with nMOS pulldown   (a)

(b)

2.9   Summary

This chapter has examined the DC characteristics of MOS transistors,
diodes, bipolar transistors and CMOS inverters. In addition the operation of
the CMOS transmission gate was reviewed. Finally, the circuit configurations of some BiCMOS inverters were surveyed. The circuits treated in this
chapter are the basis for the majority of logic and memory circuits used in
CMOS digital system design. Ensuring their correct DC operation is the first
step in constructing a correctly functioning circuit. The second step, satisfy-
ing temporal (or timing) constraints requires one to be able to estimate the
speed of a circuit. This will be treated in Chapter 4.

2.10   Exercises

1. Calculate the noise margin for the BiCMOS inverter shown in Fig.
   2.42

   a. BiCMOS → CMOS ($\beta_n = \beta_p$) and
   b. BiCMOS → pseudo-nMOS ($\beta_n/\beta_p = 4$)
      ($V_{be} = .7$ volts)

2. Calculate the noise margin for a CMOS inverter operating at $3.3V$
   with $V_{tn} = 0.7$V, $V_{tp} = -0.7$V, $\beta_p = \beta_n$. What would you do to the transistor characteristics to improve the noise margin?

3. Derive the V_{OH} and V_{OL} for the inverter shown in Fig. 2.23.

4. Derive the VI equations that predict the V_{OL} for the inverter shown in
   Fig. 2.24.

5. Design an input buffer that may be used to interface with a TTL driver ($V_{DD} = 5V$, $V_{OL} = 0.8V$, $V_{OH} = 2.0V$). Show full derivations of DC conditions.

6. Design a buffer that interfaces internal $3.3V$ logic to CMOS I/O logic operating at $5V$.

7. Does the body effect of a process limit the number of transistors that can be placed in series in a CMOS gate at low frequencies?

8. Sometimes the substrate is connected to a voltage called the substrate bias to alter the threshold of the n-transistors. If the threshold of an n-transistor is to be raised, explain to what polarity the voltage substrate would be connected. Draw a circuit diagram showing V_{DD}, V_{SS}, and substrate supplies connected to an inverter.

9. Under what voltage conditions is a p-transistor with a source connected to $V_{DD} = 5V$ ($V_{tp} = -0.7\text{V}$) a good current source?

10. Using switched current mirrors, show how you would construct a current sourced digital to analog (D/A) converter with eight distinct current level outputs.

11. Calculate the threshold implant necessary to increase the threshold voltage to $0.6V$ for the example in Section 2.1.3.1.

12. For the values given in Section 2.2.2.7 ($L_{eff} = 0.6\ \mu\text{m}$ $V_{tn} = 0.6V$) calculate the worst case substrate current as a percentage of I_{ds} for a 2-input NAND gate operated at $5V$ and $3.3V$.

2.11 Appendix—SPICE Level 3 Model

The following is a summary of the Level 3 MOS model parameters used in the HSPICE program from Meta-Software, Inc. These parameters are consistent with most SPICE implementations that are widely available. The following is reproduced in large part from the HSPICE User's Manual with the kind permission of Meta-Software. This model uses empirical values determined from processed test devices as a basis for the model equations. The basic model parameters—*LEVEL, COX, KAPPA, KP, TOX,* and *VMAX*—are reviewed in Table 2.5 in terms of a $0.5 - 1\mu$ n-well process. (*Note:* These values should be used only as a guide—check with your CMOS manufacturer for accurate model parameters.)

The Level 3 model parameters for modeling the effective width and length are given in Table 2.6.

The Level 3 model also uses some parameters that vary the threshold voltage. These are as given in Table 2.7.

The parameters related to mobility are given in Table 2.8.

TABLE 2.5 Basic Model Parameters

NAME	UNITS	TYPICAL 1μm CMOS VALUE	DESCRIPTION
LEVEL		3.0	DC model selector.
COX	F/m^2	35 – 17E–4 (100 – 200 Å)	The oxide capacitance per unit gate area. If COX is not specified, then it will be calculated from TOX.
KAPPA	1/V	0.01 – .02	Saturation field factor, used in channel-length modulation equation.
KP	amp/V^2	2.0E–5	The intrinsic transconductance parameter. If not specified, then KP is calculated as KP = UO.COX.
TOX	m	1 – 2E–8	Gate oxide thickness.
VMAX	m/s	1.5 – 2E5	Maximum drift velocity of carriers; 0.0 indicates an infinite value.

TABLE 2.6 Effective Width and Length Parameters

NAME	UNITS	TYPICAL 1μm CMOS VALUE	DESCRIPTION
DEL	m	0.0	Channel-length reduction on each side.
LD	m	.01 – .1E–6	Lateral diffusion into channel from source and drain diffusion. If LD is unspecified, but XJ is specified, then LD = 0.75 XJ.
LREF	m	0.0	Channel-length reference.
LMLT	m	1.0	Length shrink factor.
WD	m	.05 – .1E–6	Lateral diffusion into channel from bulk along width.
WMLT		0.0	Diffusion layer and width shrink factor.
WREF	m	1.0	Channel-width reference.
XJ	m	.1 – .7E–6	Metallurgical junction depth.
XL	m	0.0	Accounts for masking and etching effects.
XW	m	0.0	Accounts for masking and etching effects.

TABLE 2.7 Threshold Voltage Parameters

NAME	UNITS	TYPICAL 1μm CMOS VALUE	DESCRIPTION
DELTA		$1.0 - 1.5$	Narrow width factor for determining threshold.
ETA		$.05 - .15$	Static feedback factor for adjusting threshold.
GAMMA	$V^{0.5}$	$.2 - .6$	Body effect factor. If *GAMMA* is not specified it is calculated from $$GAMMA = \frac{\sqrt{2q\varepsilon_{Si}NSUB}}{COX}$$
ND	*1/V*	1.0	Drain subthreshold factor.
N0		1.0	Gate subthreshold factor.
LND	μ*m/V*	0.0	*ND* length sensitivity.
LN0	μ*m*	0.0	*N0* length sensitivity.
NFS	$cm^{-2}V^{-1}$	7.5E11	Fast surface state density.
NSUB	cm^{-3}	2E16	Bulk surface doping. If not specified, calculated from *GAMMA*.
PHI	*V*	.74	Surface inversion potential. If not specified it is calculated from NSUB as $$PHI = 2\frac{kT}{q}ln\left(\frac{NSUB}{Ni}\right).$$
VTO	*V*	$0.5 \rightarrow 0.7$ (N) $-0.5 \rightarrow -0.7$ (P)	Zero-bias threshold voltage. If not specified it will be calculated from other parameters.
WIC		0.0	Subthreshold model selector.
WND	μ*m/V*	0.0	*ND* width sensitivity.
WN0	μ*m*	0.0	*N0* width sensitivity.

TABLE 2.8 Mobility Parameters

NAME	UNITS	TYPICAL 1μm CMOS VALUE	DESCRIPTION
THETA	*1/V*	$0.05 - 0.15$	Mobility degradation factor.
UO	$cm^2/V.s$	600 (N) 250 (P)	Low field bulk mobility.

The drain current is calculated as follows in the Level 3 model.

Cutoff Region $V_{gs} \leq V_t$

$$i_{ds} = 0$$

On region, $V_{gs} > V_t$

$$i_{ds} = \beta \, (V_{gs} - V_t - \frac{(1 + fb)}{2} V_{de}) \, V_{de}$$

where

$$\beta \quad = KP \frac{w_{eff}}{l_{eff}}$$

$$= u_{eff} COX \frac{w_{eff}}{l_{eff}}$$

$$V_{de} = min(V_{ds}, V_{dsat})$$

and

$$fb \quad = fn + \frac{GAMMA \; fs}{4 \sqrt{PHI + V_{sb}}} \, .$$

(The 4 in this equation should be 2 but HSPICE emulates the original SPICE program and uses 2.)

The narrow width effect is included through the *fn* parameter,

$$fn = \frac{DELTA}{w_{eff}} \, .$$

The term *fs* expresses the effect of the short channel and is determined as

$$fs \quad = 1 - \frac{XJ_{scaled}}{l_{eff}} \{ \frac{LD_{scaled} + wc}{XJ_{scaled}} \sqrt{1 - (\frac{wp}{XJ_{scaled} + wp})^2} - \frac{LD_{scaled}}{XJ_{scaled}} \}$$

$$wp = xd \sqrt{(PHI + V_{sb})}$$

$$xd \quad = \sqrt{\frac{2\varepsilon_{Si}}{qNSUB}}$$

$$wc =$$

$$XJ_{scaled} \left[0.0831353 + 0.8013929 \left(\frac{wp}{XJ_{scaled}} \right) - 0.0111077 \left(\frac{wp}{XJ_{scaled}} \right)^2 \right]$$

$$XJ_{scaled} = XJ \cdot SCALM$$
$$LD_{scaled} = LD \cdot SCALM.$$

SCALM is a global scaling factor applied to all MOS models in a given HSPICE run. The effective channel length and width in the Level 3 model is determined as follows:

$$l_{eff} = L_{scaled}LMLT + XL_{scaled} - 2(LD_{scaled} + DEL_{scaled})$$

where

$$L_{scaled} = L \cdot SCALM$$
$$XL_{scaled} = XL \cdot SCALM$$
$$DEL_{scaled} = DEL \cdot SCALM$$

LMLT is a scaling factor applied on a model by model basis.

$$w_{eff} = M(W_{scaled}WMLT + XW_{scaled} - 2WD_{scaled})$$

where

$$W_{scaled} = W \cdot SCALM$$
$$XW_{scaled} = XW \cdot SCALM$$
$$WD_{scaled} = WD \cdot SCALM$$

M is a parameter that allows for multiple parallel devices. The default value is 1.

$$LREF_{scaled} = LREF_{scaled}LMLT + XL_{scaled} - 2(LD_{scaled} + DEL_{scaled})$$
$$WREF_{scaled} = M(WREF_{scaled}MLT + XW_{scaled} - 2WD_{scaled})$$

Similar to *LMLT*, *WMLT* is a model scaling factor.
The threshold voltage is calculated as follows:

$$V_{th} = V_{bi} - \frac{8.14 \times 10^{-22}}{COXl_{eff}^3}V_{ds} + GAMMAfs\sqrt{PHI + V_{sb}} + fn(PHI + V_{sb})$$

with

$$V_{bi} = V_{fb} + PHI$$

or

$$V_{bi} = VTO - GAMMA\sqrt{PHI}$$

The saturation voltage V_{dsat} is calculated as

$$V_{dsat} = \frac{V_{gs} - V_{th}}{1 + fb}$$

$$V_{dsat} = V_{sat} + V_c - \sqrt{V_{sat}^2 + V_c^2}$$

where

$$V_c = \frac{VMAX l_{eff}}{u_s}$$

If the model parameter *VMAX* is not specified, then

$$V_{dsat} = V_{sat}$$

The parameter μ_s is the normal field mobility. It is calculated as

$$u_s = \frac{UO}{1 + THETA\,(V_{gs} - V_{th})} \qquad\qquad V_{gs} > V_{th}$$

The degradation of mobility due to the lateral field and the carrier velocity saturation is determined if *VMAX* is specified.

$$u_{eff} = \frac{u_s}{1 + \dfrac{V_{de}}{V_c}}$$

The effects of channel length modulation are calculated as follows:

$$\Delta l = xd\sqrt{KAPPA.\,(V_{ds} - V_{dsat})} \qquad\qquad VMAX = 0$$

$$\Delta l = \frac{ep.xd^2}{2} + \sqrt{\left(\frac{ep.xd^2}{2}\right)^2 + KAPPAxd^2\,(V_{ds} - V_{dsat})}$$

where *ep* is the lateral electric field at the pinch off point. Its value is approximated by:

$$ep = \frac{V_c\,(V_c + V_{dsat})}{l_{eff}V_{dsat}}$$

The current in saturation is computed as

$$I_{ds} = \frac{I_{ds}}{1 - \dfrac{\Delta l}{l_{eff}}} .$$

In order to prevent the denominator from going to zero, HSPICE limits the Δl as follows:

$$\text{if } \Delta l > \frac{l_{eff}}{2}$$

$$\text{else } \Delta l = l_{eff} - \frac{(\dfrac{l_{eff}}{2})^2}{\Delta l}$$

In the subthreshold region the current is characterized by the model parameter for fast surface states, *NFS*. The modified threshold voltage, V_{on}, is determined as follows:

$$V_{on} = V_{th} + fast \qquad\qquad NFS > 0$$

where

$$fast = \frac{kt}{q}\left[1 + \frac{qNFS}{COX} + \frac{GAMMAfs\sqrt{(PHI + V_{sb})} + fn\,(PHI + V_{sb})}{2\,(PHI + V_{sb})} \right]$$

The current I_{ds} is given by

$$I_{ds} = I_{ds}(V_{on}, V_{de}, V_{sb})\, e^{\dfrac{V_{gs} - V_{on}}{fast}} \qquad\qquad V_{gs} < V_{on}$$

$$I_{ds} = I_{ds}(V_{gs}, V_{de}, V_{sb}) \qquad\qquad V_{gs} \geq V_{on}$$

The modified threshold voltage is not used in strong inversion.

2.12 References

1. L. Vadasz and A. S. Grove, "Temperature of MOS Transistor Characteristics Below Saturation," *IEEE Trans. on Electron Devices,* vol. ED-13, no. 13, 1966, pp. 190–192.

2. J. Mavor, M. A., Jack Denyer and P. B. Denyer, *Introduction to MOS LSI Design,* Reading, Mass.: Addison-Wesley, 1983, pp. 18–61 (footnote).

3. John Y. Chen, *CMOS Devices and Technology for VLSI,* Englewood Cliffs, N.J.: Prentice Hall, 1990, pp. 5–37.

4. John Y. Chen, *op. cit.,* p. 211.

5. *HSPICE User's Manual,* Campbell, Calif.: Meta-Software, 1990, pp. 7–34.

6. W. Shockley, "A unipolar field effect transistor," *Proc. IRE.,* vol. 40, Nov. 1952, pp. 1365–1376.

7. R. S. Cobbold, *Theory and Application of Field Transistors,* New York: Wiley Interscience, 1970, pp. 239–267.

8. C. T. Sah, "Characteristics of the Metal-Oxide-Semiconductor Transistor," *IEEE Trans. Ed,* ED-11, Jul. 1964, pp. 324–345.

9. L. W. Nagel, "SPICE2: A Computer Program to Simulate Semiconductor Circuits," Memo ERL-M520, Berkeley, Calif.: University of California, May 9, 1975.

10. Eric A. Vittoz, "MicroPower Techniques," in *Design of VLSI Circuits for Telecommunications,* edited by Y. Tsividis and P. Antognetti, Englewood Cliffs, N.J.: Prentice-Hall, 1985.

11. Paul R. Gray and Robert G. Meyer, *Analysis and Design of Analog Integrated Circuits, Second Edition,* New York: Wiley and Sons, 1984.

12. M. Lenzlinger and E. H. Snow, "Fowler-Nordheim tunneling into thermally grown SiO_2," *Journal of Applied Physics,* vol. 40, 1969, pp. 278–281.

13. John Y. Chen, *op. cit.,* pp. 174–232.

14. S. M. Sze, *Physics of Semiconductor Devices, Second Edition,* New York: Wiley and Sons, 1981, pp. 496–504.

15. John Y. Chen, *op. cit.,* pp. 187–199.

16. Chenming Hu, "IC Reliability Simulation," *IEEE JSSC,* vol. 27, no. 3, Mar. 1992, pp. 241–246.

17. Wen-Jay Hsu, Bing J, Sheu, Sudhir M. Gowda, and Chang-Gyu Hwang, "Advanced Integrated-Circuit Reliability Simulation Including Dynamic Stress Effects," *IEEE JSSC,* vol. 27, no. 3, Mar. 1992, pp. 247–257.

18. Takayusu Sakurai, Kazutaka Nogami, Masakazu Kakumu, and Tetsuya Iizuka, "Hot-Carrier Generation in Submicrometer VLSI Environment," *IEEE JSSC,* vol. SC-21, no. 1, Feb. 1986, pp. 187–192.

19. HSPICE User's Manual, *op. cit.,* pp. 7–34.

20. W. N. Carr and J. P. Mize, *MOS/LSI Design and Application,* New York: McGraw-Hill, 1972.

21. R. S. C. Cobbold, "Temperature Effects on MOS Transistors," *Electronic Letters,* vol. 2, no. 6, June 1966, pp. 190–192.

22. N. Weste and K. Eshraghian, *Principles of CMOS VLSI Design,* Reading, Mass.: Addison-Wesley, 1984, Edition 1, Appendix A.

23. Don Trotter, "CMOS Course Notes," Mississippi State, Miss.: Electrical and Computer Engineering Dept., Mississippi State University, 1991.

24. Adel S. Sedra and Kenneth C. Smith, *Microelectronic Circuits, Second Edition,* New York: Holt, Rinehart and Winston, 1987.

25. Mel Bazes, "Two novel fully complementary self-biased CMOS differential amplifiers," *IEEE JSSC,* vol. 26, no. 2, Feb. 1991, pp. 165–168.

26. Douglas J. Hamilton and William G. Howard, *Basic Integrated Circuit Engineering,* New York: McGraw-Hill, 1975.

27. Douglas J. Hamilton and William G. Howard, *op. cit.,* pp. 212–241.

28. Hyun J. Shin, "Performance Comparison of Driver Configurations and Full Swing Techniques for BiCMOS Logic Circuits," *IEEE JSSC,* vol. 25, no. 3, June 1990, pp. 863–865.

29. Larry Wissel and Elliot L. Gould, "Optimal Usage of CMOS within a BiCMOS Technology," *IEEE JSSC,* vol. 27, no. 3, Mar. 1992, pp. 300–306.

30. Hiroyuki Hara, Takayasu Sakurai, Makato Noda, Tetsu Nagamatsu, Katsuhiro Seta, Hiroshi Momose, Youichirou Niitsu, Hiroyuki Miyakawa and Yoshinori Watanabe, "0.5µm 2M-Transistor BiPNMOS Channelless Gate Array," *IEEE Journal of Solid State Circuits,* Vol. 26, No. 11, Nov. 1991, pp. 1615–1620.

CMOS PROCESSING TECHNOLOGY

3

The purpose of this chapter is to introduce the CMOS designer to the technology that is responsible for the semiconductor devices that might be designed. This is of importance in understanding the potential and limitations of a given technology. It also gives some background for the geometric design rules that are the interface medium between designer and fabricator.

The basics of semiconductor manufacturing are first introduced. Following this, a basic n-well CMOS process is described showing the process steps and how they relate to the design description passed from the designer to the fabrication engineer. Following this, a number of enhancements to the basic CMOS technology are described. Many of these are now required by mainstream CMOS logic and memory designers. The next section introduces the reader to layout design rules that prescribe how to manufacture the CMOS chip. The nature of CMOS latchup and the solutions to this problem are then covered. Finally, some CAD issues as they relate to process technology are covered. An appendix, Section 3.9, outlines the actual steps used in a CMOS process for those who want to get down to that level of detail.

3.1 Silicon Semiconductor Technology: An Overview

Silicon in its pure or *intrinsic* state is a semiconductor, having a bulk electrical resistance somewhere between that of a conductor and an insulator. The conductivity of silicon can be varied over several orders of magnitude by introducing *impurity* atoms into the silicon crystal lattice. These *dopants* may either supply free electrons or holes. Impurity elements that use electrons are referred to as *acceptors* since they accept some of the electrons already in the silicon, leaving vacancies or holes. Similarly, *donor* elements provide electrons. Silicon that contains a majority of donors is known as *n-type* and that which contains a majority of acceptors is known as *p-type*. When n-type and p-type materials are brought together, the region where the silicon changes from n-type to p-type is called a *junction*. By arranging junctions in certain physical structures and combining these with other physical structures, various semiconductor devices may be constructed. Over the years, silicon semiconductor processing has evolved sophisticated techniques for building these junctions and other structures having special properties.

3.1.1 Wafer Processing

The basic raw material used in modern semiconductor plants is a *wafer* or disk of silicon, which varies from 75 mm to 230 mm in diameter and is less than 1 mm thick. Wafers are cut from ingots of single-crystal silicon that have been pulled from a crucible melt of pure molten polycrystalline silicon. This is known as the 'Czochralski,' method (Fig. 3.1) and is currently the most common method for producing single-crystal material. Controlled amounts of impurities are added to the melt to provide the crystal with the

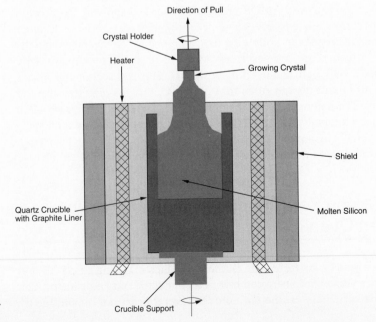

FIGURE 3.1 Czochralski method for manufacturing silicon ingots

required electrical properties. The crystal orientation is determined by a seed crystal that is dipped into the melt to initiate single-crystal growth. The melt is contained in a quartz crucible, which is surrounded by a graphite radiator. The graphite is heated by radio frequency induction and the temperature is maintained a few degrees above the melting point of silicon ($\approx 1425°C$). The atmosphere above the melt is typically helium or argon.

After the seed is dipped into the melt, the seed is gradually withdrawn vertically from the melt while simultaneously being rotated. The molten polycrystalline silicon melts the tip of the seed, and as it is withdrawn, refreezing occurs. As the melt freezes, it assumes the single crystal form of the seed. This process is continued until the melt is consumed. The diameter of the ingot is determined by the seed withdrawal rate and the seed rotation rate. Growth rates range from 30 to 180 mm/hour.

Slicing into wafers is usually carried out using internal cutting-edge diamond blades. Wafers are usually between 0.25 mm and 1.0 mm thick, depending on their diameter. Following this operation, at least one face is polished to a flat, scratch-free mirror finish.

3.1.2   Oxidation

Many of the structures and manufacturing techniques used to make silicon integrated circuits rely on the properties of the oxide of silicon, namely, silicon dioxide (SiO_2). Therefore the reliable manufacture of SiO_2 is extremely important.

Oxidation of silicon is achieved by heating silicon wafers in an oxidizing atmosphere such as oxygen or water vapor. The two common approaches are:

- Wet oxidation: when the oxidizing atmosphere contains water vapor. The temperature is usually between 900°C and 1000°C. This is a rapid process.

- Dry oxidation: when the oxidizing atmosphere is pure oxygen. Temperatures are in the region of 1200°C, to achieve an acceptable growth rate.

The oxidation process consumes silicon. Since SiO_2 has approximately twice the volume of silicon, the SiO_2 layer grows almost equally in both vertical directions. This effect is shown in Fig. 3.2 for an n-channel MOS device in which the SiO_2 (field oxide) projects above and below the unoxidized silicon surface.

3.1.3   Epitaxy, Deposition, Ion-Implantation, and Diffusion

To build various semiconductor devices, silicon containing varying proportions of donor or acceptor impurities is required. This may be achieved using epitaxy, deposition, or implantation. Epitaxy involves growing a single-crys-

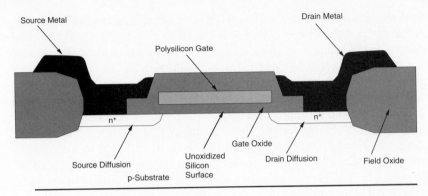

FIGURE 3.2   An nMOS transistor showing the growth of field oxide below the silicon surface

tal film on the silicon surface (which is already a single crystal) by subjecting the silicon wafer surface to elevated temperature and a source of dopant material. Deposition might involve evaporating dopant material onto the silicon surface followed by a thermal cycle, which is used to drive the impurities from the surface of the silicon into the bulk. Ion implantation involves subjecting the silicon substrate to highly energized donor or acceptor atoms. When these atoms impinge on the silicon surface, they travel below the surface of the silicon, forming regions with varying doping concentrations. At any elevated temperature (> 800°C) diffusion will occur between any silicon that has differing densities of impurities, with impurities tending to diffuse from areas of high concentration to areas of low concentration. Hence it is important once the doped areas have been put in place to keep the remaining process steps at as low a temperature as possible.

Construction of transistors and other structures of interest depends on the ability to control where and how many and what type of impurities are introduced into the silicon surface. What type of impurities are introduced is controlled by the dopant source. Boron is frequently used for creating acceptor silicon, while arsenic and phosphorous are commonly used to create donor silicon. How much is used is determined by the energy and time of the ion-implantation or the time and temperature of the deposition and diffusion step. Where it is used is determined by using special materials as masks. In places covered by the mask ion implantation does not occur or the dopant does not contact the silicon surface. In areas where the mask is absent the implantation occurs, or the predeposited material is allowed to diffuse into the silicon. The common materials used as masks include

- photoresist.
- polysilicon (polycrystalline silicon).
- silicon dioxide (SiO_2).
- silicon nitride (SiN).

The ability of these materials to act as a barrier against doping impurities is a vital factor in this process, called *selective diffusion*. Thus selective diffusion entails

- patterning *windows* in a mask material on the surface of the wafer.
- subjecting exposed areas to a dopant source.
- removing any unrequired mask material.

In the case of an oxide mask, the process used for selectively removing the oxide involves covering the surface of the oxide with an acid resistant coating, except where oxide windows are needed. The SiO_2 is removed using an etching technique. The acid resistant coating is normally a photosensitive organic material called *photoresist* (PR), which can be polymerized by ultraviolet (UV) light. If the UV light is passed through a mask containing the desired pattern, the coating can be polymerized where the pattern is to appear. The polymerized areas may be removed with an organic solvent. Etching of exposed SiO_2 then may proceed. This is called a positive resist. There are also negative resists where the unexposed PR is dissolved by the solvent. This process is illustrated in Fig. 3.3. In established processes using PRs in conjunction with UV light sources, diffraction around the edges of the mask patterns and alignment tolerances limit line widths to around 0.8 µm. During recent years, electron beam lithography (EBL) has emerged as a contender for pattern generation and imaging where line widths of the order of 0.5 µm with good definition are achievable. The main advantages of EBL pattern generation are as follows:

- Patterns are derived directly from digital data.
- There are no intermediate hardware images such as recticles or masks; that is, the process can be direct.
- Different patterns may be accommodated in different sections of the wafer without difficulty.
- Changes to patterns can be implemented quickly.

The main disadvantage that has precluded the use of this technique in commercial fabrication lines is the cost of the equipment and the large amount of time required to access all points on the wafer.

3.1.4 The Silicon Gate Process

So far we have touched on the single-crystal form of silicon used in the manufacture of wafers and the oxide used in the manufacture and operation of circuits. Silicon may also be formed in a *polycrystalline* form (not having a single-crystalline structure) called *polysilicon*. This is used as an intercon-

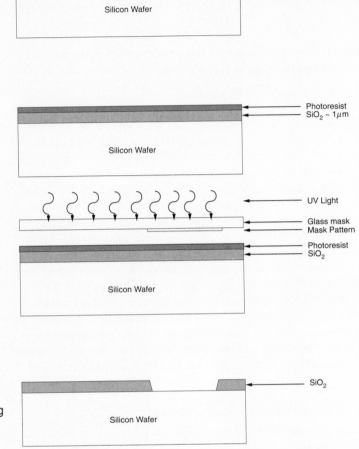

FIGURE 3.3 Simplified steps involved in the patterning of SiO_2: (a) Bare silicon wafer; (b) Wafer with SiO_2 and resist; (c) Exposing resist to UV light; (d) Final etched SiO_2

nect in silicon ICs and as the gate electrode on MOS transistors. The most significant aspect of using polysilicon as the gate electrode is its ability to be used as a further mask to allow precise definition of source and drain electrodes. This is achieved with minimum gate-to-source/drain overlap, which, we will learn, improves circuit performance. Polysilicon is formed when silicon is deposited on SiO_2 or other surfaces. In the case of an MOS transistor gate electrode, undoped polysilicon is deposited on the gate insulator. Polysilicon and source/drain regions are then normally doped at the same time. Undoped polysilicon has high resistivity. This characteristic is used to provide high-value resistors in static memories. The resistivity of polysilicon may be reduced by combining it with a refractory metal (see Section 3.2.4).

The steps involved in a typical silicon gate process entail photomasking and oxide etching, which are repeated a number of times during the processing sequence. Figure 3.4 shows the processing steps after the initial pattern-

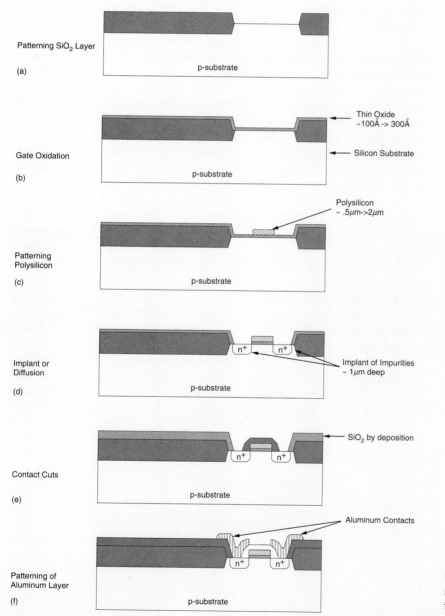

Patterning SiO$_2$ Layer

(a)

p-substrate

Gate Oxidation

(b)

Thin Oxide
~100Å -> 300Å

Silicon Substrate

p-substrate

Patterning
Polysilicon

(c)

Polysilicon
~ .5μm->2μm

p-substrate

Implant or
Diffusion

(d)

Implant of Impurities
~ 1μm deep

p-substrate

Contact Cuts

(e)

SiO$_2$ by deposition

p-substrate

Patterning of
Aluminum Layer

(f)

Aluminum Contacts

p-substrate

Figure 3.4   Fabrication steps for a silicon gate nMOS transistor

ing of the SiO_2, which was shown in Fig. 3.3. The wafer is covered with SiO_2 with at least two different thicknesses (Fig. 3.4b). A thin, highly controlled layer of SiO_2 is required where active transistors are desired. This is called the gate-oxide or thinox. A thick layer of SiO_2 is required elsewhere to isolate the individual transistors. This is normally called the field oxide. We will examine a variety of methods of achieving these two oxide thicknesses in Section 3.2.1.

Polysilicon is then deposited over the wafer surface and etched to form interconnections and transistor gates. Figure 3.4(c) shows the result of an etched polysilicon gate. The exposed gate oxide (not covered by polysilicon) is then etched away. The complete wafer is then exposed to a dopant source or is ion-implanted, resulting in two actions (Fig. 3.4d). Diffusion junctions are formed in the substrate and the polysilicon is doped with the particular type of dopant. This also reduces the resistivity of the polysilicon. Note that the diffusion junctions form the drain and source of the MOS transistor. They are formed only in regions where the polysilicon gate does not shadow the underlying substrate. This is referred to as a *self-aligned* process because the source and drain do not extend under the gate. Finally, the complete structure is covered with SiO_2 and contact holes are etched to make contact with underlying layers (Fig. 3.4e). Aluminum or other metallic interconnect is evaporated and etched to complete the final connection of elements (Fig. 3.4f). Further oxide layers, contact holes and metallization layers are normally added for extra interconnect.

Note that parasitic MOS transistors exist between unrelated transistors, as shown in Fig. 3.5. Here the source and drain of the parasitic transistor are existing source/drains and the gate is a metal or polysilicon interconnect overlapping the two source/drain regions. The "gate-oxide" is in fact the thick field oxide. The threshold voltage of this transistor is much higher than that of a regular transistor (this device is commonly called a field device) (Eq. 2.1). The high threshold voltage is usually ensured by making the field oxide thick enough and introducing a "channel-stop" diffusion, which raises

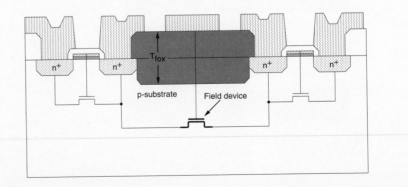

FIGURE 3.5   A parasitic MOS transistor or field device

the impurity concentration in the substrate in areas where transistors are not required, thus further increasing the threshold voltage (Section 2.1.3.1). These devices do have some useful purposes where the fact that they turn on at voltages higher than normal operating voltages may be used to protect other circuitry.

3.2 Basic CMOS Technology

CMOS (Complementary Metal Oxide Silicon) technology is recognized as the leading VLSI systems technology. CMOS provides an inherently low power static circuit technology that has the capability of providing a lower power-delay product than comparable design-rule bipolar, nMOS, or GaAs technologies. In this section we provide an overview of four dominant CMOS technologies, with a simplified treatment of the process steps. This is included primarily as a guide for better appreciation of the layout styles that may be used to implement CMOS gates.

The four main CMOS technologies are:

- n-well process.
- p-well process.
- twin-tub process.
- silicon on insulator.

In addition, by adding bipolar transistors a range of BiCMOS processes are possible.

During the discussion of CMOS technologies, process cross-sections and layouts will be presented. Figure 3.6 summarizes the drawing conventions.

3.2.1 A Basic n-well CMOS Process

A common approach to n-well CMOS fabrication has been to start with a lightly doped p-type substrate (wafer), create the n-type well for the p-channel devices, and build the n-channel transistor in the *native* p-substrate. Although the processing steps are somewhat complex and depend on the fabrication line, Fig. 3.7 illustrates the major steps involved in a typical n-well CMOS process. The mask that is used in each process step is shown in addition to a sample cross-section through an n-device and a p-device. Although we have shown a polysilicon gate process, it is of historical significance to note that CMOS was originally implemented with metal (aluminum) gates. This technology (in p-well form) formed the basis for the majority of low

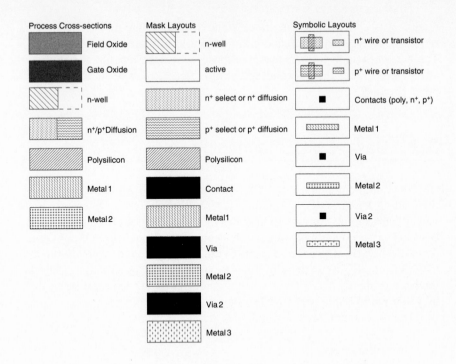

FIGURE 3.6 CMOS process and layout drawing conventions

power CMOS circuits implemented in the 1970s. The technology is robust and still in use. As can be seen from Fig. 3.7, the mask levels are not organized by component function. Rather they reflect the processing steps.

- The first mask defines the n-well (or n-tub); p-channel transistors will be fabricated in this well. Ion implantation or deposition and diffusion is used to produce the n-well (Fig. 3.7a). The former tends to produce shallower wells which are compatible with fine dimension processes. As the diffusion process occurs in all directions, the deeper a diffusion is the more it spreads laterally. This lateral spread affects how near to other structures wells can be placed. Hence, for closely spaced structures a shallow well is required. From a patterned well shape, the final well will extend outside the patterned dimension by the lateral diffusion.

- The next mask is called the "active" mask, because it defines where areas of thin oxide are needed to implement transistor gates and allow implantation to form p- or n-type diffusions for transistor source/drain regions (Fig. 3.7b). Other terms for this mask include *thinox, thin-oxide, island,* and *mesa.* A thin layer of SiO_2 is grown and covered with SiN. This is used as a masking layer for the following two steps.

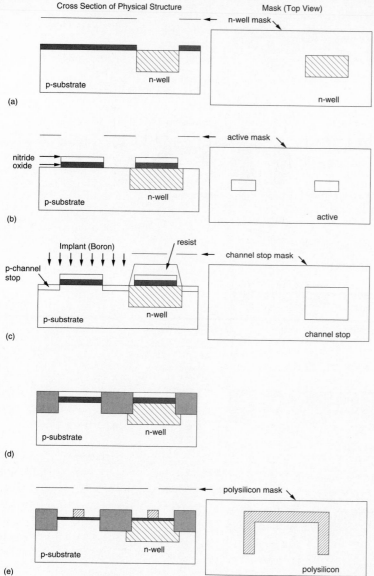

Cross Section of Physical Structure   Mask (Top View)

FIGURE 3.7 A typical n-well CMOS process

- The channel-stop implant is usually then completed. This uses the p-well mask (the complement of the n-well mask). It dopes the p-substrate in areas where there are no n-transistors p^+ using a photoresist mask (Fig. 3.7c). This, in conjunction with the thick field oxide that will cover these areas, aids in preventing conduction between unrelated transistor source/drains.

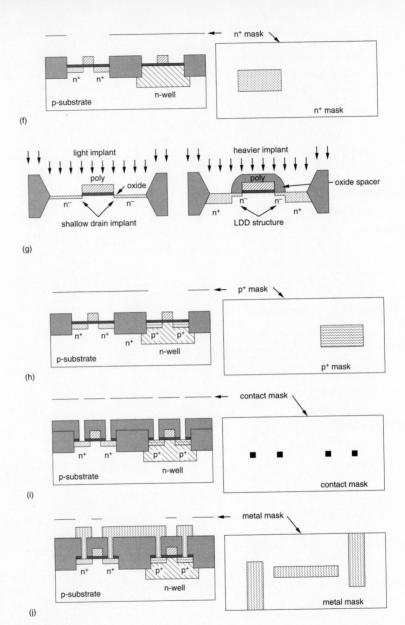

FIGURE 3.7 *(continued)*

- Following the channel-stop implant, the photoresist mask is stripped, leaving the previously masked SiO₂/SiN sandwich defining the active regions. The thick field oxide is then grown. This grows in areas where the SiN layer is absent. The oxide grows in both directions vertically and also laterally under the SiO_2/SiN sandwich (Fig. 3.7d). This lateral movement results in what is called a "bird's beak" because of the shape of the oxide encroachment under the gate oxide mask. This general oxide construction technique is called LOCOS for LOcal Oxidation Of Silicon. The oxide encroachment results in an active area that is smaller than patterned. In particular, the width dimension of a transistor will be reduced from what might be expected from the photolithography. Other techniques such as SWAMI (Side-WAll Masked Isolation)[1,2] have been developed to reduce the effect of the bird's beak. Of additional concern is the final planarity of the field oxide/gate oxide interface. If the difference in height is too great, the subsequent conductors may have "step coverage" problems in which a conductor thins and can even break as it crosses a thick to thin oxide boundary. To counter this, many planarization techniques have been developed. One such technique is to pre-etch the silicon in areas where the field oxide is to be grown by around half the final required field oxide thickness. The LOCOS oxide is then grown and the final field oxide/gate oxide interface is very planar.

- An n/p-transistor threshold voltage adjust step might then be performed using a p/n-well photoresist mask. In current fabrication processes the polysilicon is normally doped n^+. With normal doping concentrations suitable for small dimension processes, this results in threshold voltage for n-devices of around 0.5–0.7 volts. However, the p-device threshold is around –1.5 to –2.0 volts. Thus the p-device has to have its threshold voltage adjusted more than the n-device. This is done by introducing an additional negatively charged layer at the silicon/oxide interface. This moves the channel from the silicon/oxide interface further into the silicon, creating a "buried channel" device.[3] Following these two steps the gate oxide is grown.

- Polysilicon gate definition is then completed. This involves covering the surface with polysilicon and then etching the required pattern (in this case an inverted "U"). As noted previously, the "poly" gate regions lead to "self-aligned" source-drain regions (Fig. 3.7e).

- An n-plus (n^+) mask is then used to indicate those thin-oxide areas (and polysilicon) that are to be implanted n^+. Hence a thin-oxide area exposed by the n-plus mask will become an n^+ diffusion area (Fig. 3.7f). If the n-plus area is in the p-substrate, then an n-channel transistor or n-type wire may be constructed. If the n-plus area is in the n-well (not shown), then an *ohmic* contact to the n-well may be constructed. An ohmic contact is one which is only resistive in nature and is not rectifying (as in the case of a diode). In other words, there is no junction (n-type and p-type silicon abutting). Current can flow in both directions in an ohmic contact. This type of mask is sometimes called the *select* mask because it *selects* those transistor regions that are to be n-type. In modern small dimension processes, to reduce hot carrier effects, considerable effort may go into what is termed "drain engineering."[4] Rather than using one single diffusion or implantation step and mask to produce the source/drain regions, quite complicated structures are constructed. Typical of these structures is the LDD or Lightly Doped Drain structure, which is illustrated in Fig. 3.7(g). This consists of a shallow n-LDD implant that covers the source/drain region where there is no poly (i.e., the normal source/drain region). A spacer oxide is then grown over the polysilicon gate. An n^+ implant is then used to produce n^+ implants that are spaced from the edge of the original poly gate edges. The spacer is then removed, resulting in a structure that is more resistant to hot-electron effects. Current 0.25µm processes revert to a simpler self-aligned structure presumably because of the complexity of the LDD structure.

- The next step usually uses the complement of the n-plus mask, although an extra mask is normally not needed. The "absence" of an n-plus region over a thin-oxide area indicates that the area will be a p^+ diffusion or p-active. P-active in the n-well defines possible p-transistors and wires (Fig. 3.7h). A p^+ diffusion in the p-substrate allows an ohmic contact to be made. Following this step, the surface of the chip is covered with a layer of SiO_2. The LDD step is not necessarily done for p-transistors because their hot-carrier susceptibility is much less than that of n-transistors. For this reason, the drawn length dimension of p-transistors might be larger than that of the n-transistors.

- Contact cuts are then defined. This involves etching any SiO_2 down to the surface to be contacted (Fig. 3.7i). These allow metal (next step) to contact diffusion regions or polysilicon regions.

- Metallization is then applied to the surface and selectively etched (Fig. 3.7j) to produce circuit interconnections.

- As a final step (not shown), the wafer is passivated and openings to the bond pads are etched to allow for wire bonding. Passivation protects the silicon surface against the ingress of contaminants that can modify circuit behavior in deleterious ways.

The cross-section of the finished n-well process is shown in Fig. 3.8(c). The layout of the n-well CMOS transistors corresponding to this cross-section is illustrated in Fig. 3.8(b). The corresponding schematic (for an inverter) is shown in Fig. 3.8(a). From Fig. 3.8 it is evident that the p-type substrate accommodates n-channel devices, whereas the n-well accommodates p-channel devices. (Figure 3.8 also appears in color as Plate 1.)

In an n-well process, the p-type substrate is normally connected to the negative supply (V_{SS}) through what are termed V_{SS} substrate contacts, while

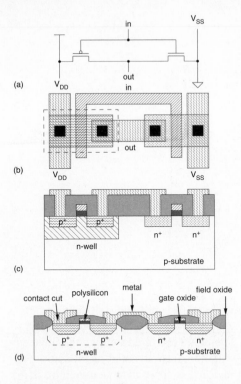

(a)

(b)

(c)

(d)

FIGURE 3.8 Cross section of a CMOS inverter in an n-well process

the well has to be connected to the positive supply (V_{DD}) through V_{DD} substrate (or well or tub) contacts. As the substrate is accessible at the top of the wafer and the bottom, connecting the substrate may be accomplished from the backside of the wafer. Topside connection is preferred because it reduces parasitic resistances that could cause latchup (see later). Substrate connections that are formed by placing n^+ regions in the n-well (V_{DD} contacts) and p^+ in the p-type substrate (V_{SS} contacts) are illustrated by Fig. 3.9(a). The corresponding layout is shown in Fig. 3.9(b). Other terminology for these contacts include "well contacts," "body ties," or "tub ties" for the V_{DD} substrate connection. We will use the term "substrate contact" for both V_{SS} and V_{DD} contacts, because this terminology can be commonly used for most bulk CMOS processes. It should be noted that these contacts are formed during the implants used for the p-channel and n-channel transistor formation.

3.2.2 The p-well Process

N-well processes have emerged in popularity in recent years. Prior to this, p-well processes were one of the most commonly available forms of CMOS. Typical p-well fabrication steps are similar to an n-well process, except that a p-well is implanted rather than an n-well. The first masking step defines the p-well regions. This is followed by a low-dose boron implant driven in by a

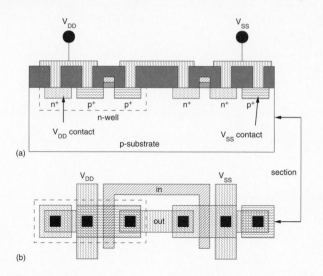

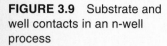

FIGURE 3.9 Substrate and well contacts in an n-well process

high-temperature step for the formation of the p-well. The well depth is optimized to ensure against n-substrate to n^+ diffusion breakdown, without compromising p-well to p^+ separation. The next steps are to define the devices and other diffusions; to grow field oxide; contact cuts; and metallization. A p-well mask is used to define p-well regions, as opposed to an n-well mask in an n-well process. A p-plus (p^+) mask may be used to define the p-channel transistors and V_{SS} contacts. Alternatively, we could use an n-plus mask to define the n-channel transistors, because the masks usually are the complement of each other.

P-well processes are preferred in circumstances where the characteristics of the n- and p-transistors are required to be more balanced than that achievable in an n-well process. Because the transistor that resides in the native substrate tends to have better characteristics, the p-well process has better p devices than an n-well process. Because p-devices inherently have lower gain than n devices, the n-well process exacerbates this difference while a p-well process moderates the difference.

3.2.3 Twin-Tub Processes

Twin-tub CMOS technology provides the basis for separate optimization of the p-type and n-type transistors, thus making it possible for threshold voltage, body effect, and the gain associated with n- and p-devices to be independently optimized.[5,6] Generally, the starting material is either an n^+ or p^+ substrate with a lightly doped *epitaxial* or *epi* layer, which is used for protection against latchup (see Section 3.5). The aim of *epitaxy* (which means "arranged upon") is to grow high-purity silicon layers of controlled thick-

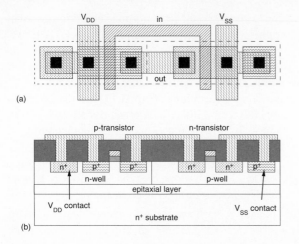

FIGURE 3.10 Twin-well CMOS process cross section

ness with accurately determined dopant concentrations distributed homogeneously throughout the layer. The electrical properties of this layer are determined by the dopant and its concentration in the silicon. The process sequence, which is similar to the n-well process apart from the tub formation where both p-well and n-well are utilized, entails the following steps:

- Tub formation.
- Thin-oxide construction.
- Source and drain implantations.
- Contact cut definition.
- Metallization.

Since this process provides separately optimized wells, balanced performance n-transistors and p-transistors may be constructed. Note that the use of threshold adjust steps is included in this process. These masks are derived from the active and n-plus masks. The cross-section of a typical twin-tub structure is shown in Fig. 3.10. The substrate contacts (both of which are required) are also included.

3.2.4 Silicon On Insulator

Rather than using silicon as the substrate, technologists have sought to use an insulating substrate to improve process characteristics such as latchup and speed. Hence the emergence of Silicon On Insulator (SOI) technologies. SOI CMOS processes have several potential advantages over the traditional CMOS technologies.[7] These include closer packing of p- and n-transistors, absence of latchup problems, and lower parasitic substrate capacitances. In the SOI process a thin layer of single-crystal silicon film is epitaxially grown on an insulator such as sapphire or magnesium aluminate spinel.[8] Alternatively, the silicon may be grown on SiO_2 that has been in turn grown on silicon. This option has proved more popular in recent years due to the compatibility of the starting material with conventional silicon CMOS fabrication. Various masking and doping techniques (Fig. 3.11) are then used to form p-channel and n-channel devices. Unlike the more conventional CMOS approaches, the extra steps in well formation do not exist in this technology. The steps used in typical SOI CMOS processes are as follows:

- A thin film (7–8 μm) of very lightly-doped n-type Si is grown over an insulator. Sapphire or SiO_2 is a commonly used insulator (Fig. 3.11a).

- An anisotropic etch is used to etch away the Si except where a diffusion area (n or p) will be needed. The etch must be anisotropic since the thickness of the Si is much greater than the spacings desired between the Si "islands" (Fig. 3.11b, 3.11c).

- The p-islands are formed next by masking the n-islands with a photoresist. A p-type dopant, boron, for example—is then implanted. It is masked by the photoresist, but forms p-islands at the unmasked islands. The p-islands will become the n-channel devices (Fig. 3.11d).

- The p-islands are then covered with a photoresist and an n-type dopant—phosphorus, for example—is implanted to form the n-islands. The n-islands will become the p-channel devices (Fig. 3.11e).

- A thin gate oxide (around 100–250 Å) is grown over all of the Si structures. This is normally done by thermal oxidation.

- A polysilicon film is deposited over the oxide. Often the polysilicon is doped with phosphorus to reduce its resistivity (Fig. 3.11f).

- The polysilicon is then patterned by photomasking and is etched. This defines the polysilicon layer in the structure (Fig. 3.11g).

- The next step is to form the n-doped source and drain of the n-channel devices in the p-islands. The n-islands are covered with a photoresist and an n-type dopant, normally phosphorus, is implanted. The dopant will be blocked at the n-islands by the photoresist, and it will be blocked from the gate region of the p-islands by the polysilicon. After this step the n-channel devices are complete (Fig. 3.11h).

- The p-channel devices are formed next by masking the p-islands and implanting a p-type dopant such as boron. The polysilicon over the gate of the n-islands will block the dopant from the gate, thus forming the p-channel devices (Fig. 3.11i).

- A layer of phosphorus glass or some other insulator such as silicon dioxide is then deposited over the entire structure.

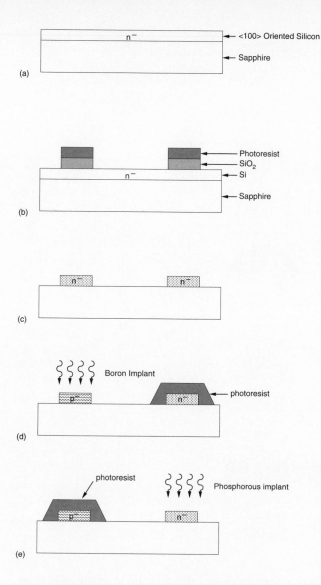

FIGURE 3.11 SOI process flow

- The glass is etched at contact-cut locations. The metallization layer is formed next by evaporating aluminum over the entire surface and etching it to leave only the desired metal wires. The aluminum will flow through the contact cuts to make contact with the diffusion or polysilicon regions (Fig. 3.11j)
- A final passivation layer of phosphorus glass is deposited and etched over bonding pad locations (not shown).

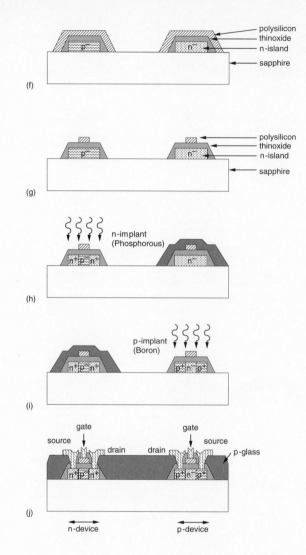

FIGURE 3.11   *(continued)*

Because the diffusion regions extend down to the insulating substrate, only "sidewall" areas associated with source and drain diffusions contribute to the parasitic junction capacitance. Since sapphire and SiO_2 are extremely good insulators, leakage currents between transistors and substrate and adjacent devices are almost eliminated.

In order to improve the yield, some processes use "preferential etch," in which the island edges are tapered. Thus aluminum or poly runners can enter and leave the islands with a minimum step height. This is contrasted to "fully anisotropic etch," in which the undercut is brought to zero, as shown in Fig. 3.12. An "isotropic etch" is also shown in the same diagram for comparison.

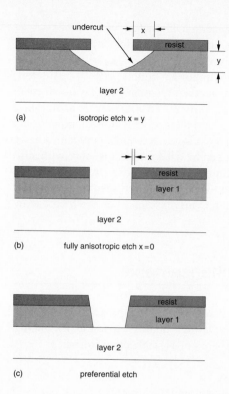

(a)   isotropic etch x = y

(b)   fully anisotropic etch x = 0

(c)   preferential etch

FIGURE 3.12 Classification of etching processes

The advantages of SOI technology are as follows:

- Due to the absence of wells, transistor structures denser than bulk silicon are feasible. Also direct n-to-p connections may be made.
- Lower substrate capacitances provide the possibility for faster circuits.
- No field-inversion problems exist (insulating substrate).
- There is no latchup because of the isolation of the n- and p-transistors by the insulating substrate.
- Because there is no conducting substrate, there are no body-effect problems. However, the absence of a backside substrate contact could lead to odd device characteristics, such as the "kink" effect in which the drain current increases abruptly at around 2 to 3 volts.[9]
- There is enhanced radiation tolerance (in fact, this is almost the sole reason the technology has been justified to date).

However, on the negative side, due to absence of substrate diodes, the inputs are somewhat more difficult to protect. Because device gains are lower, I/O structures have to be larger. Although parasitic capacitances to the substrate are reduced, the coupling capacitance between wires still exists so that the actual reduction in stray load capacitance is less than one would hope (see Chapter 4). The density advantage of SOI is not particularly important, because the density of contemporary digital processes is determined by the number and density of the metal interconnection layers. Single-crystal sapphire, spinel substrates, and silicon on SiO_2 are considerably more expensive than silicon substrates, and their processing techniques tend to be less developed than bulk silicon techniques. Recently, companies have started to produce SOI substrates that can be used interchangeably with silicon substrates in bulk CMOS fabrication lines. As the barrier to using insulating substrates is reduced, more use of them might be seen in day-to-day circuits, where the possible performance increase justifies the increase in processing cost and complexity.

3.3   CMOS Process Enhancements

A number of enhancements may be added to the CMOS processes, primarily to increase routability of circuits, provide high-quality capacitors for analog circuits and memories, or provide resistors of variable characteristics.
    These enhancements include

- double- or triple- or quadruple-level metal (or more).
- double- or triple-level poly (or more).
- combinations of the above.

We will examine these additions in terms of the additional functionality that they bring to a basic CMOS process.

3.3.1   Interconnect

Probably the most important additions for CMOS logic processes are additional signal- and power-routing layers. This eases the routing (especially automated routing) of logic signals between modules and improves the power and clock distribution to modules. Improved routability is achieved through additional layers of metal or by improving the existing polysilicon interconnection layer.

3.3.1.1   Metal Interconnect

A second level of metal is almost mandatory for modern CMOS digital design. A third layer is becoming common and is certainly required for leading-edge high-density, high-speed chips. Normally, aluminum is used for the

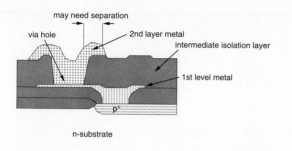

FIGURE 3.13   Two-level metal process cross section

metal layers. If some form of planarization is employed the second-level metal pitch can be the same as the first. As the vertical topology becomes more varied, the width and spacing of metal conductors has to increase so that the conductors do not thin and hence break at vertical topology jumps (*step coverage*).

Contacting the second-layer metal to the first-layer metal is achieved by a *via*, as shown in Fig. 3.13. If further contact to diffusion or polysilicon is required, a separation between the via and the contact cut is usually required. This requires a first-level metal *tab* to bridge between metal2 and the lower-level conductor. It is important to realize that in contemporary processes first-level metal must be involved in any contact to underlying areas. A number of contact geometries are shown in Fig. 3.14. Processes usually require metal borders around the via on both levels of metal although some pro-

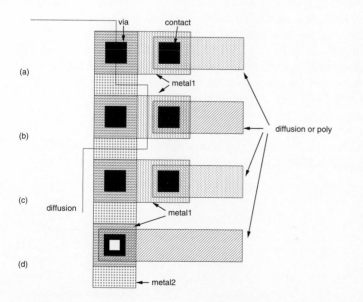

FIGURE 3.14   Two-level metal via/contact geometries

cesses require none. Processes may have no restrictions on the placement of the via with respect to underlying layers (Fig. 3.14a) or they may have to be placed inside (Fig. 3.14b) or outside (Fig. 3.14c) the underlying polysilicon or diffusion areas. Aggressive processes allow the stacking of vias on top of contacts, as shown in Fig. 3.14(d). Consistent with the relatively large thickness of the intermediate isolation layer, the vias might be larger than contact cuts and second-layer metal may need to be thicker and require a larger via overlap although modern processes strive for uniform pitches on metal1 and metal2.

The process steps for a two-metal process are briefly as follows:

- The oxide below the first-metal layer is deposited by atmospheric chemical vapor deposition (CVD).

- The second oxide layer between the two metal layers is applied in a similar manner.

- Depending on the process, removal of the oxide is accomplished using a plasma etcher designed to have a high rate of vertical ion bombardment. This allows fast and uniform etch rates. The structure of a via etched using such a method is shown in Fig. 3.13.

3.3.1.2   Polysilicon/Refractory Metal Interconnect

The polysilicon layer used for the gates of transistors is commonly used as an interconnect layer. However, the sheet resistance of doped polysilicon is between 20 and 40 Ω/square. If used as a long distance conductor, a polysilicon wire can represent a significant delay (see Chapter 4).

One method to improve this that requires no extra mask levels is to reduce the polysilicon resistance by combining it with a refractory metal. Three such approaches are illustrated in Fig. 3.15.[10] In Fig. 3.15(a), a *silicide* (e.g., silicon and tantalum) is used as the gate material. Sheet resistances of the order of 1 to 5 Ω/square may be obtained. This is called the silicide gate approach. Silicides are mechanically strong and may be dry etched in plasma reactors. Tantalum silicide is stable throughout standard processing and has the advantage that it may be retrofitted into existing process lines. Figure 3.15(b) uses a sandwich of silicide upon polysilicon,

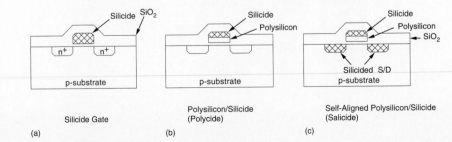

FIGURE 3.15   Refractory metal interconnect

which is commonly called the *polycide* approach. Finally, the silicide/poly-silicon approach may be extended to include the formation of source and drain regions using the silicide. This is called the *salicide* process (Self ALigned SILICIDE) (Fig. 3.15c). The effect of all of these processes is to reduce the "second layer" interconnect resistance, allowing the gate material to be used as a moderate long-distance interconnect. This is achieved by minimum perturbation of an existing process. An increasing trend in processes is to use the salicide approach to reduce the resistance of both gate and source/drain conductors.

3.3.1.3 Local Interconnect

The silicide itself may be used as a "local interconnect" layer for connection within cells.[11] As an example TiN[12] is used. Local interconnect allows a direct connection between polysilicon and diffusion, thus alleviating the need for area-intensive contacts and metal. Figure 3.16 shows a portion (p-devices only) of a six transistor SRAM cell that uses local interconnect. The local interconnect has been used to make the polysilicon-to-diffusion connections within the cell, thereby alleviating the need to use metal (and contacts). Metal2 (not shown) bit lines run over the cell vertically. Use of local interconnect in this RAM reduced the cell area by 25%. In general, local

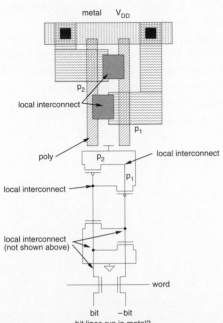

FIGURE 3.16 Local interconnect as used in a RAM cell

interconnect if available can be used to complete intracell routing, leaving the remaining metal layers for global wiring.

3.3.2   Circuit Elements

3.3.2.1   *Resistors*

Polysilicon, if left undoped, is highly resistive. This property is used to build resistors that are used in static memory cells. The process step is achieved by preventing the resistor areas from being implanted during normal processing. Resistors in the tera-Ω (10^{12} Ω) region are used.[13] A value of 3 TΩ, results in a standby current of 2μA for a 1 Mbit memory.

For mixed signal CMOS (analog and digital), a resistive metal such as nichrome may be added to produce high-value, high-quality resistors. The resistor accuracy might be further improved by laser trimming the resulting resistors on each chip to some predetermined test specification. In this process a high-powered laser vaporizes areas of the metal resistor until it meets a measurement constraint. Sheet resistance values in the KΩ/square are normal. The resistors have excellent temperature stability and long-term reliability.

3.3.2.2   *Capacitors*

Good-quality capacitors are required for switched-capacitor analog circuits while small high-value/area capacitors are required for dynamic memory cells. Both types of capacitors are usually added by using at least one extra layer of polysilicon, although the process techniques are very different.

Polysilicon capacitors for analog applications are the most straightforward. A second thin-oxide layer is required in order to have an oxide sandwich between the two polysilicon layers yielding a high-capacitance/unit area. Figure 3.17 shows a typical polysilicon capacitor. The presence of this second oxide can also be used to fabricate transistors. These may differ in characteristics from the primary gate oxide devices.

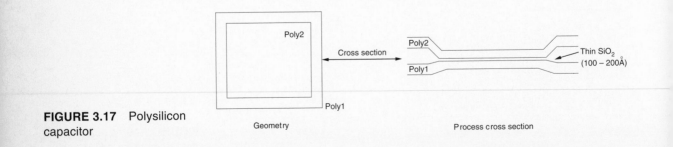

FIGURE 3.17   Polysilicon capacitor

For memory capacitors, recent processes have used three dimensions to increase the capacitance/area. One popular structure is the trench capacitor, which has evolved considerably over the years to push memory densities to 64Mbits and beyond.[14] A typical trench structure is shown in Fig. 3.18(a).[15] The sides of the trench are doped n^+ and coated with a thin 10nm oxide. Sometimes oxynitride is used because its high dielectric constant increases the capacitance. The trench is filled with a polysilicon plug, which forms the bottom plate of the cell storage capacitor. This is held at $V_{DD}/2$ via a metal connection at the edge of the array. The sidewall n^+ forms the other side of the capacitor and one side of the pass transistor that is used to enable data onto the bit lines. The bottom of the trench has a p^+ plug that forms a channel-stop region to isolate adjacent capacitors. The trench is 4µm deep and has a capacitance of 90fF. Rather than building a trench, Fig. 3.18(b) shows a fin-type capacitor used in a 64-Mb DRAM.[16,17] The storage capacitance is 20 to 30fF. The fins have the additional advantage of reducing the bit capacitance by shielding the bit lines. The fabrication of 3D-process structures such as these is a constant reminder of the skill, perseverance, and ingenuity of the process engineer.

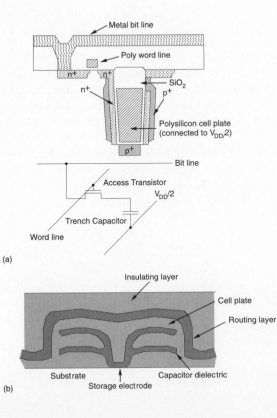

FIGURE 3.18 Dynamic memory capacitors; © IEEE 1988, © IEEE 1991.

3.3.2.3 Electrically Alterable ROM

Frequently, electrically alterable/erasable ROM (EAROM/EEROM) is added to CMOS processes to yield permanent but reprogrammable storage to a process. This is usually added by adding a polysilicon layer. Figure 3.19 shows a typical memory structure, which consists of a stacked-gate structure.[18,19] The normal gate is left floating, while a control gate is placed above the floating gate. A very thin oxide called the tunnel oxide separates the floating gate from the source, drain, and substrate. This is usually about 10 nm thick. Another thin oxide separates the control gate from the floating gate. By controlling the control-gate, source, and drain voltages, the very thin tunnel oxide between the floating gate and the drain of the device is used to allow electrons to "tunnel" to or from the floating gate to turn the cell off or on, respectively, using Fowler-Nordheim tunneling (Section 2.2.2.5). Alternatively, by setting the appropriate voltages on the terminals, "hot electrons" can be induced to charge the floating gate, thereby programming the transistor. In non–electrically alterable versions of the technology, the process can be reversed by illuminating the gate with UV light. In these cases the chips are usually housed in glass-lidded packages. (See also Section 6.3.2).

3.3.2.4 Bipolar Transistors

The addition of the bipolar transistor to the device repertoire forms the basis for BiCMOS processes. Adding an npn-transistor can markedly aid in reducing the delay times of highly loaded signals, such as memory word lines and microprocessor busses. Additionally, for analog applications bipolar transistors may be used to provide better performance analog functions than MOS alone.

To get merged bipolar/CMOS functionality, MOS transistors can be added to a bipolar process or vice versa. In past days, MOS processes always had to have excellent gate oxides while bipolar processes had to have precisely controlled diffusions. A BiCMOS process has to have both.

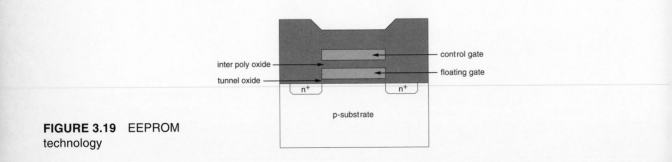

FIGURE 3.19 EEPROM technology

A mixed signal BiCMOS process[20] cross section is shown in Fig. 3.20. This process features both npn- and pnp-transistors in addition to pMOS and nMOS transistors. The major processing steps are summarized in Fig. 3.21, showing the particular device to which they correspond. The base layers of the process are similar to the process shown in Fig. 3.7. The starting material is a lightly-doped p-type substrate into which antimony or arsenic are diffused to form an n^+ buried layer. Boron is diffused to form a buried p^+ layer. An n-type epitaxial layer 4.0 μm thick is then grown. N-wells and p-wells are then diffused so that they join in the middle of the epitaxial layer. This epitaxial layer isolates the pnp-transistor in the horizontal direction, while the buried n^+ layer isolates it vertically. The npn-transistor is junction-isolated. The base for the pnp is then ion-implanted using phosphorous. A diffusion step follows this to get the right doping profile. The npn-collector is formed by depositing phosphorus before LOCOS. Field oxidation is carried out and the gate oxide is grown. Boron is then used to form the p-type base of the npn-transistor. Following the threshold adjustment of the pMOS transistors, the polysilicon gates are defined. The emitters of the npn-transistors employ polysilicon rather than a diffusion. These are formed by opening windows and depositing polysilicon. The n^+ and p^+ source/drain implants are then completed. This step also dopes the npn-emitter and the extrinsic bases of the npn- and pnp-transistors (extrinsic because this is the part of the base that is not directly between collector and emitter). Following the deposition of PSG, the normal two-layer metallization steps are completed. (*Note:* Generating the diffusions may require two distinct steps, the first being to get the impurities to the area where a diffusion is required and the second to drive the diffusion into the substrate to gain an acceptable impurity profile. These profiles have a major impact on the performance of the bipolar transistors.)

Representative of a high-density digital BiCMOS process is that represented by the cross section shown in Fig. 3.22.[21] The buried-layer–epitaxial-layer–well structure is very similar to the previous structure. However,

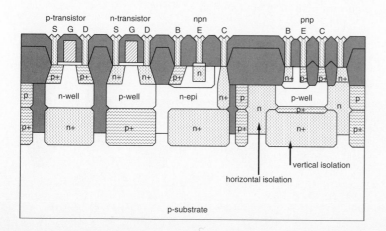

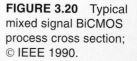

FIGURE 3.20 Typical mixed signal BiCMOS process cross section; © IEEE 1990.

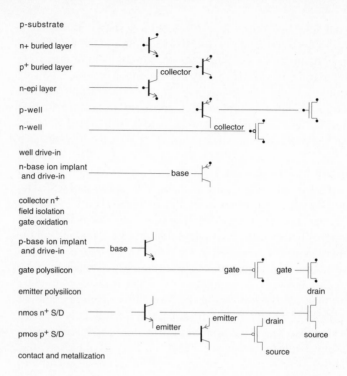

FIGURE 3.21 BiCMOS process steps for the cross section shown in Figure 3.20

because this is a 0.8μm process, LDD structures must be constructed for the p-transistors and the n-transistors. The npn is formed by a double-diffused sequence in which both base and emitter are formed by impurities that diffuse out of a covering layer of polysilicon. This process, intended for logic applications, has only an npn-transistor. The collector of the npn is connected to the n-well, which is in turn connected to the V_{DD} supply. Thus all npn-collectors are commoned. A typical npn-transistor with a 0.8μm-square emitter has a current gain of 90 and an f_t of 15 GHz.

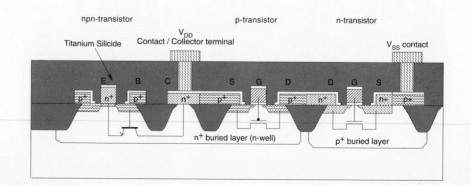

FIGURE 3.22 Digital BiCMOS process cross section; © IEEE 1991.

3.3.2.5 Thin-film Transistors

A thin-film transistor has source/drain and channel regions constructed from deposited thin films of semiconductor material. Apart from SOI processes, thin-film transistors are currently used in high-density memories and in flat-panel displays, although they have been around since the early 1960s.[22,23] Those used in memories are examples of TFTs that are added to existing CMOS processes.[24]

Representative of those transistors used in memories is the p-transistor, which is shown in Fig. 3.23(a), which is used as a load transistor in a static memory cell in a high-density SRAM.[25] In this device, third-level poly forms the gate of the device, while fourth-level poly 40nm thick forms the source, drain, and channel. The channel is separated from the gate by a 40nm

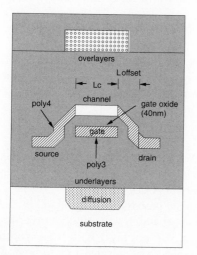

(a)

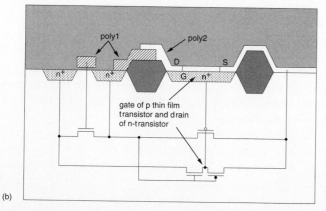

(b)

FIGURE 3.23 Examples of thin-film pMOS transistors as used in memories; © IEEE 1990.

oxide. In addition the drain is offset from the gate by the distance L_{offset}. As shown the transistor is called an "inverted staggered" thin-film transistor. The advantage of a pMOS load in memories is that the off current is of the order of 100fA compared to about 3pA for a polysilicon resistor load. For a 4Mb SRAM this results in a standby current of 0.2μA. In addition, the on current is around 10pA, which is high enough to counter any leakage current that would corrupt the data.

Another thin-film pMOS load transistor is shown[26] in Fig. 3.23(b). It is constructed from a thin film of amorphous (noncrystalline) silicon (α-silicon), 100nm thick. This film regrows crystal grains when heated to about 600°C. The larger the crystals (1–2 μm), the better the on and off characteristics of the transistor. The gate of the pMOS transistor in this instance is the source diffusion of the nMOS memory transistor. A thin gate-oxide film 40nm thick separates the "substrate" of the thin-film transistor from the gate. The pMOS transistor is 0.6μ wide and 1.4μ long. This 3D structure reduces the size of the memory cell quite considerably. As processes mature, it is highly likely that more use will be made of three-dimensional structures similar to these pMOS loads.

Thin-film CMOS transistors are also used in active-matrix LCD displays.[27] These devices have thresholds in the 4-volt range and mobilities of around 120 cm^2/Vs for the p-channel and 140 cm^2/Vs for the n-channel transistors.

3.3.3   3-D CMOS

The addition of thin-film transistors in memories effectively uses the third (vertical) dimension available on a chip. More general 3-D logic structures have been proposed and fabricated in CMOS.

One such example is shown in the process crossection in Fig. 3.24(a).[28] The substrate is an n^+ substrate upon which a p epitaxial layer is grown. Standard n-transistors are built on this epi layer with the exception of a "sinker" layer, which allows the sources of n-transistors to be down-connected to the n^+ substrate, which forms a ground plane and the V_{SS} connection. This eliminates half of the metal power wiring because V_{SS} is fed via the backside connection. A second gate oxide is grown over the n-transistor. A "seed" opening at the n-transistor drain, which allows high-quality silicon to be grown vertically and laterally, is opened. This is planarized, and a third oxide is grown on top of this epitaxially grown silicon. A polysilicon gate, used to implant self-aligned p^+ source/drains, which extend to the bottom of the epitaxially grown layer, is added on top of this structure. Planarization and metallization then are completed.

The final structure has a p- and an n-transistor with a common gate, while the p-transistor has an extra parallel gate which controls it. This basic structure allows an inverter and a 2-input multiplexer to be constructed (Fig. 3.24b). Note that there are actually two p-transistors in parallel, created

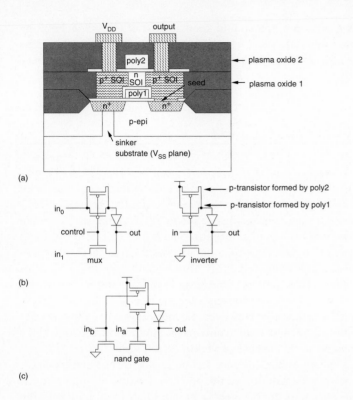

(a)

(b)

(c)

FIGURE 3.24   3-D CMOS logic technology; © IEEE 1992.

by poly1 and poly2. They both act on a common n SOI channel. By connecting the two gates together the resulting p-transistor has almost the same β as the n-transistor, thereby equalizing signal delays. By adding another series n-transistor, a 2-input NAND gate may be built (Fig. 3.24c). In the case of the inverter the p-source is connected to V_{DD}, the poly gates are commoned, and the n-source is "sinkered" to the substrate. For the 2-input multiplexer (or selector) the n- and p-sources are connected to the mux inputs while the gates are commoned to form the select line. For a NAND gate the poly gates are separately driven, as shown in Fig. 3.24(c). The diodes shown in the circuits in Fig. 3.24(b) and 3.24(c) are due to the abutting p^+ SOI drain of the p-transistor and the n^+ drain of the n-transistor. Using this novel technology, the inventors were able to design circuits that were up to 33% smaller than comparable 2-D structures.

3.3.4   Summary

This concludes the discussion of some relevant CMOS technology. Processes are constantly under development with new structures and new techniques being introduced to yield smaller, higher speed, less costly, and more reliable ICs. As a designer, you should keep abreast of CMOS technology directions because they often make previously impossible systems or ideas possible. A good forum is the annual IEEE International Electron Devices Meeting (IEDM).

3.4   Layout Design Rules

Layout rules, also referred to as *design rules,* can be considered as a prescription for preparing the photomasks used in the fabrication of integrated circuits. The rules provide a necessary communication link between circuit designer and process engineer during the manufacturing phase. The main objective associated with layout rules is to obtain a circuit with optimum yield (functional circuits versus nonfunctional circuits) in as small an area as possible without compromising reliability of the circuit.

In general, design rules represent the best possible compromise between performance and yield. The more conservative the rules are, the more likely it is that the circuit will function. However, the more aggressive the rules are, the greater the probability of improvements in circuit performance. This improvement may be at the expense of yield.

Design rules specify to the designer certain geometric constraints on the layout artwork so that the patterns on the processed wafer will preserve the topology and geometry of the designs. It is important to note that design rules do not represent some hard boundary between correct and incorrect fabrication. Rather, they represent a tolerance that ensures very high probability of correct fabrication and subsequent operation. For example, one may find that a layout that violates design rules may still function correctly, and vice versa. Nevertheless, any significant or frequent departure (*design-rule waiver*) from design rules will seriously prejudice the success of a design.

Two sets of design-rule constraints in a process relate to line widths and interlayer registration. If the line widths are made too small, it is possible for the line to become discontinuous, thus leading to an open circuit wire. On the other hand, if the wires are placed too close to one another, it is possible for them to merge together; that is, shorts can occur between two independent circuit nets. Furthermore, the spacing between two independent layers may be affected by the vertical topology of a process.

The design rules primarily address two issues: (1) the geometrical reproduction of features that can be reproduced by the mask-making and lithographical process and (2) the interactions between different layers.

There are several approaches that can be taken in describing the design rules. These include 'micron' rules stated at some micron resolution, and lambda(λ)-based rules. Micron design rules are usually given as a list of minimum feature sizes and spacings for all the masks required in a given process.

For example, the minimum active width might be specified as 1 µm. This is the normal style for industry. The lambda-based design rules popularized by Mead and Conway[29] are based on a single parameter, λ, which characterizes the linear feature—the resolution of the complete wafer implementation process—and permits first-order scaling. As a rule, they can be expressed on a single page. While these rules have been successfully used for 4–1.2 µm processes, they will probably not suffice for submicron processes.

Normally, there is some minimum grid dimension in terms of which the design rules are expressed. This is a result of the economic reality that eventually the mask has to be built and the higher the lithographic tolerance, the higher the cost of the mask. Also, historically, some mask making systems had digital accuracy limitations (i.e., 16 bits of precision). At the 1.25µ–2µ level, a minimum grid unit of .2–.25µ was adequate. In submicron processes a value of .05–.1µ is more common. In this text, we will use the λ rules to illustrate principles. Normal industry practice is to deal with the micron dimensions to ensure that the circuits built are as small as possible. Contemporary CAD tools now allow designs to migrate between compatible CMOS processes without having to resort to the linear scaling that λ rules impose.

3.4.1 Layer Representations

The advances in the CMOS processes are generally complex and somewhat inhibit the visualization of all the mask levels that are used in the actual fabrication process. Nevertheless the design process can be abstracted to a manageable number of conceptual layout levels that represent the physical features observed in the final silicon wafer. At a sufficiently high conceptual level all CMOS processes use the following features:

- Two different substrates.
- Doped regions of both p- and n-transistor-forming material.
- Transistor gate electrodes.
- Interconnection paths.
- Interlayer contacts.

The layers for typical CMOS processes are represented in various figures in terms of:

- a color scheme proposed by JPL based on the Mead-Conway colors.
- other color schemes designed to differentiate CMOS structures (e.g., the colors as used on the front cover of this book)
- varying stipple patterns.
- varying line styles.

TABLE 3.1   Layer Representations for the n-well CMOS process

LAYER	COLOR	SYMBOLIC	COMMENTS
N-well	Brown		Inside brown is n-well, outside is p-type substrate.
Thin-oxide	Green	n-transistor	Thinox may not cross a well boundary.
Poly	Red	Polysilicon	Generally n^+.
P^+	Yellow	p-transistor	Inside is p^+.
Metal1	Light blue	Metal1	
Metal2	Tan	Metal2	
Contact-cut, via	Black	Contact	
Metal3	Grey	Metal3	
Overglass			

Some of these representations are shown in Table 3.1. Where diagrams are presented, a legend will be used to indicate any different layer assignments from these defaults. At the mask level, some layers may be omitted for clarity. At the symbolic level only n- and p-transistors will be shown (i.e., no wells or select layers). The symbolic representations should be viewed as translating to the appropriate set of masks for whatever process is being considered.

The p-well and twin-tub bulk CMOS processes as well as the SOI process can be represented in a similar manner. For example, in p-well bulk CMOS the only difference in the resulting wafer structure is the reversal of the role of the well and the original substrate. Different process lines may use different combinations of the n^+, p^+, n-well, or p-well masks to define the process. It is very important to intimately understand what set of masks a particular process line uses if you are responsible for generating interface formats. For instance, an n^+ mask, which is the reverse of a p^+ mask, may be used. Thus n^+ active area denotes n-transistors, and so on. Conceptually, the mask levels in a silicon-on-insulator process are probably the simplest, The levels and visible geometry in this process correspond directly to the features that a designer has to deal with conceptually (i.e., n-regions and p-regions). Perhaps the most significant difference between SOI and bulk CMOS processes, from the designer's point of view, is the absence of wells.

3.4.2   CMOS n-well Rules

In this section we describe a version of n-well rules based on the MOSIS CMOS Scalable Rules and compare those with the rules for a hypothetical (but realistic) commercial 1μ CMOS process (Table 3.2). The MOSIS rules are expressed in terms of λ. These rules allow some degree of scaling

TABLE 3.2 CMOS Layout Rules

	λ RULE	λ/μ RULE (0.5μ)	μ RULE
A. N-well layer			
A.1 Minimum size	10λ	5μ	2μ
A.2 Minimum spacing (wells at same potential)	6λ	3μ	2μ
A.3 Minimum spacing (wells at different potentials)	8λ	4μ	2μ
B. Active Area			
B.1 Minimum size	3λ	1.5μ	1μ
B.2 Minimum spacing	3λ	1.5μ	1μ
B.3 N-well overlap of p^+	5λ	2.5μ	1μ
B.4 N-well overlap of n^+	3λ	1.5μ	1μ
B.5 N-well space to n^+	5λ	2.5μ	5μ
B.6 N-well space to p^+	3λ	1.5μ	3μ
C. Poly 1			
C.1 Minimum size	2λ	1μ	1μ
C.2 Minimum spacing	2λ	1μ	1μ
C.3 Spacing to Active	1λ	0.5μ	0.5μ
C.4 Gate Extension	2λ	1μ	1μ
D. p-plus/n-plus (p^+, n^+ for short)			
D.1 Minimum overlap of Active	2λ	1μ	1μ
D.2 Minimum size	7λ	3.5μ	3μ
D.3 Minimum overlap of Active in abutting contact (see Fig. 3.27)	1λ	0.5μ	2μ
D.4 Spacing of p^+/n^+ to n^+/p^+ gate	3λ	1.5μ	1.5μ
E. Contact			
E.1 Minimum size	2λ	1μ	0.75μ
E.2 Minimum spacing (Poly)	2λ	1μ	1μ
E.3 Minimum spacing (Active)	2λ	1μ	0.75μ
E.4 Minimum overlap of Active	2λ	1μ	0.5μ
E.5 Minimum overlap of Poly	2λ	1μ	0.5μ
E.6 Minimum overlap of Metal1	1λ	0.5μ	0.5μ
E.7 Minimum spacing to Gate	2λ	1μ	1μ
F. Metal1			
F.1 Minimum size	3λ	1.5μ	1μ
F.2 Minimum spacing	3λ	1.5μ	1μ

(continued)

TABLE 3.2   *(continued)*

	λ RULE	λ/μ RULE (0.5μ)	μ RULE
G. Via			
G.1 Minimum size	2λ	1μ	0.75μ
G.2 Minimum spacing	3λ	1.5μ	1.5μ
G.3 Minimum Metal1 overlap	1λ	0.5μ	0.5μ
G.4 Minimum Metal2 overlap	1λ	0.5μ	0.5μ
H. Metal2			
H.1 Minimum size	3λ	1.5μ	1μ
H.2 Minimum spacing	4λ	2μ	1μ
I. Via2			
I.1 Minimum size	2λ	1μ	1μ
I.2 Minimum spacing	3λ	1.5μ	1.5μ
J. Metal3			
J.1 Minimum size	8λ	4μ	4μ
J.2 Minimum spacing	5λ	2.5μ	2.5μ
J.3 Minimum Metal2 overlap	2λ	1μ	1μ
J.4 Minimum Metal3 overlap	2λ	1μ	1μ
K. Passivation			
K.1 Minimum opening		100μ	100μ
K.2 Minimum spacing		150μ	150μ

between processes as, in principal, we only need to reduce the value of λ and the designs will be valid in the next process down in size. Unfortunately, history has shown that processes rarely shrink uniformly. Thus industry usually uses the actual micron-design rules and codes designs in terms of these dimensions, or uses symbolic layout systems to target the design rules exactly. At this time, the amount of polygon pushing is usually constrained to a number of frequently used standard cells or memories, where the effort expended is amortized over many designs. Alternatively, the designs are done symbolically, thus relieving the designer of having to deal directly with the actual design rules.

The rules are defined in terms of:

- feature sizes.
- separations and overlaps.

In addition to the rules stated above, there are various spacing rules for the periphery of the chip which frequently depend on the vendor (e.g., spacing of all layers to die boundary is 20–50μ).

For each mask required in a process one needs to know whether it is "light field" or "dark field," whether light will pass through the mask to expose a photolithographic pattern or whether light will be blocked by the mask. In addition, biases are added or subtracted from the drawn dimensions of the mask to allow for varying types of processing. For instance, the active mask might be bloated to take into account the encroachment of field oxide during LOCOS. Contacts might be shrunk as etching tends to make them larger during processing. The rules in Table 3.2 are illustrated in Fig. 3.25 (and in Plate 2). The comparison between the lambda rules and micron rules reveal differences that are accentuated as process line-widths are reduced below the 1μm level. In particular, the metal widths and spacings and contact overlaps yield different pitches. For instance, the metal1 contacted pitch (contact to contact) is 4.5μ for λ = 0.5μ but 2.75μ for the equivalent micron rules. Thus the micron rules result in a 50% size reduction. The metal2 rules differ by 5μ to 2.75μ—almost a factor of 2. As many circuits are dominated by routing, this can translate almost directly to the final density of the circuit. On the other hand, the transistor pitch is generally determined by the contact-poly-contact pitch, which is 4μ for the λ rules and 3.25μ for the micron rules, which can also lead to significant layout density differences. (Note: The metal3 rules used here are extremely conservative. Most modern sub 1μ processes have equivalent metal2 and metal3 pitches.)

TABLE 3.3 Submicron CMOS Process Dimensions

LAYER		NEC[30]	HITACHI[31]	TOSHIBA[32]	HITACHI[33]	IBM[34]
Gate Oxide		15nm	13.5nm	11nm		7nm
Poly1	Width	.55μ (.65μ for p)	.6μ	.5μ	.3μ	.4μ
	Space	.55μ	.6μ	.6μ		
Poly2	Width	.55μ	.6μ	.5μ		
	Space	.55μ	.6μ	.6μ		
Poly3	Width	.55μ	.6μ	.8μ		
	Space	.55μ	.6μ	.7μ		
Poly4	Width		.6μ			
	Space		.6μ			
Contact	Size		.6μ	.6μ		
Metal1	Width	.9μ	.7μ	1.4μ	.3μ	
	Space	.55μ	.6μ	.7μ	.4μ	
Via	Size		.6μ	1.2μ		
Metal2	Width	.9μ	.7μ	1.4μ	.45μ	
	Space	.55μ	.6μ	1.2μ	.65μ	
Metal3	Width				.55μ	
	Space				.75μ	

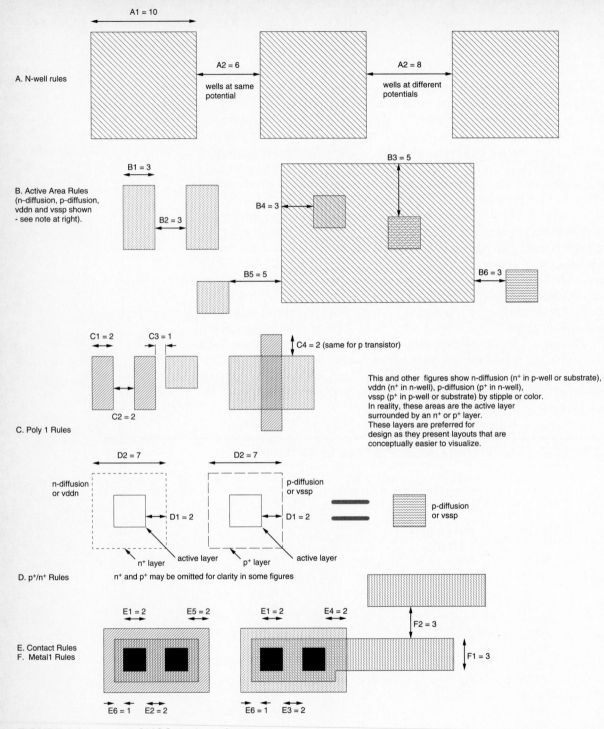

FIGURE 3.25 n-well CMOS design rules

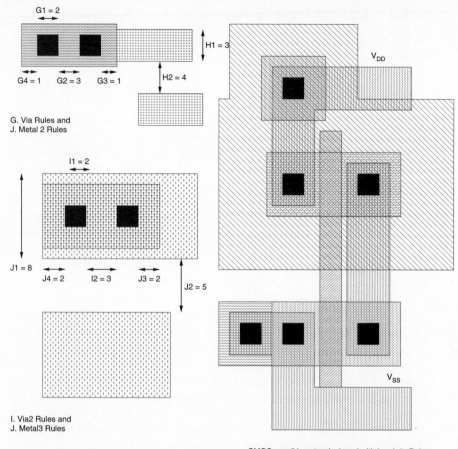

G1 = 2

H1 = 3

H2 = 4

G4 = 1 G2 = 3 G3 = 1

G. Via Rules and
J. Metal 2 Rules

I1 = 2

J1 = 8

J4 = 2 I2 = 3 J3 = 2

J2 = 5

I. Via2 Rules and
J. Metal3 Rules

V_{DD}

V_{SS}

CMOS n-well inverter designed with Lambda Rules
n^+ and p^+ layers are omitted

FIGURE 3.25 *(continued)*

Representative of processes in the 0.25–0.6μ range, the previous table (Table 3.3) summarizes the basic dimensions from published papers describing 4Mb static CMOS SRAMs and high-speed microprocessors. The RAM processes tend to have more poly layers (between 2 and 4) to enable small, dense memory cells to be constructed and the logic processes tend to have more metal layers (2 to 4) to improve routability.

These can be used as a guide to estimate sub 1μ technology rules. In particular the paper describing the IBM 0.25μ process in Table 3.3 provides a good overview of the considerations that go into a .25μm process.

3.4.3   Design Rule Backgrounder

In this section we will examine some of the reasons for the design rules listed above.

Well Rules: The n-well is usually a deeper implant compared with the transistor source/drain implants, therefore it is necessary for the outside dimension to provide sufficient clearance between the n-well edges and the adjacent n^+ diffusions. The inside clearance is determined by the transition of the field oxide across the well boundary. Some processes may permit zero inside clearance, but problems such as the 'birds-beaks' effect usually prevent this. A further point to be noted is that to avoid a shorted condition, active is not permitted to cross a well boundary. Since the n-well sheet resistance can be several KΩs per square, it is necessary to thoroughly ground the well. This will prevent excessive voltage drops due to substrate currents. Thus the rule to follow in grounding the n-well would be to put a substrate contact wherever space is available consistent with the rules outlined in Section 3.5.

Transistor rules: Where poly crosses active, the source and drain diffusion is masked by the poly region. The source, drain, and channel are thereby self-aligned to the gate. It is essential for the poly to completely cross active, otherwise the transistor that has been created will be shorted by a diffused path between source and drain. To ensure this condition is satisfied, poly is required to extend beyond the edges of the diffusion region. This is often termed the "gate extension." This effect is shown in Fig. 3.26(a) where the diffusion has increased in size and the poly has been overetched, resulting in a short. The thin oxide must extend beyond the poly gate so that diffused regions exist to carry charge into and out of the channel (Fig. 3.26b). Poly and active regions that do not meet intentionally to form a transistor should be kept separated. Both types of transistors have an active region (diffusion or implant) and a polysilicon region. A p-device has an n-well region surrounding it, whereas an n-device has an n^+ (n-plus) region surrounding it. Thin oxide areas that are not covered by n are p^+ and hence are p-devices or wires (within the n-well). Therefore a transistor is n-channel if it is inside an n^+ region; otherwise it is a p-channel device. From the above discussion it can be noted that there are two types of implant/diffusion used to form the p- and n-transistors. What is important to note is that n^+ diffusion is obtained by "logical anding" of active and n^+ (n-plus) masks, whereas p^+ diffusion is derived by "logical anding" of active and (NOT n^+) masks. Frequently, in order to simplify design the n-plus and/or p-plus masks are ignored during design and inserted automatically. A problem can occur if the orthogonal distance of n^+ (n-plus) to p^+ (p-plus) is used (Rule B.3 + B.5 for instance or B.4 + B.6). While the select layers may be added without problems for orthogonally spaced structures, diagonally positioned

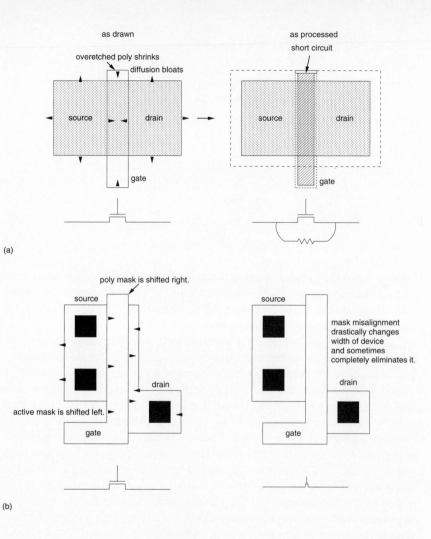

(a)

(b)

FIGURE 3.26 Effects of insufficient gate extension and source-drain extension

diffusions may violate the n-plus–p-plus spacing rules. In symbolic layout systems this frequently leads to a second set of spacings that describe diagonal constraints.

Contact Rules: There are several generally available contacts:

- Metal to p-active (p-diffusion).
- Metal to n-active (n-diffusion).
- Metal to polysilicon.
- V_{DD} and V_{SS} substrate contacts.
- Split (substrate contacts).

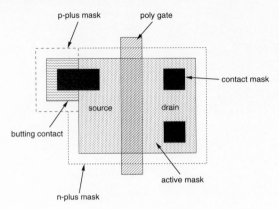

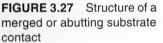

FIGURE 3.27 Structure of a merged or abutting substrate contact

Depending on the process, other contacts such as "buried" polysilicon-active contacts may be allowed. This contact allows direct connection between polysilicon and the active transistor region. Sometimes this type of contact is allowed to only one type of active area.

Because the substrate is divided into "well" regions, each isolated well must be "tied" to the appropriate supply voltage; that is, the n-well must be tied to V_{DD} and the substrate (what amounts to a p-well) must be tied to V_{SS}. This is achieved by the use of well or substrate contacts. One needs to note that every p-device must be surrounded by an n-well and that the n-well must be connected to V_{DD} via a V_{DD} contact. Furthermore, every n-device must have access to a V_{SS} contact. The split or merged contact is equivalent to two separate metal-diffusion contacts that are strapped together with metal (Fig. 3.27). This structure is used to tie transistor sources to either the substrate or the n-well. A version is also shown at the source of the n-transistor in the inverter in Fig. 3.25. Separate contacts are shown; this is consistent with modern processes, which usually require uniform contact sizes to achieve well-defined etching characteristics. Merged contact structures in older processes may have used an elongated contact rectangle (Fig. 3.27). The V_{SS} or V_{DD} merged contacts may be inset into the source of the corresponding n-transistor where wide transistors are employed. An alternative separated contact structure is shown for the V_{DD} contact for the p-transistor in Fig. 3.25. Here the n^+ well contact is separated from the p^+ source/drain diffusion.

Guard Rings: Guard rings that are p^+ diffusions in the p-substrate and n^+ diffusions in the n-well are used to collect injected minority carriers. If they are implemented in a structure, then n^+ guard rings must be tied to V_{DD}, while p^+ guard rings must be tied to V_{SS}. A p^+ diffusion with n^+ guard ring is shown in Fig. 3.28(a), while an n^+ diffusion with p^+ guard ring is shown in

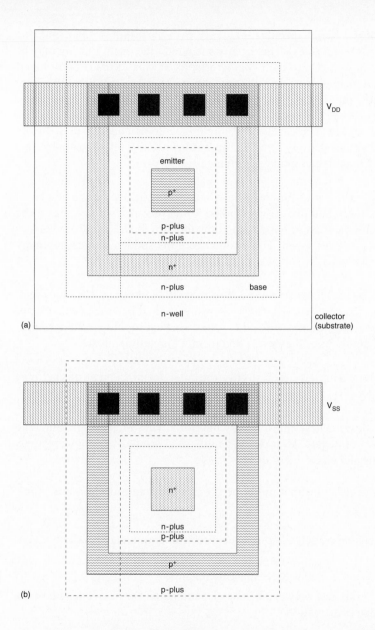

(a)

(b)

FIGURE 3.28 Guard rings

Fig. 3.28(b). Different well-enclosure rules may apply for guard-ring structures. The reason for guard rings will become more clear in Section 3.5. Incidentally, the structure shown in Fig. 3.28(a) is also that for a pnp transistor if one was required. The transistor terminals have been marked. The area of the center p^+ region is the area of the emitter. The base is the n-well and is connected via the n^+ ring. The collector is the substrate.

Metal Rules: Metal spacings may vary with the width of the metal line (so-called *fat-metal* rules). That is, at some width, the metal spacing may be increased. This is due to etch characteristics of small versus large metal wires. There may also be maximum-metal-width rules. Additionally, there may be rules that are applied to long closely spaced parallel metal lines. Some processes require a certain proportion of the chip area to be covered with metal, and in such cases metal might have to be added to chip "white space" (assuming there is some!). These rules usually relate to constraints imposed by manufacturability requirements.

Via Rules: Processes may vary in whether they allow vias to be placed over polysilicon and diffusion regions. Some processes allow vias to be placed within these areas but do not allow the vias to straddle the boundary of polysilicon or diffusion. This results from the sudden vertical topology variations that occur at sublayer boundaries.

Metal2 Rules: The possible increase in width and separation of second-level metal are conservative rules to ensure against broken conductors or shorts between adjoining wires due to the vertical topology. Modern processes frequently have the metal1 and metal2 pitches identical.

Via2 Rules: Similarly to first vias, the rules for placement of via2 may vary with process.

Metal3 Rules: These rules usually but not always increase in width and separation over metal2. Metal3 is generally used primarily for power-supply connections and clock distribution.

Some additional rules that might be present in some processes are as follows:

- Extension of polysilicon in the direction that metal wires exit a contact.
- Differing p- and n-transistor gate lengths.
- Differing gate poly extensions, depending on the device length or the device construction.

Whereas earlier processes tended to be process driven and frequently have long and involved design rules, increasingly more processes have become "designer friendly" or more specifically computer friendly because most of the mask geometries for designs are algorithmically produced. Also, system companies have created "generic" rules that span a number of different CMOS foundries that they might use. Some processes have design guidelines that feature structures to be avoided to ensure good yields. In general though, at this time, process technology is so well developed, features so small, and time to market so short that the traditional yield improvement cycle is only done for the highest volume parts. Frequently, the technology changes so fast that it is better to reimplement the circuit in the new smaller technology than worry about improving the yield on the older larger process. Of course at some time, a limit will come to how small technologies can be made and then a return to classical yield optimization will probably resurface.

Passivation or Overglass: This is a protective glass layer that covers the final chip. Openings are required at pads and any internal test points.

3.4.4   Scribe Line

The scribe line is a specifically designed structure that surrounds the completed chip and is the point at which the chip is cut with a diamond saw. The construction of the scribe line varies from manufacturer to manufacturer.

3.4.5   Layer Assignments

Table 3.4 lists the MOSIS Scalable CMOS Design-rule layer assignments for the Caltech Intermediate Form (CIF) language and Calma stream format.

TABLE 3.4   MOSIS Scalable CMOS Design-rule Layer Assignments

LAYER	CIF LAYER NAME	CALMA NUMBER
Well	CWG	14
N-well	CWN	1
P-well	CWP	2
Active	CAA	3
Select	CSG	15
P-select	CSP	8
N-select	CSN	7
Poly	CPG	4
Poly Contact	CCP	45
Poly 2 (Electrode)	CEL	5
Electrode Contact	CCE	55
Active Contact	CCA	35
Metal1	CMF	10
Via	CVA	11
Metal2	CMS	12
Via2[*]	CVB	65
Metal3[*]	CMT	14
Overglass	COG	13

[*]Author's assignment

3.4.6   SOI Rules

Usually SOI rules closely follow bulk CMOS rules except that n^+ and p^+ regions can abut. This allows some interesting multiplexer and latch circuits. A spacing rule between island edge and unrelated poly is used to ensure against shorts between the poly and island edges. This can be caused by thin or faulty oxide covering over the islands.

3.4.7   Design Rules—Summary

In commercial designs, λ rules are rarely sufficient to describe high-density, high-performance circuits. While all of these rules can be worst-cased, very inefficient designs result. A better approach is to implement systems that synthesize the correct geometry from an intermediate form. Therefore, symbolic styles of design provide a solution for creating generic CMOS circuits that can be implemented with a wide range of fabrication processes.

3.5   Latchup

If every silver lining has a cloud, then the cloud that has plagued CMOS is a parasitic circuit effect called "latchup." The result of this effect is the shorting of the V_{DD} and V_{SS} lines, usually resulting in chip self-destruction or at least system failure with the requirement to power down. This effect was a critical factor in the lack of acceptance of early CMOS processes, but in current processes it is controlled by process innovations and well-understood circuit techniques.

3.5.1   The Physical Origin of Latchup

The source of the latchup effect[35,36,37] may be explained by examining the process cross section of a CMOS inverter, shown in Fig. 3.29(a), on which is overlaid an equivalent circuit. The schematic depicts, in addition to the expected nMOS and pMOS transistors, a circuit composed of an npn-transistor, a pnp-transistor, and two resistors connected between the power and ground rails (Fig. 3.29b). Under the right conditions, this parasitic circuit has the VI characteristic shown in Fig. 3.29(c), which indicates that above some critical voltage (known as the trigger point) the circuit "snaps" and draws a large current while maintaining a low voltage across the terminals (known as the holding voltage). This is, in effect, a short circuit. As mentioned, the bipolar devices and resistors shown in Fig. 3.29(b) are parasitic, that is, an unwanted byproduct of producing pMOS and nMOS transistors. Further

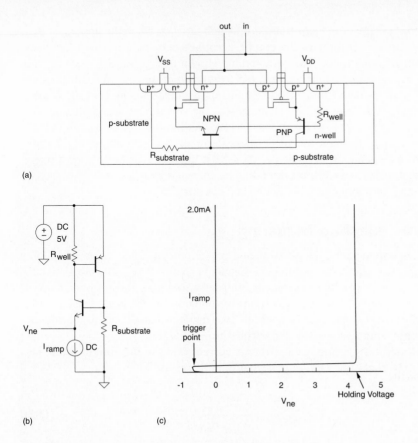

(a)

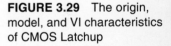

(b)   (c)

FIGURE 3.29 The origin, model, and VI characteristics of CMOS Latchup

examination of Fig. 3.29(a) reveals how these devices are constructed. The figure shows a cross-sectional view of a typical (n-well) CMOS process. The (vertical) pnp-transistor has its emitter formed by the p^+ source/drain implant used in the pMOS transistors. Note that either the drain or source may act as the emitter although the source is the only terminal that can maintain the latchup condition. The base is formed by the n-well, while the collector is the p-substrate. The emitter of the (lateral) npn-transistor is the n^+ source/drain implant, while the base is the p-substrate and the collector is the n-well. In addition, substrate resistance $R_{substrate}$ and well resistance R_{well} are due to the resistivity of the semiconductors involved.

Consider the circuit shown in Fig. 3.29(b). If a current is drawn from the npn-emitter, the emitter voltage becomes negative with respect to the base until the base emitter voltage is approximately 0.7 volts. At this point the npn-transistor turns on and a current flows in the well resistor due to common emitter current amplification of the npn-transistor. This raises the base emitter voltage of the pnp-transistor, which turns on when the pnp $V_{be} = -0.7$ volts. This in turn raises the npn base voltage causing a positive feedback condition, which has the characteristic shown in Fig. 3.29(c). At a certain npn-base-emitter voltage, called the *trigger point,* the emitter voltage suddenly "snaps back" and enters a stable state called the ON state. This state will persist as long as the voltage across the two transistors is greater than the holding voltage shown in the figure. As the emitter of the npn is the source/drains of the n-transistor, these terminals are now at roughly 4 volts. Thus there is about 1 volt across the CMOS inverter, which will most likely cause it to cease operating correctly. The current drawn is usually destructive to metal lines supplying the latched up circuitry.

3.5.2   Latchup Triggering

For latchup to occur, the parasitic npn-pnp circuit has to be triggered and the holding state has to be maintained. Latchup can be triggered by transient currents or voltages that may occur internally to a chip during power-up or externally due to voltages or currents beyond normal operating ranges. Radiation pulses can also cause latchup. Two distinct methods of triggering are possible, lateral triggering and vertical triggering.

Lateral triggering occurs when a current flows in the emitter of the lateral npn-transistor. The static trigger point is set by[38]

$$I_{ntrigger} \approx \frac{V_{pnp\text{-}on}}{\alpha_{npn} R_{well}}, \tag{3.1}$$

where

$V_{pnp\text{-}on}$ ~ 0.7 volts—the turn-on voltage of the vertical pnp-transistor

α_{npn} = common base gain of the lateral npn-transistor

R_{well} = well resistance.

Vertical triggering occurs when a sufficient current is injected into the emitter of the vertical-pnp transistor. Similar to the lateral case, this current is multiplied by the common-base-current gain, which causes a voltage drop across the emitter base junction of the npn transistor due to the resistance, $R_{substrate}$. When the holding or sustaining point is entered, it represents a stable operating point provided the current required to stay in the state can be maintained.

Current has to be injected into either the npn- or pnp-emitter to initiate latchup. During normal circuit operation in internal circuitry this may occur due to supply voltage transients, but this is unlikely. However, these conditions may occur at the I/O circuits employed on a CMOS chip, where the internal circuit voltages meet the external world and large currents can flow. Therefore extra precautions need to be taken with peripheral CMOS circuits. Figure 3.30(a) illustrates an example where the source of an nMOS output transistor experiences undershoot with respect to V_{SS} due to some external circuitry. When the output dips below V_{SS} by more than 0.7V, the drain of the nMOS output driver is forward biased, which initiates latchup. The complementary case is shown in Fig. 3.30(b) where the pMOS output transistor experiences an overshoot more than 0.7V beyond V_{DD}. Whether or not in these cases latchup occurs depends on the pulse widths and speed of the parasitic transistors.[39]

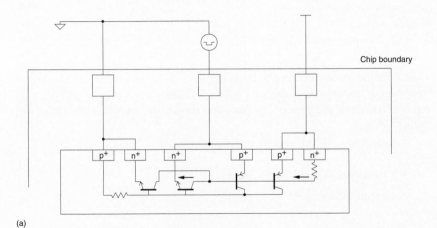

(a)

(b)

FIGURE 3.30 Externally induced latchup

3.5.3   Latchup Prevention

For latchup to occur an analysis of the circuit in Fig. 3.29(b) finds the following inequality has to be true[40]:

$$\beta_{npn}\beta_{pnp} > 1 + \frac{(\beta_{npn} + 1)\,(I_{Rsubstrate} + I_{Rwell}\beta_{pnp})}{I_{DD} - I_{Rsubstrate}}, \tag{3.2}$$

where

$$I_{Rsubstrate} = \frac{V_{be\,npn}}{R_{substrate}}$$

$$I_{Rwell} = \frac{V_{be\,pnp}}{R_{well}}$$

$$I_{DD} = \text{total supply current.}$$

This equation yields the keys to reducing latchup to the point where it should never occur under normal circuit conditions. Thus, reducing the resistor values and reducing the gain of the parasitic transistors are the basis for eliminating latchup.

Latchup may be prevented in two basic ways:

- Latchup resistant CMOS processes.
- Layout techniques.

A popular process option that reduces the gain of the parasitic transistors is the use of silicon starting-material with a thin epitaxial layer on top of a highly doped substrate. This decreases the value of the substrate resistor and also provides a sink for collector current of the vertical pnp-transistor. As the epi layer is thinned, the latchup performance improves until a point where the up-diffusion of the substrate and the down-diffusion of any diffusions in subsequent high-temperature procession steps thwart required device doping profiles. The so-called retrograde well structure is also used. This well has a highly doped area at the bottom of the well, whereas the top of the well is more lightly doped. This preserves good characteristics for the pMOS (or nMOS in p-well) transistors but reduces the well resistance deep in the well. A technique linked to these two approaches is to increase the holding voltage above the V_{DD} supply. This guarantees that latchup will not occur.

It is hard to reduce the betas of the bipolar transistors to meet the condition set above. Nominally, for a 1μ n-well process, the vertical pnp has a beta of 10–100, depending on the technology. The lateral npn-current-gain, which is a function of n^+ drain to n-well spacing, is between 2 and 5.[41] (These values are illustrative—they should be checked for the particular process being used.)

Apart from the inherent resistance to latchup of a particular process, there are a number of well-proven techniques to design CMOS layouts that are latchup resistant.

3.5.4 Internal Latchup Prevention Techniques

From Fig. 3.29(b) it may be seen that the emitter of the npn-transistor has to an nMOS transistor source returned to V_{SS}. The substrate resistor occurs between this emitter and the supply represented by a substrate contact. Clearly, if the n-transistor source is shorted to the p^+ substrate contact, much has been done to reduce $R_{substrate}$. Conversely, the well resistor occurs between the p^+ source nominally to V_{DD} and the n^+ well contact. Thus a key technique to reduce latchup is to make good use of substrate and well contacts.

In most current processes the possibility of latchup occurring in internal circuitry has been reduced to the point where a designer need not worry about the effect as long as *liberal* substrate contacts are used. The definition of "liberal" is usually acquired from designers who have completed successful designs through a given process. Modeling the parasitics is possible,[42] but the actual switching transients existent in the circuit have a great effect on any possible latchup condition. A few rules may be followed that reduce the possibility of internal latchup to a very small likelihood:

- Every well must have a substrate contact of the appropriate type.
- Every substrate contact should be connected to metal directly to a supply pad (i.e., no diffusion or polysilicon underpasses in the supply rails).
- Place substrate contacts as close as possible to the source connection of transistors connected to the supply rails (i.e., V_{SS} n-devices, V_{DD} p-devices). This reduces the value of $R_{substrate}$ and R_{well}. A very conservative rule would place one substrate contact for every supply (V_{SS} or V_{DD}) connection.
- Otherwise a less conservative rule is place a substrate contact for every 5–10 transistors or every 25–100μ.
- Lay out n- and p-transistors with packing of n-devices toward V_{SS} and packing of p-devices toward V_{DD} (see layout styles in Chapter 5). Avoid "convoluted" structures that intertwine n- and p-devices in checkerboard styles (unless you are designing in SOI which is latchup free).

3.5.5 I/O Latchup Prevention

Reducing the gain of the parasitic transistors is achieved through the use of guard rings (first encountered in Fig. 3.28). A p^+ guard ring is shown in Fig. 3.31(a) for an n^+ source/drain, while Fig. 3.31(b) shows an n^+ guard ring for a p^+ source/drain. As shown in the figures, these guard bands act as "dummy-collectors" and spoil the gain of the parasitic transistors by collecting minority carriers and preventing them from being injected into the respective bases. Carriers can still flow underneath these structures which leads sometimes to double guard banding which is illustrated in Fig. 3.31(c). While these techniques can be used on internal structures, the area penalty is usually too high except for applications such as space-borne electronics where radiation induced latchup must be avoided at all costs.

Luckily enough, as has been observed, the most likely place for latchup to occur is in I/O structures where large currents flow, large parasitics may be present, and abnormal circuit voltages may be encountered. Here the area penalty of guard rings is not at all significant. In these structures two options can be taken. The first is to use proven I/O structures designed by experts who understand the process at a detailed level. Second, rules may be applied to the design of these structures that minimize the possibility of latchup. Typical rules (n-well process) include:

- Physically separate the n- and p-driver transistors (i.e., with the bonding pad).

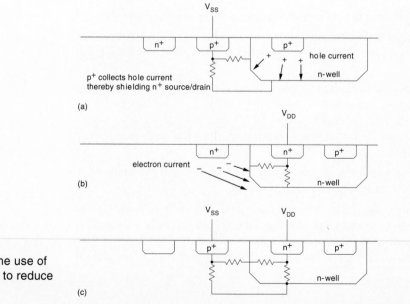

FIGURE 3.31 The use of dummy collectors to reduce latchup

- Include p^+ guard rings connected to V_{SS} around n-transistors.

- Include n^+ guard rings connected to V_{DD} around p-transistors.

- Source diffusion regions of the n-transistors should be placed so that they lie along equipotential lines when current flows between V_{SS} and the p-wells; that is, source fingers should be perpendicular to the dominant direction of current flow rather than parallel to it. This reduces the possibility of latchup through the n-transistor source, due to an effect called "field aiding."[43]

- Shorting n-transistor source regions to the substrate and the p-transistor source regions to the n-well with metallization along their entire lengths will aid in preventing either of these diodes from becoming forward-biased, and hence reduces the contribution to latchup from these components.

- The n-well should be hard-wired (via n^+) to power so that any injected charge is diverted to V_{DD} via a low-resistance path. The n-well has a relatively high sheet-resistance and is susceptible to charge injection.

- The spacing between the n-well n^+ and the p-transistor source contact should be kept to a minimum. This allows minority carriers near the parasitic pnp-transistor emitter-base junction to be collected, and reduces R_{well}. The rules for the 1μ process suggest one contact for every 10μ–50μ.

- The separation between the substrate p^+ and the n-transistor source contact should be minimized. This results in reduced minority carrier concentration near the npn-emitter-base junction. Similar spacings to those suggested above apply for processes in the 1μ range.

More details on layout and design techniques for I/O circuitry may be found in Chapter 5.

3.6 Technology-related CAD Issues

The mask database is the interface between the semiconductor manufacturer and the chip designer. Two basic checks have to be completed to ensure that this description can be turned into a working chip. First, the specified geometric design rules must be obeyed. Second, the interrelationship of the masks must, on passing through the manufacturing process, produce the correct interconnected set of circuit elements. To check these two requirements, two basic CAD tools are required, namely a Design Rule Check (DRC) program and a mask circuit-extraction program. The most common approach to implementing these tools is to provide a set of subprograms that perform general geometry operations. A particular set of DRC rules or extraction rules for a given CMOS process (or any semiconductor process) is then specified by a specification of the operations that must be performed on each mask and the intermask checks that must be completed. Accompanied by a written specification, these *run-sets* are usually the defining specification for a process.

In this section we will examine a hypothetical DRC and extraction system to illustrate the nature of these run-sets.

3.6.1   DRC—Spacing and Dimension Checks

Although we might design the physical layout of a certain set of mask layers, the actual masks used in fabrication are derived from the original specification. Similarly, when we want a program to determine what we have designed by examining the interrelationship of the various mask layers, it may be necessary to determine various logical combinations between masks.

To examine these concepts, let us posit the existence of the following functions (loosely based on the CADENCE DRACULA DRC program[44]), which we will apply to a geometric database (i.e., rectangles, polygons, paths):

- **AND** `layer1 layer2 -> layer3`
  ANDs `layer1` and `layer2` together to produce `layer3` (i.e., the intersection of the two input mask descriptions)

- **OR** `layer1 layer2 -> layer3`
  ORs `layer1` and `layer2` together to produce `layer3` (i.e., the union of the two input mask descriptions)

- **NOT** `layer1 layer2 -> layer3`
  Subtracts `layer2` from `layer1` to produce `layer3` (i.e., the difference of the two input mask descriptions)

- **WIDTH** `layer > dimension -> layer3`
  Checks that all geometry on `layer` is larger than `dimension`. Any that is not is placed in `layer3`

- **SPACE** `layer > dimension -> layer3`
  Checks that all geometry on `layer` is spaced further than `dimension`. Any that is not is placed in `layer3`

The following layers will be assumed as input:

```
nwell
active
pplus
nplus
```

```
poly
poly-contact
active-contact
metal
```

Typically, useful sublayers are first generated. First, the four kinds of active area are isolated. The set of rules to accomplish this is as follows:

```
NOT all nwell -> substrate
AND nwell active -> nwell-active
NOT active nwell -> pwell-active
AND nwell-active pplus -> pdiff
AND nwell-active nplus -> vddn
AND pwell-active nplus -> ndiff
AND pwell-active pplus -> vssp
```

In the above specification a number of new layers have been specified. For instance, the first rule states that wherever `nwell` is absent, a layer called `substrate` exists. The second rule states that all active areas within the nwell are `nwell-active`. A combination of `nwell-active` and `pplus` or `nplus` yields `pdiff` (p diffusion) or `vddn` (well tie).

To find the transistors, the following set of rules is used:

```
AND poly ndiff -> ngates
AND poly pdiff -> pgates
```

The first rule states that the combination of polysilicon and n diffusion yields the `ngates` region—all of the n-transistor gates.

Typical design rule checks (DRC) might include the following :

```
WIDTH metal > 1.25 -> metal-width-error
SPACE metal > 1.0 -> metal-space-error
```

For instance the first rule determines if any metal is narrower than 1.25μ and places the errors in the `metal-width-error` layer. This layer might then later be used with the original and an interactive mask editor to identify the errors.

A bloat command changes the dimensions of a layer.

- **BLOAT** `layer1 dimension` -> `layer2`
  Expand or contract `layer1` by `dimension` to produce `layer2`.

For instance

```
BLOAT metal 0.5 metal-exp
```
would create a layer `metal-exp` in which all metal geometries were increased in size peripherally by 0.5μ. Bloats and shrinks may be used to derive other required layers. For instance, if the gates of all p-transistors had

to be increased in length by 0.5μ, the following sequence might be used:

```
BLOAT pgates 0.25 pgates-bloat
```
The following sequence produces the nplus layer from an original specification containing only ndiff (n-transistors) and vddn (V_{DD} substrate ties).

```
AND ndiff vddn all-ndiff
BLOAT all-ndiff 2 nplus
```

3.6.2   Circuit Extraction

Now imagine that we wish to determine the electrical connectivity of a mask database. The following commands are required:

- **CONNECT** layer1 layer2
  Electrically connect layer1 and layer2
- **MOS** name drain-layer gate-layer source-layer substrate-layer
  Define an MOS transistor in terms of the component terminal layers. (This is admittedly, a little bit of magic.)

The connections between layers may be specified as follows:

```
CONNECT active-contact pdiff
CONNECT active-contact ndiff
CONNECT active-contact vddn
CONNECT active-contact vssp
CONNECT active-contact metal
CONNECT vssp substrate
CONNECT vddn nwell
CONNECT poly-contact poly
CONNECT poly-contact metal
```

The connections between the diffusions and the metal are specified by the first seven statements. The last two statements specify how the metal is connected to the poly.

Finally, the active devices are specified in terms of the layers that we have derived.

```
MOS nmos ndiff ngates ndiff substrate
MOS pmos pdiff pgates pdiff nwell
```

An output statement might then be used to output the extracted transistors in some netlist format (i.e., SPICE format). This is then used as an interface to a program that compares the connectivity that we have derived from the mask with that of, say, a circuit diagram.

It is important to realize that the above run set is manually generated. The data extracted from such a program is only as good as the input. For instance, if parasitic routing capacitances are required, then each and every layer interaction must be coded. If parasitic resistance is important in determining circuit performance, it too must be specifically included in the extraction run set. Many different coding styles exist that define the abstract layers in which the designer conceives the layout. For instance, if there are different rules that specify a well overlap for a guard structure compared with an internal structure, then a special guard layer might have to be coded in the mask database. Similar decisions have to be made concerning structures, such as resistors, that are constructed from diffusion or polysilicon.

3.7 Summary

This chapter has covered some of the more common CMOS technologies that are in current use. A representative set of n-well design rules have been introduced. These form the interface between the designer and the manufacturer. A range of process options were discussed to enhance the basic CMOS process. The important condition known as latchup has been introduced along with necessary design rules to avoid this condition in CMOS chips. Finally, some of the CAD/process interface issues were surveyed.

3.8 Exercises

1. A p-well process has the following layers:

   p-well
   active
   n-plus
   p-plus
   poly
   contact
   metal

   Draw the mask combinations for the following:

   a p-transistor
   an n-transistor
   a V_{SS} contact

a V_{DD} contact

a contact to an n-transistor source/drain

a single guard-ringed n-transistor

a double guard-ringed p-transistor

Use the design rules from Table 3.2 as appropriate.

2. Write a program that can generate a single metal CMOS inverter in an n-well technology that parametizes the widths of the p/n transistors. Use the design rules in Table 3.2.

3. Explain how the parasitic channel, which couples unrelated nMOS transistors in an n-well process, is reduced.

4. How might you use a field transistor to prevent overvoltage in a CMOS chip?

5. Explain why substrate and well contacts are important in CMOS.

6. How does a "dummy collector" prevent latchup?

7. A pad requires a pull-up resistor, which is implemented as a p-transistor that has the source connected to V_{SS}. Does this structure require any latchup protection? What about an n pull-down ($D = input$, $G = V_{DD}$, $S = V_{SS}$)?

8. A CMOS process has unequal n- and p-transistor lengths ($L_N = 0.8\mu$, $L_P = 1.0\mu$). However, a design is desired that uses the same length for each device (1.0μ). Construct a DRC run-set using the commands outlined in Section 3.6.1 that will correctly shrink all the n-transistor gates ($1.0\mu \rightarrow 0.8\mu$), and output data for the final polysilicon mask, assuming that the overall mask has to be bloated by 0.1μ.

9. Most DRC systems deal with merged "canonical" databases, where the rectangles, polygons, etc., in the geometric database are merged before geometric operations are commenced. What could happen to abutting geometric shapes if the source geometry were sized then canonicalized?

3.9 Appendix—An n-well CMOS Technology Process Flow

This section covers in gory detail the processing steps in a now old but representative n-well process developed at the University of California at Berkeley. It is described in terms of a Process Input Description Language (PIDL),[45] which can be used by a software process emulator to predict the topologies of the final structures. The steps are representative of those taken in processes today, albeit somewhat less complicated. The overall process flow gives an idea of the many steps required to produce even a simple CMOS chip.

The commands in the PIDL language are as follows:

- SUBSTRATE <NAME> (*TYPE=[P,N] IMPURITY=[ ])
  Specifies the substrate name, type, and impurity level.

- OXIDE <NAME> THICKNESS = [ ]
  Specifies oxide layer and thickness.

- DEPOSITION <NAME> (*) THICKNESS=[ ]
  Specifies a layer and thickness of a deposited layer. The (*) is followed by TYPE=[ ] IMPURITY=[ ] if the deposited layer is silicon.

- ETCH <NAME> DEPTH=[ ]
  Specifies a material and an etch depth.

- DOPE TYPE=[P, N] PEAK=[ ] DEPTH=[ ] DELTA=[ ]
  BLOCK=[ ]
  Specifies parameters necessary to define a diffusion step.

- MASK <RESIST NAME> <EXPOSED NAME> <MASK NAME>
  <POLARITY OF MASK>
  Specifies a resist layer and associated information.

The complete process input file is as follows (with abbreviations) (© IEEE 1983)[46]

```
1.  LEVEL 1
2.  SUBS SILICON TYPE=P IMPU=1e13; the substrate type and
    impurity is specified
```

Initial oxidation:

```
3.  OXIDE OXI THICK=0.1; this grows an oxide on the silicon
    surface
```

N-well definition:

```
4.  DEPO NTRD THICK=0.5; nitride is deposited over the oxide
5.  DEPO RST THICK=0.5; resist is deposited
6.  MASK RST DRST MNNL POSI; the resist is positive-masked
    (n-well)
7.  ETCH DRST DEPTH=0.6; the exposed resist is etched
8.  ETCH NTRD DEPTH=0.6; the nitride is etched
9.  ETCH RST DEPTH=0.6; the remaining resist is etched
10. OXIDE OX2 THICK=0.5; oxide is regrown
11. ETCH NTRD DEPTH=0.6
```

```
12. DOPE TYPE=N PEAK=1.5e15 DEPTH=0.0 DELTA=1 5 BLOCK=0.2
    ; well diffusion
13. ETCH OX DEPTH=0.7; oxide etched
14. OXIDE OX3 THICK=0.1; oxide regrown
```

All active area definition:

```
15. DEPO NTRD THICK=0.5; nitride deposited
16. DEPO RST THICK=0.5; resist deposited
17. MASK RST DRST MAA POSI; the resist is positive masked
    (active)
18. ETCH DRST DEPTH=0.6; the exposed resist is removed
19. ETCH NTRD DEPTH=0.6; the nitride thus exposed is etched
20. ETCH RST DEPTH=0.6; the remaining resist is removed
```

Field dope for n-channel:

```
21. DEPO RST THICK=1.0; deposit resist
22. MASK RST DRST MNWL POSI; mask
23. ETCH DRST DEPTH=1.1; etch exposed resist
24. DOPE TYPE=P PEAK=1e21 DEPTH=0.05 DELTA=0. 15
    BLOCK=0.2; diffusion step
25. ETCH RST DEPTH=1.1; remove resist
26. OXIDE OX4 THICK=0.7; grow oxide
27. ETCH NTRD DEPTH=0.6
```

Threshold adjust dope:

```
28. DOPE TYPE=P PEAK=1E2O DEPTH=0.0 DELTA=0.05 BLOCK=0.2
    ; diffusion
```

Regrow gate oxide:

```
29. ETCH OX DEPTH=0.1; remove oxide
30. OXIDE OX5 THICK=0.1; regrow oxide
```

Poly gate definition:

```
31. DEPO POLY THICK=0.30; deposit polysilicon
32. DEPO RST THICK=0.5; deposit resist
33. MASK RST DRST MSI POSI; mask resist with poly mask
34. ETCH DRST DEPTH=0.6; remove exposed resist
35. ETCH POLY DEPTH=0.6; etch exposed polysilicon
36. ETCH RST DEPTH=0.6; remove remaining resist
```

Arsenic dope for n-channel source and drain:

```
37. DEPO RST THICK=1.0; deposit resist
38. MASK RST DRST MIIN POSI; mask for n+
```

```
39. ETCH DRST DEPTH=1.1; remove exposed resist
40. DOPE TYPE=N PEAK=1e22 DEPTH=0.0 DELTA=0.2 BLOCK=0.2
    ; diffusion (or implant)
41. ETCH RST DEPTH=1.1; remove resist
```

Boron dope for p-channel source and drain:

```
42. DEPO RST THICK=1.0; deposit resist
43. MASK RST DRST MIIN NEGA; mask for p+
44. ETCH DRST DEPTH=1.1; remove exposed resist
45. DOPE TYPE=P PEAK=1e22 DEPTH=0.0 DELTA=0.2 BLOCK=0.2
    ; diffusion
46. ETCH RST DEPTH=1.1; remove remaining resist
```

LPCVD oxide (**L**iquid **P**hase Chemical **V**apor **D**eposition Oxide):

```
47. DEPO OX6 THICK=0.5; deposit oxide
```

Contact definition:

```
48. DEPO RST THICK=1.0; deposit resist
49. MASK RST DRST MCC NEGA; mask with contact mask
50. ETCH DRST DEPTH=1.1; etch exposed resist
51. ETCH OX DEPTH=1.1; etch oxide down to diffusion
52. ETCH RST DEPTH=1.1; remove resist
```

Metallization:

```
53. DEPO METL THICK=1.0; deposit metal
54. DEPO RST THICK=1.0; deposit resist
55. MASK RST DRST MME POSI; mask with metal mask
56. ETCH DRST DEPTH=1.1; remove exposed resist
57. ETCH METL DEPTH=1.1; remove exposed metal
58. ETCH RST DEPTH=1.1; remove resist
```

Some of the abbreviations are as follows:

```
NTRD  Nitride
RST   Resist
METL  Metal (Aluminum)
NEGA  Negative
POSI  Positive
MNWL  N-well mask
MAA   Thin-oxide mask
MSI   Polysilicon mask
MIIN  NPlus mask
MCC   Contact mask
MME   Metal mask
```

Using the abbreviations and language definitions, the sequence in processing may be traced. For instance, steps 31–36 deposit and etch the polysilicon layer. Step 31 deposits .3μ of polysilicon. Step 32 deposits .5μ of resist called RST. Step 33 masks this resist with a positive polysilicon mask and calls the exposed resist DRST. Step 34 etches DRST to a depth of .6μ. The exposed polysilicon is then etched to a depth of .6μ in step 35. Finally, resist RST is etched away, leaving the final polysilicon pattern. Cross sections may be generated automatically from this process file using the SIMPL-1 program.[47]

3.10   References

1. John Y. Chen, *CMOS Devices and Technology VLSI,* Englewood Cliffs, N.J.: Prentice-Hall, 1990, pp. 233–284.

2. K. Y. Ciu, J. L. Moll, and J. Manoliu, "A bird's beak free local oxidation technology for VLSI," *IEEE Trans. on Electron Devices,* ED-29, pp. 536–540.

3. John Y. Chen, *op. cit.,* pp. 5, 37, and 174–232.

4. John Y. Chen, *op. cit.,* pp. 174–232.

5. L. C. Parrillo *et al.,* "Twin-tub CMOS—a technology for VLSI circuits," *IEEE Int. Electron Devices Meeting Technical Digest,* 1980, Washington, D.C., pp. 752–755.

6. J. Agraz-Guerera, W. Bertram, R. Melin, R. Sun, and J. J. Clemens, "Twin-tub III—a third generation CMOS technology," *IEEE Int. Electron Devices Meeting Technical Digest,* 1984, Washington, D.C., p. 63.

7. H. M. Manasevit and W. I. Simpson, "Single crystal silicon on a sapphire substrate," *J. Appl. Phys.,* vol. 35, 1964, pp. 1349–1351.

8. Yasuaki Hokari, Masao Mikami, Koji Egami, Hideki Tsuya, and Masaru Kanamori, "Characteristics of MOSFET prepared on Si/Mg).Al_2O_3/SiO_2/Si structure," *IEEE JSSC,* vol. 20, no. 1, Feb. 1985, pp. 173–177.

9. Koichi Kato, Tetsunori Wada, and Kenji Taniguchi, "Analysis of kink characteristics in silicon-on insulator MOSFET's using two-carrier modeling," *IEEE JSSC,* vol. SC-20, no. 1, Feb. 1985, pp. 378–382.

10. T. P. Chow, "A review of refractory gates for MOS VLSI," *IEEE Electron Devices Meeting Technical Digest,* Dec. 1983, Washington, D.C., pp. 513–517.

11. T. Tang *et al.,* "Titanium nitride local interconnect technology for VLSI," *IEEE Trans. Electron Devices,* vol. ED-34, Mar. 1987, pp. 682–688.

12. Hiep Van Tran, David B. Scott, Pak Kuen Fung, Robert H. Haverman, Robert H. Eklund, Thomas E. Ham, Roger A. Haken, and Ashwin H. Shah, "An 8-ns 256K ECL SRAM with CMOS memory array and battery backup capability," *IEEE JSSC,* vol. 23, no. 5, Oct. 1988, pp. 1041–1047.

13. Tomohisa Wada, Toshihiko Hirose, Hirofumi Shinohara, Yuji Kawai, Kojiro Yuzuriha, Yoshio Kohno, and Shimpei Kayano, "A 34-ns 1-Mbit CMOS SRAM using triple polysilicon," *IEEE JSSC,* vol. SC-2, no. 5, Oct. 1987, pp. 727–732.

14. Koichiro Mashiko, Masao Nagatomo, Kazutami Arimoto, Yoshio Matsuda, Kiyohiro Furutani, Takayuki Matsukawa, Michihiro Yamada, Tsutomu Yoshihara, and Takao Nakano, "A 4-Mbit DRAM with folded-bit-line adaptive side-

wall-isolated capacitor (FASIC) cell," *IEEE JSSC,* vol. SC-22, no. 5, Oct. 1987, pp. 643–650.

15. Toshio Yamada, Hisakazu Kotani, Junko Matsushima, and Michihiro Inoue, "A 4-Mbit DRAM with 16-bit concurrent ECC," *IEEE JSSC,* vol. 23, no. 1, Feb. 1988, pp. 20–26.

16. Shigeru Mori, Hiroshi Miyamoto, Yoshikazu Morooka, Shigeru Kikuda, Makoto Suwa, Mitsuya Kinoshita, Atsushi Hachisuka, Hideaki Arima, Michihiro Yamada, Tsutomu Yoshihara, and Shimpei Kayano, "A 45-ns 64-Mb DRAM with a merged match-line test architecture," *IEEE JSSC,* vol. 26, no. 11, Nov. 1991, pp. 1486–1492.

17. Masao Taguchi, Hiroyoshi Tomita, Toshiya Uchida, Yasunhiro Ohnishi, Kimiaki Sato, Taiji Ema, Masaaki Higashitani, and Takashi Yabu, "A 40-ns 64-Mb DRAM with 64-b parallel data bus architecture," *IEEE JSSC,* vol. 26, no. 11, Nov. 1991, pp. 1493–1497.

18. Richard D. Jolly, Rod Tesch, Ken J. Campbell, David L. Tennant, Jay F. Olund, Robert B. Lefferts, Brendan T. Cremen, and Philip A. Andrews, "A 35-ns 64K EEPROM," *IEEE JSSC,* vol. SC-20, no. 5, Oct. 1985, pp. 971–978.

19. Koichi Seki, Hitoshi Kume, Yuzuru Ohji, Takashi Kobayashi, Atsushi Hiraiwa, Takashi Nishida, Takeshi Wada, Kazuhiro Komori, Kazuto Izawa, Toshiaki Nishimoto, Yasuroh Kubota, and Kazuyoshi Shohji, "An 80-ns 1-Mb flash memory with on-chip erase/erase-verify controller," *IEEE JSSC,* vol. 25, no. 5, Oct. 1990, pp. 1147–1152.

20. Katsumoto Soejima, Akira Shida, Hiroshi Koga, Junnichi Ukai, Hiroshi Sata, and Masaki Hirata, "A BiCMOS technology with 660MHz vertical p-n-p transistor for analog/digital ASIC's," *IEEE JSSC,* vol. 25, no. 2, Apr. 1990, pp. 410–416.

21. Ali A. Iranmanesh, Vida Ilderem, Madan Biswal, and Bami Bastani, "A 0.8mm advanced single-poly BiCMOS technology for high-density and high-performance applications," *IEEE JSSC,* vol. 26, no. 3, Mar. 1991, pp. 422–423.

22. P. K. Weimer, "The Insulated-Gate Thin-Film Transistor," in *Physics of Thin Films, Vol. 2,* New York: Academic Press, 1963, pp. 147–192.

23. Richard S. C. Cobbold, *Theory and Applications of Field-Effect Transistors,* New York: Wiley Interscience, 1970, pp. 54–64.

24. Satwinder D. S. Malhi, Hisashi Shichijo, Sanjay K. Banerjee, Ravishankar Sundaresan, Mostafa Elahy, Gordan P. Pollack, William F. Richardson, Ashwin H. Shah, Larry R. Hite, Richard H. Womack, Pallab K. Chatterjee, and Hon Wai Lam, "Characteristics and three-dimensional integration of MOSFET's in small grain LPCVD polycrystalline silicon," *IEEE JSSC,* vol. SC-20, no. 1, Feb. 1985, pp. 178–201.

25. Katsuro Sasaki, Koichiro Ishibashi, Katsuhiro Shimohigashi, Toshiaki Yamanaka, Nobuyuki Moriwaki, Shigeru Honjo, Shuji Ikeda, Atsuyoshi Koike, Satoshi Meguro, and Osamu Minato, "A 23-ns 4-Mb CMOS SRAM with 0.2mm standby current," *IEEE JSSC,* vol. 25, no. 5, Oct. 1990, pp. 1075–1081.

26. Takayuki Ootani, Shigeyuki Hayakawa, Masakazu Kakumu, Akira Aono, Masaaki Kinugawa, Hideki Takeuchi, Kazuhiro Noguchi, Tomoaki Yabe, Katsuhiko Sato, Kneji Maeguchi, and Kiyofumi Ochi, "A 4-Mb CMOS SRAM with a PMOS thin-film-transistor load cell," *IEEE JSSC,* vol. 25, no. 5, Oct. 1990, pp. 1082–1092.

27. Yutaka Takafuji, Toshihiro Yamashita, Yasunobu Akebi, Tomoaki Toichi, Takayuki Shimada, and Katsunobu Awane, "A poly-Si TFT monolithic LC data driver with redundancy," *IEEE, Proceedings of ISSCC,* Feb. 1992, San Francisco, Calif., pp. 118–119.

28. Gerhard Roos and Bernd Hoefflinger, "Complex 3D CMOS circuits based on a triple-decker cell," *IEEE JSSC,* vol. 27, no. 7, Jul. 1992, pp. 1067–1072.

29. C. A. Mead and L. A. Conway, *Introduction to VLSI Systems,* Reading, Mass.: Addison-Wesley, 1980.

30. Shingo Aizaki, Toshiyuki Shimizu, Masayoshi Ohkawa, Kazuhiko Abe, Akane Aizaki, Manabu Ando, Osamu Kudoh, and Isao Sasaki, "A 15nS 4-Mb CMOS SRAM," *IEEE JSSC,* vol. 25, no. 5, Oct. 1990, pp. 1063–1067.

31. Katsuro Sasaki, Koichiro Ishibashi, Katsuhiro Shimohigashi, Toshiaki Yamanaka, Nobuyuki Moriwaki, Shigeru Honjo, Shuji Ikeda, Atsuyoshi Koike, Satoshi Meguro, and Osamu Minato, "A 23-ns 4-Mb CMOS SRAM with 0.2mm standby current," *IEEE JSSC,* vol. 25, no. 5, Oct. 1990, pp. 1075–1081.

32. Takayuki Ootani, *et al., op. cit.*

33. Osamu Nishii, Makoto Hanawa, Tadahiko Nishimukai, Makoto Susuki, Kazuo Yano, Mitsuru Hiraki, Shohji Shukuri, and Takashi Nishida, "A 1,000 MIPS BiCMOS microprocessor with superscalar architecture," IEEE Proceedings of ISSCC, Feb. 1992, San Francisco, Calif., pp. 114–115.

34. W. S. Chang, B. Davari, M. R. Wordeman, Y. Taur, C. C. H. Hsu and M. D. Rodriquez, "A High-Performance 0.25μm CMOS Technology I—Design and Characterization," *IEEE Transactions on Electron Devices,* vol. 39, no. 4, April 1992, pp. 959–966, *and* B. Davari, W. H. Chang, K. E. Petrillo, C. Y. Wong, D. Moy, Y. Taur, M. R. Wordeman, J. Y. C. Sun, C. C. H. Hsu and M. R. Polcari, "A High-Performance 0.25μm CMOS Technology II—Technology," *IEEE Transactions on Electron Devices,* vol. 39, no. 4, Apr. 1992, pp. 967–975.

35. D. B. Estreich, "The physics and modeling of latch-up in CMOS integrated circuits," *Tech. Report No. G-2-1-9,* Integrated Circuits Laboratory, Stanford Electronics Lab., Stanford University, Nov. 1980.

36. D. B. Estreich and R. W. Dutton, "Modeling latch-up in CMOS integrated circuits and systems," *IEEE Transactions on CAD,* vol. CAD-1, no. 4, Oct. 1982, pp. 347–354.

37. R. R. Troutman, *Latch-Up in CMOS Technology: The Problem and Its Cure,* Boston, Mass.: Kluwer Academic Publishers, 1986.

38. William M. Coughran, Mark R. Pinto, and R. Kent Smith, "Computation of steady-state CMOS latchup characteristics," *IEEE Transactions on CAD,* vol. 7, no. 2, Feb. 1988, pp. 307–323.

39. John Y. Chen, *op. cit.,* pp. 285–322.

40. John Y. Chen, *op. cit.,* pp. 286–288.

41. John Y. Chen, *op. cit.,* pp. 289–290.

42. William M. Coughran *et al., op. cit.*

43. D. B. Estreich and R. W. Dutton, *op. cit.*

44. "DRACULA III," Design Rule Check Program CADENCE, Design Systems, Inc., San Jose, Calif.

45. M. A. Grimm, K. Lee, and A. R. Neureuther, "SIMPL-1 (SIMulated Profiles from the layout-version 1)," *Proc. IEDM 1983,* Dec. 1983, pp. 255–258.

46. M. A. Grimm *et al., op. cit.*

47. M. A. Grimm *et al., op. cit.*

CIRCUIT CHARACTERIZATION AND PERFORMANCE ESTIMATION

4

4.1   Introduction

In previous chapters we established that an MOS structure is created by superimposing a number of layers of conducting, insulating, and transistor-forming materials. It was further demonstrated that in a conventional silicon gate process an MOS device requires a gate-forming region and a source/drain-forming region, which consists of diffusion, polysilicon, and metal layers separated by insulating layers. Each layer has both a resistance and a capacitance that are fundamental components in estimating the performance of a circuit or system. They also have inductance characteristics that are important when considering I/O behavior but usually assumed to be negligible for most on-chip circuits.

In this section we are primarily concerned with the development of simple models that will assist us in the understanding of system behavior and that will provide the basis whereby systems performance, in terms of signal delays and power dissipation, can be estimated. The issues to be considered in this section are

- resistance, capacitance, and inductance calculations.
- delay estimations.
- determination of conductor size for power and clock distribution.

- power consumption.
- charge sharing mechanism.
- design margining.
- reliability.
- effects of scaling.

4.2   Resistance Estimation

The resistance of a uniform slab of conducting material may be expressed as

$$R = \left(\frac{\rho}{t}\right)\left(\frac{l}{w}\right) \quad \text{(ohms)},\tag{4.1}$$

where

ρ = resistivity

t = thickness

l = conductor length

w = conductor width.

The expression may be rewritten as

$$R = R_s\left(\frac{l}{w}\right) \quad \text{(ohms)},\tag{4.2}$$

where R_s is the sheet resistance having units of Ω/square. Thus to obtain the resistance of a conductor on a layer you simply multiply the sheet resistance, R_s, by the ratio of the length to width of the conductor. For example, the resistances of the two shapes shown in Fig. 4.1 are equivalent because the length-to-width ratios are the same even though the sizes are different. Table 4.1 shows typical sheet resistances that can be expected in 0.5 μm to 1 μm MOS processes. The upper metal layers have reduced resistivity because they are usually thicker. [Processes targeted at different applications may have differing metal thicknesses. For instance, a memory process might have thin metal layers to reduce vertical topology jumps, thereby improving yield. On the other hand, an ASIC process might have thick metal layers to effectively distribute power, ground and clocks.] Note that for metal having a given thickness, t, the resistivity is known, while for poly and diffusion the resistivities are significantly influenced by the concentration density of the impurities that have been introduced into the conducting regions during

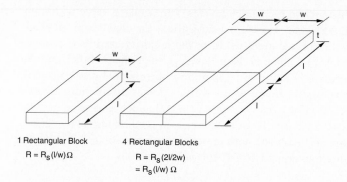

1 Rectangular Block
$R = R_S(l/w)\,\Omega$

4 Rectangular Blocks
$R = R_S(2l/2w)$
$= R_S(l/w)\,\Omega$

FIGURE 4.1   Determination of layer resistance

implantation or the extent of chemical change induced by materials such as silicides. This means that the process parameters have to be known or test structures have to be measured to accurately determine these quantities.

Although the voltage-current characteristic of an MOS transistor is generally nonlinear, it is sometimes useful to approximate its behavior in terms of a "channel" resistance to estimate performance. From Eq. (2.11), one may determine the channel resistance (in the linear region). This expression may be rewritten as

$$R_c = k\left(\frac{L}{W}\right) \tag{4.3}$$

where

$$k = \frac{1}{\mu C_{ox}(V_{gs} - V_t)}.$$

TABLE 4.1   Typical Sheet Resistances for Conductors

Material	SHEET Min	RESISTANCE Typical	Ω/SQ Max.
Intermetal (metal1-metal2)	0.05	0.07	0.1
Top-metal (metal3)	0.03	0.04	0.05
Polysilicon	15	20	30
Silicide	2	3	6
Diffusion (n^+, p^+)	10	25	100
Silicided diffusion	2	4	10
n-well	1K	2K	5K

For both the n-channel and p-channel devices, k may take a value within the range 1,000 to 30,000 Ω/sq.. Eq. (4.3) demonstrates the dependence of channel resistance on the surface mobility, μ, of the majority carriers (i.e., electrons in the n-device and holes in the p-device). Since the mobility and the threshold voltage are a function of temperature, the channel resistance and therefore switching-time parameters as well as power dissipation, change with temperature variations. The increase in the channel resistance may be approximated by +0.25% per °C for an increase in temperature above 25°C. Conductor resistances also vary with temperature from about 0.3%/°C for metal and polysilicon to around 1%/°C for well diffusions.

4.2.1   Resistance of Nonrectangular Regions

Many times during the course of a layout nonrectangular shapes are used (for instance, the corners of wires). The resistance of these shapes requires more elaborate calculation than that for simple rectangular regions. One method of calculating the resistance is to break the shape in question into simple regions, for which the resistance may be calculated.[1] Figure 4.2(a) summarizes the resistance of a number of commonly encountered shapes. Figure 4.2(b) shows some shapes that are commonly encountered in practice. Table 4.2 presents the results of a study to calculate the resistances of these shapes for different

TABLE 4.2   Resistance of Non-Rectangular Shapes

SHAPE	RATIO	RESISTANCE
A	1	1
A	5	5
B	1	2.5
B	1.5	2.55
B	2	2.6
B	3	2.75
C	1.5	2.1
C	2	2.25
C	3	2.5
C	4	2.65
D	1	2.2
D	1.5	2.3
D	2	2.3
D	3	2.6
E	1.5	1.45
E	2	1.8
E	3	2.3
E	4	2.65

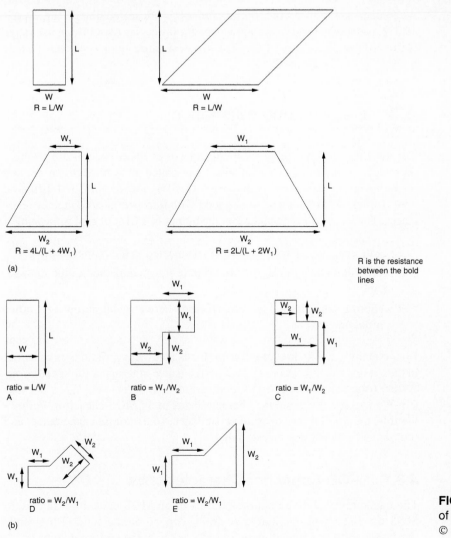

R is the resistance
between the bold
lines

FIGURE 4.2 Resistance
of nonrectangular shapes
© IEEE 1983

dimension ratios. (The resistance is measured between the bold lines.) This
shape information may also be used to estimate the effective *W/L* of odd-
shaped transistors.[2] A few precautions need to be taken, however, especially
concerning which side of a shape is the source or drain. The values shown in
Table 4.2 may be used to estimate the β of an odd-shaped transistor.

4.2.2 Contact and Via Resistance

Contacts and vias also have a resistance associated with them that is dependent on the contacted materials and proportional to the area of the contact.

As contacts are reduced in size (i.e., processes are scaled down), the associated resistance increases. Typical values for processes currently in use range from .25Ω to a few tens of Ωs. For low-resistance interlayer connections multiple contacts are used.

4.3 Capacitance Estimation

The dynamic response (e.g., switching speed) of MOS systems are strongly dependent on the parasitic capacitances associated with the MOS device and interconnection capacitances that are formed by metal, poly, and diffusion wires (often called "runners") in concert with transistor and conductor resistances. The total load capacitance on the output of a CMOS gate is the sum of

- gate capacitance (of other inputs connected to the output of the gate),
- diffusion capacitance (of the drain regions connected to the output), and
- routing capacitance (of connections between the output and other inputs).

Understanding the source of parasitic loads and their variations is essential in the design process, where system performance in terms of the speed of the system form part of the design specification.

We will first examine the characteristics of an MOS capacitor. Following this, the MOS transistor gate capacitance, source/drain capacitance, and routing capacitance will be estimated.

4.3.1 MOS-Capacitor Characteristics

The capacitance-voltage characteristics of an MOS capacitor (that is an MOS transistor without source or drain) depend on the state of the semiconductor surface. Depending on the gate voltage, the surface may be in

- accumulation.
- depletion.
- inversion.

Referring to the p-substrate structure shown in Fig. 4.3(a), an accumulation layer is formed when $V_g < 0$ ($V_g > 0$ for n-substrate). The negative charge on the gate attracts *holes* toward the silicon surface. When an *accumulation* layer is present, the MOS structure behaves like a parallel-plate capacitor. The gate conductor forms one plate of the capacitor; the high concentration of holes in a p-substrate

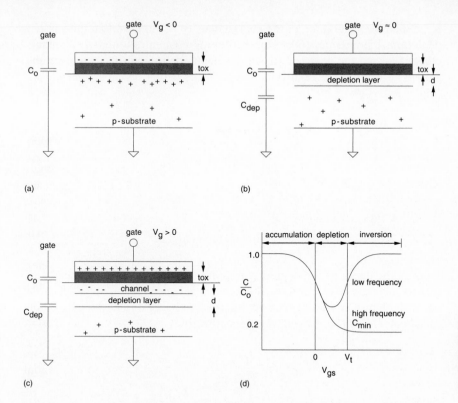

(a)

(b)

(c)

(d)

FIGURE 4.3 MOS capacitance (a) accumulation, (b) depletion, (c) inversion, (d) variation as a function of V_{gs}

(n-device) forms the second plate. Since the accumulation layer is directly connected to the substrate, the gate capacitance may be approximated by

$$C_o = \left(\frac{\varepsilon_{SiO_2} \varepsilon_0}{t_{ox}} \right) A, \tag{4.4}$$

where

A = area of gate

ε_{SiO_2} = dielectric constant (or relative permittivity of SiO_2, taken as 3.9).

ε_0 = permittivity of free space

When a small positive voltage is applied to the gate with respect to the substrate, a *depletion* layer is formed in the p-substrate directly under the gate (Fig. 4.3b). The positive gate voltage repels holes, leaving a negatively charged region depleted of carriers. A corresponding effect occurs in an n-substrate device for a small negative gate voltage.

Since the magnitude of the charge density per unit area in the surface depletion region is dependent on the doping concentration (N), electronic charge (q), and the depth of the surface depletion region (d), increasing the gate to substrate voltage also increases d. The depletion capacitance, C_{dep} (Fig. 4.3c), is given by

$$C_{dep} = (\frac{\varepsilon_0 \varepsilon_{Si}}{d}) A, \tag{4.5}$$

where

d = depletion layer depth

ε_{Si} = dielectric constant of silicon, taken as 12.

Thus as the depth of the depletion region increases, the capacitance from gate to substrate will decrease. The total capacitance from gate to substrate under depletion conditions can be regarded as being due to the gate oxide capacitance, C_o in series with C_{dep}; specifically,

$$C_{gb} = \frac{C_o C_{dep}}{C_o + C_{dep}}. \tag{4.6}$$

As the gate voltage is further increased, minority carriers (electrons for the p-substrate) are attracted toward the surface. This effectively inverts the silicon at the surface and creates an n-type channel. Surface inversion yields a relatively high conductivity layer under the gate, which restores the low-frequency capacitance to C_o. Because of the limited supply of carriers (electrons) to the inversion layer, the surface charge is not able to track fast moving gate voltages. Hence the dynamic capacitance remains the same as for the maximum depletion situation:

$$C_{gb} = C_o; \qquad\qquad low\ frequency\ (<100\text{Hz})$$

$$= \frac{C_o C_{dep}}{C_o + C_{dep}} = C_{min}; \quad high\ frequency$$

Figure 4.3(d) plots the dynamic gate capacitance as a function of gate voltage. The minimum capacitance C_{min} depends on the depth of the depletion region which depends in turn on such parameters as the substrate doping density. For instance, for a gate oxide thickness of 100–200 Å the ratio C_{min}/C_{ox} varies from .02–.3 for N_A varying from 1×10^{-14} cm^{-3} to 5×10^{-15} cm^{-3}.3

4.3.2 MOS Device Capacitances

So far, we have considered the MOS capacitor in isolation. Figure 4.4 is a diagrammatic representation of the parasitic capacitances of an MOS transistor. In this model and in the subsequent analysis the overlap of the gate over the drain and source is assumed to be zero, a simplification that is valid to a first order in self-aligned silicon gate processes.

In Fig. 4.4, the following capacitive components have been identified:

- C_{gs}, C_{gd} = gate-to-channel capacitances, which are lumped at the source and the drain regions of the channel, respectively.

- C_{sb}, C_{db} = source and drain–diffusion capacitances to bulk (or substrate) (see Section 4.3.3).

- C_{gb} = gate-to-bulk capacitance.

It is now possible to view the model in terms of circuit symbols. This is illustrated in Fig. 4.5. The total gate capacitance C_g of an MOS transistor is given by

$$C_g = C_{gb} + C_{gs} + C_{gd}. \tag{4.7}$$

The behavior of the gate capacitance of an MOS device can be explained in terms of the following simple models in the three regions of operation:

1. *Off region,* where $V_{gs} < V_t$. When the MOS device is "OFF," there is no channel, and hence $C_{gs} = C_{gd} = 0$. C_{gb} can be modeled as the series combination of the two capacitors (C_o and C_{dep}), as shown in Fig. 4.3(d).

2. *Non-saturated region,* where $V_{gs} - V_t > V_{ds}$. As a result of the formation of the channel, the gate-to-channel capacitances, C_{gs} and C_{gd}, now become significant. These capacitances are dependent on gate voltage. Their values can be conservatively estimated as

$$C_{gd} = C_{gs} = \frac{1}{2}\left(\frac{\varepsilon_0\varepsilon_{SiO_2}}{t_{ox}}\right)A. \tag{4.8}$$

C_{gb} effectively falls to zero.

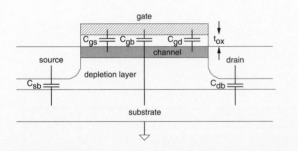

FIGURE 4.4 Process cross section showing parasitic capacitance for an MOS transistor

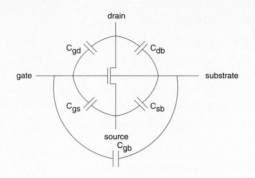

FIGURE 4.5 Circuit symbols for parasitic capacitance

3. *Saturated region,* where $V_{gs} - V_t < V_{ds}$. In this mode the channel is heavily inverted. The drain region of the channel is pinched off, causing C_{gd} to be zero. C_{gs} increases to approximately

$$\frac{2}{3}\left(\frac{\varepsilon_0 \varepsilon_{SiO_2}}{t_{ox}}\right) A.$$

The behavior of the input capacitances in the three regions of operation can be approximated as shown in Table 4.3. Experimentally, the C_{gs} and C_{gd} of a long channel n-transistor (W = 49.2μ, L = 4.5μ) is shown in Fig. 4.6(a).[4]

TABLE 4.3 Approximation of intrinsic MOS gate capacitance

	CAPACITANCE		
Parameter	Off	Non-saturated	Saturated
C_{gb}	$\dfrac{\varepsilon A}{t_{ox}}$	0	0
C_{gs}	0	$\dfrac{\varepsilon A}{2t_{ox}}$	$\dfrac{2\varepsilon A}{3t_{ox}}$
C_{gd}	0	$\dfrac{\varepsilon A}{2t_{ox}}$	0 (finite for short channel devices)
$C_g = C_{gb} + C_{gs} + C_{gd}$	$\dfrac{\varepsilon A}{t_{ox}}$	$\dfrac{\varepsilon A}{t_{ox}}$	$\dfrac{2\varepsilon A}{3t_{ox}} \rightarrow \dfrac{.9\,\varepsilon A}{t_{ox}}$ (short channel)

This graph shows the normalized capacitances varying as a function of V_{ds} for a number of $V_{gs}-V_t$ values. The variation for a short channel transistor is shown in Fig. 4.6(b). Here the length of the transistor is 0.75μ. Of particular

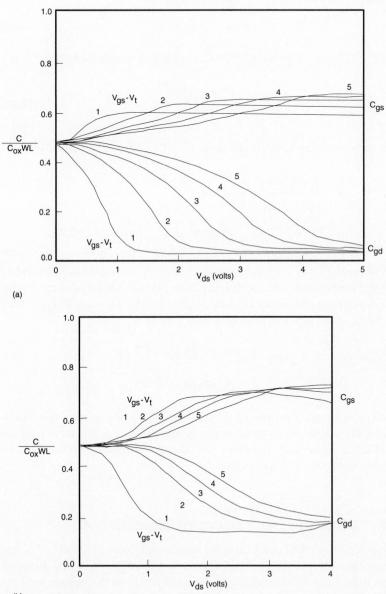

(a)

(b)

FIGURE 4.6   Total gate capacitance of an MOS transistor as a function of V_{gs} (© IEEE 1987)

importance is the finite value of C_{gd} in saturation. This is due to channel side fringing fields between the gate and drain. More accurate modeling of the MOS transistor capacitances may be achieved by using a charge based model.[5,6] For the purposes of delay calculation for digital circuits, we can conservatively approximate $C_g = C_o$.

Another way of stating this approximation is

$$C_g = C_{ox} A \qquad \text{(4.9)}$$

where C_{ox} is in the "thin-oxide" capacitance per unit area given by

$$C_{ox} = \frac{\varepsilon_0 \varepsilon_{SiO_2}}{t_{ox}}. \qquad \text{(4.10)}$$

With a thin-oxide thickness in the order of $100{\rightarrow}200$ Å, the value of C_{ox} is

$$
\begin{aligned}
C_{ox} &= \frac{3.9 \times 8.854 \times 10^{-14}}{(100 - 200) \times 10^{-8}} \\
&\approx \left(35 \rightarrow 17 \times 10^{-4} \ pF/\mu m^2 \right) \quad (t_{ox} = 100 \rightarrow 200 \ \text{Å}).
\end{aligned}
$$

Approximation of the gate capacitance may now be undertaken by simply taking the above value and multiplying it by the gate area. For example, the input (or gate) capacitance of a typical MOS transistor shown in Fig. 4.7, with $\lambda = 0.5 \ \mu m$, $W = 2 \ \mu m$, and $L = 1 \ \mu m$, $t_{ox} = 150$ Å, is

$$C_{g\,(intrinsic)} = 2 \times 25.5 \times 10^{-4} \ pF.$$

$$\approx .005 \ pF$$

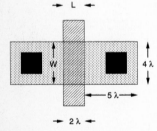

FIGURE 4.7 Physical Layout of a unit MOS Transistor for Capacitance Estimation

We will refer to this transistor as a "unit transistor"—a transistor that can be conveniently connected to metal at source and drain. It is the same width as a metal-diffusion contact. (You may elect to call a unit transistor one with the minimum width of the active region.)

4.3.3   Diffusion (source/drain) Capacitance

Shallow n^+ and p^+ diffusions form the source and drain terminals of n- and p-channel devices. Diffusion regions are also used as wires. All diffusion regions have a capacitance to substrate that depends on the voltage between the diffusion regions and substrate (or well), as well as on the effective area

fusion capacitance C_d is proportional to the total diffusion-to-substrate junction area. As shown in Fig. 4.8(a), this is a function of "base" area and also of the area of the "sidewall" periphery. The latter occurs because the diffusion region has a finite depth (X_c). It is also affected by field implants (increases capacitance) and LOCOS (reduces capacitance). Sidewall capacitance can be characterized (assuming constant depth diffusion) by a periphery-capacitance per unit length. The model generally used is shown in Fig. 4.8(b). Total C_d can be represented by

$$C_d = C_{ja} \times (ab) + C_{jp} \times (2a + 2b), \qquad \textbf{(4.11)}$$

where

C_{ja} = junction capacitance per μ^2

C_{jp} = periphery capacitance per μ

a = width of diffusion region (μ)

b = length of diffusion region (μ).

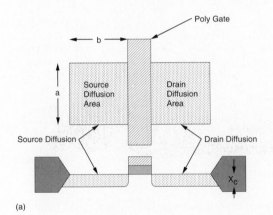

(a)

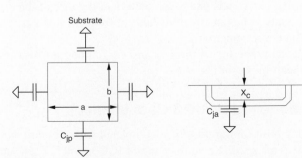

(b)

FIGURE 4.8 Area and peripheral components of diffusion capacitance

TABLE 4.4 **Typical Diffusion Capacitance Values (1μ n-well Process)**

	n-DEVICE (OR WIRE)	p-DEVICE (OR WIRE)
C_{ja}	$3 \times 10^{-4} pF/\mu m^2$	$5 \times 10^{-4} pF/\mu m^2$
C_{jp}	$4 \times 10^{-4} pF/\mu m$	$4 \times 10^{-4} pF/\mu m$

Note that the capacitance contributed by the sidewall facing the channel will be reduced somewhat by the presence of the channel-depletion region and the fact there is no field implant or LOCOS.

An obvious factor that emerges from Eq. (4.11) is that, as the diffusion area is reduced (through scaling, to be discussed later), the relative contribution of the peripheral capacitance becomes more important. Typical values for diffusion capacitances are shown in Table 4.4 for both n- and p-channel devices.

These simple capacitance calculations assume zero DC bias across the junction. Since the thickness of the depletion layer depends on the voltage across the junction, both C_{ja} and C_{jp} are functions of junction voltage, V_j. A general expression that describes the junction capacitance is

$$C_j = C_{j0} \left(1 - \frac{V_j}{V_b} \right)^{-m}, \tag{4.12}$$

where

V_j = junction voltage (negative for reverse bias)

C_{j0} = zero bias capacitance; ($V_j = 0$)

V_b = built-in junction potential ~ 0.6 volts

and m is a constant, that depends on the distribution of impurities near the junction and whether the junction is due to the bottom or the side of the diffusion. In practical models, m has an effective value of from 0.3 (for a graded junction) to 0.5 (for an abrupt junction).

The diffusion capacitance forms the C_{sb} and C_{db} components of the MOS device capacitance shown in Fig. 4.5.

4.3.4 SPICE Modeling of MOS Capacitances

Detailed modeling of circuit timing usually is completed with a circuit simulator such as SPICE. This section examines the SPICE MOSFET call and the MOSFET MODEL statement as relating to device capacitances.

The MOSFET (and the corresponding model) call in SPICE is shown below.

```
.
M1 4 3 5 0 NFET W=4U L=1U AS=15P AD=15P PS=11.5U PD=11.5U
.
.
.MODEL NFET NMOS
+ TOX=200E-8
+ CGBO=200P CGSO=600P CGDO=600P
+ CJ=200U CJSW=400P MJ=0.5 MJSW=0.3 PB=0.7
+ .....
.
.
```

This SPICE netlist fragment specifies an n-channel transistor element card `M1`, which uses an NMOS model called `NFET`. The terminal connections specify the drain connected to node 4, the gate connected to node 3, the source connected to node 5, and the substrate connected to node 0. M1 is a 4μ (`W=4U`) wide by 1μ (`L=1U`) long transistor with source and drain areas of 15μ^2 (`AS=15P  AD=15P`). The source and drain peripheries are 11.5μ (`PS=11.5U  PD=11.5U`).

The start of the MODEL statement is signified by the `.MODEL` line. The second line on the model card specifies the thin-oxide thickness (`TOX=200E-8`). This allows SPICE to calculate the voltage-dependent gate capacitance. The maximum value is

$$C_{g(intrinsic)} = W \times L \times C_{ox} = 4 \times 1 \times 17 \times 10^{-4} pF$$

$$= 0.0068 pF.$$

In the analysis in Section 4.3.2 we assumed that the gate did not overlap the source or drain. In practice this is usually not so. Even if the physical overlap is zero, fringing fields can contribute to these capacitances. To account for this, extrinsic values of C_{gso}, C_{gdo}, and C_{gbo} are added to C_{gs}, C_{gd}, and C_{gb}. These are specified in the next line of the SPICE MOSFET model by `CGSO`, `CGDO`, and `CGBO`. C_{gbo} occurs due to the polysilicon extension beyond the channel. Thus it is multiplied by the length of the transistor to yield a resulting capacitance. C_{gso} and C_{gdo} represent the gate-to-source/drain capacitance due to overlap in the physical structure of the transistor. They are multiplied by the width of the device to yield a final capacitance. Typical values for C_{gbo} range from effectively 0 to 300×10^{-12} *F/m*. Typical values for C_{gdo} and C_{gso} are 200×10^{-12} *F/m*.

In this example, the extrinsic gate capacitance for a typical MOS transistor is

$$C_{g\,(extrinsic)} = (W \times C_{gso}) + (W \times C_{gdo}) + (2L \times C_{gbo})$$

$$= \left(4 \times 3 \times 10^{-4}\right) + \left(4 \times 3 \times 10^{-4}\right) + 2 \times \left(1 \times 2 \times 10^{-4}\right)pF$$

$$= .0028\,pF.$$

In SPICE the capacitance of a source or drain diffusion is calculated as follows:

$$C_j = \left(Area \times CJ \times \left(1 + \frac{VJ}{PB}\right)^{-MJ}\right) + \left(Periphery \times CJSW \times \left(1 + \frac{VJ}{PB}\right)^{-MJSW}\right), \quad \textbf{(4.13)}$$

where

CJ = the zero-bias capacitance per junction area

$CJSW$ = the zero-bias-junction capacitance per junction periphery

MJ = the grading coefficient of the junction bottom

$MJSW$ = the grading coefficient of the junction sidewall

VJ = the junction potential

PB = the built-in voltage (~0.4–0.8 volts)

$Area = AS$ or AD, the area of the source or drain

$Periphery = PS$ or PD, the periphery of the source or drain.

PB, CJ, CJSW, MJ, and MJSW are specified in the model card. AS, AD, PS, and PD are specified by the element card. VJ is built in and VB depends on circuit conditions. At $VJ = 2.5$ volts (half rail),

$$C_{jdrain} = \left[15 \times 10^{-12} \times 2 \times 10^{-4}(1 + 2.5/0.7)^{-0.5}\right] +$$

$$\left[11.5 \times 10^{-6} \times 4 \times 10^{-10}(1 + 2.5/0.7)^{-0.3}\right]pF$$

$$= \left(15 \times 2 \times 10^{-4} \times .47\right) + \left(11.5 \times 4 \times 10^{-4} \times .63\right)pF$$

$$= .0014pF + .0029pF$$

$$= .0043pF.$$

Summarizing these capacitances then,

$$C_{gtotal} = 0.015 + .0052 = .02pF$$

$$C_{drain} = C_{source} = .0043pF\ (@\ 2.5\ \text{volts}).$$

Thus, in this process, the gate capacitance of an n-channel device is about 4.5 times the source/drain capacitance. Bearing in mind that the fan-out of a gate might range from 1 to 10, it can be seen that the gate capacitance dominates the loading in current CMOS technologies. On a historical note, in the first edition of this book, in the example given above, the source/drain capacitance was about two times the gate capacitance. This was primarily because the gate oxides were around 500 Å and the diffusions were deeper and thus had much higher peripheral capacitance contributions. Always check the particular technology in which you are designing to become familiar with the relative importance of the stray capacitance terms!

Note: Some designers prefer to set the MOS diffusion capacitances to zero (i.e., $AD = AS = PS = PD = 0$) and model each source/drain as an appropriately dimensioned diode. This is done to have more control over the area and also to model the effects of leakage.

4.3.5   Routing Capacitance

4.3.5.1 Single Wire Capacitance

Routing capacitances between metal and poly layers and the substrate can be approximated using a parallel-plate model ($C = (\varepsilon/t)A$), where A is area of the parallel-plate capacitor, t is the insulator thickness, and ε is the permittivity of the insulating material between the plates. The parallel-plate approximation, however, ignores fringing fields that occur at the edges of the conductor due to its finite thickness. In addition, a conductor can exhibit capacitance to an adjacent conductor on the same layer. These are shown in Fig. 4.9. The effect of fringing fields is to increase the effective area of the plates. A detailed analysis of the field lines using field theory can yield the actual capacitance of a given structure. Due to the computational burden of calculating this for large numbers of conductors, a number of authors have proposed approximations to this calculation.[7] One approximation treats the

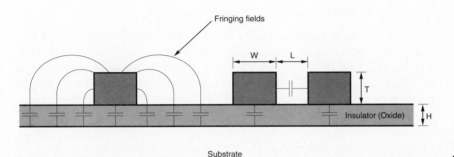

FIGURE 4.9 Effect of fringing fields on capacitance

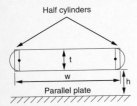

FIGURE 4.10
Simple Capacitance
Model to Account for
Fringing Fields

conductor as a rectangular middle section with two hemispherical end caps, as shown in Fig. 4.10.[8] The total capacitance is assumed to be the sum of a parallel-plate capacitor of width $w - t/2$ and a cylindrical capacitor of radius $t/2$. This results in an expression for the capacitance as follows:

$$C = \varepsilon \left[\frac{w - \dfrac{t}{2}}{h} + \frac{2\pi}{ln\left\{ 1 + \dfrac{2h}{t} + \sqrt{\dfrac{2h}{t}\left[\dfrac{2h}{t} + 2\right]} \right\}} \right], \qquad (4.14)$$

where

w = the width of the conductor

h = the insulator thickness

t = the conductor thickness

ε = the permittivity of the insulator.

This formula is accurate within 10% for $w \geq t/2$ and $t \approx h$.

An empirical formula that is computationally efficient and relatively accurate[9] is given by

$$C = \varepsilon\left[\left(\frac{w}{h}\right) + 0.77 + 1.06\left(\frac{w}{h}\right)^{0.25} + 1.06\left(\frac{t}{h}\right)^{0.5} \right]. \qquad (4.15)$$

As a result of the contribution of fringing fields to the capicitance of a conductor and mutual capacitance, poly and metal lines will actually have a higher capacitance than that predicted by the parallel-plate model. Methods for more accurately computing the fringing factor can be found in Ruehli et al.[10]

4.3.5.2 Multiple Conductor Capacitances

Modern CMOS processes usually have multiple routing layers and the capacitance interactions between layers can become quite complex. Three-dimensional field simulators are used to accurately compute the capacitance of conductor structures.[11] However, these simulations are too complex to perform on the many signal nets that might be in a CMOS chip. Thus empirical formulae are sought that model the capacitance of structures to a close approximation. This section will describe one such set of formulae based on the work of Chern, Huang et al.[12]

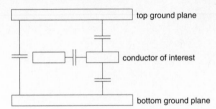

FIGURE 4.11   Multilevel-layer capacitance model

The model for the routing structure is shown in Fig. 4.11. It consists of three potential layers:

- A top ground plane.
- The conductor of interest.
- A bottom ground plane.

The capacitance of the middle layer (conductor of interest) is divided into three components:

- The line-to-ground capacitance.
- The line-to-line capacitance.
- The crossover capacitance.

In Fig. 4.12, the capacitance of middle layer 2 to ground (C_2) consists of the capacitance to layer 3 (C_{23}), the capacitance to layer 1 (C_{21}), and the capacitance between other parallel conductors on layer 2 (C_{22}). Thus

$$C_2 = C_{21} + C_{23} + C_{22}. \tag{4.16}$$

The capacitances C_{21} and C_{23} will be given by formulae for crossover capacitance. The line to line capacitance C_{22} is affected by the presence or absence of layer-1 and layer-3 ground planes. For this reason C_{22} is constructed from the weighted sum of these two conditions:

$$C_{22} = AC_{(line\text{-}to\text{-}line,\ 2\ ground\ planes)} + BC_{(line\text{-}to\text{-}line,\ isolated)}, \tag{4.17}$$

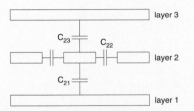

FIGURE 4.12   Specific capacitances in a three-layer-metal system

where $A + B = 1$. The weighting factors are given as

$$A = \frac{R}{(P_1 + P_3)} \tag{4.18a}$$

$$B = \frac{O}{(P_1 + P_3)}, \tag{4.18b}$$

where

$P_1 = (W_1 + S_1)$, the layer-1 pitch, (W_n is the width on layer n, S_n is the spacing on layer n),

$P_3 = (W_3 + S_3)$, the layer-3 pitch,

and R and O are determined by

$$R = W_1 + 2T_1 + W_3 + 2T_3 \qquad \text{for } S_1 \geq 2T_1 \text{ and } S_3 \geq 2T_3$$

$$R = W_1 + S_1 + W_3 + 2T_3 \qquad \text{for } S_1 < 2T_1 \text{ and } S_3 \geq 2T_3$$

$$R = W_1 + 2T_1 + W_3 + S_3 \qquad \text{for } S_1 \geq 2T_1 \text{ and } S_3 < 2T_3$$

$$R = W_1 + S_1 + W_3 + S_3 \qquad \text{for } S_1 < 2T_1 \text{ and } S_3 < 2T_3 \ (T_n \text{ is the thickness of the conductor on layer } n).$$

R is the measure of the ground-plane coverage, which is dependent on W_1 and W_3 with a sidewall contribution due to T_1 and T_3.

$$O = S_1 - 2T_1 + S_3 - 2T_3 \qquad \text{for } S_1 \geq 2T_1 \text{ and } S_3 \geq 2T_3$$

$$O = S_3 - 2T_3 \qquad \text{for } S_1 < 2T_1 \text{ and } S_3 \geq 2T_3$$

$$O = S_1 - 2T_1 \qquad \text{for } S_1 \geq 2T_1 \text{ and } S_3 < 2T_3$$

$$O = 0.0 \qquad \text{for } S_1 < 2T_1 \text{ and } S_3 < 2T_3 \text{ (continuous upper and lower ground planes)}$$

The term O measures the amount of space and hence the capacitance to a ground plane. The capacitance is corrected for sidewall contributions when the conductors are widely spaced.

The capacitance formulae are as follows:

Line-to-ground Capacitance

One-ground plane:

$$\frac{C}{\varepsilon} = \frac{W}{H} + 3.28 \left(\frac{T}{T + 2H} \right)^{0.023} \left(\frac{S}{S + 2H} \right)^{1.16} \tag{4.19}$$

Two-ground planes:

$$\frac{C}{\varepsilon} = \frac{W}{H} + 1.086\left(1 + 0.685e^{\frac{-T}{1.343S}} - 0.9964e^{\frac{-S}{1.421H}}\right) \times$$

$$\left(\frac{S}{S+2H}\right)^{0.0476} \times \left(\frac{T}{H}\right)^{0.337}$$

(4.20)

Line-to-line Capacitance

One-ground plane:

$$\frac{C}{\varepsilon} = 1.064\left(\frac{T}{S}\right) \times \left(\frac{T+2H}{T+2H+0.5S}\right)^{0.695}$$

$$+ \left(\frac{W}{W+0.8S}\right)^{1.4148} \times \left(\frac{T+2H}{T+2H+0.5S}\right)^{0.804}$$

$$+ 0.831\left(\frac{W}{W+0.8S}\right)^{0.055} \times \left(\frac{2H}{2H+0.5S}\right)^{3.542}$$

(4.21)

Two-ground planes:

$$\frac{C}{\varepsilon} = \frac{T}{S}\left(1 - 1.897e^{\frac{-H}{0.31S} - \frac{-T}{2.474S}}\right.$$

$$+ 1.302e^{\frac{-H}{0.082S}} - 0.1292e^{\frac{-T}{1.326S}}\left.\right)$$

$$+ 1.722\left(1 - 0.6548e^{\frac{-W}{0.3477H}}\right)e^{\frac{-S}{0.651H}}$$

(4.22)

Crossover Capacitance

$$\frac{C}{\varepsilon} = \frac{W_1 W_2}{H}$$

$$+ 0.9413 \, FC(T_1, S_1) \, 2W_2\left(\frac{S_1}{S_1 + 0.01H}\right)^{0.2}$$

$$+ 0.9413 \, FC(T_2, S_2) \, 2W_1\left(\frac{S_2}{S_2 + 0.01H}\right)^{0.2}$$

(continued)

$$+ 1.14 \ FC(T_1, S_1) \ (S_2 S_1)^{0.5} \left(\frac{W_2}{H} \right)^{0.182}$$

$$+ 1.14 \ FC(T_2, S_2) \ (S_2 S_1)^{0.5} \left(\frac{W_1}{H} \right)^{0.182},$$

where

$$FC(T, S) = \left(1 - 0.326 e^{\frac{-T}{0.133 S}} - 0.959 e^{\frac{-S}{1.966 H}} \right) \tag{4.23}$$

C/ε = the normalized capacitance (per unit length of conductor)

ε = the dielectric permittivity of the insulator between the conductors

W = the width of a metal line

T = the thickness of a metal line

H = the thickness of the dielectric between conductors

S = the clear space between parallel conductors

numerical subscripts refer to two crossing conductor layers.

The valid range for these formulae is

$$0.3 \leq \frac{W}{H} \leq 10$$

$$0.3 \leq \frac{S}{H} \leq 10 \rightarrow \text{cutoff } S/H = 10$$

$$0.3 \leq \frac{T}{H} \leq 10.$$

For each layer in the process, the above formulae can be applied to estimate the capacitance of a given conductor. Consider the simplified process cross section shown in Fig. 4.13 for a two-level-metal process. The following dielectric and conductor thicknesses have been used:

Thin-oxide	200 Å
Field-oxide	6000 Å
Polysilicon	3000 Å
M1-poly–oxide	6000 Å
Metal1	6000 Å
M1-M2–oxide	6000 Å

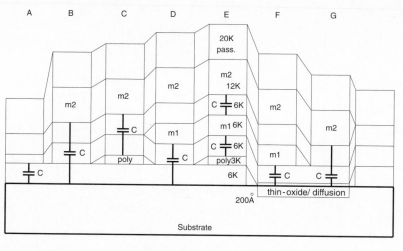

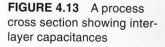

FIGURE 4.13 A process cross section showing inter-layer capacitances

Metal2	12000 Å
Passivation	20000 Å

The shielding conditions for the cross section views are shown in Table 4.5.

The capacitance of routing layers may be calculated directly from the parameters and the appropriate formulae for each piece of geometry in a layout. However, many times even this is too much computation to undertake on a large circuit. Thus it is common to calculate the area of a conductor and then apply a weighting factor dependent on the conductor size and routing density. The weighting factor can be determined from a graph such as the one shown in Fig. 4.14,[13] which is in turn computed from the formulae above and the appropriate conductor thicknesses and separations. Usually chip manufacturers will supply area capacitance and perimeter capacitance figures for each layer that are backed up by measurement of capacitance test

TABLE 4.5   Parasitic Capacitance Table

CONDITION	LAYER	LINE-TO-GROUND EQUATION	LINE-TO-LINE EQUATION
A	Poly-substrate	4.19	4.21
B	Metal2-substrate	4.19	4.21
C	Poly-metal2	4.20	4.22
D	Metal1-substrate	4.20	4.22
E	Metal1-poly	4.20	4.22
E	Metal1-metal2	4.20	4.22
F	Metal1-diffusion	4.20	4.22
G	Metal2-diffusion	4.19	4.21

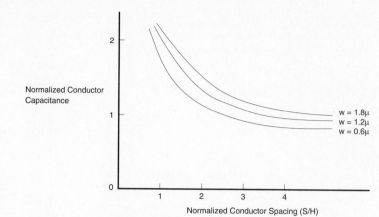

FIGURE 4.14 Typical conductor capacitances as a function of spacing; © IEEE 1992

structures. It is often prudent to include test structures on chips that enable the designer to independently calibrate a process to a set of design tools. These structures might range from simple ring-oscillators using inverters, NAND and NOR gates to structures that drive long, closely coupled lines in antiphase (which maximizes the effective coupling capacitance).

Other approaches to capacitance modeling may be found in the literature.[14,15,16,17,18] A recent technique that recognizes the type of structures predominantly found in today's CMOS ICs uses a technique called the "missing neighbor model."[19] This technique uses a table-lookup method to detect the presence or absence of adjacent conductor segments in regular routing channels. Because these channels form the majority of interconnect on all gate-array, standard-cell, and most current custom chips, the model has wide applicability. The basic numerical data is calculated by using a two-dimensional Poisson solver that finds the capacitance of a conductor in the vicinity of other conductors that are grounded.

When calculating stray capacitance values, one must keep in mind the actual dimensions that layers assume on the chip, rather than the drawn dimensions. For instance, a metal wire might be drawn at 1μ but end up being etched to 0.75μ. In addition the manufacturing tolerance on the width might be ±0.25μ. Also, the dielectric thicknesses will vary within manufacturing tolerances. One normally uses the worst-case value (i.e., the maximum width and thinnest dielectrics) for delay and dynamic power calculations and the minimum width and maximum thickness dielectrics for race calculations. On the other hand when calculating *RC* delays, the minimum width of a conductor might be used.

4.3.6 Distributed RC Effects

The propagation of a signal along a wire depends on many factors, including the distributed resistance and capacitance of the wire, the impedance of the driving source, and the load impedance. For very long wires with appreciable sheet resistance propagation delays caused by distributed resistance capacitance (*RC*) in the wiring layer can dominate. This transmission-line effect is particularly severe in poly wires because of the relatively high resistance of this layer but can be of equal importance in silicide wires and heavily loaded metal wires. A long wire can be represented in terms of several *RC* sections, as shown in Fig. 4.15.

The response at node V_j with respect to time is then given by

$$C\frac{dV_j}{dt} = (I_{j-1} - I_j)$$

$$= \frac{(V_{j-1} - V_j)}{R} - \frac{(V_j - V_{j+1})}{R}.$$ (4.24)

As the number of sections in the network becomes large (and the sections become small), the above expression reduces to the differential form:

$$rc\frac{dV}{dt} = \frac{d^2V}{dx^2},$$ (4.25)

where

x = distance from input

r = resistance per unit length

c = capacitance per unit length.

The form of this relation is that of the well-known diffusion equation. The solution for the propagation of a voltage step along the wire shows that the rise/fall delay, t_x, along a wire of length x is

$$t_x = kx^2,$$ (4.26)

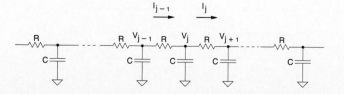

FIGURE 4.15
Representation of long wire in terms of distributed *RC* sections

where k is a constant. Alternatively, a discrete analysis of the circuit shown in Fig. 4.15 yields an approximate signal delay of

$$t_n = 0.7 \times \frac{RCn\,(n+1)}{2}, \qquad (4.27)$$

where

n = number of sections.

(The 0.7 factor accounts for a rise/fall delay to half rail.)

As n becomes very large (i.e., as the individual sections become very small), this reduces to

$$t_1 = 0.7\frac{rcl^2}{2}, \qquad (4.28)$$

where

r = resistance per unit length

c = capacitance per unit length

l = length of the wire.

The l^2 term in Eq. (4.28) shows that signal delay will be totally dominated by this RC effect for very long signal paths.

To illustrate the delays that can occur, we first consider the problem of running a long polysilicon wire in a single-metal process. (This was a frequent requirement in these processes for signals such as word lines in memories. In two-level-metal processes, poly should never be used for more than local interconnect or very slow global interconnect.) To optimize speed of a long poly line, one possible strategy is to segment the line into several sections and insert buffers within these sections. Figure 4.16 shows a poly bus of length 2 mm that has been divided into two 1-mm sections. For $r = 20\ \Omega/\mu m$ and $c = 4 \times 10^{-4}\ pF/\mu m$, Eq. (4.28) yields

$$t_1 = 0.7 \times 4 \times 10^{-15}\, l^2$$

for the delay of a 1-mm section.

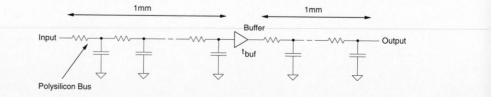

FIGURE 4.16
Segmentation of an RC line using a buffer

Assuming that the delay associated with the buffer is τ_{buf}, the total delay for this bus is

$$t_p = 2.8 \times 10^{-15} \, (1000)^2 + \tau_{buf} + 2.8 \times 10^{-15} \, (1000)^2$$
$$= 2.8 \, ns + \tau_{buf} + 2.8 \, ns$$
$$= 5.6 \, ns + \tau_{buf}.$$

This may be contrasted with the situation in which the buffer is missing, which yields

$$t_p = 11.2 \, ns.$$

Thus by keeping τ_{buf} small, significant gain can be obtained through appropriately segmenting the bus. The buffer delay, τ_{buf}, does, in fact, depend on the resistance of the first section of the bus and on the capacitance of the second section of the bus. The relative importance of these two terms depends on other circuit parameters, such as final load capacitance. In some situations, it may be preferable to use a wide poly wire to reduce overall series resistance at the expense of capacitance.

The advent of two-level-metal processes and silicided polysilicon have reduced the need for interconnections in native polysilicon. Usually, only metal is used for interconnect in these processes. However, in structures such as RAMs, silicided word lines might be used to reduce the size of the layout. As circuit speeds have increased, even metal connections can give rise to RC-delay effects, especially in heavily loaded clock lines. Consider a $50pF$ clock load distributed over a 10 mm chip in 1μ metal. Assuming the clock travels along two edges with the clock buffer in one corner, the total clock length might be 20 mm. Assuming the $50pF$ is distributed along the line, the delay to the end is ($r = .05 \, \Omega/\mu m$, c = $50pF/20mm$)

$$t_p = 0.7 \times 6.25 \times 10^{-17} \times (20000)^2$$
$$= 17.5 \, ns.$$

Buffers might solve this problem, but a more straightforward method to solve this skew problem is to widen the clock line and distribute the clock line from the top center of the chip. This decreases r, while increasing c a tiny bit. It also reduces l, thus reducing the skew to a manageable number. For instance, reducing l to 10 mm and increasing the clock line width to 20 μm, results in

$$t_p = 0.7 \times .625 \times 10^{-17} \times (10000)^2$$
$$= .44 \, ns.$$

Clock distribution is an important problem in high-speed, high-density chips. The above calculations are typical of the kind that have to be made in high-speed circuits to ensure correct temporal operation of a chip.

A model for the distributed RC delay, which takes driver and receiver loading into account, is shown in Fig. 4.17. R_s is the output resistance of the driver; C_l is the receiver input capacitance; R_t and C_t are the total, lumped resistance and capacitance of the line; and τ is the RC delay calculated using Eq. (4.28)–($rcl^2/2$). Such a model yields results that are very economical in terms of computation and, more importantly, are accurate enough for most design purposes. The approximations described in this section can (and should) be verified via simulation to check the accuracy of any critical RC delay problem.

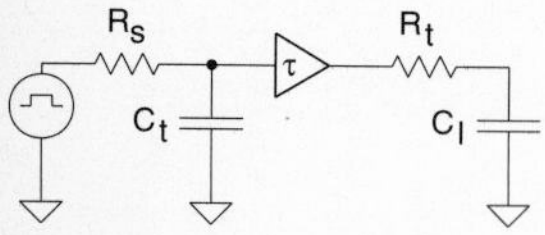

FIGURE 4.17 Simple model for RC delay calculation

4.3.7 Capacitance Design Guide

As a guide to the design process and, in particular, to the choice of routing layers, Table 4.6 is provided. It shows representative capacitance values (no fringing) for a 1 µm ($\lambda = 0.5$ µm) n-well CMOS process. As an approximation, the area of the routing capacitance can be doubled to account for fringing capacitance.

TABLE 4.6 Typical 1µm CMOS capacitances

PARAMETER	CAPACITANCE (ATTO FARADS (10^{-18})/µm^2)	COMMENTS
C_{jan}	300	n-diffusion area—varies widely with process
C_{jpn} (aF/µm)	400	n-diffusion periphery—varies widely with process
C_{jap}	500	p-diffusion area—varies widely with process
C_{jpp} (aF/µm)	400	p-diffusion periphery—varies widely with process
C_{gs}	1800	Gate capacitance—increases as t_{ox} thins
C_p	50	Poly-over field oxide
C_{m1}	30	Metal1-over field oxide
C_{m1p}	60	Metal1 to poly
C_{m1d}	60	Metal1 to diffusion
C_{m2}	20	Metal2 to substrate
C_{m2m1}	50	Metal2 to metal1
C_{m2p}	30	Metal2 to poly
C_{m2d}	30	Metal2 to diffusion
C_{m3}	10	Metal3 to substrate
C_{m3m2}	30	Metal3 to metal2
C_{m3m1}	15	Metal3 to metal1
C_{m3p}	12	Metal3 to poly
C_{m3d}	10	Metal3 to diffusion

It is important to be able to estimate capacitances before any detailed layout is completed. For each process, it is useful to have an approximate figure for the gate capacitance of a unit-size n- and p-transistor, the capacitance of 100 μm of poly wire, etc. In this way, bus loadings and other critical parasitics can be estimated to a first order without having to complete the design first. (Hint: In sub 0.5μ processes, metal wiring capacitance becomes quite dominant.)

Example

A register that fits in a data-path is 25 μm tall (the direction of repetition). A metal2 clock line runs vertically to link all registers in an n-bit register. The register has 30 μm of 1 μm metal1, 20 μm of 1 μm poly (over field), and 16 μm of 1 μm gate capacitance.

1. Calculate the per-bit clock load and the load for a 16-bit register.

2. What would be the *RC* delay to the register from a clock buffer using 5 mm of 1μ metal2 (.05 Ω/sq)?

3. How wide would the clock line have to be to keep the skew below .5 *ns* if a register file containing 32 16-bit registers was fed with the same 5 mm metal2 wire?

   1. The parasitics are as follows:

      $C_{m1} = 30 \times 30 = 900 \; aF$

      $C_p = 20 \times 50 = 1000 \; aF$

      $C_{gs} = 16 \times 1800 = 28,800 \; aF$

      $C_{reg1} = 900 + 1000 + 28,800 \; aF = .030pF$

      $C_{reg16} = 16 \times C_{reg1} = 0.48pF$

   2. $R_{metal2} = 5000 \times .05$

      $\qquad\qquad = 250 \; ohms$

      Because the capacitance load is at the end of the wire, we can approximate the *RC* delay by adding the metal2 track capacitance to the load capacitance and performing a simple *RC* calculation.

      $C_{total} = 0.48 + C_{metal2} \; pF$

      $\qquad\quad = 0.48 + (5000 \times 20 \times 10^{-6})pF$

      $\qquad\quad = 0.58pF$

      $RC \quad = 250 \times .58 \times 10^{-12} \; S$

      $\qquad\quad = .145 \; ns$

   3. We now have 32 registers, so the load capacitance of the registers is

      $C_{regfile} = 32 \times C_{reg16}$

      $\qquad\quad = 15.36pF.$

The RC for a 1 μm-wide clock feed is

$$= 3.84 \ ns.$$

Hence the clock line has to be widened by 3.84/0.5 or 7.68. For safety one might choose a 10μ wire.

Now

$$
\begin{aligned}
C_{total} &= 15.36 + C_{metal2} \ pF \\
&= 15.36 + (5000 \times 10 \times 20 \times 10^{-6})pF \\
&= 16.36pF \\
RC &= 25 \times 16.36 \times 10^{-12} \ s \\
&= 0.41 \ ns.
\end{aligned}
$$

Assuming the capacitance values are worst-cased, this width would be adequate. If typical values were used, a larger safety margin to cater for process variations might be prudent (i.e., double the calculated width in the example above). It is important to note that if a decision does not materially affect performance or density, it is always wise to make decisions which yield the highest operating margin (i.e., if it doesn't cost you, make it big!).

4.3.8   Wire-Length Design Guide

For the purposes of timing analysis, an electrical node may be defined as that region of connected paths in which the delay associated with signal propagation is small in comparison with gate delays. For sufficiently small wire lengths, RC delays can be ignored. Wires can then be treated as a single electrical node and modeled as simple capacitive loads. It is therefore useful to define simple electrical rules that can be used as a guide in determining the maximum length of communication paths for the various interconnect levels. To do this we require that wire delay and gate delay satisfy the following condition:

$$\tau_w \ll \tau_g \tag{4.29}$$

On substituting Eq. (4.28) into Eq. (4.29), we obtain the result

$$l \ll \sqrt{\frac{2\tau_g}{rc}} \ . \tag{4.30}$$

This establishes an upper bound on the allowable length of lightly loaded interconnects where the above approximations are valid. For example, for a

TABLE 4.7   Guidelines for Ignoring RC Wire Delays

LAYER	MAXIMUM LENGTH (λ)
Metal3	10000
Metal2	8000
Metal1	5000
Silicide	600
Polysilicon	200
Diffusion	60

minimum-width aluminum wire, assuming a gate delay of 200 *ps*,

$$l \ll \sqrt{\frac{2 \times .2 \times 10^{-9}\,(\tau_g)\,\lambda^2}{.05\,(r) \times 30 \times 10^{-18}\,(c)}}$$

$$\approx 16000\lambda.$$

So, conservatively,

$$l < 5000\lambda.$$

What this illustrates is that in a 1μ process, the *RC* delay of any minimum-width aluminum wire above 2.5 mm long should be taken into account, particularly for clock lines.

The electrical rules governing interconnect paths for a typical CMOS process are illustrated in Table 4.7 in terms of λ (as used in specification of design rules). This table assumes gate delays of the order of 100 *ps* to 500 *ps* and the signals are lightly loaded. Heavily loaded signals such as clocks should *always* be checked for *RC* skew problems. The significant factor that emerges from the table is the difference in tolerable communication distance between the metal layers and the polysilicon and diffusion layers. The rules shown in Table 4.7 should be recalculated for a given process.

4.4   Inductance

Although on-chip inductances are normally small, bond-wire inductance can cause deleterious effects in large, high-speed I/O buffers. Also, as processes shrink, it is likely that on-chip inductance might have to be taken into account.

The inductance of a cylindrical wire above a ground plane is given by

$$L = \frac{\mu}{2\pi} ln\left(\frac{4h}{d}\right), \tag{4.31}$$

where

μ = the magnetic permeability of the wire (typically 1.257×10^{-8} H/cm)

h = the height above the ground plane

d = the diameter of the wire.

This equation is appropriate for calculating the inductance of bonding wires and the pins on packages. For calculating the inductance of a conductor on a chip the following expression is approximately accurate (it assumes that thickness is negligible and $w < h$),

$$L = \frac{\mu}{2\pi} ln\left(\frac{8h}{w} + \frac{w}{4h}\right), \tag{4.32}$$

where

w = the width of the conductor

h = the height above the substrate (distance to backplane).

In the case of package inductance, values are normally supplied by the manufacturer (normally in the range from 3–15 nH). The inductance of a bond wire is of importance when calculating the inductive spike that occurs when a large current is drawn through a wire in a short period of time. The voltage change is as follows:

$$dV = L\frac{dI}{dt} \tag{4.33}$$

In high-speed designs it is important for power-supply connections to keep the inductance to a level where the change in voltage does not disturb the behavior of the chip. For more extensive treatment of the issues concerning packaging and interconnect, see Bakoglu.[20]

At the chip level, the inductance of on-chip wires may be estimated from the equation above:

$$L = \frac{1.257 \times 10^{-8}}{2\pi} ln\left(\frac{8000}{1} + \frac{1}{4000}\right) \quad \text{(assuming chip thickness = 1 mm)}$$

$$= 1.7 \times 10^{-8} \text{ Henrys/meter}$$

$$= 1.7 \times 10^{-2} \text{ nH/mm}$$

Apart from the very highest performance chips, these values are not of great importance.

4.5 Switching Characteristics

In this section, we develop analytic and empirical models that describe the switching characteristics of a CMOS inverter. These models are of use to understand the parameters that affect CMOS delays. More detailed analysis or simulation is usually required to yield models that accurately predict the performances of today's processes.

The switching speed of a CMOS gate is limited by the time taken to charge and discharge the load capacitance C_L. An input transition results in an output transition that either charges C_L toward V_{DD} or discharges C_L toward V_{SS}.

Before proceeding, however, we need to define some terms. Referring to Fig. 4.18:

- *Rise time, t_r* = time for a waveform to rise from 10% to 90% of its steady-state value.

- *Fall time, t_f* = time for a waveform to fall from 90% to 10% of its steady-state value.

- *Delay time, t_d* = time difference between input transition (50%) and the 50% output level. (This is the time taken for a logic transition to pass from input to output.)

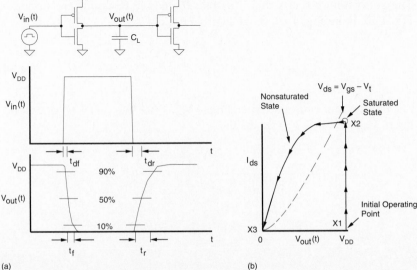

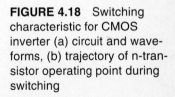

FIGURE 4.18 Switching characteristic for CMOS inverter (a) circuit and waveforms, (b) trajectory of n-transistor operating point during switching

Furthermore a differentiation is made between t_{df}, the high-to-low delay (input rising), and t_{dr}, the low-to-high delay (input falling).

We will first develop a simple analytic model to predict the delay of a CMOS inverter in order to understand the parameters that affect this delay.

4.5.1 Analytic Delay Models

4.5.1.1 Fall Time

Figure 4.18(a) shows the familiar CMOS inverter with a capacitive load, C_L, that represents the load capacitance (input of next gates, output of this gate and routing). Of interest is the voltage waveform, $V_{out}(t)$, when the input is driven by a step waveform, $V_{in}(t)$, as shown in Fig. 4.18(a). Figure 4.18(b) shows the trajectory of the n-transistor operating point as the input voltage, $V_{in}(t)$, changes from zero volts to V_{DD}. Initially, the n-device is cut off and the load capacitor, C_L, is charged to V_{DD}. This is illustrated by $X1$ on the characteristic curve. Application of a step voltage (i.e., $V_{gs} = V_{DD}$) at the input of the inverter changes the operating point to $X2$. From there onwards, the trajectory moves on the $V_{gs} = V_{DD}$ characteristic curve toward point $X3$ at the origin. From the switching characteristics shown in Fig. 4.18, it is evident that the fall time, t_f, consists of two intervals:

1. t_{f1} = period during which the capacitor voltage, V_{out}, drops from 0.9 V_{DD} to $(V_{DD} - V_{tn})$.
2. t_{f2} = period during which the capacitor voltage, V_{out}, drops from $(V_{DD} - V_{tn})$ to 0.1 V_{DD}.

The equivalent circuits that illustrate the above behavior are shown in Fig. 4.19. From Fig. 4.19(a), while in saturation

$$C_L \frac{dV_{out}}{dt} + \frac{\beta_n}{2} (V_{DD} - V_{tn})^2 = 0. \tag{4.34}$$

Integrating from $t = t_1$, corresponding to $V_{out} = 0.9 \, V_{DD}$, to $t = t_2$ corresponding to $V_{out} = (V_{DD} - V_{tn})$ results in

$$t_{f1} = 2 \frac{C_L}{\beta_n (V_{DD} - V_{tn})^2} \int_{V_{DD} - V_{tn}}^{0.9 V_{DD}} dV_{out}$$

$$= \frac{2C_L (V_{tn} - 0.1 V_{DD})}{\beta_n (V_{DD} - V_{tn})^2} \tag{4.35}$$

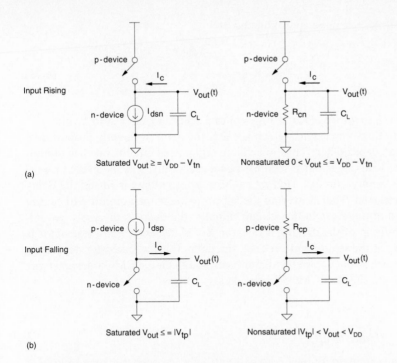

(a)

(b)

FIGURE 4.19 Equivalent circuits for fall- and rise-time determination

When the n-device begins to operate in the linear region, the discharge current is no longer constant. The time, t_{f2}, taken to discharge the capacitor voltage from $(V_{DD} - V_{tn})$ to $0.1\ V_{DD}$ can be obtained as before, giving

$$t_{f2} = \frac{C_L}{\beta_n (V_{DD} - V_{tn})} \int_{0.1 V_{DD}}^{V_{DD} - V_{tn}} \frac{dV_{out}}{\dfrac{V_{out}^2}{2(V_{DD} - V_{tn})} - V_{out}}$$

$$= \frac{C_L}{\beta_n (V_{DD} - V_{tn})} ln\left(\frac{19 V_{DD} - 20 V_{tn}}{V_{DD}}\right)$$

$$= \frac{C_L}{\beta_n V_{DD} (1 - n)} ln\,(19 - 20n),$$

(4.36)

with $n = V_{tn}/V_{DD}$.

Thus the complete term for the fall time, t_f is

$$t_f = 2\frac{C_L}{\beta_n V_{DD} (1 - n)}\left[\frac{(n - 0.1)}{(1 - n)} + \frac{1}{2} ln\,(19 - 20n)\right].$$

(4.37)

The fall time, t_f, can be approximated as

$$t_f \approx k \times \frac{C_L}{\beta_n V_{DD}},\tag{4.38}$$

where $k = 3$ to 4 for values of $V_{DD} = 3$ to 5 volts and $V_{tn} = .5$ to 1 volt.

From this expression, we can see that the delay is directly proportional to the load capacitance. Thus to achieve high-speed circuits one has to minimize the load capacitance seen by a gate. Secondly, it is inversely proportional to supply voltage. That is, as the supply voltage is raised the delay time is reduced. Thus, lowering the supply voltage on a circuit will reduce the speed of the gates in that circuit. Finally, the delay is inversely proportional to the β of the driving transistor. So, as the width of a transistor is increased or the length is decreased, the delay for that transistor decreases. These three "knobs" form the major basis by which the CMOS designer can optimize the speed of CMOS logic gates.

4.5.1.2   Rise Time

Due to the symmetry of the CMOS circuit, a similar approach may be used to obtain the rise time, t_r (Fig. 4.19b). Thus

$$t_r = 2\frac{C_L}{\beta_p V_{DD}(1-p)}\left[\frac{(p-0.1)}{(1-p)} + \frac{1}{2}ln\,(19-20p)\right]\tag{4.39}$$

with $p = |V_{tp}|/V_{DD}$.

As before, Eq. (4.39) may be approximated by

$$t_r \cong 3 \rightarrow 4\frac{C_L}{\beta_p V_{DD}}.\tag{4.40}$$

For equally sized n- and p-transistors, where $\beta_n = 2\beta_p$,

$$t_f = \frac{t_r}{2}.\tag{4.41}$$

Thus the fall time is faster than the rise time, primarily due to different carrier mobilities associated with the p- and n-devices (i.e., $\mu_n = 2\mu_p$). Therefore, if we want to have approximately the same rise and fall time for an inverter, we need to make

$$\frac{\beta_n}{\beta_p} = 1. \tag{4.42}$$

This implies that the channel width for the p-device must be increased to approximately two to three times that of the n-device, so

$$W_p = 2\text{–}3\ W_n. \tag{4.43}$$

Note that to accurately specify the width ratio required to achieve equal rise and fall times, an accurate ratio of β_n and β_p must be known. These, in turn, depend on the parameters of the process being used.

4.5.1.3 Delay Time

In most CMOS circuits, the delay of a single gate is dominated by the output rise and fall time. The delay is approximately given by

$$t_{dr} = \frac{t_r}{2} \tag{4.44}$$

and

$$t_{df} = \frac{t_f}{2}. \tag{4.45}$$

An alternative formulation is given by

$$t_{df} = A_N \frac{C_L}{\beta_n}, \tag{4.46}$$

where A_N is a process constant for a specific supply voltage.[21] A_N has been derived as

$$A_N = \frac{1}{V_{DD}(1-n)} \left[\frac{2n}{1-n} + ln\left(\frac{2(1-n) - V_O}{V_O} \right) \right],$$

where

$$n = \frac{V_{tn}}{V_{DD}}$$

$$V_O = \frac{V_{out}}{V_{DD}}.$$

For $V_{tn} = 0.7$ volts, $V_{DD} = 5$ volts, $V_{out} = 2.5$ volts, A_N is .283.
Similarly,

$$t_{dr} = A_p \frac{C_L}{\beta_p}, \tag{4.47}$$

where A_P is a process constant for a specific supply voltage. A_P has been derived as

$$A_p = \frac{1}{V_{DD}(1+p)} \left[\frac{-2p}{1+p} + ln \left(\frac{2(1+p) - V_O}{V_O} \right) \right],$$

where

$$p = \frac{V_{tp}}{V_{DD}}.$$

For $V_{tp} = -0.7$, $V_{DD} = 5$, $V_{out} = 2.5$, A_P is .283.
The average gate delay for rising and falling transitions is

$$t_{av} = \frac{t_{df} + t_{dr}}{2}. \tag{4.48}$$

Figure 4.20 illustrates a SPICE simulation of a step input applied to an inverter driving a capacitive load. The process parameters for the simulation were as follows:

$V_{tn} = .767$ volts, $V_{tp} = -.938$ volts, $\beta_n = 4.04 \times 10^{-4}$, $\beta_p = 3.48 \times 10^{-4}$, $V_{DD} = 5.0$, $C_L = .5\ pF$, substituting into Eq. (4.39):

$t_r = 1.04\ ns$, compared with $1.14\ ns$ from SPICE (level 1).

Substituting into Eq. (4.37):

$t_f = .83\ ns$, compared with $.89\ ns$ for SPICE (level 1).

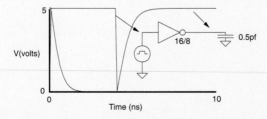

FIGURE 4.20 SPICE simulation of CMOS inverter transient response

If we examine the delays t_{dr} and t_{df}, we see that

$t_{dr} = .5$ *ns*, compared with .52 *ns* for SPICE.

$t_{df} = .4$ *ns*, compared with .45 *ns* for SPICE.

For the delay times, the error is 7% to 10%. The equations for delay times developed in this section have the limitation that only first-order MOS equations were used to calculate the drain currents flowing in the transistors. Unfortunately, today's processes have quite complicated modeling equations. While more complex equations may be incorporated into a similar analysis, most designers (and CAD programmers) have found it more feasible to take an empirical approach to calculating delay values.

4.5.2 Empirical Delay Models

In an empirical delay model, a circuit simulator is used to model the inverter or gate in question and then the measured values are backsubstituted into appropriate delay equations. For instance, one can backsubstitute into Eqs. (4.46) and (4.47) to obtain values for A_N and A_P. For the simulation shown in Fig. 4.20 ($W_p = 2W_n$),

$$A_P = t_{dr-spice} \frac{\beta_p}{C_L} = .52 \times 10^{-9} \times \frac{3.48 \times 10^{-4}}{0.5 \times 10^{-12}} = .36 \, (.31 \text{ calc}) \quad \textbf{(4.49a)}$$

$$A_N = t_{df-spice} \frac{\beta_n}{C_L} = .45 \times 10^{-9} \times \frac{4.04 \times 10^{-4}}{0.5 \times 10^{-12}} = .36 \, (.29 \text{ calc}). \quad \textbf{(4.49b)}$$

These constants may now be used to predict delay values for a wide range of gates. That is, for gates with $W_p = 2W_n$,

$$t_{dr} = .36 \frac{C_L}{\beta_p} \quad \textbf{(4.50a)}$$

$$t_{df} = .36 \frac{C_L}{\beta_n}. \quad \textbf{(4.50b)}$$

One may also couch these equations in terms of the width of the transistor. Notice that these equations now represent the delay in terms of an *RC* delay where the effective resistance of the transistor is given by .36/β. We will use this in Section 4.5.4.

4.5.3   Gate Delays

The delay of simple gates may be approximated by constructing an "equivalent" inverter. This is an inverter where the pull-down n-transistor and the pull-up p-transistor are of a size to reflect the effective strength of the real pull-down or pull-up path in the gate. For instance, in the 3-input NAND gate shown in Fig. 4.21, $W_p = W_n$ for all transistors. When the pull-down path is conducting, all of the n-transistors have to be turned on. The effective β of the n-transistors is given by

$$\beta_{neff} = \cfrac{1}{\cfrac{1}{\beta_{n1}} + \cfrac{1}{\beta_{n2}} + \cfrac{1}{\beta_{n3}}} \quad \text{(summation of series conductances)} \qquad \textbf{(4.51)}$$

For $\beta_{n1} = \beta_{n2} = \beta_{n3}$

$$\beta_{neff} = \frac{\beta_n}{3}.$$

For the pull-up case, only one p-transistor has to turn on to raise the output. Thus,

$$\beta_{peff} = \beta_p.$$

For $\beta_p = 0.3\, \beta_n$

$$t_r = k\frac{C_L}{0.3\beta_n V_{DD}}, \; t_f = k\frac{C_L}{\dfrac{\beta_n}{3} V_{DD}}$$

$$\frac{t_r}{t_f} \approx 1.$$

For a more graphical understanding of this, a series transistor connection is illustrated in Fig. 4.22(a). Consider the three n-transistors in series. Imagine the gates of three transistors of the same width and length in series being brought closer together. Finally, when the gate regions abut (illegal, but for purposes of illustration), the resulting transistor has a length of $3L$ (Fig. 4.22b). Thus

$$\beta_{series} = \frac{\beta_n}{3}.$$

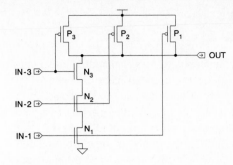

FIGURE 4.21 A 3-input NAND gate

Hence,

$$\tau_{series} = k\frac{C_L}{\frac{\beta_n}{3}V_{DD}}$$

which is three times the delay time for one transistor.

In general, the fall time t_f is mt_f for m n-transistors in series. Similarly the rise time t_r for k p-transistors in series is kt_r. In comparison, the fall time t_f for a parallel connection of transistors is t_f/m for m transistors in parallel, if all the transistors are turned on simultaneously. For k p-transistors in parallel, the rise time is t_r/k for k devices in parallel if all transistors are turned on simultaneously. These times are important if the fastest delay through a gate has to be evaluated. For other delay approaches, see also Vemuru and Thorbjornsen,[22] Dhar and Franklin,[23] and Sakurai and Newton.[24]

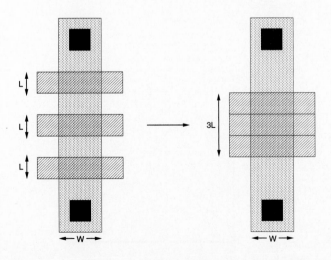

FIGURE 4.22 Graphical illustration of the effect of series transistors

4.5.4 Further Delay Topics

4.5.4.1 Input Waveform Slope

The analytical expression for the delay in an inverter was calculated under certain conditions with idealized VI equations. In addition to differences in device occurrences, real circuits have other effects that lead to discrepancies in timing values in those circuits. For instance, the input waveform was assumed to be a step function. The slope of the input waveform can modify the delay of a gate. When the input rises or falls rapidly, the delay of the charge or discharge path is determined by the rate at which the transistors in the path can charge or discharge the capacitors in the tree. When the input changes slowly, it will contribute to the output delay. This effect is shown in the SPICE simulation shown in Fig. 4.23, where two identical circuits are driven by waveforms of differing slope. The results are tabulated below in Table 4.8.

Signal y incurs an extra .35 ns in rise time and .3 ns in fall time from the slower changing input signal b. For the fast changing input (a), the p- and n-transistors are still in saturation when the input reaches its final value.

Hedenstierna and Jeppson[25] provide the following modification to Eq. (4.47) to take account of rise time:

$$t_{dr} = t_{dr\text{-}step} + \frac{t_{input\text{-}fall}}{6}(1 - 2p),$$

(4.52)

where

$t_{dr\text{-}step}$ = the step-input rise time calculated in Eq. (4.47)

$t_{input\text{-}fall}$ = the input fall time.

$$p = \frac{V_{tp}}{V_{DD}}$$

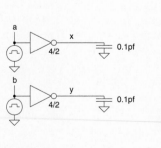

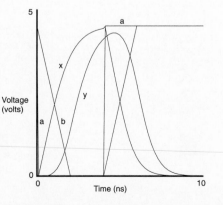

FIGURE 4.23 Effect of input rise and fall time on inverter delays

TABLE 4.8   Effect of Input Rise Time on Inverter Delay

INPUT	OUTPUT	RISE TIME	RISE DELAY	FALL DELAY
a	x	0.1 ns	1.06 ns	0.94 ns
b	y	2 ns	1.41 ns	1.24 ns
b	y	5 ns	1.87 ns	1.67 ns

Similarly,

$$t_{df} = t_{df\text{-}step} + \frac{t_{input\text{-}rise}}{6}(1 + 2n) \qquad (4.53)$$

$$n = \frac{V_{tn}}{V_{DD}}$$

This is valid for input rise or fall times that satisfy the following criteria:

$$\frac{t_{input\text{-}rise}\beta_p V_{DD}}{C_L} < \frac{6p}{(1-p)^3} \qquad (4.54a)$$

and

$$\frac{t_{input\text{-}fall}\beta_n V_{DD}}{C_L} < \frac{6n}{(1-n)^3}. \qquad (4.54b)$$

4.5.4.2   Input Capacitance

In the derivation of the inverter delay the input capacitance was assumed to be constant. In practice, we have seen that as the voltage changes on the gate of a transistor, so does the capacitance of the gate terminal. Also an effect known as *bootstrapping* can modify the effective input capacitance of an inverter or logic gate. This variation can lead to further error in the simple model presented in Section 4.4. Bootstrapping may be understood by examining Fig. 4.24(a). The inverter has the normal C_{in} (C_{gs}) and load capacitance. C_{gd}, the gate to drain capacitance, has been added. In the case where the input is rising (that is, the output is high), the effective input capacitance is $C_{gs} + C_{gd}$. When the output starts to fall, the voltage across C_{gd} changes, requiring the input to supply more current to charge C_{gd}. This effect is seen

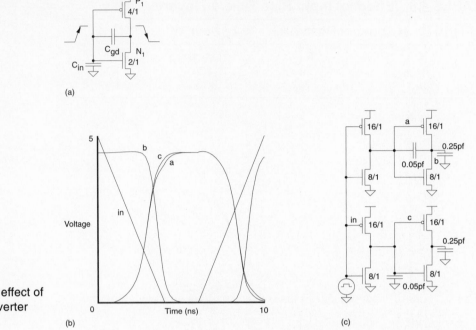

FIGURE 4.24 The effect of bootstrapping on inverter delay

in Fig. 4.24(b) for the circuit shown in Fig. 4.24(c). As waveform b falls (rises), waveform a slows down as the extra capacitance is charged. Because C_{gd} is small, this is usually a small effect (as evidenced in Fig. 4.24b).

If the inverter is biased in its linear region, the C_{gd} may appear multiplied by the gain of the inverter. This is known as the Miller effect but is seldom of importance in digital circuits because the input passes rapidly through the linear region. It is, however, of major importance in analog circuits.

4.5.4.3   Switch-Level RC Models

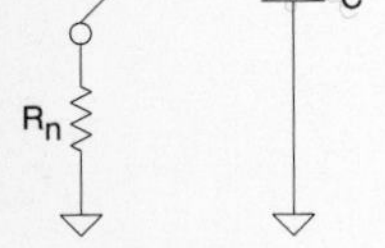

Resistance-capacitance (RC) modeling techniques[26,27] represent transistors as a resistance discharging or charging a capacitance, as shown in Fig. 4.25. A variety of timing models have been developed to estimate the delays of logic gates, using the switch behavior of the transistors involved in the gate. These models include the following:

- Simple RC delay.
- Penfield-Rubenstein Model.
- Penfield-Rubenstein Slope Model.

FIGURE 4.25
A switch-level RC model

In the simple *RC* model the total resistance of the pull-up or pull-down path is calculated and all the capacitance of nodes involved in switching are lumped onto the output of the gate.[28,29,30] For instance, in Fig. 4.26, the fall delay for any input would be calculated as follows:

$$t_{df} = \Sigma R_{pulldown} \times \Sigma C_{pulldown\text{-}path}$$
$$= (R_{N1} + R_{N2} + R_{N3} + R_{N4}) \times (C_{out} + C_{ab} + C_{bc} + C_{cd}), \qquad \textbf{(4.55)}$$

while the rise delay (from node A) would be calculated as

$$t_{dr} = R_{P4} \times C_{out}. \qquad \textbf{(4.56)}$$

An *effective* resistance is used for each transistor type (n or p), size, and state (on or off). Other factors, such as circuit style, may lead to different effective resistance values. The effective resistance is multiplied by the *W/L* ratio of the transistor to arrive at a final value for the resistance.

The *RC* delay calculation tends to pessimize the delay because it assumes that all the internal capacitance has to be discharged or charged to switch the gate.

The Penfield-Rubenstein model[31] was developed to calculate delays in generalized *RC* trees. For a group of transistors in series (as in a NAND gate), this formulation simplifies to the Elmore delay[32] for an *RC* ladder, which is

$$t_d = \sum_i R_i C_i \qquad \textbf{(4.57)}$$

where R_i is the summed resistance from point *i* to power or ground and C_i is the capacitance at point *i*. For instance in the 4-input NAND gate shown in

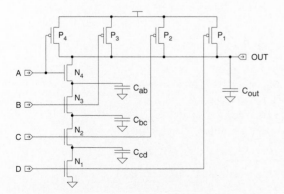

FIGURE 4.26   A 4-input NAND gate showing parasitic capacitances

Fig. 4.26, this would result in

$$t_{df} = (R_{N1} \times C_{cd}) + [(R_{N1} + R_{N2}) \times C_{bc}] + [(R_{N1} + R_{N2} + R_{N3}) \times C_{ab}]$$
$$+ (R_{N1} + R_{N2} + R_{N3} + R_{N4}) \times C_{out.}$$

This model may be improved by taking into account the rise or fall time of the input waveform. The *Slope Model* defines the intrinsic rise time (or fall time) as the rise time that would occur if the input was driven by a step function.[33,34] The actual input rise time is then divided by this value to arrive at a rise-time-ratio, which indicates the degree to which the switched transistor is turned on. Combining the slope model with the Penfield-Rubenstein delay model results in the Penfield-Rubenstein Slope Delay Model, which is widely used in transistor-level-timing analyzers and switch-level simulators. In order to use such delay models one requires tables of transistor resistance values from which to calculate delays. A method of deriving the effective resistance values for such models is to use SPICE with test circuits similar to that shown in Fig. 4.27. Here, we measure the rise and fall time of a specific-

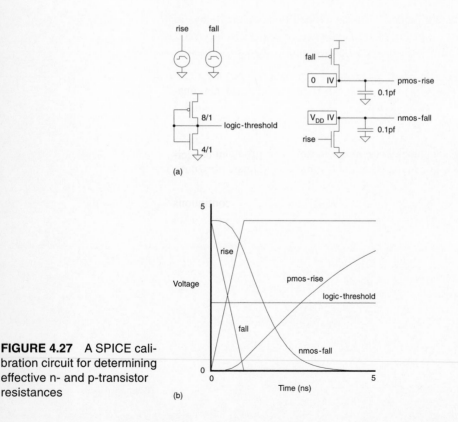

FIGURE 4.27 A SPICE calibration circuit for determining effective n- and p-transistor resistances

sized transistor (or range of sizes) for a variety of input (fall) rise times generated by the pulse generators. A self-biased inverter (with a p/n ratio that is representative of the logic library being used) establishes a nominal switching voltage level or logic threshold. The delay between the input and output at the logic threshold is measured and, from the known load capacitance, the effective resistance may be calculated. For instance in the example shown, a minimum-sized n-transistor switches $.1pF$ in $1.1\ ns$. Hence the effective resistance is

$$R_n = \frac{t_{df}}{C}$$

$$= \frac{1.1 \times 10^{-9}}{.1 \times 10^{-12}}$$

$$= 11K\Omega$$

Once this resistance has been calculated for a variety of transistor widths and rise and fall times, the effective pull-up or pull-down resistance of a gate may be found by interpolation.

4.5.4.4   Macromodeling

The approach of macromodeling involves deriving a set of accurate formulae to calculate gate capacitance and logic gate behavior based on the device equations.[35] In this approach, the circuit is divided into gates and memory elements. These modules are characterized by power, input and output capacitance and waveforms. A typical model along with the timing model is shown in Fig. 4.28. Here t_{swin} is the input waveform, C_{in} is the input capacitance, t_{beout} is the delay through the gate, t_{swout} is the output waveform, and C_L is the output capacitance. The waveforms are typically represented by linear ramps with exponential tails. Macromodeling techniques can be discontinuous in the first derivative, posing problems for optimization programs. Other

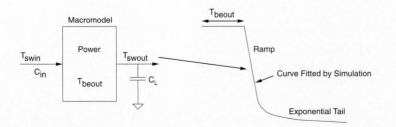

FIGURE 4.28   Model used in macromodeling approach

well-behaved analytical models have also been developed for CMOS gate delays.[36]

Another approach (and more common in the ASIC community) treats logic gates as simple delay elements. Each gate type is simulated with a circuit simulator, and an equation of the following type is used to determine the delay of a particular gate (for both rising and falling inputs):

$$t_d = t_{internal} + k \times t_{output} \tag{4.58}$$

Here the delay is divided into a fixed internal delay, $t_{internal}$, and an output delay, t_{output}, that is proportional to the output loading, k. The output loading and t_{output} are related in such a way as to arrive at an appropriate delay. Figure 4.29 shows a typical SPICE circuit used to calibrate delay equations. It uses three input NAND gates driving load capacitances of zero, $0.5pF$, and $1pF$ to determine the internally loaded delay and the delay at two capacitance values. Table 4.9 summarizes the data gained from this simulation and the values of the gate delays that would be placed in a data sheet.

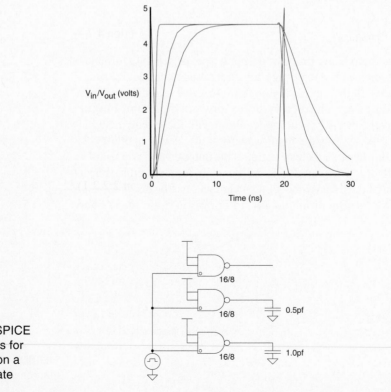

FIGURE 4.29 SPICE circuit and results for delay modeling on a 3-input NAND gate

TABLE 4.9   NAND3 SPICE Delays

Time	LOAD CONDITIONS		
	$C = 0$	$C = .5pF$	$C = 1pF$
t_{dr} (ns)	0.255	1.32	2.38
t_{df} (ns)	0.42	2.36	4.27
t_{output} rise (ns/pF)		2.12	2.12
t_{output} fall (ns/pF)		3.82	3.82

Thus for this gate the delay equations would be

$$t_{dr} = .255 + k \times 2.12 \ ns \ (k \text{ is in } pF)$$

$$t_{df} = .42 + k \times 3.82 \ ns \ (k \text{ is in } pF)$$

This might be completed for each input to the gate, or only the worst-case input-to-output delay (speed and power) might be used. In the worst-case slow process, the temperature (high) and the voltage (low) of the process would be used to determine the slowest speed operation of a circuit. In the worst-case fast process, the temperature (low) and the voltage (high) of the process should be used for logic race or power checks (but see Section 4.7— with static logic the power dissipation should not change markedly over process and temperature). For an explanation of these terms see Section 4.10.

4.5.4.5   Body Effect

Body effect is the term given to the modification of the threshold voltage, V_t, with a voltage difference between source and substrate. Specifically, $\Delta V_t \propto \gamma \sqrt{V_{sb}}$, where γ is a constant, V_{sb} is the voltage between source and substrate, and ΔV_t is the change in threshold voltage (see Section 2.2.2.1). For instance, in the 4-input NAND gates shown in Fig. 4.30(a), the n-transistor at the output will switch slower if the source potential of this transistor is not the same as the substrate. The SPICE simulation in Fig. 4.30(b) illustrates how this occurs. In the upper NAND gate the lower transistors are initially turned on while transistor N_{4A} is turned off. This results in the source of N_{4A} being at ground when the input on N_{4A} rises. The result is seen in Fig. 4.30(b) in the form of waveform CD, which rises to about 1.7 volts before being discharged to ground through the four-series n-devices. In the lower NAND gate, the upper transistors are turned on initially, while transistor N_{1B} is turned off. Hence, the nodes cd, bc, and ab are at an n-threshold below V_{DD} (~3.1 volts). When N_{1B} turns on, nodes ab, bc, and cd are pulled to ground in that order. This slows the output transition, as can be seen in the

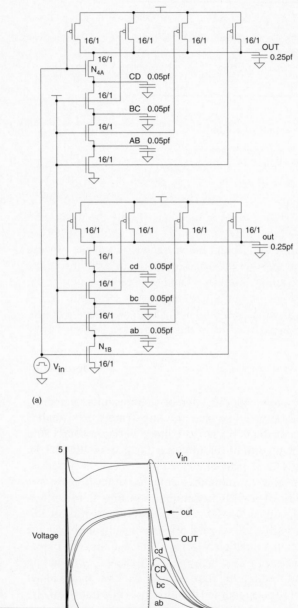

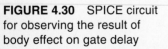

FIGURE 4.30 SPICE circuit for observing the result of body effect on gate delay

SPICE plot (in this case about .4 *ns*). When the load capacitance is much greater than the internal capacitance of the gates, this effect is minimized; however, for performance optimized circuits it can be significant. To minimize this effect, gate design should minimize "internal" node capacitance and take into account the relative body effect of the two types of transistor.

Given that a number of series transistors may be required in a gate, a further optimization may be made. As the body effect is essentially a dynamic problem involving the charging of parasitic capacitances, we can use the natural time sequencing of signals to offset the body effect. The first strategy is to place the transistors with the latest arriving signals nearest the output of a gate. The early signals, in effect, "discharge" internal nodes, and the late-arriving signals have to switch transistors with minimum body effect. The other strategy mentioned previously is to minimize the capacitance of internal nodes. Thus if a diffusion wire had to be used to minimize the geometric topology of a gate, we would try to use it at the output of a gate rather than on some internal node. In the same vein, connections on internal nodes should be completed in metal or local interconnect, if available. The diffusion attached to transistors should be optimized to reduce its area and periphery contributions to parasitic capacitance.

4.5.5 Summary

While much effort has been directed at analytically modeling CMOS inverter and gate delays, the most pragmatic approach is either to use the Penfield-Rubenstein or Penfield-Rubenstein-Slope models for transistor-level modeling or simulate gates with SPICE and measure the appropriate delays. With a good programmable CAD system these tasks can be highly automated. In fact, at least one commercial semiconductor vendor can automatically create a new standard cell data book and mask library automatically using a highly automated symbolic layout system. These methods are fast and accurate when the delays are derived from a circuit simulator that is known to accurately model a given process.

Precise process calibration requires that

- the transistors are modeled accurately.
- the parasitic capacitances are modeled accurately.

The modeling of transistors may be checked by including individual transistors of appropriate widths as probe-accessible test structures on a chip (most manufacturers include these in their own PCMs (Process Control Monitors)—a bit of detective work can usually locate and identify these). Such transistors may be probed using microprobes and their DC characteristics compared to that of the circuit simulator. Sometimes, the output transistors in I/O buffers may be accessed in such a way that their characteristics may be measured in cases where no test structures are available.

The modeling of capacitance may be checked by probing test capacitance structures. However, it is usually easier to measure the delay of a number of gates in a known path and reverse-engineer the capacitance by comparing the measured delay with that of a simulator with known good DC MOS models. At least two distinct types of stray capacitance should be measured. The first is in tightly packed, locally connected structures such as the internals of datapaths (i.e., adder carry chains). The second type of routing is where a single gate drives a large routing capacitance to a number of gates spread across a chip. A good choice here might be a lightly loaded signal and a heavily loaded signal such as a clock. The reason for these two measurements is that in the closely packed case, loading is dominated by the intrinsic load of the gates, while in the second case the loading is dominated by routing area capacitance and fringing capacitance.

The calibration of a given set of CAD tools usually requires at least one pass through a given CMOS fabrication line. Manufactured devices may also be compared to simulations using SEM circuit probing, which is probably one of the best ways of debugging and calibrating CAD tools. In this technique, an electron beam is raster-scanned across an exposed chip (under vacuum) and the reflected electrons are measured to estimate the circuit potential. A TV image may be constructed that shows the chip surface with the voltage levels being represented by gray levels. Sampling techniques allow time plots of signals to be measured and stored.

Whatever the technique, accurate process calibration is the key to predictable performance estimation.

4.6 CMOS-Gate Transistor Sizing

4.6.1 Cascaded Complementary Inverters

The discussions so far have led us to believe that if we want to have approximately the same rise and fall times for an inverter, for current CMOS processes, we must make

$$W_p \approx (2{\rightarrow}3) \times W_n, \qquad \qquad \textbf{(4.59)}$$

where W_p is the channel width of the p-device and W_n is the channel width of the n-device. This, of course, increases layout area and, as we shall see later, dynamic power dissipation. In some cascaded structures it is possible to use minimum or equal-sized devices without compromising the switching response.

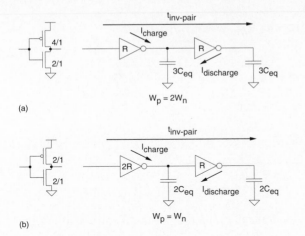

(a)

(b)

FIGURE 4.31   CMOS inverter pair timing response

This is illustrated in the following analysis, in which the delay response for an inverter pair (Fig. 4.31a) with $W_p = 2W_n$ is given by

$$t_{inv\text{-}pair} \propto t_{fall} + t_{rise}$$

$$\propto R3C_{eq} + 2\frac{R}{2}3C_{eq}$$

$$\propto 3RC_{eq} + 3RC_{eq}$$

$$\propto 6RC_{eq}, \tag{4.60}$$

where R is the effective "on" resistance of a unit-sized n-transistor and $C_{eq} = C_g + C_d$ is the capacitance of a unit-sized gate and drain region. The inverter pair delay, with $W_p = W_n$ (Fig. 4.31b), is

$$t_{inv\text{-}pair} \propto t_{fall} + t_{rise}$$

$$\propto R2C_{eq} + 2R2C_{eq}$$

$$\propto 6RC_{eq}. \tag{4.61}$$

Thus we find similar responses are obtained for the two different conditions.

TABLE 4.10   Variation in V_{inv} with β_n/β_p ratio

V_{DD}	V_{tn}	V_{tp}	β_n	β_p	V_{inv}
5	.7	−.7	1	1	2.5
5	.7	−.7	.5	1	2.8
5	.7	−.7	1	.5	2.2
3	.5	−.5	1	1	1.5
3	.5	−.5	.5	1	1.67
3	.5	−.5	1	.5	1.32

It is important to remember that changes in the β ratio also affect inverter threshold voltage, V_{inv}. From Eq. (2.22), the relation defining V_{inv} is given by

$$
V_{inv} = \frac{V_{DD} + V_{tp} + V_{tn}\sqrt{\dfrac{\beta_n}{\beta_p}}}{1 + \sqrt{\dfrac{\beta_n}{\beta_p}}}.
$$

Table 4.10) summarizes V_{inv} for a range of values of V_{DD}, V_{tp}, V_{tn}, β_p, and β_n. This shows less than 15% variation in V_{inv} for these β ratios. Based on these results it is evident that, if necessary, in self-loaded circuits minimum-sized devices may be used to reduce power dissipation and increase circuit packing density. When the circuits have to drive any significant routing load, this optimization does not apply and the n- and p-transistors should be sized to yield equal rise and fall times.

4.6.2   Cascaded Pseudo-nMOS Inverters

A simple timing model of the pseudo-nMOS inverter introduced in Chapter 2 is shown in Fig. 4.32. This uses the 3:1 transistor-width ratios determined in that chapter. The approximate delay for a pair of inverters is

$$
t_{inv\text{-}pair} \propto 6R\,(C_g + 2C_d) + R\,(C_g + 2C_d)
$$

$$
\propto 7RC_{eq},
$$

(4.62)

where

$$
C_{eq} = C_g + 2C_d.
$$

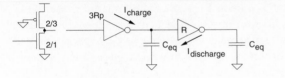

FIGURE 4.32   Pseudo-nMOS inverter pair timing response

Note that this speed may be improved by sacrificing noise margin (i.e., making the pull-up stronger).

4.6.3   Stage Ratio

Often it is desired to drive large load capacitances such as long buses, I/O buffers, or, ultimately, pads and off-chip capacitive loads. This is achieved by using a chain of inverters (or perhaps other logic gates) where each successive inverter is made larger than the previous one until the last inverter in the chain can drive the large load in the time required. The optimization to be achieved here is to minimize the delay between input and output while minimizing the area and power dissipation. The ratio by which each stage is increased in size is called the stage ratio.

Following the derivation given in Mead and Conway,[37] consider the circuit shown in Fig. 4.33. It consists of n-cascaded inverters with stage-ratio a, driving a capacitance C_L. Thus inverter inv-1 is a minimum-sized inverter

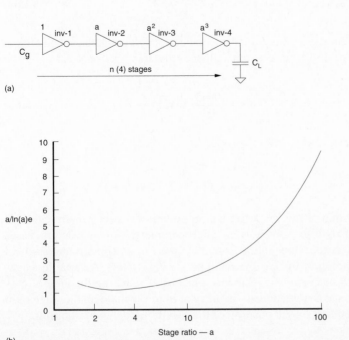

FIGURE 4.33   Stage ratio (a) circuit; (b) graph

driving inverter inv-2, which is a times the size of a minimum inverter. Similarly, inverter inv-2 drives inverter inv-3, which is a^2 the size of a minimum inverter. The delay through each stage is at_d, where t_d is the average delay of a minimum-sized inverter driving another minimum-sized inverter (actually the delay through any inverter driving an identically sized inverter). Hence the delay through n stages is nat_d. If the ratio of the load capacitance to the capacitance of a minimum inverter, C_L/C_g, is R, then $a^n = R$. Hence $ln(R) = nln(a)$. Thus the total delay is

$$Total\ Delay\ =\ nat_d\ =\ ln\,(R)\,\frac{a}{ln\,(a)}t_d. \tag{4.63}$$

In this equation t_d is a constant and $ln(R)$ depends on the ratio of internal to external load and is constant for a given load and process. The variable part of Eq. (4.63) is graphed in Fig. 4.33(b) for various values of a from 1 to 100. The y scale is normalized to e. The graph shows that for this simple analysis the stage ratio minimizes the total delay when the stage ratio equals e (~2.7).

More detailed analysis that accounts for the contribution of the intrinsic output capacitance of the inverter illustrates that this ratio varies from 3 to 5 depending on the process.[38] The optimum stage ratio may be determined from

$$a_{opt}\ =\ e^{\frac{k+a_{opt}}{a_{opt}}}, \tag{4.64}$$

where k is the intrinsic output load capacitance and input gate capacitance of an inverter. For the 1μ process capacitances given in Section 4.3.4

$$k\ =\ \frac{C_{drain}}{C_{gate}}\ =\ \frac{.0043}{.02}\ =\ .215, \tag{4.65}$$

which yields

$$a_{opt} = 2.93.$$

In the first edition of this book, for a 2.5μ process $k = 3.57$, which results in $a_{opt} = 5.32$. This illustrates how a design parameter can vary as processes advance. In practice, stage ratios from 2 to 10 are quite common in practical circuits depending on speed, area, and power constraints. A variable-stage-ratio approach has been suggested as a means of reducing the area of cascaded inverters at a slight penalty in delay.[39] In this technique, the stage ratio is varied depending on the position of the inverter in the overall buffer.

Although we have considered the delay through cascaded inverters, the concept of maintaining a good stage ratio is also of importance for a cascaded path through any logic gates where high-speed designs are involved. A variety of software packages have been developed to aid in the optimization of transistor sizes in cascaded CMOS gates.[40]

4.7 Power Dissipation

There are two components that establish the amount of power dissipated in a CMOS circuit. These are:

- Static dissipation due to leakage current or other current drawn continuously from the power supply.
- Dynamic dissipation due to
  —switching transient current.
  —charging and discharging of load capacitances.

4.7.1 Static Dissipation

Considering a complementary CMOS gate, as shown in Fig. 4.34, if the input = '0,' the associated n-device is "OFF" and the p-device is "ON." The output voltage is V_{DD} or logic '1.' When the input = '1,' the associated n-channel device is biased "ON" and the p-channel device is "OFF." The output voltage is 0 volts (V_{SS}). Note that one of the transistors is always "OFF" when the gate is in either of these logic states. Since no current flows into the gate terminal, and there is no DC current path from V_{DD} to V_{SS}, the resultant quiescent (steady-state) current, and hence power P_s, is zero.

However, there is some small static dissipation due to reverse bias leakage between diffusion regions and the substrate. In addition, subthreshold conduction can contribute to the static dissipation. We need to look at a simple model that describes the parasitic diodes for a CMOS inverter in order to have an understanding of the leakage involved in the device. The source-drain diffusions and the n-well diffusion form parasitic diodes. This can be represented in the profile of an inverter shown in Fig. 4.35. In the model, a parasitic diode is shown between n-well and substrate. Since parasitic diodes are reverse-biased, only their leakage current contributes to static power dissipation. The leakage current is described by the diode equation

$$i_o = i_s(e^{qV/kT} - 1), \tag{4.66}$$

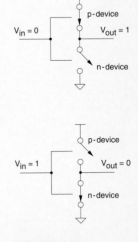

FIGURE 4.34
CMOS inverter model for static power dissipation evaluation

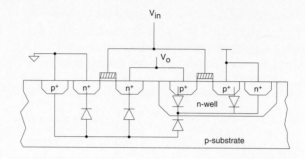

FIGURE 4.35 Model describing parasitic diodes present in a CMOS inverter

where

  i_s = reverse saturation current

  V = diode voltage

  q = electronic charge (1.602×10^{-19} C)

  k = Boltzmann's constant (1.38×10^{-23} J/K)

  T = temperature.

The static power dissipation is the product of the device leakage current and the supply voltage. A useful estimate is to allow a leakage current of 0.1nA to 0.5nA per device at room temperature. Then total static power dissipation, P_s, is obtained from

$$P_s = \sum_{1}^{n} \text{leakage current} \times \text{supply voltage}, \qquad \textbf{(4.67)}$$

where

       n = number of devices.

  For example, typical static power dissipation due to leakage for an inverter operating at 5 volts is between 1 and 2 nanowatts.

  Of course, static dissipation can occur in gates such as pseudo-nMOS gates, where there is a direct path between power and ground. If such gates are used, their static dissipation must be factored into the total static power dissipation of the chip.

Example

For a process with β_p of 30 $\mu A/V^2$ and a β_n of 85 $\mu a/V^2$ ($V_{tn} = |V_{tp}| = 0.7V$, $V_{DD} = 5V$), calculate the static power dissipation of a 32 × 32 ROM which contains a 1:32 pseudo-nMOS row decoder and pMOS pull-ups on the 32-bit lines. The aspect ratio of all pMOS pull-ups (W/L) is 1. Each pMOS load can source $(\beta(V_{gs} - V_t)^2)/2$ of current.

$$I_{load} = \left(30\,(5 - 0.7)^2 \right)/2\mu A = 277\mu A$$

$$P_{load} = 1.4\ mW = (277\mu A \times 5V)$$

Assuming that one row decoder is on and 50% of the bit lines are on at any one time yields

$$P_{total} = 17 \times 1.4\ mW$$

$$= 23.6\ mW.$$

4.7.2 Dynamic Dissipation

During transition from either '0' to '1' or, alternatively, from '1' to '0,' both n- and p-transistors are on for a short period of time. This results in a short current pulse from V_{DD} to V_{SS}. Current is also required to charge and discharge the output capacitive load. This latter term is usually the dominant term. The current pulse from V_{DD} to V_{SS} results in a "short-circuit" dissipation that is dependent on the input rise/fall time, the load capacitance and gate design. This is of relevance to I/O buffer design. Figure 4.36 shows three inverters with varying loads from $0pF$ to $.2pF$ with voltage sources to measure currents in SPICE. The output voltage waveforms are shown at the top of the diagram. The currents flowing in the n- and p-transistors are shown beside each inverter. With no loading, the short-circuit current is quite evident. As the capacitive load is increased, the discharge or charge current starts to dominate the current drawn from the power supplies. Appropriate simulations would show that slow rising or falling edges would increase the short circuit current.

The dynamic dissipation can be modeled by assuming that the rise and fall time of the step input is much less than the repetition period. The average dynamic power, P_d, dissipated during switching for a square-wave input, V_{in}, having a repetition frequency of $f_p = 1/t_p$, is given by

$$P_d = \frac{1}{t_p} \int_0^{t_p/2} i_n(t)\, V_{out} dt + \frac{1}{t_p} \int_{t_p/2}^{t_p} i_p(t)\, (V_{DD} - V_{out})\, dt,$$

$$\text{(4.68)}$$

where

i_n = n-device transient current

i_p = p-device transient current.

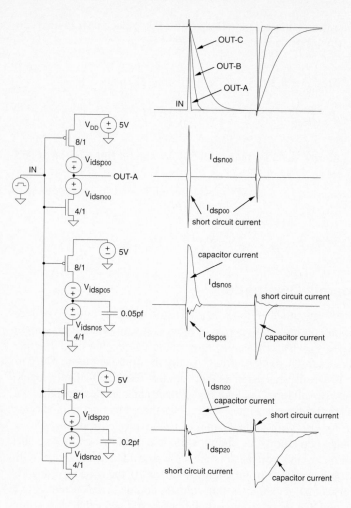

FIGURE 4.36 SPICE circuits and results showing dynamic short-circuit current and capacitive current for a CMOS inverter for varing load capacitances (the 0V voltage sources are used to measure currents)

For a step input and with $i_n(t) = C_L \, dV_{out}/dt$ (C_L = load capacitance)

$$P_d = \frac{C_L}{t_p} \int_0^{V_{DD}} V_{out} dV_{out} + \frac{C_L}{t_p} \int_{V_{DD}}^0 (V_{DD} - V_{out}) \, d(V_{DD} - V_{out})$$

$$= \frac{C_L V_{DD}^2}{t_p}$$

(4.69)

with $f_p = 1/t_p$,

resulting in

$$P_d = C_L V_{DD}{}^2 f_p. \tag{4.70}$$

Thus for a repetitive step input the average power that is dissipated is proportional to the energy required to charge and discharge the circuit capacitance. The important factor to be noted here is that Eq. (4.70) shows power to be proportional to switching frequency but independent of the device parameters.

4.7.3 Short-Circuit Dissipation

The short-circuit power dissipation is given by

$$P_{sc} = I_{mean} \cdot V_{DD}.$$

For the input waveform shown in Fig. 4.37(a), which depicts the short circuit (Fig. 4.37b) in an unloaded inverter,

$$I_{mean} = 2 \times \left[\frac{1}{T} \int_{t_1}^{t_2} I(t)\, dt + \frac{1}{T} \int_{t_2}^{t_3} I(t)\, dt \right] \tag{4.71}$$

assuming that $V_{tn} = -V_{tp}$ and $\beta_n = \beta_p$ $(=\beta)$ and that the behavior is symmetrical around t_2.

$$= 2 \times \frac{2}{T} \int_{t_1}^{t_2} \frac{\beta}{2} \left(V_{in}(t) - V_t \right)^2 dt,$$

with

$$V_{in}(t) = \frac{V_{DD}}{t_r} t$$

$$t_1 = \frac{V_t}{V_{DD}} t_r$$

$$t_2 = \frac{t_r}{2}.$$

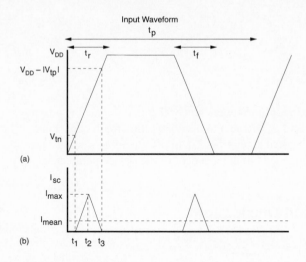

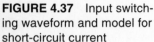

FIGURE 4.37 Input switching waveform and model for short-circuit current

Thus for an inverter without load, assuming that $t_r = t_f (= t_{rf})$,

$$P_{sc} = \frac{\beta}{12}(V_{DD} - 2V_t)^3 \frac{t_{rf}}{t_p}. \qquad (4.72)$$

where t_p is the period of the input waveform. This derivation is for an unloaded inverter. It shows that the short-circuit current is dependent on β and the input waveform rise and fall times. Slow rise times on nodes can result in significant (20%) short-circuit dissipation for loaded inverters. Thus it is good practice to keep all edges fast if power dissipation is a concern. Further discussion may be found in Veendrick.[41] As the load capacitance is increased the significance of the short-circuit dissipation is reduced by the capacitive dissipation P_d.

4.7.4   Total Power Dissipation

Total power dissipation can be obtained from the sum of the three dissipation components, so

$$P_{total} = P_s + P_d + P_{sc} \qquad (4.73)$$

When calculating the power dissipation, a rule of thumb is to add all capacitances operating at a particular frequency and calculate the power. Then the power from other groups operating at different frequencies may be summed. The dynamic power dissipation may be used to estimate total power consumption of a circuit and also the size of V_{DD} and V_{SS} conductors to minimize transient-induced voltage drops.

For a complex circuit it is often impractical to calculate the power dissipation in a detailed manner. The following are some approximations of increasing accuracy.

- Calculate the total capacitance driven by gate outputs in the circuit. Estimate the percentage activity of the circuit operating at the maximum clock frequency (say, 50%). Use Eq. (4.69) to calculate the dynamic power as follows:

$$P_d = \frac{percentage\text{-}activity \times C_{Total} V_{DD}^2}{t_p} \tag{4.74}$$

- Partition the circuit into smaller parts where the activity factor may be calculated more accurately and repeat the above calculation.
- Some simulators (especially switch-level) have the ability to be modified to sum the total capacitance switched by each switch on each node over the course of a simulation run. After any simulation is run, the total number of clock cycles that have been simulated are used in conjunction with the capacitance as follows:

$$P_d = \frac{C_{TOTAL\text{-}SWITCHED} V_{DD}^2}{TOTAL\text{-}NUMBER\text{-}OF\text{-}CYCLES \times t_p} \tag{4.75}$$

- Device-level timing simulators can sum the current drawn from both power supplies over the course of a simulation, thus yielding a current waveform that may be used to estimate power dissipation (and *IR* drop in conductors, noise, etc.).

4.7.5   Power Economy

In large projects, where many designers are involved in the design of modules that go into a large chip or for low-power applications, each module is usually given a power budget. This is a power dissipation that the module can not exceed. It is then the job of the designer to meet this constraint (in addition to all the other normal constraints).

Minimizing power may be achieved in a number of ways. DC power dissipation may be reduced to leakage by only using complementary logic gates. The leakage in turn is proportional to the area of diffusion, so the use of minimum-sized devices is of advantage. (A process with low leakage helps too!) Dynamic power dissipation may be limited by reducing supply voltage, switched capacitance, and the frequency at which logic is clocked. Supply voltage tends to be a system-design consideration, and low-power systems use 1.5 to 3 volt supplies. Minimizing the switched capacitance again tends to favor using minimum-sized devices and optimal allocation of resources such as adders and registers. Manual layout techniques are also of use to minimize routing capacitance. Another big gain can be made by only operating the minimum amount of circuitry at high speeds or having a variable clock depending on how much computation has to be completed.

The fundamental factors that affect power dissipation have been presented in this section. There are many ingenious methods of manipulating architecture, circuit, and layout to achieve low-power, high-speed goals.

4.8   Sizing Routing Conductors

Metal power-carrying conductors have to be sized for three reasons:

- Metal migration.
- Power supply noise and integrity (i.e., satisfactory power and signal voltage levels are presented to each gate).
- *RC* delay.

Metal migration or electromigration is the transport of metal ions through a conductor resulting from the passage of direct current. It is caused by a modification of the normally random diffusion process to a directional one caused by charge carriers. This can result in the deformation of conductors and subsequent failure of circuitry. Factors that influence the electromigration rate are

- current density.
- temperature.
- crystal structure.

In determining the minimum size of conductors, particularly those for V_{DD} and V_{SS}, it is necessary to estimate the current density in the conductor. If the current density, J, of a current-carrying conductor exceeds a threshold value, we find that the conductor atoms begin to dislocate and move in the direction of the current flow. If there is a constriction in the conductor, the conductor atoms move at a faster rate in the region of the constriction. This results in a weakening of the constriction, which eventually blows like a fuse. For example, the limiting value for 1 μm-thick aluminum is

$$J_{Al} \approx 1 \rightarrow 2 \text{ mA/μm}.$$

As a rule of thumb, 0.4 mA/μm to 1.0 mA/μm of metal width should be used for both V_{DD} and V_{SS} lines (although check with the process you are using).

Apart from electromigration, voltage drops can occur on power conductors due to IR drop during charging transients. Poor V_{DD} or V_{SS} levels can lead to poor logic levels which reduce the noise margin of gates and cause incorrect operation of gates. While electromigration usually sets the minimum width of conductors, the need to supply correct V_{DD} and V_{SS} is often the driving consideration. Sometimes the supply conductors can not be increased to the desired width. For such circumstances, other techniques such as adding extra supply pins to distribute the current flow could be considered.

For a discussion of the importance of sizing conductors to minimize RC delay see Section 4.3.5.

4.8.1   Power and Ground Bounce

As a module is clocked, the current drawn from the power-supply leads tends to rise as the clock transitions. The current reflects various stages of logic triggered by values changing due to the clock transition. As any gates may change close to the clock, large current spikes may occur. These lead to what is termed "ground bounce" for the ground lead and "power bounce" for the power lead. Careful power supply routing should insure that these spikes do not interfere with the operation of any circuitry. If the threshold of logic gates is around 2.5 volts, spikes to approximately 1 volt are tolerable when complementary logic is used. Where dynamic logic or logic with low noise margins are used one must be particularly careful about noise on the power supplies.

Ground bounce can also occur in I/O pads when the pad drives an outside load. Generally in pad design, separate power and ground buses are routed to the I/O buffers so that the ground bounce does not flow through internal circuitry.

Clock buffers can also cause considerable ground bounce in their supply leads because they usually drive a large capacitance. Very careful attention must be paid to the design of the power-supply connections to large clock buffers. Often in high-performance designs, on-chip "bypass" capacitors are added between the power bus and the substrate. These normally utilize the gate capacitance of large n-transistors placed under the power buses.

Example

What would be the conductor width of power and ground wires to a 50 MHz clock buffer that drives $100pF$ of on-chip load to satisfy the metal-migration consideration ($J_{AL} = 0.5$ mA/μ)? What is the ground bounce with the chosen conductor size? The module is 500μ from both the power and ground pads and the supply voltage is 5 volts. The rise/fall time of the clock is 1 *ns*.

1. $P = CV_{DD}^2 f$
   $$= 100 \times 10^{-12} \times 25 \times 50 \times 10^6$$
   $$= 125 \text{ mW}$$
   $$I = 25 \text{ mA}$$

   Thus the width of the clock wires should be at least 60μ. A good choice would be 100μ.

2. $R = 500/100 \times .05$
   $$= 5 \text{ squares} \times .05 \ \Omega/\text{sq.}$$
   $$= .25\Omega$$

   $$IR = \frac{C_d V}{dt} R = \frac{100 \times 10^{-12} \times 5}{1 \times 10^{-9}} \times .25$$

   $$= 125 \text{ mV (also see Section 5.5.16)}$$

4.8.2   Contact Replication

Often, a single run of a conductor can not be made to supply all circuits or modules in a design. In these cases a layer change may be necessary. Because this involves the use of interlayer contacts or vias, the resistance and current-carrying capacity of these structures must be taken into account for the effects mentioned in this section.

The current density in a contact (window, cut) periphery must be kept below about 0.1 mA/μm. We find that, due to current crowding around the perimeter of a window, a chain of small windows, suitably spaced, generally provides just as much current-carrying capacity as a single long, narrow contact. The direction of the current flow after passing through a contact can also influence the current-carrying capacity. If the current flow turns at right angles or reverses, a square array of contents is generally required, while if the flow is in the same direction, fewer contacts may be used. Figure 4.38 illustrates these points.

4.9   Charge Sharing

In many structures a bus can be modeled as a capacitor, C_b, as shown in Fig. 4.39. Sometimes the voltage on this bus is sampled (latched) to determine the state of a given signal. Frequently, this sampling can be modeled by the two capacitors, C_s and C_b, and a switch. In general, C_s is in some way related to the switching element. The charge associated with each of the capacitances prior to closing the switch can be described by

$$Q_b = C_b V_b \tag{4.76}$$

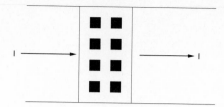

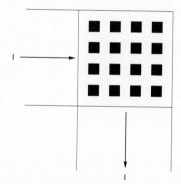

FIGURE 4.38   Contact structures for linear and orthogonal joints

and

$$Q_s = C_s V_s.$$

The total charge Q_T is then given by

$$Q_T = C_b V_b + C_s V_s \qquad \text{(4.77)}$$

The total capacitance C_T is given by

$$C_T = C_b + C_s. \qquad \text{(4.78)}$$

Therefore, when the switch is closed, the resultant voltage V_R (not shown in Fig. 4.39) is

$$V_R = \frac{Q_T}{C_T} = \frac{C_b V_b + C_s V_s}{C_b + C_s}. \qquad \text{(4.79)}$$

For example, if

$$V_b = V_{DD}$$

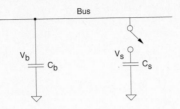

FIGURE 4.39 Charge-sharing mechanism

and

$$V_b \gg V_s,$$

then

$$V_R = V_{DD} \left[\frac{C_b}{C_b + C_s} \right]. \tag{4.80}$$

To ensure reliable data transfer from C_b to C_s, it is necessary to ensure $C_s \ll C_b$. A useful rule to follow is $C_b > 10C_s$. Charge sharing does not necessarily occur only on buses. Most frequently, problems involving charge sharing occur in dynamic logic gates (see Chapter 5).

Example

A precharge bus has a loading of $10pF$. At a point in the clock cycle, 64 registers with transmission gates on their inputs turn on. The input load of each register (after the transmission gate) is $.1pF$. Calculate the change in precharge voltage.

What would be an alternative approach?

1. Here $C_b \; = 10pF$

$\quad\quad C_s \; = 64 \times .1pF = 6.4pF$

$\quad\quad V_{DD} = 5V$

Hence

$$V_R \; = 5 \times \frac{10}{10 + 6.4}$$

$$= 3.05 \text{ volts (change in voltage is 1.9V).}$$

2. The most obvious approach to alleviating the problem is to use buffer inverters on the input of each register. (The above example would probably point to a very suspect design approach!)

One must always be aware of charge sharing problems any time charge is stored on a node. As mentioned, this most frequently occurs in dynamic logic (see Chapter 5) and dynamic memories.

4.10   Design Margining

So far when considering the various aspects of determining a circuit's behavior, we have only alluded to the variations that might occur in this behavior given different operating conditions. In general, there are three different sources of variation, two environmental and one manufacturing. These are

- operating temperature.
- supply voltage.
- process variation.

One must aim to design a circuit that will reliably operate over all extremes of these three variables. Failure to do so invites circuit failure, potentially catastrophic system failure, and a rapid decline in reliability (not to mention a loss of customers).

4.10.1   Temperature

In Chapter 2, the temperature dependence of the drain current was found to be proportional to $T^{-1.5}$. That is, as the temperature is increased, the drain current is reduced for a given set of operating conditions. This variation is shown in Fig. 4.40. For commercially specified parts, the ambient temperature range is usually specified from 0°C to 70°C. Industrially specified parts are required to operate over a spread of –40°C to 85°C, while military parts need to operate from –55°C to 125°C.

The die temperature is specified as follows:

$$T_j = T_a + \theta_{ja} \times P_d \qquad (4.81)$$

where

T_j = the junction temperature in °C (temperature of the chip itself)

T_a = the ambient temperature in °C (temperature of surrounding air)

θ_{ja} = the package thermal impedance, expressed in °C/watt

P_d = the power dissipation.

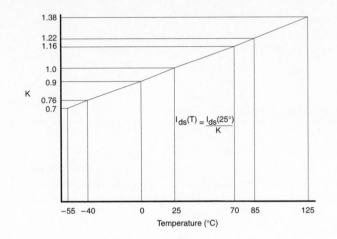

FIGURE 4.40 I_{ds} versus temperature (© LSI Logic, 1987)

For instance, if we have a package with a θ_{ja} of 30°C/watt and we are dissipating 1 watt, the junction temperature for an ambient of 85°C would be 115°C.

The lowest industrial temperature would be –10°C. Using the graph in Fig. 4.40, the current variation would be approximately .8 to 1.3 of that at 25°C.

Processes usually specify an absolute maximum temperature, below which the device characteristics are guaranteed not to drift with time. This is of the order of 70–125°C.

Apart from transistors, capacitors and resistors will have thermal coefficients, that is, a variation with temperature. These variations are not very important for digital circuits but are of great importance for analog circuits.

4.10.2   Supply Voltage

The basic supply voltage for current digital CMOS systems is 5 volts. Smaller dimension processes will use a lower voltage, initially 3.3 volts and then maybe lower, while portable CMOS systems using batteries might have 1–3 volt power supplies. Component tolerances, temperature variation, or battery condition all combine to alter these nominal supply voltages. Thus when specifying a part, a variation on the supply voltage accompanies the data sheet. Normally, this is ±10%. Thus for a nominal 5-volt power-supply, the lowest expected power-supply voltage is 4.5 volts and the highest is 5.5 volts. Similarly, for 3.3 volts, the values are 3.0 volts and 3.6 volts. Some environments might require even larger variations in power supply.

For analog circuits, the designer must consider the voltage coefficient of each integrated device (i.e., transistor, resistor, capacitor), that is, the variation in that component's value with the operating voltage. This drives process engineers to search for voltage insensitive structures and circuit designers to search for components with opposite voltage coefficients so that one component's variation will be cancelled by another.

4.10.3   Process Variation

The fabrication process is a long sequence of chemical reactions that result in device characteristics that follow a normal or Gaussian distribution, as shown in Fig. 4.41. Retaining parts with a 3σ distribution will result in .26% of parts being rejected. A 2σ retention results in 4.56% parts being rejected, while a 1σ results in 31.74% of parts being rejected. Obviously, keeping parts that are within 1σ of nominal would waste a large number of parts. A 3- or 2-σ limit is normal. A manufacturer with a commercially viable CMOS process should be able to supply a set of device parameters that are guaranteed to yield 2 or 3σ.

The variations in device performance can be caused by variations in doping densities, implant doses and variations in the width and thickness of active diffusion and oxide layers and passive conductors. When considering

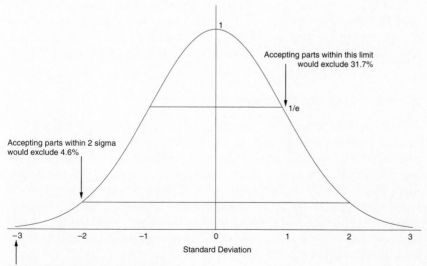

FIGURE 4.41 The distribution of process parameters

transistors the following terms are used to describe the boundary cases of performance:

- nominal.
- fast.
- slow.

That is, transistors that are of nominal speed, faster that nominal and slower than nominal. In CMOS, because there are at least two types of transistors, the characteristics of which are somewhat independent, one can have each of the above speeds ascribed to each type of transistor. Thus we might have the following boundary combinations:

- Fast-n fast-p.
- Fast-n slow-p.
- Slow-n slow-p.
- Slow-n fast-p.

In some processes, the gains of the p- and n-transistors track but the threshold voltages might not. In these cases one might see process corners such as the following:

- Slow-n, low-V_{tp}.
- Low-V_{tn}, slow-p.

In addition to the transistor variation, conductors can vary in width and thickness, thus giving rise to variations in stray capacitance and resistance. Compared with the device variations, these tend to be smaller. They are usually insignificant for digital circuits but can be significant for analog circuits.

4.10.4   Design Corners

When combined with the lowest temperature and the highest operating voltage that the circuit will encounter, the fast-n/fast-p processing corner is usually called the *worst-power* or *high-speed* corner (the term *corner* refers to an imaginary box that surrounds the guaranteed performance of the transistors). The slow-n/slow-p combination will have the slowest speed. When combined with the highest temperature and lowest operating voltage, it is usually called the *worst-speed* corner. The other combinations such as slow-n/fast-p or slow-n/low-V_{tp} are of importance when designing ratioed circuits such as pseudo-nMOS.

Careful design involves simulating circuits at all appropriate corners to ensure correct operation and adequate performance. For instance, in a digital circuit, one would simulate or timing-analyze the circuit at the worst-speed corner at the minimum clock period required. This corner would also be used to check external setup times while the worst power corner would be used to check hold time constraints (see Chapter 5). Depending on the constraints, many engineers operate the clock 10–20% higher in frequency at this corner during simulation or timing verification to give some extra margin. Then the circuit would be simulated and/or timing analyzed at the worst-power corner, checking for timing hazards and clock races. The power dissipation would also be checked at this corner (although this corner only changes the static dissipation due to ratioed logic).

4.10.5 Packaging Issues

Package selection can be very important because packages vary widely in cost and thermal impedance. Usually the more expensive a package for a given number of pins, the better the thermal impedance. The thermal impedance is a measure of the effectiveness with which a package can conduct heat away from the die. Ceramic pin-grid arrays (PGAs) have thermal impedances in the range 15–30°C/watt, while plastic quad flat packs (PQFPs) might range 40–50°C/watt. Some packages have finned metal heatsinks, spreaders, or embedded metal slugs to improve the thermal impedance, while keeping the cost low. Figure 4.42 shows some typical packages. Packages are normally rated in still air and for a given rate of air flow over the package. High-cost packaging might include forced air or liquid cooling through tiny ducts in the package. Usually the designer is bound by some constraints such as cost or maximum die temperature, and the design evolves accordingly. Some computers, notably the Cray, are largely defined by unique cooling technology. Packages also have a wide variation in lead inductance, with ceramic pin-grid arrays having the lowest values and cheap plastic packages the highest. Often in high-speed parts high-power-dissipation capability and low-lead inductance requirements go hand in hand.

FIGURE 4.42 Typical packages used for CMOS chips

TABLE 4.11   CMOS Digital System Checks (Commercial)

PROCESS	TEMP	VOLTAGE	TESTS
Fast-n/fast-p	0°C	5.5V (3.6V)	Power dissipation (DC), clock races, hold time constraints
Slow-n/slow-p	125°C	4.5V (3.0V)	Circuit speed, setup time constraints
Slow-n/fast-p	0°C	5.5V (3.6V)	Pseudo-nMOS noise margin, level shifters, memory write/read, ratioed circuits
Fast-n/slow-p	0°C	5.5V (3.6V)	Memories, ratioed circuits, level shifters

4.10.6   Power and Clock Conductor Sizing

Power supply, noise on power supplies, and clock conductor size should be checked for metal migration problems at the worst-power corner.

4.10.7   Summary

In summary then, any CMOS digital system should have completed the checks shown in Table 4.11.

Usually the main corner exercised is the slow-n/slow-p, high-temperature, low-voltage corner because this affects speed, which is usually the dominant design goal. Of course, remember the "published paper corner" (fast-n fast-p, low-temp, high-voltage), which makes all your circuits seem about three times faster than a manufacturable part.

4.11   Yield

An important issue in the manufacture of VLSI structures is the yield.[42] Although yield is not a performance parameter, it is influenced by such factors as

- technology.
- chip area.
- layout.[43]

Once the silicon has been processed, other manufacturing yield factors, such as scribe yield and packaging yield, also contribute to the overall yield of a device.

Yield is defined as

$$Y = \frac{No.\ of\ Good\ Chips\ on\ Wafer}{Total\ Number\ of\ Chips} \tag{4.82}$$

and may be described as a function of the chip area and defect density. Two common equations are used:

Seeds's model,[44] which is given by

$$Y = e^{-\sqrt{AD}} \tag{4.83}$$

where

A = chip area

D = defect density (defined as lethal defects per cm^2).

This model is used for large chips and for yields less than about 30 percent. Murphy's model,[45] which is described by

$$Y = \left[\frac{1 - e^{-AD}}{AD}\right]^2. \tag{4.84}$$

This model is used for small chips and for yields greater than 30 percent.

A more recent generalized model is as follows[46]:

$$Y = \prod_{i=1}^{N}\left(1 + \sum_{j}\frac{A_j D_i P_{ij}}{c_i}\right)^{-c_i} \tag{4.85}$$

where

i = the ith type of defect

j = the jth module

P_{ij} = the probability that an i defect will cause a fault in the jth area

c_i = the constant relating to the density of a ith type of defect.

From these relations it is obvious that yield decreases dramatically as the area of the chip is increased. The latter two models account for the clustering of defects (i.e. they are not independent). One can easily encounter a situation in which all of the chips on a wafer are found to be defective. Modern fabrication lines using dry etching techniques generally yield a D value of around 1 to 5 defects/cm^2. In order to improve yield it is possible to incorporate redundancy into the structure. In random logic, yield improvement is minimal due to increase in area. However, in memory structures, it is possible to gain dramatic improvement in yield through incorporation of redundant cells. The parametric yield is related to the number of chips that fail performance tests. In general, well-designed digital CMOS chips should not encounter parametric yield problems. For further information see Cox et al.[47]

4.12 Reliability

Designing reliable CMOS chips involves careful circuit design and processing with attention directed to the following potential reliability problems:

- "Hot electron" effects.
- Electromigration.
- Oxide failure.
- Bipolar transistor degradation.
- Package/chip power dissipation (die temperature).
- ESD protection.

Currently chips are subjected to a process called accelerated life testing where packaged chips are subjected to overvoltage and overtemperature in an effort to emulate the aging process. Any failures may then be used to estimate the actual lifetime of the part. This process is time consuming and comes right at the end of the project. Current research[48,49] attempts to build simulators that can estimate the reliability of a chip at the earliest possible point in the design cycle. These simulators are sure to be increasingly used as designers aim to increase the reliability of their chips and systems.

4.13 Scaling of MOS-transistor Dimensions

So far in this chapter, we have examined some electrical design issues and formulated some electrical design rules that should be taken into account when building high-performance circuits with current CMOS processes. As CMOS processes are improved and device dimensions are reduced, these rules will change. In this section, we take a look at the effect that these reduced dimensions will have on electrical circuit behavior.

4.13.1   Scaling Principles

First-order "constant field" MOS scaling theory is based on a model formulated by Dennard et al.[50,51] This indicates that the characteristics of an MOS device can be maintained and the basic operational characteristics preserved if the critical parameters of a device are scaled in accordance to a given criterion. With a constant field scaling, the scaled device is obtained by applying a dimensionless factor α to

- all dimensions, including those vertical to the surface.
- device voltages.
- the concentration densities.

In practice, alternative scaling methods have been used over the last few years. The first is *constant voltage scaling,* where the V_{DD} voltage is kept constant, while the process is scaled. Another is *lateral scaling,* where only the gate length is scaled (this is commonly called a "gate-shrink" because it can be easily done to an existing mask database for a design). There are clearly other types of scaling that can be applied.

The resultant effect of these three types of scaling is illustrated in Table 4.12.

For constant field scaling, Table 4.12 shows that if device dimensions (which include channel length, L; channel width, W; oxide thickness, t_{ox}; junction depth, X_j; applied voltages; and substrate concentration density, N) are scaled by the constant parameter α, then the depletion layer thickness, d, the threshold voltage, V_t, and the drain-to-source current, I_{ds}, are also scaled. One of the important factors to be noted is that since the voltage is scaled, electric field, E, in the device remains constant. This has the desirable effect that many nonlinear factors essentially remain unaffected.

With constant voltage scaling, E increases, which has led to process development to reduce the deleterious effects of high fields (i.e., L_{DD} structures).

The depletion regions associated with the pn junctions of the source and drain determine how small we can make the channel. As a rule, the source-drain distance must be greater than the sum of the widths of the depletion layers to ensure that the gate is able to exercise control over the conductance of the channel. Thus in order to reduce the length of the channel you need to reduce the width of the depletion layers. This is accomplished by increasing the doping level of the substrate silicon. In constant field scaling as we scale device dimensions by $1/\alpha$, the drain-to-source current, I_{ds}, per transistor decreases to $1/\alpha$, the number of transistors per unit area; that is, circuit density scales up by α^2, which subsequently results in the current density scaling linearly with α. Constant-voltage scaling exacerbates this problem

TABLE 4.12   Influence of Scaling on MOS-Device Characteristics

PARAMETER	SCALING MODEL		
	Constant field	Constant voltage	Lateral
Length (L)	$1/\alpha$	$1/\alpha$	$1/\alpha$
Width (W)	$1/\alpha$	$1/\alpha$	1
Supply voltage (V)	$1/\alpha$	1	1
Gate-oxide thickness (t_{ox})	$1/\alpha$	$1/\alpha$	1
Current ($I = (W/L)(1/t_{ox})V^2$)	$1/\alpha$	α	α
Transconductance (g_m)	1	α	α
Junction depth (X_j)	$1/\alpha$	$1/\alpha$	1
Substrate doping (N_A)	α	α	1
Electric Field across gate oxide (E)	1	α	1
Depletion layer thickness (d)	$1/\alpha$	$1/\alpha$	1
Load Capacitance ($C = WL/t_{ox}$)	$1/\alpha$	$1/\alpha$	$1/\alpha$
Gate Delay (VC/I)	$1/\alpha$	$1/\alpha^2$	$1/\alpha^2$
	RESULTANT INFLUENCE		
DC power dissipation (P_s)	$1/\alpha^2$	α	α
Dynamic power dissipation (P_d)	$1/\alpha^2$	α	α
Power-delay product	$1/\alpha^3$	$1/\alpha$	$1/\alpha$
Gate Area ($A = WL$)	$1/\alpha^2$	$1/\alpha^2$	$1/\alpha$
Power Density (VI/A)	1	α^3	α^2
Current Density	α	α^3	α^2

(increasing the current density by α^3), while a gate shrink increases the current density by α^2. Thus proportionately wider metal power conductors are necessary for more densely packed structures. This is usually solved by adding metal layers that are used solely for power and ground.

Another characteristic illustrated in Table 4.12 is power density. For constant-field scaling, both the static power dissipation, P_s, and frequency-dependent dissipation, P_d, decrease by $1/\alpha^2$ as the result of scaling. (The P_d value assumes that the frequency of operation is 1/Gate Delay.) However, since the number of devices per unit area increases by α^2, the resultant effect is that the power density remains constant. The power density for constant-voltage scaling increases by α^3, while lateral scaling increases the power density by α^2.

An estimation of the limit in power density is derived from the thermodynamic relationship given by Equation 4.81, that is

$$T_j = T_a + \theta_{ja}Pd$$

Generally, the thermal resistance is expressed as $\Delta°C$ per watt, which means one watt of heat energy will raise the temperature by $\Delta°C$. For a 144-pin pin-grid-array (PGA) package, this value is in the range of 20°C (ceramic) to 40°C (plastic) per watt. If we assume an ambient temperature of 70°C, and the maximum allowed silicon junction temperature is about 110°C, then the maximum power dissipation that does not require special cooling is

$$P_{max} = \frac{T_j - T_{amb}}{\theta_{ja}} = \frac{110 - 70}{(20 \to 40)} \qquad \textbf{(4.86)}$$

$$= 1 \to 2 \; \text{watts(plastic-ceramic)}$$

Specialized packages with forced cooling can handle power dissipations in the 10s of watts and above (at a cost!). The increase in power density for constant-voltage scaling and lateral scaling has forced manufacturers to develop and designers to use new package solutions.

As the temperature increases, the carrier mobility falls, thus reducing the gain of devices. This, in turn, would reduce the speed of circuits. If high-temperature, high-speed circuits are required, then special consideration during design is necessary. The current density increases for constant-field and constant-voltage scaling, necessitating better metalization (usually more layers) to deal with metal migration problems.

It is necessary to recognize that the variables shown in Table 4.12 are only first-order approximations. A more rigorous analysis would modify some of the values. For example, scaling of the substrate doping level by α causes the mobility to decrease slightly. Therefore the propagation delay, as a rule, does not improve by as much as the predicted factor of $1/\alpha$. However, power dissipation will decrease by somewhat more than the expected value of $1/\alpha^2$. Thus the power-speed product remains at $1/\alpha^3$.

One of the limitations of first-order scaling is that it gives the wrong impression of being able to scale proportionally to zero dimension, or to zero threshold voltages. In reality, both theoretical and practical considerations do not permit such behavior.

4.13.2   Interconnect-Layer Scaling

Although scaling gives a number of improvements, there are a number of circuit parameters—such as voltage drop, line propagation delay, current density, and contact resistance—that exhibit significant degradation with scaling. For example, scaling the thickness and width of a conductor by α reduces the cross-sectional area by α^2. The scaled-line resistance, R', is given by

$$R' = \frac{\rho}{\frac{t}{\alpha}} \left[\frac{\frac{L}{\alpha}}{\frac{W}{\alpha}} \right] \tag{4.87}$$

$$= \alpha R,$$

where ρ is the conductivity term, which is related to sheet resistance by $Rs' = \rho/\alpha t$, and t is conductor thickness. The voltage drop along such a line can now be expressed as a constant field scaling

$$V_d' = (I/\alpha)(\alpha R) \tag{4.88}$$

$$= IR.$$

In a similar manner, we can derive the line-response time as

$$t_s' = (\alpha R)(C/\alpha) \tag{4.89}$$

$$= RC,$$

which is a constant. The influence of scaling on interconnection paths if the routing is scaled by α and the current increases by $1/\alpha$, is summarized in Table 4.13.

For a constant chip size, many of the communication paths do not scale. That is, they still traverse the width or length of a chip (which usually does not vary). Thus the actual RC delays and voltage drops that are seen are greater than those predicted by Table 4.13.

The significance of this result is that it is somewhat difficult to take full advantage of the higher switching speeds inherent in scaled devices when signals are required to propagate over long paths. Thus the distribution and organization of clocking signals becomes a major problem as geometries are scaled. In addition, metal lines must carry a higher current with respect to

TABLE 4.13 Influence of Scaling on Interconnect Media (Constant Field)

PARAMETERS	SCALING FACTOR
Line resistance (r)	α
Line response (rc)	1
Voltage drop	1

cross-sectional area; thus electron migration becomes a major factor to consider. This is another reason that as processes have developed, more metal layers have been added before scaling the gate dimensions drastically.

As the level of integration increases, the average line length on a chip tends to increase also, thereby increasing the capacitance. In addition, the resistance of wires increases and becomes more important relative to transistor resistance. However, the power dissipation per gate decreases, which diminishes the ability of gates driving wiring capacitances. Under such conditions, average gate delay is determined by the interconnection rather than the gate itself.

4.13.3 Scaling in Practice

In the time since this book was first published, Table 4.14 shows a (personal) scaling history that has been observed (with an idea of the chips designed).

What this shows is that over the last ten years, a constant-voltage scaling approach has been followed as the chips get more complex and faster. As this book is being written a real move to 3.3V is being seen. The problems that dominate design today are metal migration and *RC* delays (in metal wires!) just as was predicted by the scaling theory years ago.

4.14 Summary

In this chapter we have developed models to allow us to estimate circuit timing performance, power dissipation, and circuit yield. The principles of environmental and process-based design margining were also introduced. Combined with the models, design margining applied to any CMOS design

TABLE 4.14   A Scaling History Since 1980 (Personal)

YEAR	TECH	CHIP	SIZE (Tr)	SIZE (mm^2)	SPEED (MHz)	V_{DD}	TYPE OF SCALING
1980–1984	3.5μ	16-bit datapath and RAM	12K	25	5	5	
1985	2.0μ	Lisp μProcessor	250K	225	5	5	Constant voltage
1987	1.5μ	Lisp μProcessor	250K	144	8	5	Constant voltage
1989	1.2μ	Lisp μProcessor	250K	100	12	5	Constant voltage
1990	1.0μ	Ghost canceller	500K	54	56	3–5	Constant voltage
1992	0.8μ	Video decoder	1.2M	120	40	5	Constant voltage
1993+	0.5μ	??	>1M	100+	100+	3.3	Scaled V_{DD} and gate length

method are the basis for designing reliable, well-engineered systems. The chapter concluded with an example of scaling theory that can be used to evaluate what a particular process scaling approach might yield.

4.15   Exercises

1. Explain the types of simulation you would carry out to DC- and AC-margin a CMOS chip that employed a mix of complementary and pseudo-nMOS logic.

2. A 6-in. wafer (1 defect/cm^2) costs $1000 to process. The function that is required for a CMOS chip can occupy 10 mm $\times$10 mm for a single chip or take 4 identical 5.5 $\times$ 5.5 mm chips. The package cost for the 10mm chip is $15.00, whereas the package cost of the smaller chip is $2.00. The testing cost for each die (prior to packaging) is $1.50. What is the cheapest solution?

3. Design an output buffer that will buffer an external load of $50pF$ in 5 ns (internal driver $W_p = 4\mu$, $W_n = 4\mu$, $L_n = L_p = 1\mu$, $C_g = .0017\ pF/\mu^2$, assume source, drain and other stray capacitance is equal to gate capacitance of stage).
   (Use $\mu_n \varepsilon/t_{ox} = 90\mu A/V^2$, $\mu_p \varepsilon/t_{ox} = 30\mu A/V^2$, $V_{DD} = 3V$.)

   a. Calculate the current drawn by 16 such buffers that are simultaneously driven at a clock rate of 20 MHz.

   b. How many power and ground pads would be required for the 16 buffers if the $V_{DD} + V_{SS}$ feed to each pad is 100μ?

4. A single lead package of inductance $20nH$ is in series with power and ground pads for the clock with the buffer designed above. What is the inductive spike due to this package inductance? What could you do to reduce this?

5. Explain how the shape of the input waveform to a CMOS logic gate alters the delay through the gate.

6. Calculate the approximate dynamic and short-circuit power dissipated in a chip operating with a V_{DD} of $5V$ at 100 MHz with an internal switched capacitance of $300pF$ (the average rise/fall time is 200 ps). How does the short-circuit dissipation change if the average rise/fall time is 500 ps?

7. A clock buffer takes an external TTL clock at 50 MHz and has to drive $300pF$ of on-chip load. Design a circuit for a buffer that minimizes the skew between the input clock and the on-chip clock. Cal-

culate the necessary power and ground bus-widths to keep the ground bounce below .25V with a 5V supply. (Assume the length of the power + ground-supply wires is 200μ.) What width wire should exit the clock buffer to drive the on-chip load? How many contacts are required in the via that is necessary in this connection? (Ignore inductive effects)

8. Derive the scaled values for speed and power density for a process option that scales the voltage and the gate length.

9. You are designing logic intensive devices on a 2-level-metal, single poly CMOS process and you have the option of using a new process step that adds silicided polysilicon or metal3. Which would you choose, and why?

10. Explain why different criteria might be used to size transistors in tightly coupled small-fan-out circuits versus widely spaced high-fan-out circuits.

11. A silicided word line (1μ wide, 1 mm long, 4 Ω/square) is used for a RAM memory. If the per bit capacitance of each RAM cell is 8000 aF (transistors and routing) and there are 64 memory cells in a row, what is the worst-case word-line delay for a driver with $\beta_p = 2.5$mA/V^2 and $\beta_n = 3$mA/V^2? How would you improve this speed?

4.16 References

1. M. Horowitz and R. W. Dutton, "Resistance extraction from mask layout," *IEEE Transactions on CAD,* vol. CAD-2, no. 3, Jul. 1983, pp. 145–150.
2. E. F. Girczyc and A. R. Boothroyd, "A one-dimensional DC model for nonrectangular IGFETs," *IEEE Journal of Solid State Circuits,* vol. SC-18, no. 6, Dec.1983, pp. 778–784.
3. Lance A. Glasser and Daniel W. Dobberpuhl, *The Design and Analysis of VLSI Circuits,* Reading, Mass.: Addison-Wesley, 1985, pp. 78–79.
4. Bing J. Sheu and Ping-Keung Ko, "Measurement and modeling of short-channel MOS transistor gate capacitances," *IEEE JSSC,* vol. SC-22, no. 3, June 1987, pp. 464–472.
5. Steve Shoo-Shiun Chung, "A charge-based capacitance model of short-channel MOSFET's," *IEEE Transactions on CAD,* vol. 8, no. 1, Jan. 1989, pp. 1–7.
6. Mehmet A. Cirit, "The Meyer model revisited: why is charge not conserved?" *IEEE Transactions on CAD,* vol. 8, no. 10, Oct. 1989, pp. 1033–1037.
7. Erich Barke, "Line-to-ground capacitance calculation for VLSI: a comparison," *IEEE Transactions on CAD,* vol. 7, no. 2, Feb. 1988, pp. 295–298.
8. C. P. Yuan and T. N. Trick, "A simple formula for the estimation of the capacitance of two-dimensional interconnects in VLSI circuits," *IEEE Electronic Device Letters,* vol. EDL-3, 1982, pp. 391–393.

9. N. v.d. Meijs and J. T. Fokkema, "VLSI circuit reconstruction from mask topology," *Integration,* vol. 2, no. 2, 1984, pp. 85–119.

10. A. E. Ruehli and P. A. Brennan, "Accurate metallization capacitances for integrated circuits and packages," *IEEE JSSC,* vol. SC-8, no. 4, Aug. 1973, p. 289.

11. J. H. Chern, J. T. Maeda, L. A. Arledge, and P. Yang, "SIERRA: A 3-D device simulator for reliability modeling," *IEEE Transactions on CAD,* vol. 8, no. 5, 1989, pp. 516–527.

12. Jue-Hsien Chern, Jean Huang, Lawrence Arledge, Ping-Chung Li, and Ping Yang, "Multilevel metal capacitance models for CAD design synthesis systems," *IEEE Electron Device Letters,* vol. 13, no. 1, Jan. 1992, pp. 32–34.

13. Chern et al., 1992, *op. cit.*

14. A. E. Ruehli, "Survey of computer-aided electrical analysis of integrated circuits interconnections," *IBM Journal of Research and Development,* vol. 23, 1979, pp. 626–639.

15. A. E. Ruehli and P. A. Brennan, "Efficient capacitance calculations for three-dimensional multiconductor systems," *IEEE Transactions on Microwave Theory Techniques,* vol. MTT-21, Feb. 1973, pp. 76–82.

16. A. E. Ruehli and P. A. Brennan, "Capacitance models for integrated circuit metallization wires," *IEEE JSSC,* vol. SC-10, Dec. 1975, pp. 530–536.

17. Zhen-qiu Ning and Patrick M. DeWilde, "SPIDER: capacitance modelling for VLSI interconnections," *IEEE Transactions on CAD,* vol. 7, no. 12, Dec. 1988, pp. 1221–1228.

18. A. H. Zemanian, Reginald P. Tewarson, Chi Ping Ju, and Juif Frank Jen, "Three-dimensional capacitance computations for VLSI/ULSI interconnections," *IEEE Transactions on CAD,* vol. 8, no. 12, Dec. 1989, pp. 1319–1326.

19. W. Richard Smith, Scott Powell, and George Persky, "A 'missing neighbor model' for capacitive loading in VLSI interconnect channels," *IEEE JSSC,* vol. SC-22, no. 4, Aug. 1987, pp. 553–557.

20. H. B. Bakoglu, *Circuits, Interconnections and Packaging for VLSI,* Reading, Mass.: Addison-Wesley, 1990, chapters 4, 5, and 6.

21. M. I. Elmasry, "Digital MOS integrated circuits: a tutorial," in *Digital MOS Integrated Circuits,* edited by M. I. Elmasry, New York: IEEE Press, 1981, pp. 4–27.

22. Srinivasa R. Vemuru and Arthur R. Thorbjornsen, "Variable-taper CMOS buffer," *IEEE JSSC,* vol. 26, no. 9, Sept. 1991, pp. 1265–1269.

23. Sanjay Dhar and Mark A. Franklin, "Optimum buffer circuits for driving long uniform lines," *IEEE JSSC,* vol. 26, no. 1, Jan. 1991, pp. 32–40.

24. Takayasu Sakurai and A. Richard Newton, "Delay analysis of series-connected MOSFET circuits," *IEEE JSSC,* vol. 26, no. 2, Feb. 1991, pp. 122–131.

25. Nils Hedenstierna and Kjell O. Jeppson, "CMOS circuit speed and buffer optimization," *IEEE Transactions on CAD,* vol. CAD-6, no. 2, Mar. 1987, pp. 270–281.

26. J. Rubenstein, P. Penfield, Jr., and M. A. Horowitz, "Signal delay in RC networks," *IEEE Transactions on CAD,* vol. CAD-2, Jul. 1983, pp. 202–211.

27. M. Horowitz, "Timing models for MOS circuits," Ph.D. Dissertation, Center for Integrated Systems, Stanford University, 1983.

28. C. Mead and L. Conway, *Introduction to VLSI Systems,* Reading, Mass.: Addison-Wesley, 1980.

29. John K. Ousterhout, "Switch-level delay models for digital MOS VLSI," *Proc. 21st IEEE/ACM Design Automation Conference,* Alberquerque, N.M., June 1984, pp. 542–548.

30. C. J. Terman, "Simulation Tools for VLSI," in *VLSI CAD Tools and Applications* (Wolfgang Fichtner and Martin Morf, eds.), Boston, Mass.: Kluwer Academic, 1987.

31. J. Rubenstein et al., *op. cit.*

32. W. C. Elmore, "The transient response of damped linear networks with particular regard to wide-band amplifiers," *Journal of Applied Physics,* vol. 19, no. 1, Jan. 1948, pp. 55–63.

33. D. J. Pilling and J. G. Skalnik, "A circuit model for predicting transient delays in LSI logic systems," *Proc. 6th Asilomar Conference on Circuits and Systems,* 1972, pp. 424–428.

34. John K. Ousterhout, *op. cit.*

35. M. D. Matson and L. A. Glasser, "Macromodeling and optimization of digital MOS VLSI circuits," *IEEE Transactions on CAD,* vol. CAD-5, Oct. 1986, pp. 659–678.

36. Berhhard Hoppe, Gerd Neuendorf, Doris Schnitt-Landsiedel, and Will Specks, "Optimization of high-speed CMOS logic circuits with analytical models for signal delay, chip area, and dynamic power dissipation," *IEEE Transactions on CAD,* vol. 9, no. 3, Mar. 1990, pp. 236–247.

37. C. Mead and L. Conway, *op. cit.*

38. Nils Hedenstierna and Kjell O. Jeppson, *op. cit.*

39. Srinivasa R. Vemuru and Arthur R. Thorbjornsen, *op. cit.*

40. J. Fishburn and A. Dunlop, "TILOS: A polynomical programming approach to transistor sizing," *Proc. IEEE International Conference on Computer Aided Design (ICCAD),* Nov. 1985, pp. 326–328.

41. Harry J. M. Veendrick, "Short-circuit—dissipation of static CMOS circuitry and its impact on the design of buffer circuits," *IEEE JSSC,* vol. SC-19, no. 4, Aug. 1984, pp. 468–473.

42. K. Saito and E. Arai, "Experimental analysis and new modeling of MOS LSI yield associated with the number of elements," *IEEE JSSC,* vol. SC-17, no. 1, Feb. 1982, pp. 28–33.

43. R. D. Rung, "Determining IC layout rules for cost minimization," *IEEE JSSC,* vol. SC-16, no. 1, Feb. 1981, pp. 35–43.

44. R. B. Seeds, "Yield and cost analysis of bipolar LSI," paper 1.1, *Proc. IEEE International Electron Devices Meeting,* Oct. 1967.

45. B. T. Murphy, "Cost-size optima of monolithic integrated circuits," *Proc. IEEE,* vol. 52, Dec. 1964, pp. 1537–1545.

46. Charles Kooperberg, "Circuit layout and yield," *IEEE JSSC,* vol. 23, no. 4, Aug. 1988, pp. 887–892.

47. P. Cox, P. Yang, S. S. Mahant-Shetti and P. K. Chatterjee, "Statistical modeling for efficient parametric yield of MOS VLSI circuits," *IEEE JSSC,* vol. SC-20, no. 1, Feb. 1985, pp. 391–398.

48. Chenming Hu, "IC reliability simulation," *IEEE JSSC,* vol. 27, no. 3, Mar. 1992, pp. 241–246.

49. Wen-Jay Hsu, Bing J. Sheu, Sudhir M. Gowda, and Chang-Gyu Hwang, "Advanced integrated-circuit reliability simulation including dynamic stress effects," *IEEE JSSC,* vol. 27, no. 3, Mar. 1992, pp. 247–257.

50. R. H. Dennard et al. in *Semiconductor Silicon Electrochemical Society,* (H. R. Huff and R. R. Burgess, eds.), 1973.

51. R. H. Dennard et al., *IEEE JSSC,* vol. SC-9, 1974.

CMOS
CIRCUIT AND
LOGIC DESIGN

5

5.1 Introduction

In Chapter 1, CMOS logic was introduced with the assumption that MOS transistors act as simple switches. We have seen in subsequent chapters that certain limitations pertain to MOS transistors that detract from this idealized viewpoint. Furthermore, we have only considered fully complementary logic structures and the ratioed CMOS inverter.

In this chapter we examine alternative CMOS logic configurations to the fully complementary CMOS logic gate. The objective in doing this is to present a range of options that may be useful in a given design situation. While 95% of all design in CMOS can be accommodated by complementary CMOS gates, high-speed, low-power-dissipation, or density restrictions may force another solution: one should always use the circuit that satisfies the application that is the safest and easiest to design and verify. Clocking strategies and storage are also treated. We examine the effects of nonideal switch behavior on circuits. Since we may be interested in designing physical layouts and creating performance optimized designs, two areas have to be addressed in order to achieve a prescribed behavior:

1. Circuit (structural) design.
2. Layout (physical) design.

As we will see, these two phases of design are intimately meshed. The density, behavior, and power dissipation of circuits may have a direct impact on any high-level architectural decisions and may allow or preclude options based on the selection of a logic style or clocking method. For this reason it is important for the system designer to have some idea of low-level circuit options.

The final section in this chapter deals with the important area of I/O circuits—the interface between the internals of a chip and the outside world.

5.2   CMOS Logic Gate Design

To achieve correct operation of any integrated logic gate, both functional and temporal (timing) constraints have to be satisfied. For CMOS logic the following effects can result in incorrect functioning of a gate:

- Incorrect or insufficient power supplies, or power supply noise (noise on ground and/or power signals).
- Noise on gate inputs.
- Faulty transistors.
- Faulty connections to transistors.
- Incorrect ratios in ratioed logic.
- Charge sharing or incorrect clocking in dynamic gates.

It is important to note that when supplied with good power supplies, a correctly fabricated complementary CMOS gate will *always* function correctly (even in the presence of a good deal of noise and with low power-supply voltages). This "safeness" of function is one good reason CMOS technology is a cornerstone of modern system-level IC technology—there is little guesswork. For ratioed gates and dynamic gates, the function of gates may be compromised by poor design, sloppy layout, and unforeseen noise. This is why these styles of gates are not generally used in ASIC logic designs.

When it comes to optimizing the temporal aspects of CMOS logic, there are many more options from which to choose. Accordingly, a fair proportion of the design cycle of a performance CMOS IC might be spent optimizing the speed of the design. As we found in Chapter 4, the rise/fall delay time of a CMOS logic gate may be approximated by

$$k \frac{C_{load}}{\beta_{eff(rise/fall)} V_{DD}}, \tag{5.1}$$

where

k is a constant ($\approx$ 2–4)

$\beta_{eff(rise/fall)}$ is the effective β of the pull-up or pull-down chain in a gate

C_{load} is the load capacitance seen by the gate

V_{DD} is the power supply voltage.

In turn, β_{eff} is affected by

- the number and size of transistors in series (or parallel) in the pull-down (fall time) or pull-up (rise time).

C_{load}, the load capacitance seen by the gate is dependent on

- the size of the transistors in the gate (self-loading)
- the size and number of transistors to which the gate is connected.
- the routing capacitance between a gate and the ones it drives.

Furthermore, not reflected in this simple derivation, the speed of a gate may be affected by the rise or fall time of the input signal.

In many designs there will be many logic paths that do not require any conscious effort when it comes to speed. However, usually there will be a number of paths, called the *critical paths,* that require attention to timing details. These may be recognized by experience or timing simulation, but most designers use a timing analyzer, which is a design tool that automatically finds the slowest paths in a logic design (see Chapter 6). The critical paths can be affected at four main levels:

- The architectural level.
- The RTL/logic gate level.
- The circuit level.
- The layout level.

The most leverage is achieved by completing a good architecture. That is, designing the overall function in the most efficient manner at the highest level possible. This usually requires broad knowledge of the algorithims that implement the function and chip parameters, such as how many gate delays fit in a clock cycle, how fast addition occurs, or how fast memories access.

The next level of timing optimization comes at the RTL/logic level where pipelining, the type of gates (i.e., INVERTER/BUFFER, NAND/ AND, NOR/OR, XOR) and the fan-in and fan-out of the gates (see Section 5.2.1) are designed. This transformation from function to logic and registers may be done by experience, by experimentation, or, increasingly, by logic synthesis. Remember, however, that no amount of skillful logic design can overcome a poor architecture.

Once the logic level has been decided, the circuit level of design can be used to optimize a critical speed path. This may be done by sizing transistors or using other styles of CMOS logic (see later in this chapter).

Finally, one can affect the speed of a set of logic by rearranging the physical layout.

Depending on the style of CMOS design, some of these knobs may be tweaked to various extents. For instance, in semicustom gate-array design the size of transistors is fixed, and many times the layout is produced automatically. For this reason critical paths are fed to the placement and routing programs so that critical nets are routed with the minimum interconnect length between them. At the other extreme, custom design allows for the maximum of flexibility in optimization.

For the remainder of this section we will examine some of the first-order logic design trade-offs required in the design of CMOS logic to meet certain timing constraints. Following this, some second-order effects that are of importance when designing for maximum performance will be covered.

5.2.1   Fan-in and Fan-out

The *fan-in* of a logic gate is the number of inputs the gate has in the logic path being exercised. Figure 5.1(a) illustrates the fan-in of a number of gates. For instance, a 4-input NAND gate has a fan-in of 4, while a 2-input NOR gate has a fan-in of 2.

The *fan-out* of a logic gate is the total number of gate inputs that are driven by a gate output. This is usually expressed in terms of some default gate size. For instance, one might express the loading of a minimum-sized inverter (for the technology, library, etc.) as unity. In the circuit shown in Fig. 5.1(b), the 2-input NAND gate has a fan-out of 4.

The *stage ratio* is the increase in transistor size in successive logic stages. Correct selection of this ratio can markedly affect timing in cascaded logic stages.

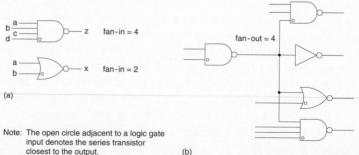

FIGURE 5.1   (a) Fan-in of CMOS gates; (b) Fan-out of CMOS gates

Note: The open circle adjacent to a logic gate input denotes the series transistor closest to the output.

The fan-in of a gate affects the speed of the gate in the following manner. In Chapter 4 we found that if two identical transistors are connected in series, the rise (or fall) time will be approximately double that for a single transistor with the same capacitive load.

When gates with large numbers of inputs have to be implemented, the best speed-performance may be obtained by using gates where the number of series inputs ranges between about 2 and 5. To illustrate this point, a very simple analysis will be presented. We will consider t_{dr} the worst-case rise delay time for an m-input NAND gate (one p-device turning on), to be (ignoring body effect - see Section 4.5.4.5),

$$t_{dr} = \frac{R_p}{n}(mnC_d + C_r + kC_g), \tag{5.2a}$$

where

R_p = the effective resistance of p-device in a minimum-sized inverter

n = width multiplier for p-devices in this gate

k = the fan-out (number of inputs connected to gate output, say, in units of minimum-sized inverters)

m = fan-in of gate

C_g = gate capacitance of a minimum-sized inverter

C_d = source/drain capacitance of a minimum-sized inverter (Note: the p and n contributions are added as an approximation (worst case) to model the effect of internal diffusion regions loading the gate as well as diffusions connected to the output)

C_r = routing capacitance.

This can be reformulated as

$$t_{dr} = \frac{R_p}{n}(mnrC_g + q(k)C_g + kC_g)$$

$$= \frac{R_pC_g}{n}(mnr + q(k) + k)$$

$$= R_pC_g mr + \frac{R_pC_g}{n}q(k) + \frac{R_pC_g}{n}k, \tag{5.2b}$$

where

$r = C_d/C_g$, the ratio of the intrinsic drain capacitance of an inverter to the gate capacitance, and

$q(k)$ = a function of the fan-out representing the routing capacitance as a multiplier times the gate capacitance.

The latter function might be used prior to an actual layout being available. The function varies for different technologies and different types of circuit layout types. A starting approximation for standard cell or gate-array layouts (usually control logic) would be $q(k) = k$; in other words, the routing capacitance adds as much routing capacitance as there is gate capacitance. In these cases it might pay to increase the size of the driving transistors because the relative effect of the routing capacitance would be reduced. In custom-designed data paths, $q(k)$ might be $.1 \rightarrow .2k$ where the circuit is dominated by self-loading. In these cases there might be no advantage to increasing the size of the transistors because this only increases the routing capacitance while the self-loading remains constant and the area increases.

The above equation is of the form

$$t_{dr} = t_{internal\text{-}r} + k \times t_{output\text{-}r}, \tag{5.3}$$

which was a gate-delay approximation introduced in Chapter 4 (Eq. 4.58) with

$$t_{internal\text{-}r} = R_p C_g m r$$

and

$$t_{output\text{-}r} = \frac{R_p C_g}{n}\left(1 + \frac{q(k)}{k}\right).$$

Similarly, the fall delay time, t_{df} is approximated by

$$t_{df} = m\frac{R_n}{n}\left(mnrC_g + q(k)C_g + kC_g\right) \tag{5.4a}$$

$$= R_n C_g m^2 r + mk\frac{R_n C_g}{n}\left(1 + \frac{q(k)}{k}\right) \tag{5.4b}$$

$$= t_{internal\text{-}f} + k \times t_{output\text{-}f} \tag{5.4c}$$

where

R_n = the effective resistance of an n-device in a minimum-sized inverter.

The previous equations assume an "equal-sized"-gate strategy often used in standard cells and gate arrays where the p- and n-transistors in gates are fixed in size with relation to each other. This condition is usually enforced to automate the layout in a straightforward manner. Another equally valid strategy would be to use an "equal-delay" method, where the rise and fall times are equalized. This may allow a somewhat smaller gate. In the above example of a NAND gate, the n–pull-down chain would normally be the slowest. Hence with

$$t_{dr} = t_{df}$$

$$\frac{R_p}{n}(mnrC_g + q(k)C_g + kC_g) = m\frac{R_n}{n}(mnrC_g + q(k)C_g + kC_g)$$

$$R_p = mR_n.$$

Thus

$$\beta_p W_p = \frac{\beta_n W_n}{m}.$$

Hence, using this method, the p-devices would be made $\beta_n/m\beta_p{}^{\text{th}}$ the width of the n-devices.

The equations for an m-input NOR gate for the equal-sized option are similar in nature:

$$t_{dr} = m\frac{R_p}{n}(mnrC_g + q(k)C_g + kC_g) \tag{5.5}$$

$$t_{df} = \frac{R_n}{n}(mnrC_g + q(k)C_g + kC_g) \quad \text{[one n-device turning on]} \tag{5.6}$$

5.2.2 Typical CMOS NAND and NOR Delays

Figure 5.2 shows the delay for a family ($W_n = 6.4\mu$, $L_n = 1\mu$, $W_p = 12.8\mu$, $L_p = 1\mu$, $t_{input\text{-}rise/fall} = .1ns$, $C_L = 0{\rightarrow}1pF$) of NAND and NOR gates measured by simulation with SPICE at the worst-speed-process corner for a particular CMOS process. Table 5.1 summarizes the data shown in Fig. 5.2 in terms of Eq. (5.1).

From these graphs we can calculate effective resistances for the transistors. From Eq. (5.2) with the following process parameters:

$n = 4$ (the n-transistors are four times the minimum size that they can be)

$kC_g = C_L$

$q(k) = 0$ (all load lumped into the kC_g term above)

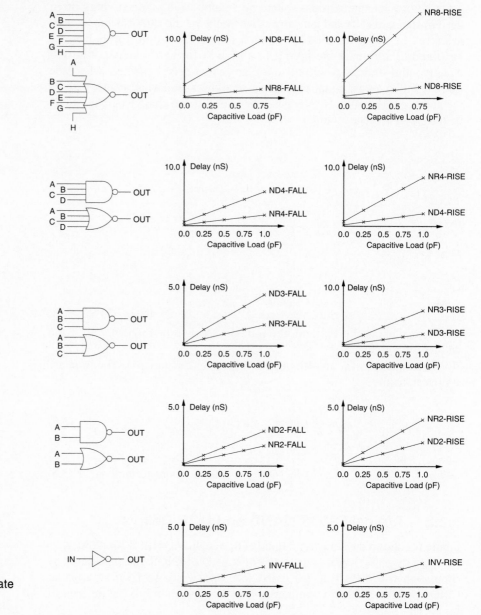

FIGURE 5.2 CMOS gate delays

$$rC_g \ (= C_d) = .005pF \ (C_g = .003pF, \ r = 1.7); \ W_p = 2W_n$$

$$t_{df\text{-}nand} = m\frac{R_{n\text{-}nand}}{4}(m \times 4 \times .005 + C_L)$$

TABLE 5.1   NAND- and NOR-Gates Delays Measured with SPICE

GATE	$t_{internal-f}$ (ns)	$t_{output-f}$ (ns/pF)	$t_{internal-r}$ (ns)	$t_{output-r}$ (ns/pF)
INV	.08	1.7	.08	2.1
ND2	.2	3.1	.15	2.1
ND3	.41	4.4	.2	2.1
ND4	.68	5.7	.25	2.1
ND8	2.44	10.98	.38	2.2
NR2	.135	1.75	.25	4.1
NR3	.14	1.83	.52	6.2
NR4	.145	1.88	.9	8.2
NR8	.19	1.8	3.35	16.4

with

$$R_{n-nand} = \frac{4 \times t_{df-nand}}{m\,(.02 \times m + kC_g)}$$

Similarly, values for the n-transistor resistance in NOR gates and p-transistor resistance may be calculated. The results for the gates shown in Fig. 5.2 are tabulated in Table 5.2.

Table 5.1 and Fig. 5.2 show that, for a given size of transistor, NAND gates are generally a better choice than NOR gates in complementary CMOS logic. If NOR gates are used, the fan-out should be limited. In general, any large fan-out should be driven with an inverter.

TABLE 5.2   Effective Resistance Values for a Typical 1μ CMOS Process (m = 1 – 4)

GATE	R_n (Ω)	R_p (Ω)
INV	7.1K	8.5K
ND2	6.3K	8.6K
ND3	6.0K	8.7K
ND4	5.9K	8.8K
NR2	7.3K	8.4K
NR3	7.4K	8.4K
NR4	7.5K	8.4K

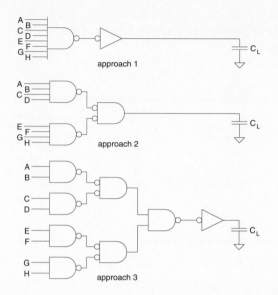

FIGURE 5.3 8-input AND gate construction

Example

As an example of a simple logic decision consider the implementation of an 8-input AND gate driving a $1pF$ load (for instance, a row decoder in a RAM or ROM), we may use the following (Fig. 5.3):

- Approach 1—An 8-input NAND and an inverter.
- Approach 2—Two 4-input NANDs and a 2-input NOR.
- Approach 3—Four 2-input NANDs, two 2-input NORs, a 2-input NAND, and an inverter.

Using the values in Table 5.1, the delays shown in Table 5.3 may be calculated (for a rising output). (*Note:* The effect of the input rise or fall time of a gate is accounted for by adding $.44t_{f/r}$ (Eq. 4.53) to the step-input delay of the stage. SPICE values are shown in parentheses under the calculated delay value.)

The result shows a classic CMOS trade-off. The approach with the most number of stages provides the best result. (For completeness, the SPICE fall times are 2.7*ns*, 2.3*ns*, and 2.6*ns*, respectively).

Note that any series resistance inserted in series with the charging or discharging path of a gate will affect switching speed. For instance, if an n-transistor is connected to the V_{SS} supply by a long resistive wire, the gate will be slower than necessary. Therefore, you should watch long resistive connections in gates. This also includes connecting power supplies to gates via resistive polysilicon, diffusion, or insufficient contacts.

TABLE 5.3   Comparison of Approaches to Designing an 8-input AND Gate

APPROACH	DELAY STAGE 1 *ns*	DELAY STAGE 2 *ns*	DELAY STAGE 3 *ns*	DELAY STAGE 4 *ns*	TOTAL DELAY (SPICE) *ns*
1 ND8-> INV	2.82 ND8 falling	3.37 INV rising			6.2 (6.5)
2 ND4-> NR2	.88 ND4 falling	4.36 NR2 rising			5.24 (5.26)
3 ND2-> NR2-> ND2-> INV	.31 ND2 falling	.4 NR2 rising	.31 ND2 falling	2.17 INV rising	3.19 (3.46)

5.2.3   Transistor Sizing

In Chapter 4 we discussed the notion of the stage ratio of an inverter chain where inverters are sized progressively to drive a large capacitive load. As the example above shows, certain logic circuits can also have signals that have large capacitive loads due to large fan-out. Clocks and reset signals are common examples, but other examples, such as row line decoders, frequently arise. In these cases, it may be advantageous to size logic gates to improve the delay between stages. The level at which this can be exercised depends on the design style that is being used. In a gate array one may increase the gate size by a fixed size that is usually that of the unit-sized inverter. Standard cell designs might have a wider range of size options for gates. Custom designs allow continuous variability of each transistor size in each gate. The latter freedom is rarely needed in all but the most stringent designs.

In the 8-input AND gate example above, the transistors were a uniform size (as would be the case in a gate-array or maybe a standard cell library). The total area of each implementation is proportional to the total size of the transistors used in the approach. Thus the areas for the implementations in Fig. 5.3 are summarized in Table 5.4.

In a row decoder application, there would be some advantage to reducing the area and the fan-in to the smallest possible value because multiple row drivers are needed to drive different rows in the memory. An improvement may be achieved by initially sizing the transistors to balance the rise and fall times of the gates. For instance, by reducing the size of the parallel p-transistors in the NAND gates, reducing the size of the parallel n-transistors in the NOR gates, and grading the transistor sizes in Approach 3, the delays shown in Table 5.5 were achieved.

TABLE 5.4   Areas for 8-input AND Gate Implementations

APPROACH	AREA
1	216
2	216
3	360

(*Note:* In a real row decoder, some of the gates may be shared between different row decoders.)

The rise times have been improved and the areas improved at the expense of the fall times. This may be an appropriate trade-off for a row decoder, where the rise time determines the access time of the memory.

The ability to arbitrarily size transistors to achieve optimal delays is often limited by other layout constraints. For instance, Approach 1 would allow a simple software layout generator to be constructed to cater for a generalized n-bit decoder, while the other approaches may be more difficult. Other considerations might drive the selection, such as minimizing the area for power dissipation considerations.

When designing circuits at any level (gate or circuit), one must balance the time to optimize such gates versus the overall effect on the system. Quite often, the performance gained might be in a part of a circuit where the improvement will not be reflected as a gain in the system as a whole. For this reason, a good starting point is to use minimum-sized devices throughout and then optimize paths from a critical-path-timing analysis. Minimum size means different things in different design technologies. In a custom design it might mean a transistor that is the minimum size that geometric design rules allow. In a gate array system it might be a transistor pair 5 to 10 times the size of the smallest pair that can be fabricated.

5.2.4   Summary

From the discussion in this section it may be seen that when designing CMOS complimentary logic with speed as a concern, there are some basic

TABLE 5.5   Delays for 8-input AND Gate with Some Transistor Sizing

APPROACH	t_r	t_f	AREA	ORIGINAL FAN-IN	NEW FAN-IN
1	4.7	4.5	120	12	6
2	4.8	4.9	136	12	6
3	3.4	3.7	124	12	4

guidelines:

- Use NAND structures where possible.
- Place inverters (or at worst, small fan-in NAND gates) at high fan-out nodes, if possible.
- Avoid the use of NOR structures in high-speed circuits, especially with a fan-in greater than four and where the fan-out is large.
- Use a fan-out below 5–10.
- Use minimum-sized gates on high fan-out nodes to minimize the load presented to the driving gate.
- Keep rising and falling edges sharp.
- When designing with power or area as a constraint, remember that large fan-in complementary gates will always work given enough time.

5.3 Basic Physical Design of Simple Logic Gates

In this section we will examine the physical layout of CMOS gates in a general sense to examine the impact of the physical structure on the behavior of the circuit. This section begins with an outline of different inverter layout forms. (To simplify layouts, "unit"-sized transistors will generally be shown. In actual layouts, the correct dimension transistors would be arrived at via detailed circuit design. P-transistors will often be shown double the "unit" size. A symbolic layout style is used to show most layouts. This omits select layers and wells and uses stylized contacts [no surrounds]. Wires and transistors are arranged on a grid. Actual layouts would space the grid proportionately to design rules.)

5.3.1 The inverter

By examining the circuit diagram for the inverter (Fig. 5.4a), we should be able to effect a physical layout by substituting layout symbols for the schematic symbols. In a schematic, lines drawn between device terminals represent connections. Any nonplanar situation is dealt with by simply crossing two lines (i.e., the connection between the drain of the n-transistor and the drain of the p-transistor). However, in a physical layout, we have to concern ourselves with the interaction of physically different interconnection layers. We know from our consideration of the fabrication process, that the source and drain of the n-transistor are n-diffusion regions, while the p-transistor uses p-diffusion regions for these connections. Additionally, in a bulk CMOS process, we can not make a direct connection from n-diffusion to

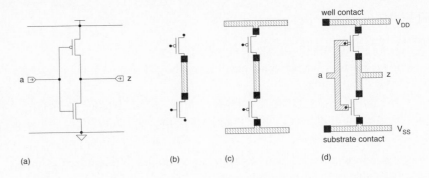

FIGURE 5.4 A sequence of steps to create the physical layout of an inverter

p-diffusion. Thus we have to implement the simple interdrain connection in the structural domain as at least one wire and two contacts in the physical domain. Assuming that the process does not have local interconnect or buried contacts, this connection has to be in metal. Substituting layout symbols, the partial inverter shown in Fig. 5.4(b) results. By similar reasoning, the simple connections to power, V_{DD}, and ground, V_{SS}, could be made using metal wires and contacts (Fig. 5.4c). Power and ground are usually run in metal (for low resistance from circuit to power supply). The common gate connection may be a simple polysilicon wire. Finally, we must add substrate contacts that are not implied in the schematic. The resulting symbolic schematic is shown in Fig. 5.4(d). Converting this to a symbolic layout yields the arrangement shown in Fig. 5.5(a). An alternative layout is shown in Fig. 5.5(b), where the transistors are aligned horizontally.

Note that there are some topology variations that may be used to enable nonplanar connection schemes to be implemented. For instance, if a metal line has to be passed through the middle of the cell from the left end of the cell to the right end, the layout shown in Fig. 5.5(c) could be used. Here, horizontal metal straps connect to a vertical metal2 or polysilicon line, which in turn connects the drains of the transistors. Alternatively, if a metal line is to be passed from left to right at the top or bottom of the cell, the power and ground connections to the transistors may be made in the appropriate diffusion layer (Fig. 5.5d). This, in effect, makes the inverter transparent to horizontal metal connections that may have to be routed through the cell. From the considerations that affect performance, the previous deviations from the original layout have little effect. In the case of a vertical polysilicon drain connection, an extra connection resistance is incurred. This would be approximately $2R_{contact} + R_{poly}$, where $R_{contact}$ is the resistance of a metal-polysilicon contact and R_{poly} is the resistance of the polysilicon runner. In addition, a slight extra capacitance may be incurred. Usually the result of both of these effects would be inconsequential. For the power and ground diffusion connections, the penalty is a series-connection resistance and

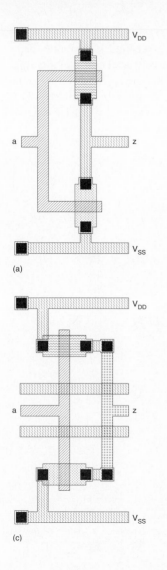

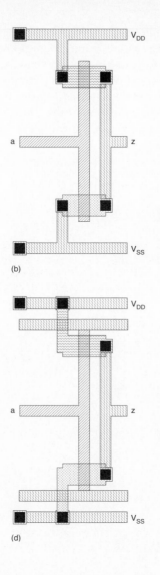

FIGURE 5.5   Symbolic layouts for the CMOS inverter

increased capacitance. As a rule of thumb, the resistance should be kept an order of magnitude below the transistor "on" resistance. The capacitance on supply connections does not normally affect performance and in some cases may be intentionally increased to reduce on-chip power supply noise. Running a polysilicon connection from left to right must be completed below or above the transistors, with the transistors using metal connections to power and ground. Polysilicon passing from left to right through the middle of the cell requires a metal strap. These layouts are also shown in Plate 3(a).

The addition of a second layer of metal allows more interconnect freedom with the two other interconnect layers. The second-level metal may be used to run V_{DD} and V_{SS} supply lines. Alternatively, second-level metal may be used to strap polysilicon in a parallel connection style to reduce delays due to long poly runs. In these cases, the layouts remain approximately the same, with the exception of the added metal2 wires and metal1 connection stubs. Some options are shown in Figs. 5.6(a) and 5.6(b).

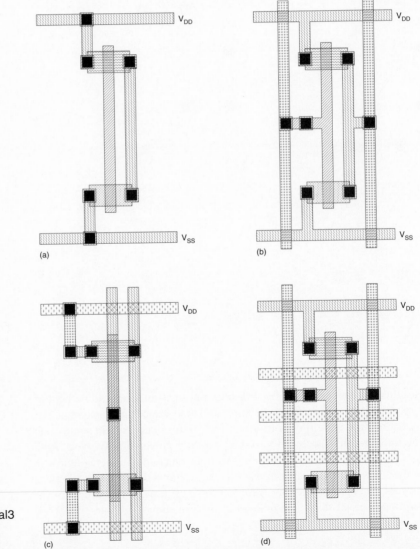

FIGURE 5.6 Metal2/Metal3 symbolic layouts for the CMOS inverter

A third level of metal is usually used for power and ground connections. Figures 5.6(c) and 5.6(d) show some alternative layouts. A strict vertical poly, horizontal metal1, vertical metal2, horizontal metal3 is of use for sea-of-gates structures (Fig. 5.6d). See also Plate 3(b).

Note that in addition to increasing the size of the transistors (Fig. 5.7a) a large inverter may be constructed from many smaller inverters connected in parallel. This is symbolically shown in Fig. 5.7(b). In large transistors, the source and drain regions should be "stitched" with the contacts and metal to reduce source-drain resistance. Placing transistors back to back (Fig. 5.7b)

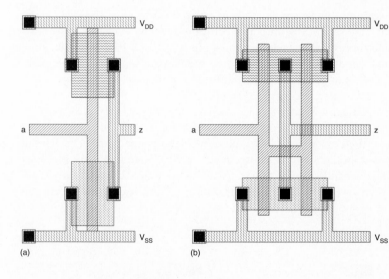

FIGURE 5.7 Various methods for creating large inverters by paralleling small inverters and changing transistor shape

yields a more optimum drain capacitance because of the smaller merged diffusion regions. This results from the fact that the drain area does not increase in size much but the gain of the transistors (β) is doubled. A further reduction in drain capacitance is achieved by using the donut ("round transistor") connection, shown in Fig. 5.7(c). Here the β of the transistors is almost quadrupled, while the drain area is substantially the same as for a single minimum-sized inverter. Plate 3(c) shows these layouts in color. In essence, these variations represent some "forms" for an inverter (and to an extent, other gates) that will be used in various situations in this text.

5.3.2   NAND and NOR Gates

Similar reasoning can be applied to converting the 2-input NAND schematic to a layout. Figure 5.8(a) shows a direct translation of a schematic. By orienting the transistors horizontally, the layout in Fig. 5.8(b) is possible. Note that in the case of the NAND gate, the latter layout is much cleaner (and smaller). This is in general true for multiple-input static gates, and we will adopt a style where transistors are oriented horizontally and polysilicon gate signals run vertically. Where departures are made from this style, the reasons for doing so will be given. Note, of course, that the gate could be rotated 90° to obtain vertical metal and horizontal polysilicon connections. The 2-input NOR gate symbolic layout is shown in Fig. 5.9(a). Note that there is a variation of the connection to the two transistors in parallel. The alternative layout is shown in Fig. 5.9(b). The latter connection, in common with the

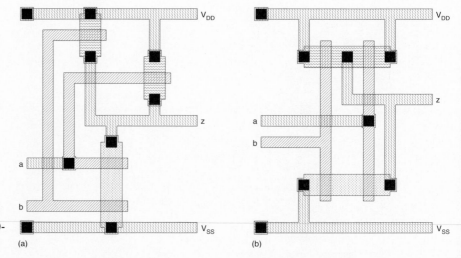

FIGURE 5.8 Typical NAND-gate symbolic layouts

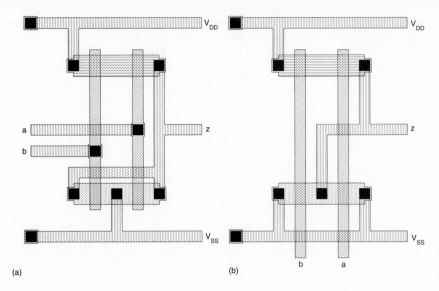

(a)   (b)

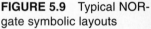

FIGURE 5.9   Typical NOR-gate symbolic layouts

paralleled inverters, has less drain area connected to the output. This results in a faster gate. The same variation may be applied to the NAND gate. This will be further discussed in Section 5.5. Complex gates are an extension of the gates so far treated.

5.3.3   Complex Logic Gates Layout

All complementary gates may be designed using a single row of n-transistors above or below a single row of p-transistors, aligned at common gate connections. Most "simple" gates may be designed using an unbroken row of transistors in which abutting source-drain connections are made. This is sometimes called the "line of diffusion" rule, referring to the fact that the transistors form a line of diffusion intersected by polysilicon gate connections.

If we adopt this layout style, it has been shown that there are techniques for automatically designing such gates.[1] Those automated techniques that are applicable to static complementary gates are reviewed here. The CMOS circuit is converted to a graph where (1) the vertices in the graph are the source/drain connections, and (2) the edges in the graph are transistors that connect particular source-drain vertices. Two graphs, one for the n-logic tree, and one for the p-logic tree, result. Figure 5.10 shows an example of the graph transformation. The connection of edges in the graphs mirror the series-parallel connection of the transistors in the circuits. Each edge is named with the gate signal name for that particular transistor. Thus, for instance, the p-graph has four vertices: Z, I_1,

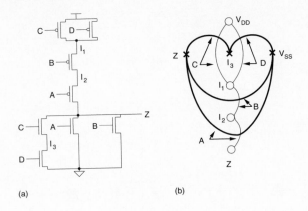

FIGURE 5.10   CMOS-logic-gate graph representation

(a)

(b)

I_2, and V_{DD}. It has four edges, representing the four transistors in the p-logic structure. Transistor A (A connected to gate) is an edge from vertex Z to I_2. The other transistors are similarly arranged in Fig. 5.10(b). Note that the graphs are the dual of each other as the p- and n-trees are the dual of each other. The n-graph (dark lines and crosses) overlays the p-graph in Fig. 5.10(b) to illustrate this point. If two edges are adjacent in the p- or n-graph, then they may share a common source-drain connection and may be connected by abutment. Furthermore, if there exists a sequence of edges (containing all edges) in the p-graph and n-graph that have identical labeling, then the gate may be designed with no breaks. This path is known as an Euler path. The main points of the algorithm[2] are as follows:

1. Find all Euler paths that cover the graph.
2. Find a p- and an n-Euler path that have identical labeling (a labeling is an ordering of the gate labels on each vertex).
3. If the paths in step 2 are not found, then break the gate in the minimum number of places to achieve step 2 by separate Euler paths.

In the example shown in Fig. 5.10, the original graph with a possible Euler path is shown in Fig. 5.11(a). The sequence of gate signal labels in the Euler path is (A,B,C,D). Note that the graph for the n- and p-graph allow this labeling. To complete a layout the transistors are arranged in the order of the labeling n- and p-transistors in parallel rows, as shown in Fig. 5.11(b). Vertical polysilicon lines complete the gate connections. Metal routing wires complete the layout. This procedure may be followed when manually designing a gate.

A variation of the single line of n- and p-transistors occurs in logic gates where a signal is applied to the gates of multiple transistors. In this case,

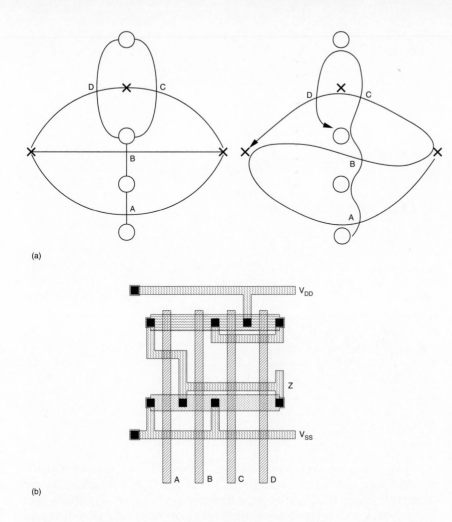

(a)

(b)

FIGURE 5.11   Euler paths in a CMOS gate and the corresponding layout (symbolic)

transistors may be stacked on the appropriate gate signal. This also occurs in cascaded gates that cannot be constructed from a single row of transistors. A good example of this is the complementary XNOR gate. The schematic for this gate is shown in Fig. 5.12(a). According to the style of layout that we have used to date, two possible layouts are shown in Fig. 5.12(b) and Fig. 5.12(c). The layout shown in Fig. 5.12(b) uses the single row of n- and p-transistors, with a break, and that in Fig. 5.12(c) uses a stacked layout. The selection of the styles would depend on the overall layout—whether a short, fat, or long thin cell were needed. Note that the gate segments that are maximally connected to the supply and ground rails should be placed adjacent to these signals.

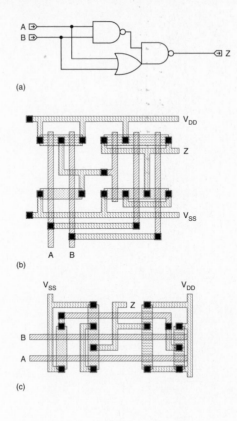

FIGURE 5.12 Complementary CMOS XNOR gate—alternative layout styles

An automatic approach to achieve this style of layout that uses a graph-theoretic approach has been proposed.[3–7] The approach is based on the use of interval graphs to optimally place transistors on vertical polysilicon lines in a gate matrix style (see Chapter 6). The layout style is similar to that used so far, with vertical polysilicon lines and horizontally arranged transistors. Power and ground run at the top and bottom of the cell. The approach is summarized in Fig. 5.13.

- Transistors are grouped in strips to allow maximum source/drain connection by abutment. To achieve better grouping, polysilicon columns are allowed to interchange to increase abutment.

- The resultant groups are then placed in rows with groups maximally connected to the V_{SS} and V_{DD} rails placed toward these signals. Row placement is then based on the density of other connections.

- Routing is achieved by vertical diffusion or manhattan (horizontal and vertical) metal routing. This normally would require a maze router (see Chapter 6).

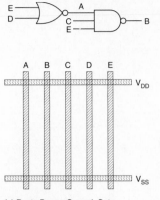

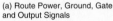

(a) Route Power, Ground, Gate
and Output Signals

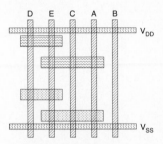

(b) Order Gate and Outputs to
optimize horizontal transistor connectivity

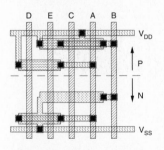

(c) Rearrange vertical strip ordering to
optimize power routing internal gate connection
and output connections

FIGURE 5.13 Outline of
automated approach to
CMOS-gate layout

5.3.4 CMOS Standard Cell Design

When designing standard cells or polycells, geometric regularity is often
required while maintaining some common electrical characteristics between
cells in the library. A common physical limitation is to fix the physical height
of the cell and vary the width according to the function. A typical standard cell
is shown in Fig. 5.14. It is composed of a row of n-transistors of maximum

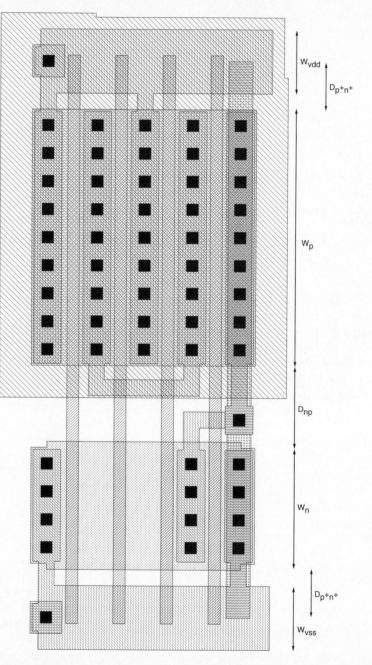

FIGURE 5.14 Typical CMOS standard-cell mask layout

height W_n and a row of p-transistors of maximum height W_p, separated by a distance D_{np}, the design-rule separation between n- and p-active areas. Power (V_{DD}) and ground (V_{SS}) busses traverse the cell at the top and bottom. The internal area of the cell is used for routing the transistors of specific gates.

A design objective in which W_p and W_n are selected may take into account such parameters as power dissipation, propagation delay, noise immunity, and area. Kang provides a good summary of the approach in the selection of W_p and W_n.[8] The basic steps are as follows:

1. Identify a sample selection of gates (i.e., INVERTER, NAND, NOR) and compute an "average" delay time.

2. Calculate an objective function that relates worst-case propagation time to the ratio of W_p/W_n.

3. Calculate an objective function relating the noise immunity to W_p/W_n.

4. Select an appropriate ratio that balances the required objective functions.

   (*Hint:* For normal CMOS gates in current processes $W_p = W_n$ is widely used.)

Techniques may then be employed to automatically generate these gates in a process-independent manner from fairly straightforward intermediate forms.[9] Note that in the above gate structure, all transistors of similar type were assumed to be the same size. One may further optimize a parameter such as noise immunity by adjusting individual transistor sizes to that below the maximum width allowed.

Figure 5.15 shows a few representative examples of standard cell layouts. Figure 5.15(a) shows a style which compresses the series n-transistors to reduce internal capacitance. Figure 5.15(b) is a fairly standard gate while Fig. 5.15(c) straps the polysilicon with metal2. Figure 5.15(d) shows a standard cell in which the substrate connection is separated from the V_{SS} line. This style is sometimes used in mixed-signal (analog + digital) chips to reduce the amount of current injected into the substrate. This current can affect analog performance.

5.3.5 Gate Array Layout

Standard-cell chips require all mask levels to construct a chip. A gate-array chip uses a fixed "image" of under layers with a set of discretionary wiring layers providing the personalization of the array. Typically, the well, diffusion, and polysilicon layers are fixed, and contact, metal1, via, and metal2 are programmed. Figure 5.16(a) shows a typical "site" consisting of three transistor pairs. A programmed site is shown in Fig. 5.16(b). These may be arrayed in rows, as shown in Fig. 5.16(c). Routing tracks are placed in the spaces between rows of transistors. Design decisions involve the size of the transistors, the connectivity of the polysilicon, and the number of tracks allowed in a routing channel.

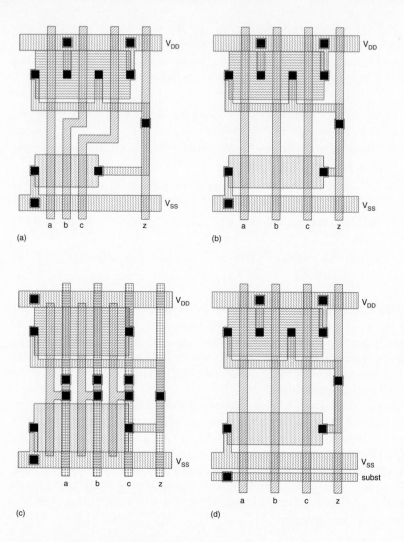

FIGURE 5.15 More standard-cell layout styles (symbolic)

5.3.6 Sea-of-Gates Layout

The general layout style used for the gate array may be generalized to build a type of circuit called *sea-of-gates* or a CMOS cell array.[10] In this array, continuous rows of n- and p-diffusion are run across the master chip. These in turn are arrayed regularly in the Y dimension without regard to routing channels. A logic gate is "isolated" from a neighboring logic gate by tying the gate terminal of the end transistors to V_{SS} (n) or V_{DD} (p). Routing channels are routed across rows of unused transistors as required. This results in a much more general-purpose array, the decision about the number of routing tracks per routing row having been finessed. The basic array architecture is shown in Fig. 5.17. A variety of gate personalizations are shown in Fig. 5.18. Figure 5.18(a) shows a 3-input NAND gate, and Fig. 5.18(b) shows two

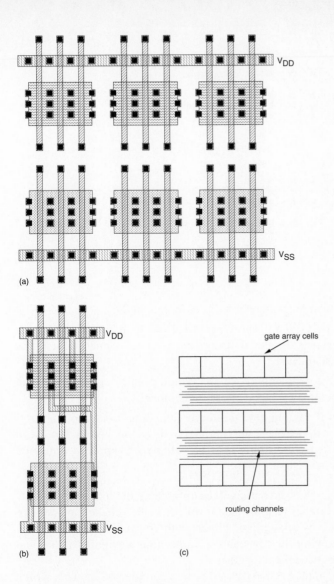

gate array cells

routing channels

FIGURE 5.16 Gate array layout: (a) unprogrammed base array; (b) personalized cell; (c) routing strategy

inverters driving a 2-input NOR gate. Other gate-array and sea-of-gate styles are discussed in Chapter 6.

5.3.7 General CMOS Logic-Gate Layout Guidelines

From the considerations given to the layout of complementary gates, the following general layout guidelines may be stated:

1. Complete the electrical gate design, taking into account the factors mentioned in Section 5.2.

2. Run V_{DD} and V_{SS} in metal at the top and bottom of the cell.

287

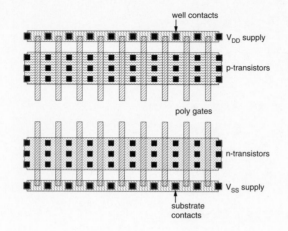

FIGURE 5.17 Array architecture of sea-of-gates layout style

3. Run a vertical polysilicon line for each gate input.

4. Order the polysilicon gate signals to allow the maximal connection between transistors via abutting source-drain connections. These form gate segments.

5. Place n-gate segments close to V_{SS} and p-gate segments close to V_{DD}, as dictated by connectivity requirements.

6. Connections to complete the logic gate should be made in polysilicon, metal, or, where appropriate, in diffusion. (As in the case of connections to the supply rails or outputs.) Keep capacitance on internal nodes to a minimum.

Note that the style of layout involves optimizing the interconnection at the transistor level rather than the gate level. As a rule, smaller and perhaps faster layouts result by taking logic blocks with 10- to 100-transistor complexities and following the rules above, rather than designing individual gates and trying to piece them together.

This improvement in density is due to a number of factors, which include the following:

1. Better use of routing layers—routes can occur over cells.

2. More "merged" source-drain connections.

3. More usage of "white" space (blank areas with no devices or connections) in sparse gates.

4. Use of optimum device sizes—the use of smaller devices leads to smaller layouts.

Improvements gained by optimizing at this level over a standard-cell approach can be up to 100% to 300% or more in area. Furthermore, cells can

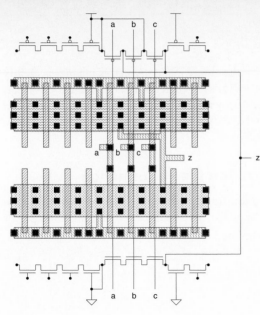

(a)

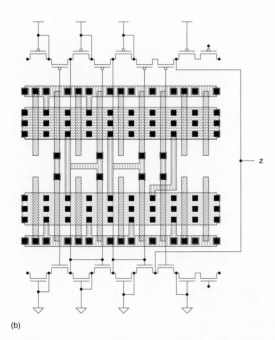

(b)

FIGURE 5.18 Various personalized sea-of-gates layouts: (a) 3-input NAND gate; (b) two inverters driving a 2-input NOR gate

be designed in such a way to provide "transparent routing" for cell-to-cell communication. This greatly reduces the global wiring problem. The problem with such approaches is that, unless automated, they can be quite labor intensive. These days it is probably only worth investing manual effort in highly repetitive and reused structures such as data paths and widely used

289

standard cells. Implementing control logic manually in this manner is clearly a mistake because this type of logic often changes and the manual effort has to be continually spent to keep up with the changes. With metal3 standard cell layouts, where it is easy to have over-the-cell routing, very dense layouts may be automatically constructed.

5.3.8   Layout Optimization for Performance

In this section a potpourri of optimization techniques will be presented. One technique that has been demonstrated to increase the speed of gates consisting of

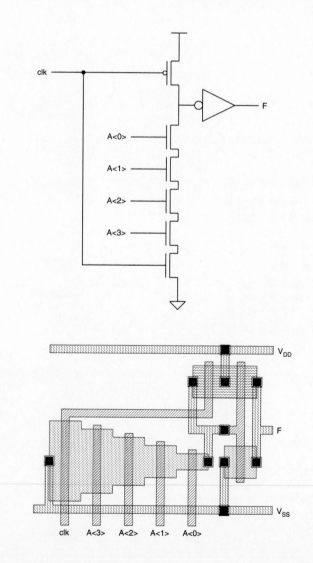

FIGURE 5.19   Grading-series transistors in an AND gate to reduce delay

series combinations of transistors for older technologies, is to vary the size of the transistor according to the position in the series structure.[11] This is shown in Fig. 5.19 for a 4-input AND gate. The transistor closest to the output is the smallest, with transistors increasing in size the nearer they are to V_{SS}. The decreased switching times are attributed to the dominance of the capacitance term in the RC time constant of the gate. In older technologies, increases in performance of 15% to 30% have been demonstrated. More recent experience tends to suggest that in submicron technologies, where the source/drain capacitances are less, this improvement is limited to 2% to 4% and thus is hardly worthwhile.[12]

Another effect that leads to less than ideal gates occurs in the case of parallel connected transistors. This effect was encountered in the construction of parallel inverters and was also demonstrated in the 2-input NOR gate constructed previously. In the NOR schematic, the output is connected to one p-transistor drain and two n-transistor drains. However, in one of the NOR layouts (Fig. 5.9b), the drain connection between the two n-transistors is merged. This effectively means that only two drain connections are connected to the output, thus reducing the capacitance at the output. The parallel connection of two sources to the ground rail adds capacitance to the ground rail but does not affect the output switching speed. Another example is seen in the gate that implements the function $F = \overline{(A + B + C) \cdot D}$ (Fig. 5.20a). The n-transistor connection for this gate is shown in Fig. 5.20(b). The ground connection may be made at point 1 or 2. Point 1 would be preferred, because this connects three of the source regions to ground (V_{SS}). Actually, by merging the source-drain connections, only two V_{SS} connections are made. In general, as a result of this effect and the body effect, we try to assemble the most capacitive nodes closest to the supply and ground rails. Symbolic layouts for the function in Fig. 5.20(a) are shown in Fig. 5.20(c) and Fig. 5.20(d), illustrating two approaches to implementing the gate. The gate in Fig. 5.20(c) has one "unit" output p-drain capacitance, one output n-drain capacitance, and four "internal" drain capacitances (two n and two p). Figure 5.20(d) has four output drain capacitances and four internal drain capacitances. Thus the layout in Fig. 5.20(c) improves diffusion capacitance by at least one n and one p. Where this capacitance dominates, the optimized layout would result in a faster circuit.

Note that these strategies may coincide. For instance, in the complex gate shown in Fig. 5.20, the signal A may be delayed with respect to signals B, C, and D. This also indicates that connection-point 1 should be grounded. If there is some doubt regarding the organization of a gate (i.e., signal B arrived first), a simulation should be done on the gate with appropriately timed inputs.

5.3.9   Transmission-Gate Layout Considerations

In the case of complementary gates, there is one point of contact between the n-transistors and the p-transistors. As we have seen, this can be completed

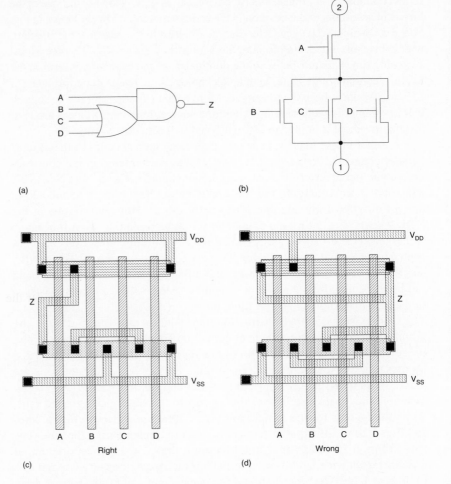

(a)

(b)

(c)

Right

(d)

Wrong

FIGURE 5.20 Optimization of CMOS gate layout involving multiple source-drain connections

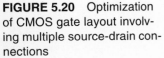

with a metal strap or a combination of metal and polysilicon straps where metal routing transparency is required. When considering a transmission gate, the source and drain terminals of the p- and n-transistors are paralleled. According to the layout strategy presented, the layouts shown in Fig. 5.21 would be suitable. Note that in Fig. 5.21(a), no metal lines can pass from left to right. The layout shown in Fig. 5.21(b) is longer but has horizontal metal transparency. The decision on which layout is more suitable would depend on the circuit being designed. For instance, in a shift-register-delay line, Fig. 5.21(a) might be preferred due to its small size. In a data path, where bus lines may have to pass horizontally, Fig. 5.21(b) would be preferred. Figure 5.21(c) shows a metal2 version.

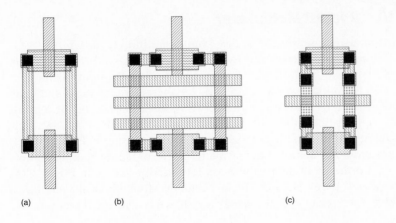

FIGURE 5.21   Transmission-gate layouts

The transmission gate has to be supplied with a switching signal and its complement. These signals may be generated at some distance and may have to be routed to the transmission gate (i.e., in an array of registers controlled by a common signal). In these cases, it is necessary to consider the routing of the gate signals to the transmission gates. Three possibilities are shown in Fig. 5.22. In Fig. 5.22(a) the control inputs are run horizontally in metal, outside the transistors. Note that in this case, polysilicon can be passed horizontally between the n- and p-transistors. In Fig. 5.22(b) the control signals are routed vertically in polysilicon. In this case, the transistors are offset to allow the passage of the vertical control lines. Figure 5.22(c) shows another layout using metal2, in which the poly is run vertically but is strapped by metal1.

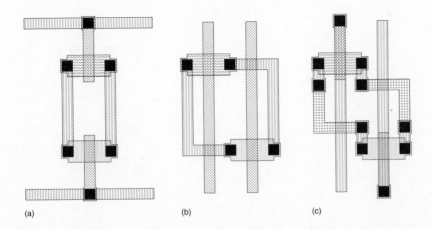

FIGURE 5.22   Routing to transmission gates

5.3.10 2-input Multiplexer

Apart from use in multiplexers, the 2-input multiplexer in Fig. 5.23(a) is used frequently in latches and registers. Possible layouts are shown in Figs. 5.23(b) and 5.23(c). Note that in Fig. 5.23(c) the control lines are crossed in the middle of the latter cell. An alternative layout is shown in Fig. 5.23(d). If the mux was to be a stacked structure, the number of contacts in each control line should be equalized, as shown in Fig. 5.23(e). This equalizes any delay that might arise due to contact resistance. Figure 5.23(f) does not cross the control lines. An alternative to the manhattan layouts shown is illustrated in the partial mask level design shown in Fig. 5.23(g) which uses 45° wires. Because this is a common structure, manhattan-based symbolic systems can treat the signal switch as a special crossover symbol.

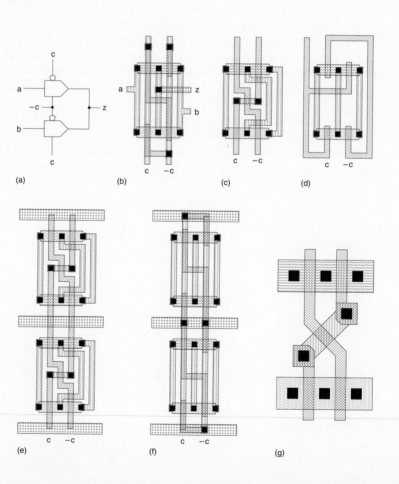

FIGURE 5.23 Two-input multiplexer: (a) circuit; (b) metal strapped select lines; (c) poly select lines (with metal crossover); (d) poly select lines (no crossover); (e) stacked mux (poly); (f) stacked mux (metal); (g) 45° crossover mux

5.4   CMOS Logic Structures

In some situations, the area taken by a fully complementary static CMOS gate may be greater than that required, the speed may be too slow, or the function may just not be feasible as a purely complementary structure (such as in the case of a large PLA). In these cases, it is desirable to implement smaller and faster gates at a cost of increased design and operational complexity and, possibly, decreased operational margin. There are a number of alternate CMOS logic structures that can be used. These structures will be summarized in this section.

5.4.1   CMOS Complementary Logic

For review, the complementary CMOS inverter, NAND, and NOR gates are shown in Fig. 5.24. All complementary gates may be designed as ratioless circuits (that is, there does not have to be a fixed ratio in size between pull-

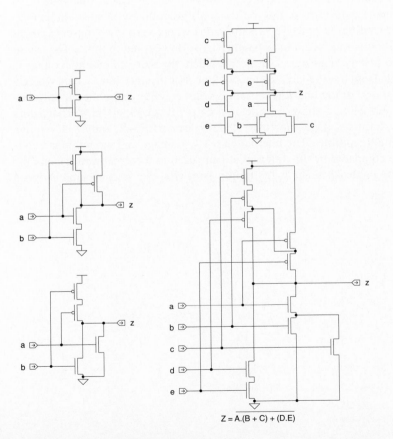

$$Z = \overline{A.(B + C) + (D.E)}$$

FIGURE 5.24   CMOS complementary gates (review)

up and pull-down structures). This feature is used in certain layout design strategies such as gate arrays and sea-of-gates. If all transistors are the same size the circuit will function correctly (compared to some other MOS logic families where this is not the case). A complex gate that will form the basis for comparison between logic families is shown. It implements the function

$$Z = \overline{A.(B+C) + (D.E)}$$

Apart from varying the ratio of the transistors in a complementary CMOS gate to vary input threshold or speed, the supply voltage can be increased or decreased to achieve higher noise immunity, decrease the power dissipated in the circuit, or meet a system-supply voltage constraint. The supply voltage can generally be increased within some safety margin (1.5–2.0 volts) of where the source-drain diodes break down (for instance, in most 5-volt CMOS processes the break-down voltage is around 7 volts). High-voltage CMOS processes might allow 15- to 30-volt supply voltages while newer high-density processes might only allow 2.5 to 3.5 volts. In applications such as watches, 1.0 to 1.5 volts only may be available. A conventional 5-volt-process CMOS gate will operate with very low supply voltages by virtue of subthreshold conduction, albeit very slowly. Usually at some point the leakage current from the source-drain junctions will cause the gate to cease operating. In power-down situations where very low quiescent power dissipation is required, this capability is worth keeping in mind. The use of on-chip voltage regulators is desirable where the required on-chip voltage is different from the available system voltage. For instance, this occurs where 5 volts is used at the board level, but 3 volts is required by the on-chip circuitry. Figure 5.25 shows a typical on-chip voltage regulator.[13] Transistors P_6–P_{10} form a voltage reference, while transistors P_1–P_4 and N_1–N_2 form a differential amplifier. The internal chip V_{DD} voltage and the reference voltage are compared by the differential pair (used as a current mirror), and the resulting control voltage is fed to transistor (P_5) that is connected between

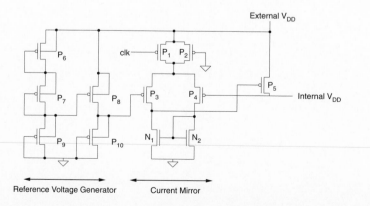

FIGURE 5.25 A voltage regulator for reducing the on-chip V_{DD} supply

the internal supply and the external supply. In this particular example, the *clk* signal was generated ahead of where the chip required a large amount of current so that the internal supply did not droop.

As we have seen, the CMOS complementary gate has two function-determining blocks—an n-block and a p-block. There are normally 2n transistors in an n-input gate. Variations from the complementary CMOS gate include the following techniques:

- reducing the noise margin of the gate

  and/or

- reducing the function-determining transistors to one polarity.

5.4.2   BiCMOS Logic

The output drive capability of a CMOS gate can be enhanced if bipolar transistors are available as circuit elements, as is the case in a BiCMOS process. A BiCMOS NAND gate is shown in Fig. 5.26(a). Transistors N_1 and N_2 supply the pull-down npn-transistor with base current when the input is high. N_3 clamps the pull-down when the output is high. In common with the related BiCMOS inverter, P_1 or P_2 supply base current to the pull-up npn-transistor. Another type of BiCMOS gate is shown in Fig. 5.26(b). In this gate the nMOS transistors of the NAND are replicated in the pull-down path of the output in an effort to get a good V_{OL} level. Clearly the series transistors should be limited if the pull-down speed is to match the pull-up speed. Another approach to building BiCMOS gates is to simply use CMOS gates for logic and then use any of the driver structures demonstrated in Chapter 2 as output stages.

Two schools of thought currently apply to the use of BiCMOS for digital-only chips. The first embraces the technology as a speed-enhancing option, especially for highly automated design techniques such as gate arrays. The second line of reasoning favors finer-line CMOS processes with

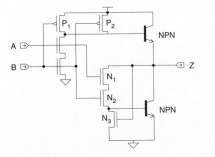

(a)

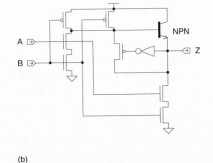

(b)

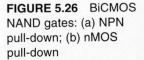

FIGURE 5.26 BiCMOS NAND gates: (a) NPN pull-down; (b) nMOS pull-down

their lower production costs and economies of scale. There is definitely a trade-off to be made, and time will determine which course designers, managers and their customers take.

The most useful place for BiCMOS drivers is as bus drivers, I/O drivers, and in other applications where a high drive capability is required.[14–26] Other uses are in memory-sense amplifiers. In mixed-signal chips, bipolar transistors are of extreme utility in designing simple, high performance op-amps and other linear circuits.

5.4.3 Pseudo-nMOS Logic

A pseudo-nMOS gate is shown in Fig. 5.27(a). It is the extension of the inverter dealt with in Chapter 2. Here the load device is a single p-transistor, with the gate connected to V_{SS}. Alternatively, the p-load may be connected as a constant-current source to provide better process tracking and optimized pull-down sizes. The gain ratio of the n-driver transistors to p-transistor load, $\beta_{driver}/\beta_{load}$, has to be selected to yield sufficient gain to generate consistent high and low logic levels. The design of this style of gate thus involves ratioed transistor sizes to ensure correct operation. That is, the effective β_n/β_p ratio has to be consistent with the values predicted in Section 2.4 for all combinations of input values. The main problem with the gate is the static power dissipation that occurs whenever the pull-down chain is turned on. As the p-load is always turned on, when the n pull-down is on, current flows in the gate structure. There are $n + 1$ transistors in an n-input pseudo-nMOS gate. In a complementary gate, the capacitive load on each input is at least two unit-gate loads (the gate input capacitance of a unit-sized transistor). In this type of gate, the minimum load can be one unit-gate load as a result of using only one transistor for each term of the input function. However, if minimum-sized driver transistors are used, the gain of the pull-up has to be decreased to provide adequate noise margins. This, in turn, slows the rise time of the gate. A gate so implemented should have a density advantage over a fully complementary gate. Figure 5.27(b) shows a circuit that may be used to ensure the V_{OL} noise margin of the pseudo-nMOS inverter. Transistors P_1–P_2 and N_1–N_3 form a bias generator that tracks process and V_{DD} changes. The drain of transistor N_2 is maintained at V_{tn} above ground. This is stabilized by the pass transistor N_3 and the feedback inverter N_1–P_2. This inverter is ratioed so that the V_{IL} is around V_{tn} (by making N_1 large). The combination of N_2 and N_3 causes a current to flow in the current mirror, P_1. By ratioing the transistor sizes used in pseudo-nMOS gates, the V_{OL} level may be set. For instance, if $2W_{N_2} = W_{N_4} = W_{N_5}$ and $W_{P_2} = 0.5W_{P_1}$, then the V_{OL} will be approximately $0.5V_{tn}$. Figure 5.27(c) shows some typical SPICE waveforms for a pseudo-nMOS inverter using the biasing scheme. The V_{bias} line should be liberally bypassed (with on-chip capacitance—i.e., transistor gates) if it travels large distances so that it does not bounce with clock changes.

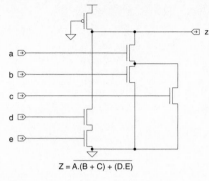

$$Z = \overline{A.(B + C) + (D.E)}$$

(a)

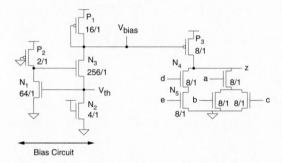

(b)

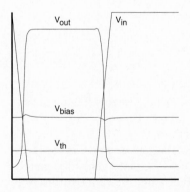

(c)

FIGURE 5.27 Pseudo-nMOS logic: (a) circuit; (b) bias circuit; (c) SPICE waveforms

An alternate way to represent the pseudo-nMOS gate is shown in Fig. 5.28. This is called CMOS Multidrain Logic by the inventors.[27] It represents, in layout and logic style, bipolar integrated injection logic or merged transistor logic.[28] The gates formed are electrically identical to the gate shown in Fig. 5.27, but the gates are combined in an open-drain manner,

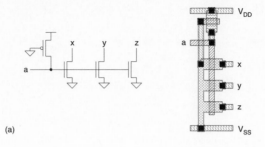

(a)

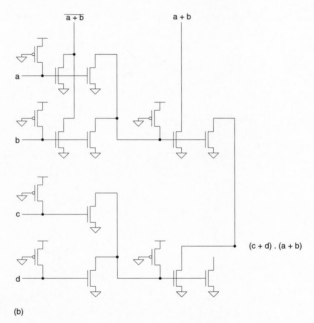

(b)

FIGURE 5.28 CMOS multi-drain logic: (a) basic gate; (b) symbolic layout; (c) typical use

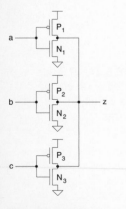

FIGURE 5.29
Ganged CMOS logic

which may have some benefits in automated layout systems. Figure 5.28(a) shows the basic gate and a representative layout. Figure 5.28(b) shows a variety of circuit connections and logic functions.

Another manifestation of the pseudo-nMOS gate is shown in Fig. 5.29. This is called a symmetric CMOS NOR gate[29] or, more generally, ganged CMOS.[30] As with the normal pseudo-nMOS gate, the ratios of the p- to n-transistors have to be selected to ensure correct operation. For instance if $\beta_{driver}/\beta_{load}$ has to be greater than ratio R to ensure correct operation for an n-input pseudo-nMOS gate, then $\beta_{driver}/\beta_{load}$ has to be greater than $R/n-1$ for a symmetric n-input NOR gate ($n \geq 2$). The gate structure is best used for gates with fan-in less than 4 and in these applications has been shown to be about 1.4–1.6 times as fast as the pseudo-nMOS NOR gate. One surprising attribute of the NOR gate shown in Fig. 5.29 is the ability for it to operate as a NAND gate by suitably ratioing the p-transistors to be able to overcome the n-transistors. While this usually leads to slower NAND gates than regular CMOS implementations, it does lead to some interesting topologies. Moreover ganged CMOS implementations have been demonstrated with internal quaternary (four-valued) nodes.[31]

5.4.4 Dynamic CMOS Logic

A basic dynamic CMOS gate is shown in Fig. 5.30. It consists of an n-transistor logic structure whose output node is precharged to V_{DD} by a p-transistor and conditionally discharged by an n-transistor connected to V_{SS}. (Alternatively, an n-transistor precharged to V_{SS} and a p-transistor discharge to V_{DD} and a p logic-block may be used.) *clk* is a single-phase clock. The precharge phase occurs when $clk = 0$. The path to the V_{SS} supply is closed via the n-transistor "ground switch" during $clk = 1$ (evaluate phase). The input capacitance of this gate is the same as the pseudo-nMOS gate. The pull-up time is improved by virtue of the active switch, but the pull-down time is

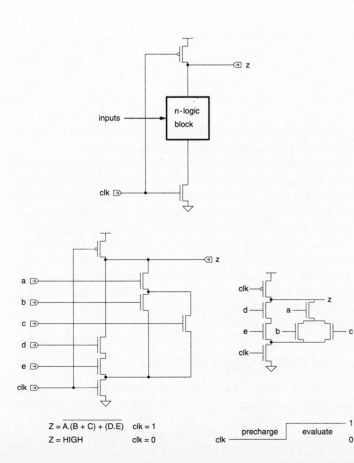

$Z = \overline{A.(B + C) + (D.E)}$ clk = 1
$Z = $ HIGH   clk = 0

FIGURE 5.30 Basic CMOS dynamic gate

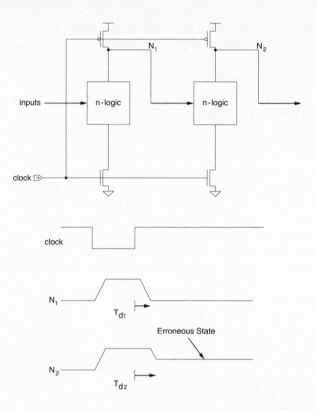

FIGURE 5.31 Erroneous evaluation in cascaded dynamic CMOS gates

increased due to the ground switch. Note that the ground switch may be omitted if the inputs are guaranteed to be zero during precharge.

A number of problems are manifest in this structure. Firstly, the inputs can only change during the precharge phase and must be stable during the evaluate portion of the cycle. If this condition is not met, charge redistribution effects can corrupt the output node voltage. Simple single-phase dynamic CMOS gates can not be cascaded. For instance, consider the circuit in Fig. 5.31. When the gates are precharged, the output nodes are charged to V_{DD}. During the evaluate phase, the output of the first gate will conditionally discharge. However, some delay will be incurred due to the finite pull-down time. Thus the precharged node (N_1) can discharge the output node of the following gate (N_2) before the first gate is correctly evaluated. Modifications to these basic dynamic gates to correct this problem are demonstrated in Section 5.4.7, 5.4.8, and 5.5.11.

5.4.5 Clocked CMOS Logic (C²MOS)

A clocked CMOS gate is shown in Fig. 5.32. This form of gate was originally used to build low-power-dissipation CMOS logic.[32] The reasons for

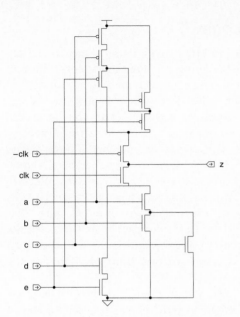

FIGURE 5.32   A clocked CMOS gate (C^2MOS)

the reduced dynamic power dissipation stem mainly from metal gate CMOS layout considerations and are not particularly relevant in today's technologies. The main use of such logic structures at this time is to form clocked structures that incorporate latches or that interface with other dynamic forms of logic (see Section 5.4.8). The gates have the same input capacitance as regular complementary gates but larger rise and fall times due to the series clocking transistors. The series clock transistors can either be at the output of the gate or at the power supply ends. Clocked CMOS circuitry is one recommended remedy for "hot electron" effects, because it places an additional n-transistor in series with the logic transistors.[33] In this application, because the clock normally is the last changing input, the clock transistor has to be placed at the bottom of the n-logic tree. This is at odds with placement at the center, which yields a faster gate.

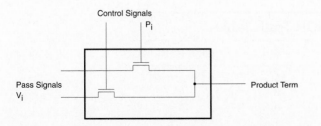

FIGURE 5.33   Model for pass transistor logic

5.4.6 Pass-Transistor Logic

One form of logic that is popular in nMOS-rich circuits is pass-transistor logic, the simplest example probably being a 2-input multiplexer. Formal methods for deriving pass-transistor logic have been presented for nMOS.[34] They are based on the model shown in Fig. 5.33, where a set of *control* signals are applied to the gates of n-transistors. Another set of *pass* signals are applied to the sources of the n-transistors. In the notation given by Radhakrishnan et al.,[35] product terms P_i consist of a number of n-transistors in series controlled by control variables and fed with a pass variable. Thus $F = P_1(V_1) + P_2(V_2) + \ldots + P_n(V_n)$, where V_i are the pass variables. When P_i is true, V_i is passed to the output. Pass variables can take the values $\{0, 1, X_i, -X_i, Z\}$, where X_i and $-X_i$ are the true and complement of the ith input variable and Z is the high-impedance state. Design of pass-transistor networks using a Karnaugh map involves constructing the cells in the Karnaugh map in the normal manner. For instance, in the case of a 2-input XNOR gate, the truth table is shown in Table 5.6.

The pass-network Karnaugh map is augmented with the possible pass variables that can be passed to the output to yield the function. This is shown in Table 5.7.

Now instead of grouping '1's, as one would with a normal-logic gate, any variable may be cast as a pass variable or control variable and grouped together. For instance, in the above Karnaugh map, by grouping the $-B$ columns when A is 0, and the B columns when A is 1, the function could be implemented, using A as a control variable and B as a pass variable, as

$$F = -A(-B) + A(B).$$

This may be implemented using complementary switches or n-transistors as shown in Figs. 5.34(a) and 5.34(b). In this circuit when the control variable A is true, the pass variable B is passed to the output. When A is false, $-B$ is passed to the output. The circuit in Fig. 5.34(c) may also be used for an XNOR/XOR function. It is left as an exercise to cast this in terms of control and pass variables.

TABLE 5.6 XNOR Truth Table

A	B	$\overline{A \oplus B}$	PASS FUNCTION
0	0	1	$-A + -B$
0	1	0	$A + -B$
1	0	0	$-A + B$
1	1	1	$A + B$

TABLE 5.7   Modified Karnaugh Map

		A		A	
		0		**1**	
B	**0**	−A		−A	
			−B		B
	1	A		A	
			−B		B

Each cell in the Karnaugh map must be covered by the resulting expression. Note that groupings that pass both true and false input variables to the output are not allowed to prevent undefined states. In addition, if a complementary implementation is required, the p-pass function that is the dual of the n-structure must also be constructed. In a complementary version, pass variables of value 0 require only an n-transistor network, while pass variables of 1 only require a p-transistor network.

A popular use of pass-transistor logic is in the construction of a Boolean function unit, shown in Fig. 5.35.[36,37] This implements all Boolean combinations of inputs A and B, depending on the function inputs P_4–P_1 as summarized in Table 5.8. For instance Table 5.9 illustrates some of the functions that may be implemented.

The nMOS-only structure is shown in Fig. 5.35(a). In CMOS, this structure can be replicated, as shown in Fig. 5.35(b), by using a full-transmission gate for each original n-transistor. A more realizable layout is possible by using the circuit shown in Fig. 5.35(c). This alleviates many direct n- to p-transistor connections. A dynamic version is shown in Fig. 5.35(d). In terms of speed, the nMOS version has the fastest fall time and the comple-

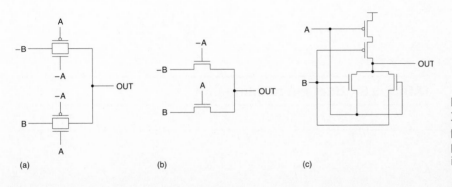

(a)   (b)   (c)

FIGURE 5.34   Two-input XNOR gate implemented in pass-transistor logic: (a) complementary; (b) single-polarity; (c) cross-coupled

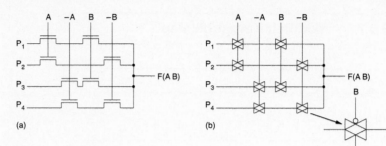

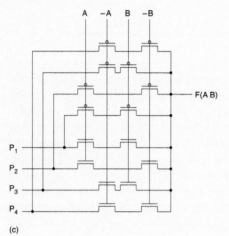

FIGURE 5.35 Boolean Function Unit: (a) an nMOS structure; (b) a full-transmission-gate implementation; (c) a complementary version with improved layout; (d) a dynamic (or static) version (static with feedback p-transistor connected)

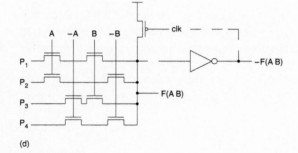

TABLE 5.8 Boolean function unit

		B	
		0	1
A	0	P_4	P_3
	1	P_2	P_1

TABLE 5.9 Some Functions Implemented by the Boolean Function Unit

OPERATION	P1	P2	P3	P4
NOR (A,B)	0	0	0	1
XOR (A,B)	0	1	1	0
NAND (A,B)	0	1	1	1
AND (A,B)	1	0	0	0
OR (A,B)	1	1	1	0

mentary version has the fastest rise time. Using larger p-transistors decreases the rise time but increases the fall time. The dynamic version is roughly the same speed as the nMOS version but requires a precharge period that may extend clock-cycle times. An alternative to the dynamic approach is to include a buffer that is fed back to a p-transistor pull-up. This then yields a static gate with zero DC power dissipation. The p-transistor pull-up and n-transistor pull-downs must be ratioed to allow the output buffer to switch (over all process corners). In this case and in the precharge case, the terms in the logic expression do not need to guarantee a '1' at the output of the gate because the p pull-up ensures this condition if no n pull-downs are turned on.

The apparent advantages of pass-transistor networks in CMOS should be studied carefully and judiciously utilized. A few points detract from the use of pass networks. To achieve good logic levels, complementary pass-networks are desirable but incur extra pull-down delays. In comparison with regular gates, the merging of source and drain regions is difficult, leading to higher internal node capacitances. Finally, true and complement control variables are required. The best use that can be made of such networks is when an efficient pass-gate structure can be found for a particular function. The Boolean function unit is a good example of this kind of a structure. Transmission gates can have a significant speed advantage when a few stages are cascaded in a circuit block. The delay characteristics are some-what like the *RC* delay lines treated in Chapter 4, where there is a square-law relationship between the number of stages and the delay. The number of transistors may be reduced if the output nodes can be precharged, or if the static implementation, as illustrated in the function block implementation shown in Fig. 5.35(d), can be used. Note that the complementary versions might be of use with preplaced transistor sites such as those found in gate arrays. The effectiveness of any pass-transistor network must be assessed for any given situation by simulation and layout. Note that the pass networks derived here may be used with the CVSL logic covered in Section 5.4.9. Pass networks, due to their lower stray capacitance, have a good future in low-power, high-performance systems.

5.4.7 CMOS Domino Logic

A modification of the clocked CMOS logic allows a single clock to precharge and evaluate a cascaded set of dynamic logic blocks. This involves incorporating a static CMOS inverter into each logic gate, as shown in Fig. 5.36(a).[38] During precharge ($clk = 0$), the output node of the dynamic gate is precharged high and the output of the buffer is low. As subsequent logic stages are fed from this buffer, transistors in subsequent logic blocks will be turned off during the precharge phase. When the gate is evaluated, the output will conditionally discharge, allowing the output of the buffer to conditionally go high. Thus each gate in sequence can make at most one transition (1 to 0). Hence, the buffer can only make a transition from 0 to 1. In a cascaded set of logic blocks, each state evaluates and causes the next stage to evaluate—in the same way a line of dominos fall. Any number of logic stages may be cascaded, provided that the sequence can evaluate within the evaluate clock phase. A single clock can be used to precharge and evaluate all logic gates within a block.

Some limitations are evident with the structure. First, each gate must be buffered (maybe this is an advantage!). Second, only noninverting structures are

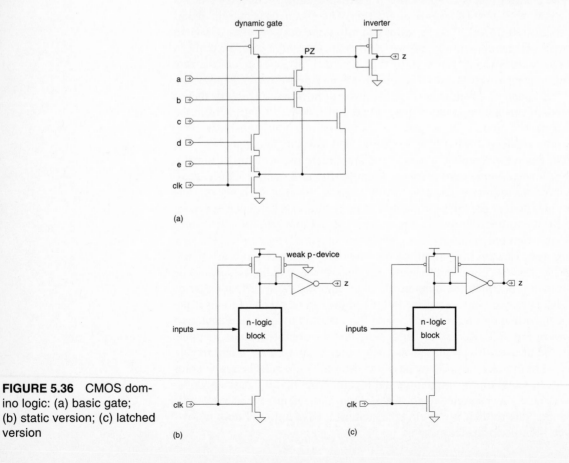

FIGURE 5.36 CMOS domino logic: (a) basic gate; (b) static version; (c) latched version

possible. Finally, in common with all-dynamic–CMOS, charge redistribution can be a problem. Depending on the situation, the effect of these problems can be minimized. For example, in complex logic circuits, such as arithmetic logic units, the necessary XOR gates may be implemented conventionally (as complementary gates) and driven by the last domino circuit. The buffer is often needed from circuit-loading considerations, and would be needed in any case.

The domino gate may be made static by including a weak p-transistor, as shown in Fig. 5.36(b). A weak p-transistor is one that has low gain (small W/L ratio). It has to have a gain such that it does not fight the pull-down transistors, yet can balance the effects of leakage. This will allow low frequency or static operation when the clock is held high. In this case, the pull-up time could be an order of magnitude slower than the pull-down speed. In addition, the current drawn by the gate during evaluation should be small enough that the static power dissipation of a circuit would not be impacted. Note that the precharge transistor may be eliminated if the time between evaluation phases is long enough to allow the weak pull-up to charge the output node (i.e., a switched pseudo-nMOS gate). The inclusion of the weak p pull-up does little to aid high-frequency operation because the transistor does not have enough time to operate. In addition, the additional capacitance can slow the gate. Note that the gate may also be made latching by including a weak p feedback transistor, as shown in Fig. 5.36(c). Figure 5.37 shows what can happen if intermediate nodes are precharged inappropriately. In Fig. 5.37(a), the clocked n-transistor has been placed closest to the output. Thus if capacitances C_2–C_7 are charged low, input A_0 is low and inputs A_{5-1} are high, upon asserting the clock, the charge stored in C_1 is dumped into C_2–C_7. Depending on the ratio of the capacitances, this level could erroneously go low, thus triggering the output inverter. If n_1 is charged to V_{DD}, then the voltage after evaluation is given by

$$V_{n_1} = \frac{C_1}{\sum\limits_{i=2}^{7} C_i + C_1} V_{DD}.$$

If $C_1 = 3 \times C_2$ and $C_2 = C_3 = C_4 = C_5 = C_6 = C_7$ then

$$V_{n_1} = \frac{3C_2}{7C_2 + 2C_2} V_{DD}$$

$$= .3 V_{DD}$$

$$= 1.5V,$$

which is below the threshold of the buffering inverter.

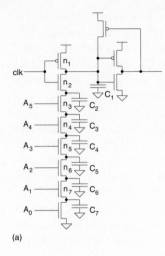

(a)

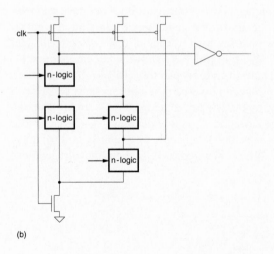

(b)

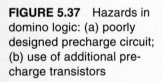

FIGURE 5.37 Hazards in domino logic: (a) poorly designed precharge circuit; (b) use of additional precharge transistors

The solution here is to place the clocked n-transistor at the bottom of the AND tree. Another example is shown in Fig. 5.37(b), where intermediate nodes in a complex domino gate have been provided with their own precharge transistor.

5.4.8 NP Domino Logic (Zipper CMOS)

A further refinement of the domino CMOS is shown in Fig. 5.38(a). Basically, the domino buffer is removed, while cascaded logic blocks are alternately composed of p- and n-transistors.[39,40,41] In the circuit in Fig. 5.38(a), when *clk* = 0, the first stage (with n-transistor logic) is precharged high. The

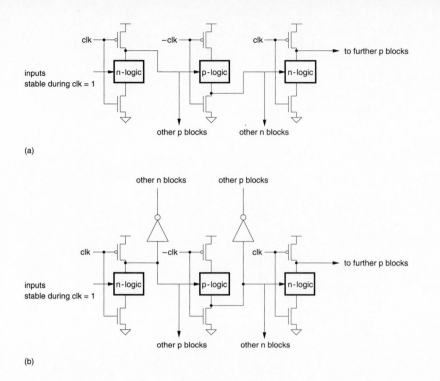

(a)

(b)

FIGURE 5.38 NP domino logic: (a) basic gate; (b) domino connections of an NP domino gate

second stage is precharged low and the third stage is precharged high. As the second logic stage is composed of p-transistors, these will all be turned off during precharge. Also, as the second stage is precharged low, the n-transistors in the third logic state will be off. Domino connections are possible, as shown in Fig. 5.38(b).

Common advantages of the dynamic logic styles are as follows:

- Smaller area than fully static gates.
- Smaller parasitic capacitances, hence higher speed.
- Glitch free operation if designed carefully.

The last point is the catch. If you want to use dynamic circuits, you must be prepared to invest the extra design effort to ensure correct operation under all circuit conditions (process corners, timing sequences, noise sensitivity).

5.4.9 Cascade Voltage Switch Logic (CVSL)

The basic form of this style of CMOS logic is depicted in Fig. 5.39(a).[42] It is a differential style of logic requiring both true and complement signals to be routed to gates. Two complementary nMOS switch structures are constructed and then connected to a pair of cross-coupled p pull-up transistors.

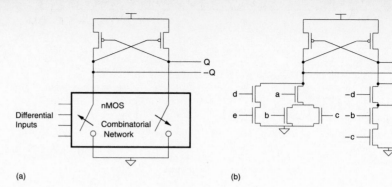

(a)

(b)

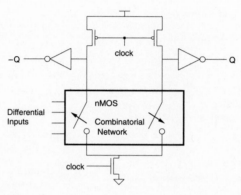

(c)

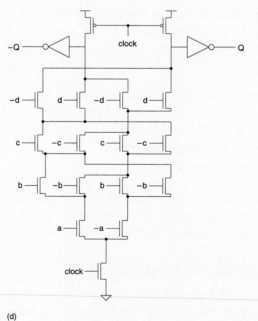

FIGURE 5.39 Cascade voltage switch logic: (a) the basic gate; (b) a particular function; (c) clocked version; (d) a four-way XOR gate implemented in CVSL

(d)

When the inputs switch, nodes Q and $-Q$ are pulled either high or low. Positive feedback applied to the p pull-ups causes the gate to switch. The logic trees may be further minimized from the full differential form using logic minimization algorithms. This version, which might be termed a "static" CVSL gate, is slower than a conventional complementary gate employing a p-tree and n-tree. This is because during the switching action, the p pull-ups have to "fight" the n pull-down trees. Figure 5.39(b) shows the implementation of the example gate. In isolation, this is not a very efficient implementation of this gate; however, in certain cases, such as multiple input XOR gates, the implementation is quite reasonable.

Further refinement leads to a clocked version of the CVSL gate (Fig. 5.39c). This is really just two "domino" gates operating on true and complement inputs with a minimized logic tree. The advantage of this style of logic over domino logic is the ability to generate any logic expression, making it a complete logic family (as noted in Section 5.4.7, domino logic can only generate noninverted forms of logic). This is achieved at the expense of the extra routing, active area, and complexity associated with dealing with double-rail logic. However, the ability to generate any logic function is of advantage where automated logic synthesis is required. A four-way XOR gate is shown in Fig. 5.39(d).[42] The performance of the dynamic CVSL gate may be improved with the addition of a latching sense amplifier as shown in Fig. 5.40.[43] This variation is called Sample-Set Differential Logic (SSDL). It works slightly differently from dynamic CVSL. When $clk = 0$, P_1, P_2, and N_1 are turned on. One output will be at V_{DD} and the other will be slightly below V_{DD} because a path exists to V_{SS} through one of the n trees. When $clk = 1$, the latching sense amplifier forces the lower output

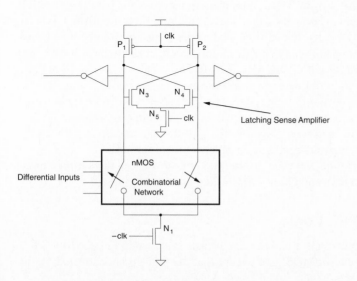

FIGURE 5.40 A latching sense amplifier for use with dynamic CVSL (SSDL logic)

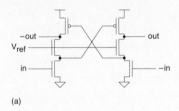

(a)

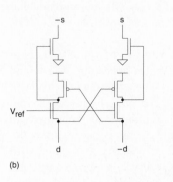

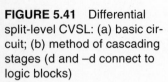

FIGURE 5.41 Differential split-level CVSL: (a) basic circuit; (b) method of cascading stages (d and –d connect to logic blocks)

(b)

quickly to V_{SS}. Thus the pull-down time is now determined by a single pull-down rather than the series connection pull-downs as in conventional CVSL.

Another form of CVSL logic is shown in Fig. 5.41.[44] Here the p-load is replaced with cross-coupled current-controlled cascoded n- and p-transistors. The complementary n pull-down trees are retained in common with CVSL. This form is called a *differential split-level gate*. A reference voltage, V_{ref}, is set to an n threshold above $V_{DD}/2$. This sets the voltage at the n pull-downs to roughly $V_{DD}/2$. The complementary nodes at the p pull-ups are full V_{DD} level signals. The gate gains speed by having the n pull-downs only pull-down $V_{DD}/2$ volts. In addition, as the p pull-ups are isolated by the cascode n-transistors, they can be made larger, thus increasing the pull-up speed from the conventional CVSL gate. Finally, by repartitioning the gate and using open-drain outputs, the reduced voltage swing nodes can be passed between gates, thus improving speed. The logic style was originally invented to allow the use of n pull-downs with smaller gate dimensions than the cascode n-transistor and p pull-ups. As the pull-downs operate at a lower V_{ds} potential "hot electron" reliability problems are minimized.

Design techniques for the differential pull-down structures utilize modified Karnaugh map and Quine-McCluskey tabular methods.[45]

5.4.10 SFPL Logic

A Source Follower Pull-up Logic (SFPL) gate is shown in Fig. 5.42(a).[46] It is similar to a pseudo-nMOS gate except that the pull-up is controlled by the

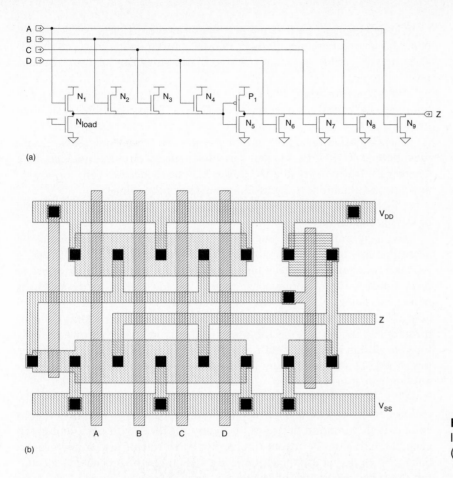

(a)

(b)

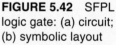

FIGURE 5.42 SFPL logic gate: (a) circuit; (b) symbolic layout

inputs. In turn this can lead to the use of smaller n pull-downs. The gate of the p load is driven by a parallel source follower, consisting of drive transistors N_1–N_4 and load transistor N_{load}. The combination of any driver turning on and the load are ratioed to provide about 2–3 volts at the input of the inverter formed by P_1 and N_5. This voltage tends to turn P_1 off, which allows smaller n pull-downs to be used. This reduces the self-loading of the output and improves the speed of the gate. The gate style shows a marked advantage in high fan-in gates, albeit at some DC power cost. Figure 5.42(b) shows a typical symbolic layout of the gate in Fig. 5.42(a), which demonstrates a compact layout.

5.4.11 Summary

A large number of additional options have been presented in this section. Where should one use what gate?

Complementary logic is the best option in the majority of CMOS circuits. It is noise-immune, dissipates no DC power, and is fast, and its creation may be highly automated. Large fan-in gates can lead to excessive levels of logic.

BiCMOS gates should be used in mixed-signal situations or perhaps in high-speed applications over finer-line CMOS if the economics justify the use.

Pseudo-nMOS logic is of most utility in large fan-in NOR gates. Examples include ROMs, PLAs, and carry look-ahead circuits in adders (see Chapter 8). If necessary, the DC power may be reduced to zero for power-down or test situations by controlling the gate of the p-load.

Clocked CMOS logic is of possible benefit in "hot electron"-susceptible processes and conditions.

By using transmission gate logic, significant speed advantage may be accrued if structures are limited to a few series-transmission gates. The style is of use for complex Boolean functions where the size and/or power has to be minimized. There is not much commercial CAD support for the synthesis of transmission gate designs. One should compare density, speed, power, and ease of design of any circuit designed with that of the corresponding complementary CMOS circuit to justify use. (In other words, do your own analysis for your design problem. Don't rely on generalized examples.) Sometimes low-threshold n-transistors are provided in processes specifically for transmission-gate use.

CMOS domino logic should be used for low-power or high-speed applications. Be careful of charge redistribution effects. If you do not wish to exhaustively simulate the gates at the circuit level with back-annotated capacitances from the layout (including the effects of power and ground bounce) then do not use them. Remember that often the precharge time will rob the speed advantage over static designs in poorly designed clocking schemes. Many novices (and pros too!) have been caught by not understanding all of the problems that can arise when this logic is used.

CVSL logic is potentially of use in fast gates using cascode CVSL or SSDL. The gates are generally synthesizable, which may be an advantage in some CAD environments. Size, design complexity, and reduced noise immunity may lead one to avoid this logic family. Some designers regard the logic highly while others are still seeking a use for it.

Many times it may be possible to create a hybrid gate that merges two of the styles of logic design covered in Section 5.2. Chapter 8 illustrates this with respect to an adder carry-chain by mixing static logic and transmission gates.

As a general rule, the more you can make a gate look like an inverter, the faster it will operate. Small numbers of cascaded transmission gates are also fast.

Remember, you have a set of switches with which to implement a given logic function. In certain cases it may be worthwhile to spend some time to compose and optimize such a hybrid gate.

5.5 Clocking Strategies

5.5.1 Clocked Systems

In this chapter we have discussed various alternative forms of CMOS logic. Although we have studied logic gates in isolation, no global clocking strategy has been suggested. Virtually every useful VLSI system must store some state, implying some form of storage elements. As an example, Fig. 5.43(a) shows what is termed a finite-state machine (FSM) which is composed of a set of logic inputs feeding a block of combinational logic resulting in a set of logic outputs. Some of the outputs are fed back to the inputs via storage devices that are clocked by a system clock (or clocks). The machine operates by determining the "next state" as a function of the "current state" and the external inputs. The outputs are a function of the "current state" and perhaps the external inputs. When the clock transitions (let us assume 0→1), the

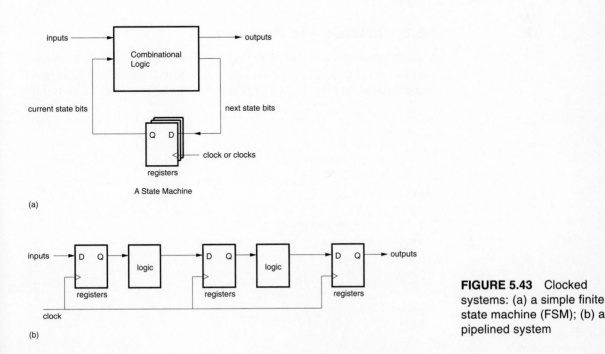

FIGURE 5.43 Clocked systems: (a) a simple finite state machine (FSM); (b) a pipelined system

"next state" bits are transferred to the "current state" bits and the "current state" bits and inputs trickle through the combinational logic to the outputs and the "next state" bits. When the outputs are stable the system may be clocked again. The minimum time in which the outputs and "next state" bits settle determines the maximum frequency that the clock may operate. The design of FSMs is covered in Chapter 8.

A pipelined system is shown in Fig. 5.43(b). Pipelined systems use storage devices to capture the output of each processing stage at the end of each clock period, and in general have no feedback. The majority of VLSI systems are a combination of pipelined and finite-state machines. More examples of pipelined systems may be found in Chapter 8.

The storage devices used in FSMs or pipelined systems are in turn defined in terms of a set of clock waveforms used to store and access the state of each storage element. The selection of a particular clocking strategy influences how many transistors are used per storage element and how many clock signals need to be routed throughout the chip. These decisions impact the size of the chip and the power dissipated by the chip. Hence, one of the most important decisions that may be made at the commencement of a design is the selection of the clocking strategy. In this section we will first examine the use of a single clock and then explore multiphase clocking techniques. Suitable memory and logic elements for each clocking strategy will be summarized. Some layout guidelines are also given.

5.5.2 Latches and Registers

A single-phase clock is shown in Fig. 5.44, in conjunction with the timing waveforms for a storage element called a positive edge-triggered register (sometimes called a flip-flop). The behavior of the register is as follows: If the

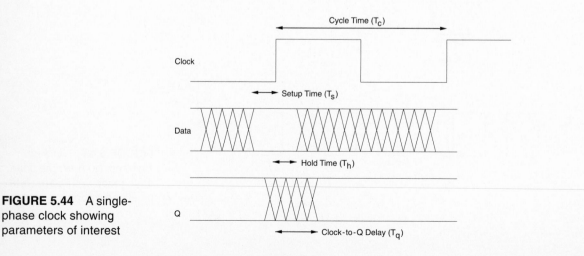

FIGURE 5.44 A single-phase clock showing parameters of interest

signal at the data input (commonly called the D input) is stable within a window around the positive transition of the clock, then some time later that D value will propagate to the output of the register (commonly called the Q output). The time before the clock edge that the D input has to be stable is called the *setup time* (T_s) and the time after the clock edge that the D input has to remain stable is called the *hold time* (T_h). The delay from the positive clock input to the new value of the Q output is called the *clock-to-Q delay* (T_q). The time between successive positive clock transitions is called the *cycle time* (T_c).

The level-sensitive latch. The first step in building an edge-triggered register is to build a level-sensitive latch. In Chapter 1, a latch storage element was developed using two inverters and a multiplexer. This structure is shown in Fig. 5.45(a), with one inverter merged into an inverting mux. This is a negative level-sensitive latch because the D input is passed to the output when the clock (*clk*) is low. The D input must be stable for a short time before and after the positive clock transition. Note that this is not an edge-triggered storage element because the output changes in sympathy with the input while the clock is low. A positive level-sensitive latch is shown in Fig. 5.45(b). The Q output reflects the input when the clock is high.

The edge-triggered register. By combining two level-sensitive latches, one positive-sensitive and one negative-sensitive, a designer can construct an edge-triggered register as shown in Fig. 5.45(c). By convention the first latch stage is called the *master* and the second is called the *slave*.

While the clock is low, the master negative level-sensitive-latch output (QM) follows the D input while the slave positive-latch holds the previous value. When the clock transitions from 0 to 1, the master latch ceases to sample the input and stores the D value at the time of the clock transition. The slave latch opens, passing the stored master value (QM) to the output of the slave latch (Q). The D input is prevented from affecting the output because the master is disconnected from the D input. When the clock transitions from 1 to 0, the slave latch locks in the master-latch output and the master starts sampling the input again. This sequencing is shown in Fig. 5.45(d).

Thus this device is a positive edge-triggered register (also called a D register or D flip-flop) by virtue of the fact that it samples the input at the rising edge of the clock. By reversing the latch polarities, a negative edge-triggered register may be constructed. Figure 5.45(e) shows a CMOS implementation of a D register. Apart from the D latch/register a number of other storage circuits are popular.

The RS latch. The reset-set, or *RS*, latch is shown in Fig. 5.46. It consists of cross-coupled NAND (or NOR) gates. In the implementation shown, Q changes to 1 when S is 1, and changes to 0 when R is 1. Q is undefined for $S = 1$ and $R = 1$. The latch maintains its state for $S = 0$ and $R = 0$.

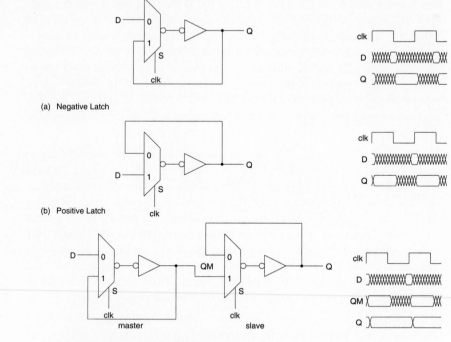

(a) Negative Latch

(b) Positive Latch

(c) Positive edge-triggered register (single-phase clock)

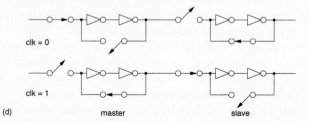

(d)   master   slave

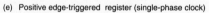

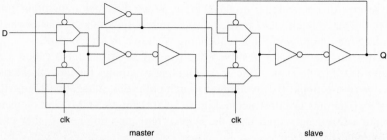

(e) Positive edge-triggered register (single-phase clock)

FIGURE 5.45 Static CMOS latches and registers: (a) negative level-sensitive latch; (b) positive level-sensitive latch; (c) positive edge-triggered register; (d) operation of register; (e) CMOS circuit implementation

The T register. A toggle register, or "*T* latch," toggles the *Q* output as the *T* input is varied. When *clk* = 1, the *Q* output is complemented, whereas when *clk* = 0, *Q* is unchanged. This gives a divide-by-2 function. A *T* latch which is shown in Fig. 5.47 may be used as a low-gate-count counter element (it is in fact a *D* register, with –*Q* tied to the *D* input). However, the counter bits ripple through the cells, which may be undesirable and, if not resettable, may pose testability problems (see Chapter 7).

The JK register. A *JK* register is shown is Fig. 5.48. When *J* and *K* are 0, the *Q* output is maintained. If *K* = 1, *J* = 0, *Q* is set to 0; when *K* = 0, *J* = 1,

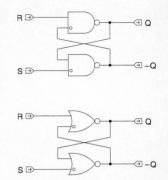

FIGURE 5.46 An *RS* latch (a) NAND; (b) NOR

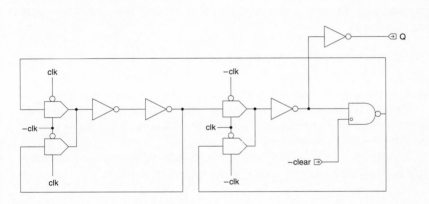

FIGURE 5.47 A *T* register

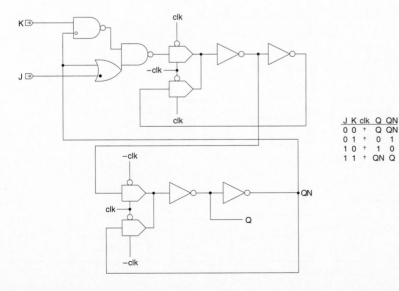

J	K	clk	Q	QN
0	0	↑	Q	QN
0	1	↑	0	1
1	0	↑	1	0
1	1	↑	QN	Q

FIGURE 5.48 A *JK* register

Q is set to 1. When J and K are both set to 1, the output Q toggles. The JK register is in essence the combination of an RS and a T latch.[47] The JK register was a popular TTL structure because, when combined with an asynchronous set and reset, it combines set, reset, T-register and D-register functionalities. For this reason it has carried over to gate-array libraries. Today the D register or a register combined with multiplexers is almost always used as a replacement.

5.5.3 System Timing

Latches and registers may be used in a variety of ways to implement clocked systems. For instance, Fig. 5.49(a) shows a typical pipelined system with input and output registers separated by combinational logic. Here the cycle time is given by

$$T_c = T_q + T_d + T_s \tag{5.7}$$

where T_d is the worst-case delay through the combinational logic block.

Pipeline registers are placed in series with the logic flow to attain some desired cycle time in situations where the overall logic delay, T_d, is greater than the required cycle time, T_c.

An alternate structure is shown in Fig. 5.49(b) where the combinational logic is bounded by latches. This structure can be extended, as shown in

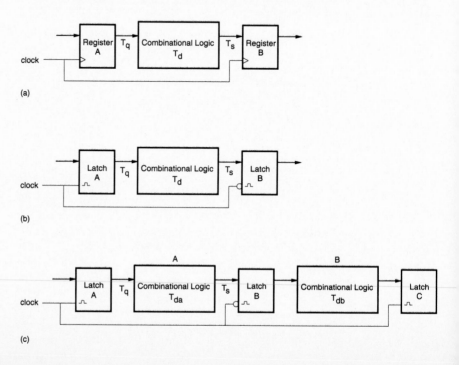

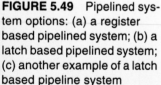

FIGURE 5.49 Pipelined system options: (a) a register based pipelined system; (b) a latch based pipelined system; (c) another example of a latch based pipeline system

Fig. 5.49(c), by alternating positive and negative level-sensitive latches with combinational logic. A saving has been made in the number of latches per combinational block compared with the registered configuration in Fig. 5.49(a). However, in the latch case the logic at the output of latch A receives data T_q later than the $0 \rightarrow 1$ transition and must deliver it to latch B T_s before the $1 \rightarrow 0$ transition of the clock. Thus the logic delay of block A (T_{da}) must satisfy the following inequality:

$$T_{da} < T_{c1} - T_{qa} - T_{sb} \text{ (assuming a 50\% duty cycle)} \qquad \textbf{(5.8)}$$

where

T_{qa} = the clock-to-Q time of latch A

T_{sb} = the setup time of latch B.

Similarly,

$$T_{db} < T_{c0} - T_{qb} - T_{sa}. \qquad \textbf{(5.9)}$$

In the limit $T_{da} = T_c/2 - T_{qa} - T_{sb}$ and $T_{db} = T_c/2 - T_{qb} - T_{sa}$

$T_{c1} = T_{da} + T_{qa} + T_{sb}$

and,

$T_{c0} = T_{db} + T_{qb} + T_{sa}$

$T_c = T_{da} + T_{db} + [2(T_q + T_s)]$ (assuming that latch A and B are identical).

The register pipelining strategy is the simplest to think about because it is edge based; that is, all state changes occur at the rising (or falling) clock edge.

5.5.4   Setup and Hold Time

The setup and hold time of a register are deviations from an ideal register caused by finite circuit delays. The hold time relates to the delay between the clock input to the register and the storage element. That is, the data has to be held for this period while the clock travels to the point of storage. The setup time is the delay between the data input of the register and the storage element. As the data takes a finite time to travel to the storage point, the clock can not be changed until the correct data value appears. In most CMOS registers, these delays are very small and each type of register has characteristic setup and hold times due to the circuit construction.

In a synchronous system, if the data input to a register does not obey the setup and hold-time constraints, then potential *clock race* problems may occur. These races result in erroneous data being stored in registers. For instance, imagine that the data violates the hold-time violation. The data changes to a new value before the clock can change. With a setup time constraint the clock changes before the data assumes the correct value.

Assuming a perfectly synchronous system with perfect clocks, zero hold-time registers, and clock-to-Q time greater than the setup time, no clock race problems should occur. However, at the chip level this might be hard to ensure. Consider the block diagram shown in Fig. 5.50 where two modules are interconnected. Here a delay has been included in series with the data and the clock lines to the modules. The earliest that data appears at the input of register M_2 is at time $T_{c1} + T_{q1}$, assuming zero delay in the logic block. The clock appears at register M_2 at time T_{c2}. Assuming zero internal setup and hold times in the registers, if T_{c2} lags the data change ($T_{c2} > (T_{c1} + T_{q1})$), the module M_2 will store the data from the current cycle rather than the previous cycle. This is a hold-time violation and may be caused in practice by T_{c1} and T_{q1} being close to zero while a delay is introduced into the T_{c2} clock line. This might be due to *RC* delay or clock-buffer delay. If the delay $(T_{c1} + T_{q1}) - T_{c2}$ is larger than the cycle time, T_c, then the data will arrive late at M_2. This will cause a setup-time violation. This occurs when the circuit is too slow for the clock cycle used. While T_{c2} may be artificially increased to allow more time for the data to set up, the constraint $T_{c2} < (T_{c1} + T_{q1})$ becomes harder to meet and data delays may have to be artificially added to meet the constraint. In general, this type of temporal tight-rope walking should be avoided in products where time to market is short and first-time correctness is important. A recent style of design called *wave pipelining* takes this to the extreme by using the delay of logic stages as the delay elements in the circuit.[48]

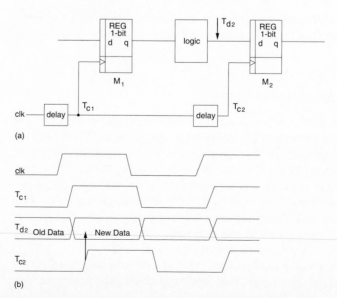

FIGURE 5.50 Clock skew and the relation to setup and hold times

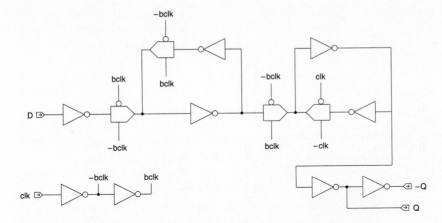

FIGURE 5.51 A typical gate array or standard-cell edge triggered *D* register

5.5.5 Single-Phase Memory Structures

The simplest clocking methodology is to use a single clock in conjunction with the register shown in Fig. 5.51. In this register, the necessary clocks are locally generated within the register and the *Q* and *–Q* are buffered. This is the normal kind of register used in gate arrays and standard cell designs. In custom designs it is desirable to reduce the number of transistors in the basic register. The clock buffers are candidates to be replaced by a global clock buffer supplying *bclk* and *–bclk*. However, because these signals now travel across the chip and may be heavily loaded, they may develop a skew in relation to each other. Consider *–clk* delayed with respect to *clk* as shown in Fig. 5.52(a). We see that the first-transmission-gate n-transistor can be turned on at the same time as the second-transmission-gate n-transistor. Hence the value on the input

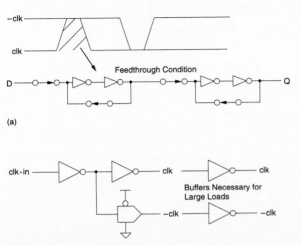

FIGURE 5.52 Clock skew in a *D* register: (a) effect; (b) balanced delay clock driver

can ripple through the two transmission gates, leading to invalid data storage. This problem means that close attention must be paid to the clock distribution to minimize the clock skew. One method for achieving this is shown in Fig. 5.52(b).[49] A conventional clock buffer consisting of two inverters is shown. The buffered true clock will always be delayed with respect to the buffered inverted clock. To reduce this undesirable delay, the *clk* signal may be passed through a transmission gate to equalize delay with respect to *–clk*. The transmission gate should use similar-sized (though slightly smaller) transistors as those used in the inverters (verify this by simulation). Subsequent buffers may also be used to keep the initial buffers small. Routing load must also be balanced between the two phases of the clock. Another method of avoiding excessive clock skews is to use a local buffer for every n-bit register.

Realizing that single-phase registers are composed of two latches operating on complementary clocks, a variety of latches will be reviewed (Fig. 5.53). The feedback transmission gate may be eliminated by using a weak trickle inverter as the feedback inverter, as shown in Fig. 5.53(a). Here the trickle

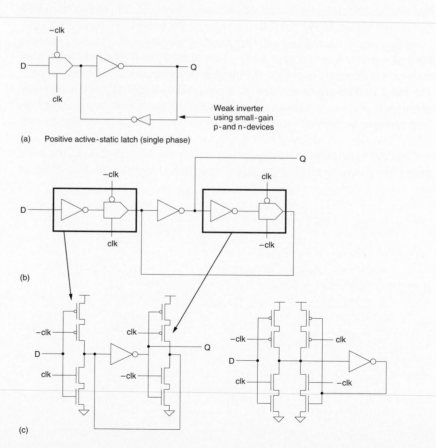

FIGURE 5.53 Various CMOS static latches: (a) a "jamb" latch; (b) a transmission-gate latch; (c) a tristate-buffer latch with the clocks at the center of the tristates

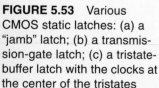

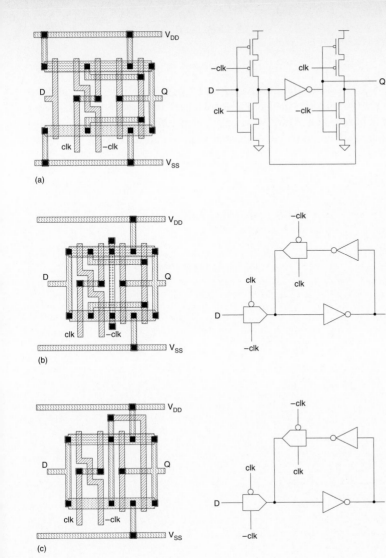

FIGURE 5.54 Typical latch symbolic layouts

inverter is constructed with low-gain n- and p-transistors. This is achieved by employing transistors with a length (*L*) greater than the minimum value. Alternatively, the *W/L* of the driving transmission-gate transistors may be made larger than those in the feedback inverter. The transmission gate (and associated source-driver circuitry) must be capable of overdriving the trickle inverter for all process corners and circuit conditions. When the transmission gate is turned off, the trickle inverter locks in the stored state in the latch.

Figure 5.53(b) shows the latch in Fig. 5.45 with a buffering input inverter. By eliminating the connections at the confluence of the inverter and the transmission gate, the latch in Fig. 5.53(c) may be constructed without loss of function. This eliminates a metal connection, yielding a smaller latch. Figure 5.54 illustrates some symbolic layouts for some of the latches treated so far.

327

Figure 5.55(a) shows latches based on a CVSL structure. An n and a p version are shown that are cascaded to form a register. Figure 5.55(b) shows a latch that is based on a static RAM cell. These latches are complementary in nature, have reduced noise margin, and require careful design. However,

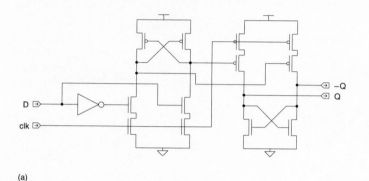

(a)

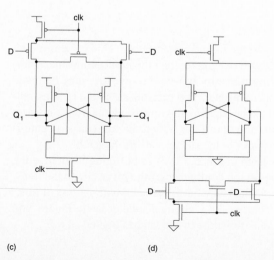

(b)

FIGURE 5.55 More static registers: (a) a fast static CVSL style register; (b) a latch based on a RAM cell; (c) a "double-edge" triggered register

(c)

(d)

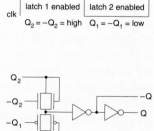

(e)

they are small and can be very fast. Figure 5.55(c) shows latches that may be used to clock data on both edges of the clock—so called Double-Edge-Triggering.[50,51]

Two final designs are shown in Fig. 5.56(a) and Fig. 5.56(b). The first design uses only one clock phase and a gated RS latch. It uses 14 transistors, compared with 8–10 transistors in the preceding flip-flops. The second design shows this circuit extended to build a master-slave design with set and reset.

It is often desirable (and in many design styles mandatory) to be able to set the state of a storage element. This can be done either synchronously with the clock (i.e., by loading a zero into a loadable register) or asynchronously (i.e., without regard to the state of the clock). Asynchronous settable and resettable registers are shown in Fig. 5.57. Smaller versions of these registers may be built by "jambing" the state of the master via a single transistor

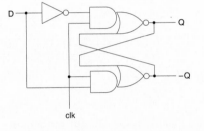

Positive active-static latch (single phase)

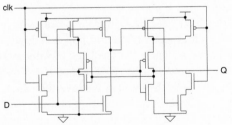

(a)

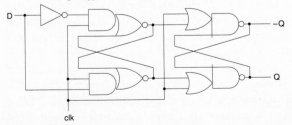

Positive edge-triggered static register (single phase)

(b)

FIGURE 5.56 Logic gate based latches: (a) positive level-sensitive latch; (b) positive edge-triggered register

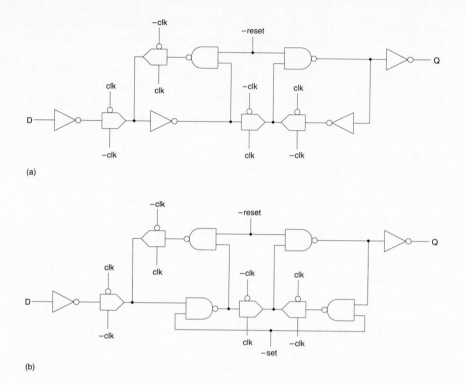

FIGURE 5.57
Asynchronously settable
and resettable registers

connected to power or ground. This would usually draw DC current and in some circumstances may be undesirable.

So far the latches and registers that have been described have been static; that is, they store their state when the clock is stopped and power is maintained. To reduce the number of transistors in a latch, the feedback inverter and transmission gate may be eliminated, as shown in Fig. 5.58(a) and (b). Now the latched value is stored on the capacitance of the input of the inverter, which is composed predominantly of gate capacitance. A word of caution here: The clock-to-Q delay in this style of latch can be very small and the designer must be very careful to ensure that the latches are not transparent.[52] Particular attention must be paid to providing the latches with sharp antiphase clocks, particularly if used in shift registers where there is no logic between storage elements. Internal inversion of the clock is often the only way of ensuring these clock constraints at high clock speeds (>40 MHz). The dynamic latch with a "tristate" inverter is shown in Fig. 5.58(c). The corresponding registers are shown in Figs. 5.58(d) and 5.58(e).

The requirements of implementing a single clocked register with the minimum number of transistors has led to the structures shown in Fig. 5.59. Figure 5.59(a) shows a representative clocking method used in DEC's ALPHA microprocessor.[53] It consists of a latch L_1 which is transparent

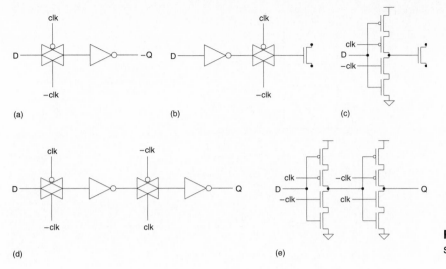

(a)   (b)   (c)

(d)   (e)

FIGURE 5.58 Dynamic single clock latches

when the clock is high and opaque when the clock is low. A complementary latch L_2 is transparent when the clock is low and opaque when the clock is high. Logic is interposed between L_1 and L_2 and between L_2 and L_1 similar to the clocking scheme shown in Fig. 5.49(c). Specific implementations of an L_1 latch are shown in Fig. 5.59(b), while implementations of an L_2 latch are shown in Fig. 5.59(c). In Figs. 5.59(b) and 5.59(c), unbuffered and buffered versions of the latches are shown. The operation of the unbuffered L_1 latch in Fig. 5.59(b) is as follows: When *CLK* rises, node X is either pulled low (D high) through N_2 and N_1 or high (D low) through P_1. Transistors N_3, N_4, and P_2 similarly act as an inverter to produce Q from node X. When the clock is low, transistors N_2 and N_4 are turned off. Assuming D was high when the clock was high, then X is initially low and Q is high. With the clock low, a high-to-low transition on D causes X to go high, which turns P_2 off, holding the value at the Q output. If D was low when the clock is asserted, then X is high and Q is low. If D transitions high when the clock is low, P_1 is turned off, holding node X high (tristated). Thus Q is in turn held low. The P_3 feedback transistor is added to counteract noise sources and leakage that tend to reduce the voltage on node X.

The P_2 latches shown in Fig. 5.59(c) operate similarly with N_3 providing the feedback function. Figure 5.59(d) and 5.59(e) show the latches without feedback transistors while Fig. 5.59(f) and 5.59(g) show registers implemented using these latches. These are the versions as proposed originally by Yuan and Svensson.[88] Logic may be integrated into the first stage of the latches as shown in Fig. 5.59(h) and Fig. 5.59(i). In Fig. 5.59(h) transistors N_1 and N_5 form the pull-down path of a two-input buffered NAND gate,

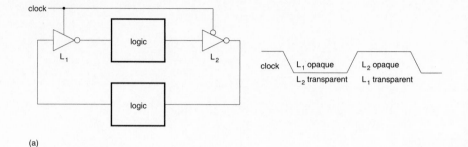

(a)

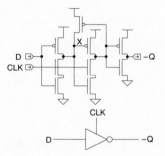

(b)

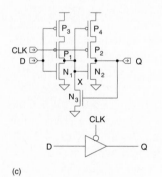

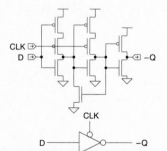

(c)

FIGURE 5.59 Single-phase dynamic latch clocking: (a) clocking method; (b) clock active high latches; (c) clock active low latches; (d) latch b) without feedback and buffer; (e) latch c) without feedback and buffer; (f) register (positive edge); (g) register (negative edge); (h) NAND gate/ latch combination; (i) OR gate/latch combination

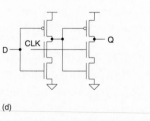

(d)

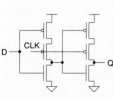

(e)

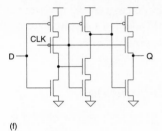

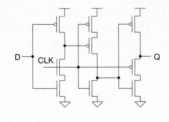

(f)

(g)

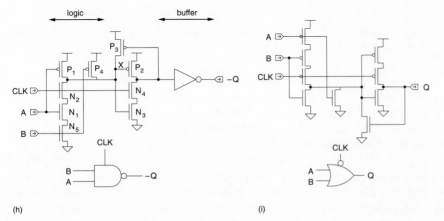

(h)

(i)

Figure 5.59 *(continued)*

while P_1 and P_4 form the pull-up path. Figure 5.59(i) shows an unbuffered two-input OR gate. Careful design of the latches and clocking is required when using the latches shown in Fig. 5.59. DEC's ALPHA designers carefully characterized race through susceptibility by simulating various combinations of latches, with varying clock rise and fall times, and voltage, temperature and process extremes. In the ALPHA case, clock rise and fall times below 0.8 *ns* caused no failures while a value of 1.0 *ns* showed some sign of failure. A value of 0.5 *ns* was set for the clock rise and fall time to prevent latch failure. The clock distribution used in ALPHA is discussed in Section 5.6.4.

While dealing with dynamic storage nodes some guidelines should be noted. The period that charge will stay on a storage capacitor is usually determined by the leakage of the diffusions (sources and drains) connected to the gate. This is highly dependent on temperature, but assuming a leakage current of 1 *nA* and a storage capacitance of .02*pF*, then $C(\Delta V/\Delta i) = (.02 \times 10^{-12} \times 5)/10^{-9} = 100$ μ*s*. Thus this node has to be refreshed (clocked with the old or new state) roughly every 100 μ*s*. Dynamic nodes should not be left floating for long periods of time even if the storage of the correct state is unimportant (i.e., in a power-down mode). The leakage characteristics of the storage node may cause the node to assume a level that causes the inverter to draw significant current. Dynamic nodes should always be refreshed or clamped to a known state when in standby or low-power mode.

5.5.6   Phase Locked Loop Clock Techniques

Phase Locked Loops (PLL)s are used to generate internal clocks on chips for two main reasons:

- To synchronize the internal clock of a chip with an external clock.
- If it is desired to operate the internal clock of a chip at a higher rate that the external clock input.

While PLLs have been used for some time to regenerate clocks from data (as in modems, etc.), it is only relatively recently that PLLs have been used to aid system-clocking issues. This has occurred because the on-chip clock frequencies have increased to the point where having to allow for small skews at the board level can decrease the overall speed of the system drastically. A PLL allows an internal clock to be generated that is in phase with an externally delivered clock. Figure 5.60(a) shows an example of a chip that receives an external clock that is internally buffered. This buffered clock is distributed across the chip and feeds an output register, which in turn feeds an output buffer. The delay time from the clock input to a new valid output data value is comprised of the clock-buffer delay, the RC delay to the register, the clock-to-Q delay of the register, and the output buffer delay. Consider the scheme in Fig. 5.60(b). Here a PLL senses the internal clock at the input of the register (or some other convenient place) and feeds this to a PLL, which also receives the input clock. The PLL generates a clock that is in phase with the input clock. Thus the clock-buffer delay and the RC clock-line delay is eliminated from the input clock to output data delay time. By including a divider in the PLL loop, the on-chip frequency may be increased by the divider ratio. In this case a division by 4 results in an internal clock that runs four times faster than the input clock. Chapter 9 demonstrates an example of this type of PLL-clocking system. A further system example is shown in Fig. 5.60(d), where a PLL that drives a high-speed tristate bus is used in each chip. This ensures that the output-enables of chips are synchronized with each other, which reduces tristate fights and improves overall timing.

A block diagram of a charge-pump PLL is shown in Fig. 5.61(a). It consists of a phase detector, a charge pump, a loop filter, and a Voltage Controlled Oscillator (VCO). The phase detector detects the difference between the reference clock and the VCO clock and applies charge-up or charge-down pulses to the charge pump. These pulses are used to switch voltage or

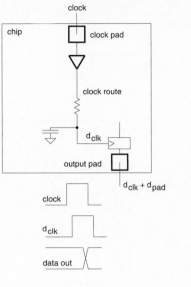

(a)

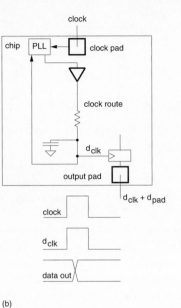

(b)

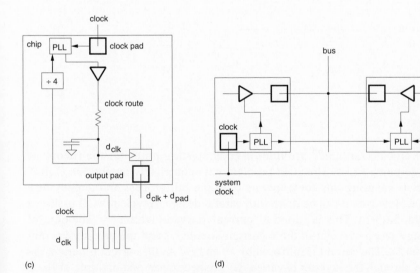

(c)

(d)

FIGURE 5.60 Phase locked loops for clock synchronization: (a) a chip without a PLL and a potential skew problem; (b) a PLL-clock-generator solution to clock skew; (c) a clock-multiplying PLL; (d) another use of PLL clocks to synchronize data transfers between chips

current sources, which charge or discharge a capacitor. The loop filter filters these pulses and applies the resulting control voltage to the VCO. The VCO changes oscillation frequency depending on the control voltage. Thus the total system forms a feedback system where the VCO is locked to the refer-

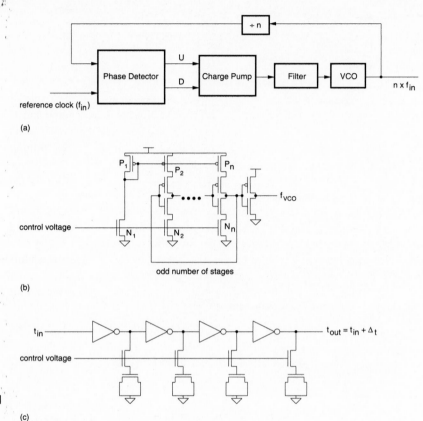

FIGURE 5.61   A charge pump PLL: (a) basic PLL block diagram; (b) typical CMOS VCO circuit; (c) typical CMOS VCDL circuit

ence clock. Gardner[54] provides a good starting point for the theory of charge-pump PLLs. The main problem with completely monolithic CMOS PLLs is ensuring that they operate over the full process and temperature range. However, it can be done with careful design. A typical VCO is shown in Fig. 5.60(b). This is called a "current-starved inverter."[55] The control voltage sets a currrent in the n current-source N_1 and the inverter-current source N_2. The current is mirrored by P_1 and P_2. As the current is varied, the delay through the inverter is varied. By connecting an odd multiple of these stages in series, an oscillator is constructed. As the control voltage is varied, the oscillation frequency changes (over quite a large range). For further representative circuits see Chapter 9 or Jeong et al.[56] and Young et al.[57]

As an alternative to using a VCO, designers have used a Voltage-Controlled Delay Line (VCDL).[58] This uses the circuit shown in Fig. 5.60(c). This uses a control voltage on an n-transistor to vary the amount of load capacitance seen by an inverter. A number of these stages are cascaded to form the delay line.

5.5.7 Metastability and Synchronization Failures

If the data and clock do not satisfy the setup and hold-time constraints of a register, a synchronization failure may occur.[59] This is a failure that is due to the inherent analog nature of the storage elements used in all electronic circuits. A latch with clock deasserted is normally a bistable device; that is, it has two stable states (one and zero). Under the right conditions the latch may enter a metastable state. Here the output is in an indeterminate state between 0 and 1. At the CMOS circuit level for the latches and registers shown previously, this means that the sampled input to an inverter that is responsible for determining the state is close to the inverter threshold voltage. Thus, in effect, the latch is perfectly balanced between making a decision to resolve a one or a zero. In practice, noise (switching and/or thermal) or a slight initial imbalance eventually pushes the latch output one way or the other. However, the output decision is arbitrary and the interpretation of this signal may cause a synchronization failure. The problem occurs when logic looking at the output of the latch interprets the resultant value differently due to the delay caused by the metastability. Figure 5.62(a) shows a SPICE simulation example in which the data signal in a latch is moved toward the clock edge. The top waveform shows the circuit inputs consisting of a fast-falling clock signal and a slow-rising data signal. The delay time of the rising data (from 0) is varied from 2.2 to 2.4 *ns*. At 2.2 *ns* the *Q* output of the latch makes a low to high transition. At a delay of 2.3 *ns* as the data is moved closer to the clock edge, *Q* still makes a low-to-high transition, but only after passing through an operating point that includes some time spent at the transition point of the inverters (i.e., a delay). With a delay of 2.4 *ns* the *Q* stays low. The extra time spent in the metastable region can lead to subsequent circuitry interpreting the output of the latch as two different values in the same clock cycle.

Within a synchronously clocked system, as long as the clock-to-*Q* delays are longer than the setup times, synchronization failures can not occur (no clock skew, clock cycle long enough). However, at the boundary of two independently clocked systems or a clocked system and the asynchronous real world, synchronization problems may occur. To deal with these asynchronous interfaces, circuits called synchronizers are used at the interface between independently clocked systems. These lower the risk of synchronizer failure to an acceptable level.

The probability of a synchronizer failure has been shown to be[60]:

$$F(t) = \frac{T_o}{\delta} \, e^{-\frac{t}{\tau_r}} \quad \text{for } t \text{ is greater than some time } h, \qquad (5.10)$$

where

δ, T_o, τ_r and h are parameters of the latch design.

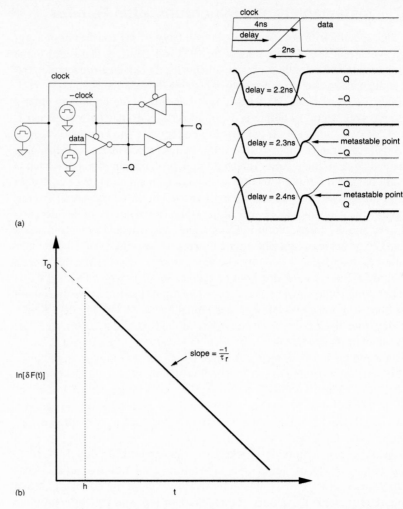

FIGURE 5.62 Metastability: (a) test circuit showing register entering metastable state; (b) graph showing probability of latch entering metastable state

δ is the range of time over which clock-to-data time varies (assumes a uniform distribution); τ_r is commonly called the time constant of resolution of the latch. This parameter is related to the overall gain-bandwidth of the latch amplifiers (i.e., as the gain-bandwidth of the inverters increases, this parameter becomes smaller). This value may be estimated by observing the small signal frequency response of the inverters or observing the time constant of the latch when exiting the metastable condition. T_o is related to the efficiency of converting a time difference in the signals at the input of the latch to an initial condition (voltage difference) at the metastable resolving node within the latch. Glasser and Dobberpuhl[61] estimate this as $(V_{IH} - V_{IL})$

$\times (dV_{in}/dt)^{-1}$, where dV_{in}/dt is the rate of charge of the data input to the latch and V_{IH} and V_{IL} are the input high- and low-noise margins respectively. In practice, this appears to be optimistic and values of 10–100 times or more of this value are observed.[62]

Equation 5.10 may be rexpressed as

$$ln\,[\delta F\,(t)\,] \;=\; -\frac{1}{\tau_r}\,(t) + ln\,(T_o) \qquad t > h, \tag{5.11}$$

which is graphed in Fig. 5.62(b). Measurements carried out on actual latches can be used to determine these values, or they may be estimated analytically or by simulation.[63–68] T_o is the extrapolated intercept at $t = 0$, while $-1/\tau_r$ is the slope of the characteristic.

From probability theory, the mean time between the failure of the output to be resolved within some time (t_f) ($MTBU$) is given by

$$MTBU\,(t_f) \;=\; \frac{1}{f_c f_d}\, e^{-\frac{t_f}{\tau_r}} \tag{5.12}$$

where

t_f = the time after the change in the clock by which the latch output must be resolved

f_c = the frequency of the clock

f_d = the frequency of the data.

Example:

To assesss the $MTBU$ of a typical system, consider the following conditions:

f_c = 50 MHz

f_d = 100 KHz

t_f = 10 ns (assume that half the cycle is taken by real logic and half may be taken by potential synchronizer delay)

Assume that

T_o = .1s

τ_r = .2 ns

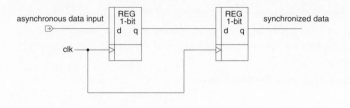

FIGURE 5.63 A typical register based synchronizer

$$MTBU(t_f) = \frac{1}{f_c f_d T_o e^{-\frac{t_f}{\tau_r}}}$$

$$= \frac{1}{50 \times 10^6 \times 100 \times 10^3 \times .1 \times e^{\frac{-10}{.2}}}$$

$$= 1 \times 10^{10} \text{ seconds.}$$

To deal with synchronizer problems circuits such as that shown in Fig. 5.63 are used. This consists of two cascaded registers which allow a whole clock cycle for the output of the first register to resolve. More registers may be cascaded to improve the metastability characteristics of the circuit at the cost of increased latency through the synchronizer. As far as good synchronizer register (latch) design is concerned, the general principle that applies is to keep the resolving circuit fast. This minimizes the exponential term in τ_r in Eq. (5.10). This involves minimizing parasitics by careful circuit and layout design.

The main point to realize is the following: If you have an asynchronous input that enters your chip, you should calculate the *MTBU* and design for an acceptable value. Use a synchronizer such as that shown in Fig. 5.62 before using the signal for any internal use. Failure to do so will surely cause system problems that will be both difficult and expensive to track down.

Problems with interfacing synchronous circuits with real-world asynchronous events has lead researchers to propose *self-timed* systems. (For instance see Chapter 7 in Mead and Conway.[69])

5.5.8   Single-phase Logic Structures

Conventional static logic may be used with single-phase clocking. In addition, domino nMOS logic may be used to improve speed, reduce area, and reduce dynamic power consumption. However, it is difficult to pipeline such logic stages while using a single clock and complement. A logic family termed N-P CMOS dynamic logic (Figs. 5.64 and 5.38) may be used to optimize speed and density at the expense of more detailed circuit and system design.[70]

N-P CMOS dynamic logic combines N-P sections of domino logic with a C^2MOS latch as the output stage. We can build *clk* blocks (Fig. 5.64a), which resolve (or evaluate) during *clk* = 1, and −*clk* blocks (Fig. 5.64b), which resolve during −*clk* = 1. Cascading these N-P blocks is achieved using the structure in Fig. 5.64(c). This yields a pipelined structure in which *clk* sections are precharged and −*clk* sections are evaluated when *clk* = 0 and −*clk* = 1. Information to −*clk* sections is held constant by the clocked CMOS latch in the output of *clk* sections. When *clk* = 0 and −*clk* = 1, *clk* sections are evaluated and −*clk* sections are precharged. Often it is desired to mix N-P dynamic sections with static logic or to connect N-P sections with domino

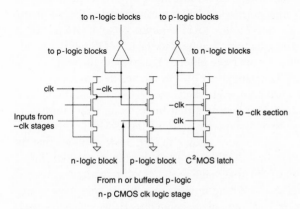

(a)

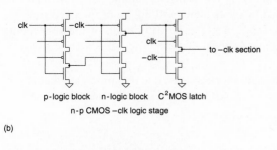

(b)

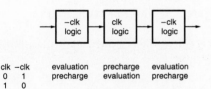

(c)

FIGURE 5.64 Cascaded NP Logic

sections. If this is done, two problems must be avoided: First, self-contained sections must be internally race free. Second, when different sections are cascaded to form pipelined systems, clock skew should result in no deleterious effects. We will examine some rules that have been proposed to deal with both problems.[70]

In the case of internal races, the basic rules for dynamic domino must be obeyed:

1. During precharge, logic blocks must be switched off.
2. During evaluation, the internal inputs can make only one transition.

For complete dynamic blocks, either alternate n-p logic gates may be used or n-n or p-p blocks may be cascaded with buffer inverters between sections following the domino rules. Static logic structures may be used. Where this is done, it is best to keep the logic static up to the C^2MOS latch, because the static structures generally can create glitches that violate the second condition mentioned above. When using the C^2MOS latches in conjunction with N-P logic sections, an additional rule guarantees race-free operation, even in the presence of clock skew. This requires that there be an even number of static inversions between the final dynamic gate and the C^2MOS output latch.

The ability to pipeline sections, as shown in Fig. 5.64, assumes that the output of a logic block (*clk* or *–clk*) does not glitch (due to precharging or input variations) the input of the next stage in the pipeline while it is resolving its output. Under perfect clocking conditions, this assumption is met merely by following the *clk,–clk* logic block sequence. However, in the presence of clock skew, an early output transition on one block may glitch the next stage while it is still actively resolving its inputs. An additional rule ensures that glitches caused by clock skew will not be propagated from the output of one logic block through to the C^2MOS latch of the succeeding logic block. This rule states that either:

- There exists in each logic block at least one dynamic gate that is separated from the previous C^2MOS output stage by an even number of inversions;

  or

- The total number of inversions between the C^2MOS stage and the previous C^2MOS stage is even. Figure 5.65 illustrates these rules.

The logic/latches shown in Fig. 5.59 are also appropriate for single clock systems.

Figure 5.66 illustrates one final method of clocking using a single clock. This combines CVSL gates of both polarities with *RS* flip-flops on the outputs of each gate.[71] This allows a single clock to be used at the expense of circuit and layout complexity.

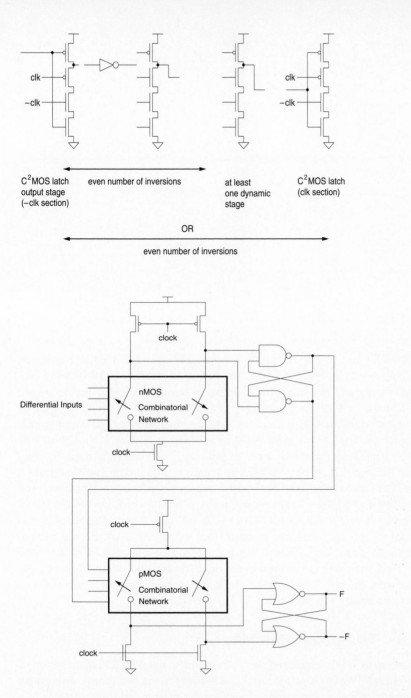

FIGURE 5.65 NP logic design rules

C^2MOS latch output stage (−clk section)

even number of inversions

at least one dynamic stage

C^2MOS latch (clk section)

OR

even number of inversions

FIGURE 5.66 Another single clock clocking scheme

While clock skew of various kinds can be deleterious to the operation of a system, it can also be used to advantage. Figure 5.67 shows a sequence of pipelined registers, each of which is supplied with a delayed clock. Because

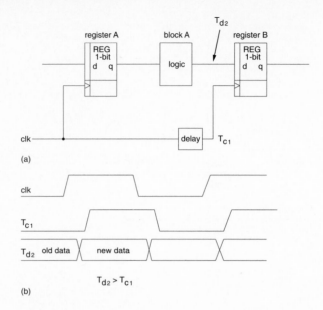

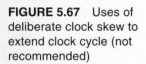

FIGURE 5.67 Uses of deliberate clock skew to extend clock cycle (not recommended)

the clock feeding register B is delayed with respect to that feeding register A, the data from logic block A has a little longer to stabilize prior to being registered by register B. The problem is now to ensure that logic outputs do not race to register B, that is, that there is a lower bound on the delay time of logic block A outputs. This clocking scheme must be designed with extreme care with the appropriate tools (such as a timing analyzer). It should only be used as a last resort when all other techniques have been exhausted. In cases where the skewed clocks are used to equalize logic activity throughout a cycle (as in mixed analog/digital chips), the race condition may be countered by using the clocking scheme shown in Fig. 5.68(a), which latches the inputs to register B with the clock to register B. In this case, the timing advantage is lost but each logic block can operate from a delayed clock. Figure 5.68(b) shows a safe clocking scheme that prevents races but robs from the cycle time. Here the clock is buffered in the opposite direction to the flow of data. If this were the way in which the clock was to be distributed, then this is a preferable strategy to that shown in Fig. 5.68(a).

5.5.9 Two-phase Clocking

A problem in single-phase registers can be the generation and distribution of near-perfectly overlapping clocks. This has been commonly solved by employing two nonoverlapping clocks for the master and slave sections of a register. Thus we can have between one and four clock lines to route around a chip. Usually, two main clocks would be distributed with buffers to generate local clocks. A typical set of clock waveforms and a simple register

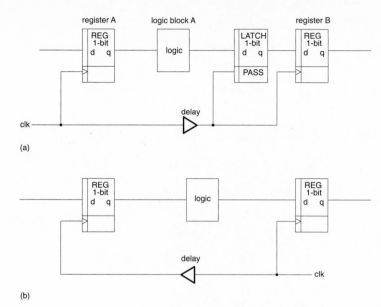

(a)

(b)

FIGURE 5.68 Two methods of avoiding clock skew problems in situations where clock skew is present: (a) latched outputs; (b) contra-data-direction clock

(DFF1) are shown in Fig. 5.69(a). Note that $phi_1(t) \cdot phi_2(t) = 0$ for all t. The operation of the register is illustrated in Fig. 5.69(b). During $phi_1 = 1$, the master transmission gate is closed, thereby storing the input level on the gate capacitance of the inverter and the output capacitance of the transmission gate (C_1). The state of the slave is stored on a similar capacitance, C_2. During $phi_2 = 1$, the stage-1 transmission gate opens and the inverse of the stored value on C_1 is placed on C_2.

The selection of the actual clock relationships depends on the circuit. Some guidelines would be as follows. If phi_1 is used as a precharge clock, then it has to be of a duration to allow precharge of the worst-case node in the circuit. Typically, this might be on a RAM bit line. The delay between clocks has to be chosen to ensure that for the combination of worst-case conditions, the two clocks do not overlap. Clock skew can occur in two forms. The first is shown in Fig. 5.69(c), where the clocks applied to a register have travelled through different delay paths to arrive at the latch. The skew occurs while both clocks are simultaneously HIGH, causing the two transmission gates in the register to be transparent, similar to the single-phase skew examples. Another type of skew can occur even if the clocks are perfectly overlapping. This is shown in Fig. 5.69(d). Here, the rise and fall times are so slow that the period of the transition region causes the latch transmission gates to couple. Both of these conditions can lead to incorrect values being stored on the C_1 and C_2 capacitances. Thus the period of the clocks must allow for the worst-case logic propagation time in combinational blocks that are to be latched.

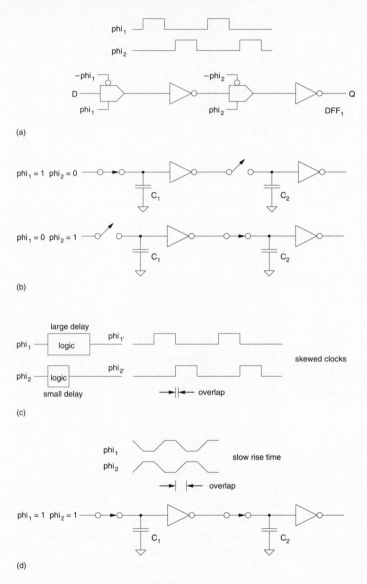

FIGURE 5.69 Two-phase clocking: (a) dynamic register and clock waveforms; (b) operation; (c) failure due to clock skew; (d) failure due to slow-rise-time clocks

5.5.10 Two-phase Memory Structures

Two-phase registers are usually replications of single-phase structures with phi_1 feeding the master and phi_2 feeding the slave of the register. Examples are shown in Fig. 5.70.

Considering DFF1, two representative layouts are shown in Fig. 5.71(a) and Fig. 5.71(e). A layout representing the DFF2 configuration are shown in

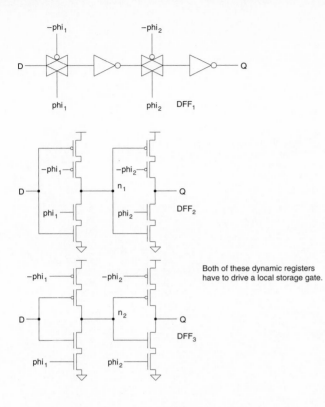

Both of these dynamic registers have to drive a local storage gate.

FIGURE 5.70 Two-phase dynamic registers

Fig. 5.71(b), Fig. 5.71(c) and Fig. 5.71(d). Note that routing clocks in poly-silicon may lead to clock-delay problems if long unbuffered clock lines are used. An example of DFF2 may be found in Fig. 8.70.

A reduction in the number of clock lines can be accommodated if only n-transistors are used in the transmission gate, as would be the case in an n-MOS design (Fig. 5.72a). Two effects occur in this configuration. First, the '1' level transferred to the input of the inverter is degraded to approximately $V_{DD} - V_{tn}$. This has the effect of slowing down the low transition of the inverter. Furthermore, the high-noise margin (N_{MH}) of the inverter is degraded. It also has the possible effect of causing static power dissipation. For instance, if $|V_{tp}| < V_{tn(body\text{-}affected)}$, then the p-transistor in the inverter will be turned on when the inverter output is in the low state, thus causing current to flow through the inverter. This is consistent with the reduced N_{MH}. Although this is not catastrophic, it must be taken into account when calculating total power dissipation. Figure 5.72(b) shows the addition of p feedback transistors to provide fully restored logic levels. Figure 5.72(c) shows a single clock version of this register that uses p and n pass transistors. The rising transition at the output of the inverter in Fig. 5.72(a) and 5.72(b) is faster

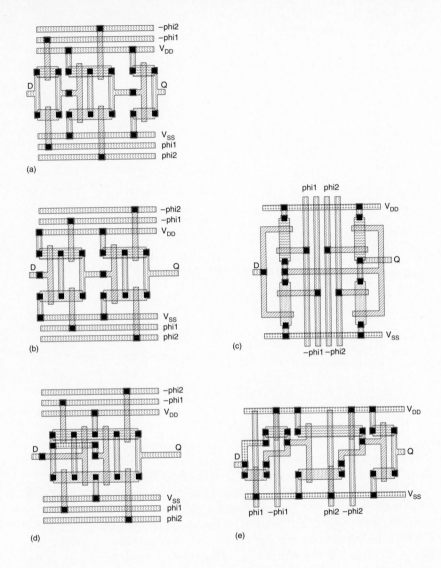

FIGURE 5.71 Various layouts for two-phase dynamic registers

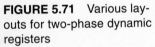

because the capacitance at the output is reduced due to the absence of the p-transistor. There is no hard and fast rule for when the various flip-flop configurations should be used. Of course, if density is crucial the last mentioned flip-flop could be used, provided that the speed was suitable and that static power dissipation was not a problem. This can only be reconciled by worst-case simulation and power-dissipation calculations.

Clock distribution techniques for two-phase clocks may differ depending on the design. One technique is to globally distribute the two clocks with

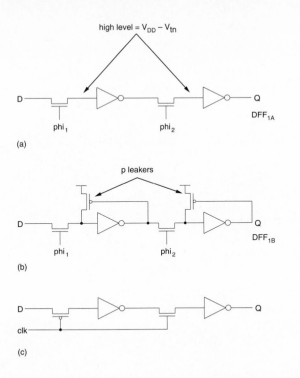

(a)

(b)

(c)

FIGURE 5.72 Two-phase registers with single-polarity clocks: (a) nMOS pass transistors; (b) adding p leakers; (c) a single clock version; (c) pass gate

or without their complements. Alternatively, a single clock can be distributed and local two-phase clocks generated in individual modules. Regardless of the technique, somewhere a two-phase clock generator is required. Figure 5.73 shows the basic circuit that is normally used; it is based on a cross-coupled *RS* flip-flop. This basic design is modified to meet drive considerations and overlap requirements. Methods for conditionally loading two-phase registers are shown in Fig. 5.74. Figure 5.74(a) uses a multiplexer on the front of a two-phase register. Alternatively, the clocks to the master may be gated, as shown in Fig. 5.74(b). Gating the clock can lead to delayed clocks, which in turn can lead to clock-skew problems in registers that are susceptible to skew. Another clock qualification method is shown in Fig. 5.74(c). This uses

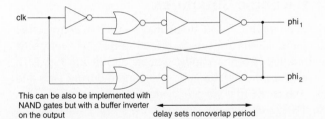

This can be also be implemented with NAND gates but with a buffer inverter on the output

delay sets nonoverlap period

FIGURE 5.73 Two-phase clock generator

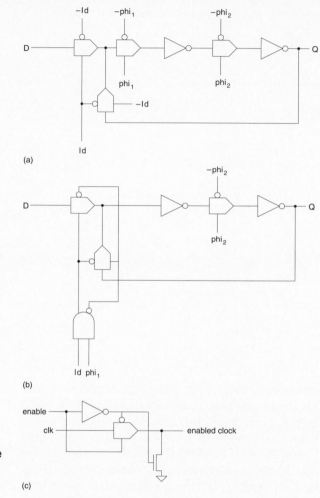

FIGURE 5.74   Clock-enable circuits: (a) mux; (b) clock gating; (c) pass gate

a transmission gate as an AND function. The n-pulldown transistor ensures that the falling edge of the enabled clock is fast.

5.5.11   Two-phase Logic Structures

For two-phase systems, conventional static logic may be used in conjunction with the memory elements that have been described in the last section. If dynamic logic is required, the two-phase logic scheme outlined in Fig. 5.75 may be used. In this scheme, the first stage is precharged during phi_1 and evaluated during phi_2. While the first stage is evaluated, the second stage is precharged and the first-stage outputs are stored on the second-stage inputs.

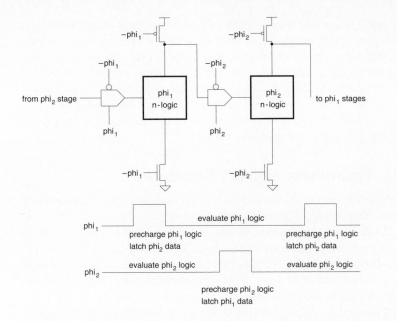

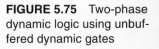

FIGURE 5.75 Two-phase dynamic logic using unbuffered dynamic gates

During phi_1, the second stage is evaluated and latched in a succeeding phi_1 stage.

Domino n-MOS gates may also be employed. A typical gate is shown in Fig. 5.76. Here, a single clock (phi_1 or phi_2) is used to precharge and evaluate the logic block. The succeeding stage is operated on the opposite clock phase, as illustrated in Fig. 5.76. The difference between this logic structure and that previously shown in Fig. 5.75 is that in the domino logic, a number of logic stages may be cascaded before latching the result.

5.5.12 Four-phase Clocking

The dynamic logic that has been described has a precharge phase and an evaluate phase. The addition of a "hold" phase can simplify dynamic-circuit-

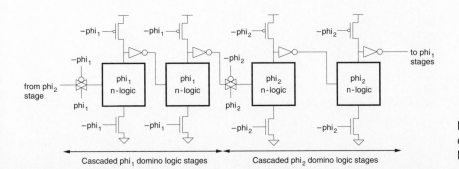

FIGURE 5.76 Two-phase dynamic logic using domino logic

logic design. This primarily results from the elimination of charge sharing in the evaluation cycle. Four-phase clocking schemes have been historically very popular for a variety of reasons, including circuit size, clocking safeness, and the ability to generate a wide variety of clocks using the abundance of edges available. Modern designs tend to minimize the number of clock phases used, and employ self-timed circuits to generate special clocks. A disadvantage of four-phase logic is the number of clocks that may have to be generated and distributed throughout the chip.

5.5.13   Four-phase Memory Structures

A four-phase flip-flop is shown in Fig. 5.77(a) with its corresponding clock waveforms. During $clk_1 = 0$, node n_1 precharges. When $clk_2 = 1$ and $clk_1 = 1$, node n_1 conditionally discharges. When clk_2 falls to 0, this value is held on node n_1 regardless of the state of the input D. During $clk_3 = 0$, Q is precharged; during $clk_4 = 1$, $clk_3 = 1$, this node is conditionally discharged according to the state of node n_1. This configuration can still have charge-sharing problems because the intermediate nodes in the inverters (inv_1 and inv_2) may be corrupted due to charge sharing with outputs n_1 and Q, respectively. This is solved by altering the clock waveforms so that clk_2 is actually clk_{12} and clk_4 is clk_{34}, as shown in Fig. 5.77(b). With these clocking waveforms the intermediate nodes are precharged uniformly.

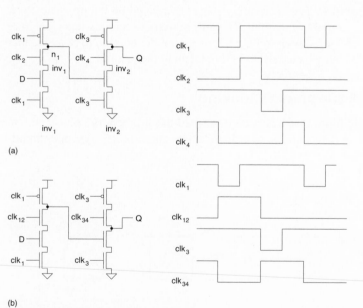

FIGURE 5.77 Four-phase registers: (a) type 1; (b) type 2

5.5.14 Four-phase Logic Structures

Historically, the main purpose in adopting a four-phase clocking strategy was to enable the four-phase logic gates to be built (although static gates may be also used). Illustrative gates are shown in Figs. 5.78, 5.79, and 5.80.

Arguments for using such a clocking strategy include the fact that no more clock lines are needed than for two-phase clocking if certain four-phase structures are used. In addition, a strict ratioless circuit technique may be applied, which can lead to very regular layouts.

Improvements on the dynamic structure in Fig. 5.30 use the forms of two- and four-phase logic that have been developed for earlier types of MOS design.[72] These gates add a sample-and-hold clock phase to the precharge and evaluate cycles. Figure 5.78(a) shows one version of a gate implemented using the clock relationships shown in Fig. 5.78(b). The composite clocks clk_{12} and clk_{23} are used in this example. During clk_1, node P_z is precharged, while node z is held at its previous value. When clk_2 is true, node P_z remains precharged and, in addition, the transmission gate turns on, thus precharging node z. When clk_3 is asserted, the gate evaluates and node P_z conditionally discharges. Node z follows node P_z because the transmission gate remains on. Finally, when clk_4 is true, node z will be held in the evaluated state. The state of node P_z is immaterial. There are four types of gates characterized by the phase in which evaluation occurs. When using such logic gates, they must be used in the appropriate sequence. The allowable connections between types are shown in Fig. 5.79. Note that four levels of logic may be evaluated per bit time. Alternatively, a two-phase logic scheme may be employed by using type-4 gates and type-2 gates or type-1 gates and type-3 gates.

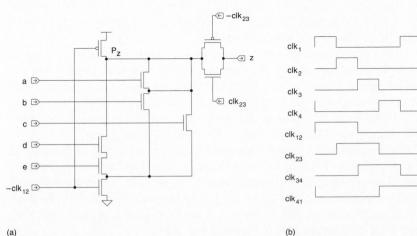

(a)   (b)

FIGURE 5.78 Four-phase logic—type A

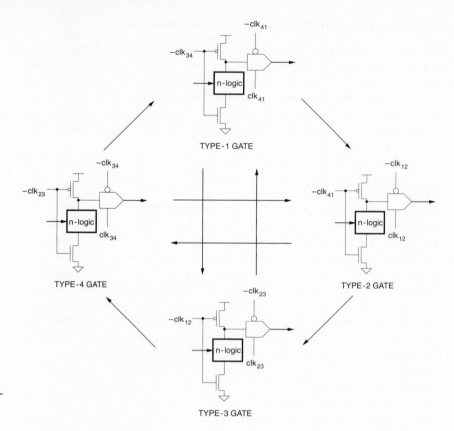

FIGURE 5.79 Allowable logic gate interconnections—type A logic

An alternate four-phase structure is shown in Fig. 5.80(a).[73] The clocking waveforms are shown in Fig. 5.80(b). This gate type is more restrictive than the previous gate, but the circuit is simpler and the number of clocks reduced, and the layout would be smaller.

The number of transistors required for such logic gates is either $n + 4$ or $n + 3$ for an n-input gate. A problem that occurs with such gates is that the clock frequency must be long enough to allow for the slowest gate to evaluate. Thus fast gates tend to evaluate quickly and the remainder of the cycle is "dead time." Other system-design problems arise when trying to distribute four or more clocks and synchronize them around a large chip.

It is also possible to use a four-phase clock as a general clocking technique for domino circuits. By using the appropriate logic gate, any combination of phases may be generated locally for circuits requiring different clocking strategies: clk_1 may be used as a slave latch clock, clk_2 the first-level logic evaluation, clk_3 as the master latch clock, and clk_4 as the second-level logic evaluation. This is shown in Fig. 5.81.

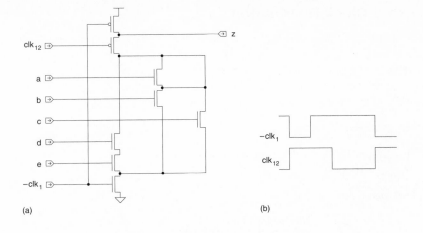

(a)   (b)

FIGURE 5.80 Four-phase
logic—type B

5.5.15 Recommended Clocking Approaches

For first-time designs, where mostly static logic is to be used, the single-phase clocking scheme is probably preferable using fully self-contained static registers. For standard-cell and gate-array designs this will usually be the only option permitted. The clock-routing problem is minimal, especially in data-path designs. A two-phase clocking strategy is a little easier to work in the timing for RAMs, ROMs, and PLAs. In the past, the two-phase scheme was popular because it guaranteed latch behavior and worked well with small dynamic latches. In today's processes and circuits, cycle times are so short that guaranteeing the nonoverlap time for a two-phase clocking scheme over all process corners can cut significantly into the cycle time. In addition, the CMOS processes are extremely dense, thus obviating the need for the smallest latch possible. All of these trends lead one to consider only single-phase clocking for complex, high-speed CMOS circuits. Special clocks are normally generated using self-timed logic circuits. Alternative clocking schemes may be of utility in special situations.

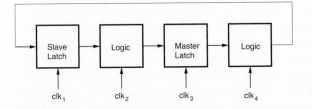

FIGURE 5.81 Four-phase
logic clocking method

5.5.16   Clock Distribution

Assuming that either a one- or two-clock system is chosen, the problem still exists of how to distribute the clock. If one counts up all the capacitance in the registers in a large CMOS design, it may well add up to over a $1000pF$. If this has to be driven in a small time and at a high repetition rate, the peak transient current and average dynamic current can be in the amp range. For example,

$$V_{DD} = 5V$$

$$C_{register} = 2000pF \text{ (20K register bits @ .1pF)}$$
$$T_{clock} = 10 \text{ ns}$$

$$T_{rise/fall} = 1 \text{ ns}$$

$$I_{peak} = C\frac{dv}{dt} = \frac{2000 \times 10^{-12} \times 5}{1.0 \times 10^{-9}} = 10A$$

$$P_d = CV_{DD}^2 f = 2000 \times 10^{-12} \times 25 \times 100 \times 10^6$$
$$= 5 \text{ watts.}$$

Two main techniques are used:

- a single large buffer.
- a distributed-clock-tree approach.

With the first approach, a single buffer (i.e., cascaded inverters) is used to drive a global clock that feeds all modules. The geometric aspects of this approach involve ensuring that a low-skew clock is fed to all modules on the chip. Approaches for achieving this are discussed in the next section. The distributed-clock-tree method constructs a tree of clock buffers with some suitable geometry such that modules that communicate with each other receive well-defined and well-behaved clocks. For instance, Fig. 5.82 shows an example of a clock tree that drives a datapath (e.g., a FIR filter). The leaves of the clock tree feed n-bit datapaths. Either the delay to each datapath can be carefully simulated and matched or the distribution of the clock can be arranged so that any RC delay occurs in a safe slew direction (i.e., opposite to the direction of data flow).

The first approach (i.e., a single buffer) is preferred in designs that have a large number of diverse modules that have no discernable structured routing approach (i.e., a microprocessor or digital signal processor). Some examples of clock buffer layout options for this style are given in the next section. The second approach is more suitable for highly structured DSP structures such as FIR filters. In general, one is just trading one style of design for another. In the end the design must be engineered in a detailed manner—there is no such thing as a design-free clocking strategy in today's high-performance processes.

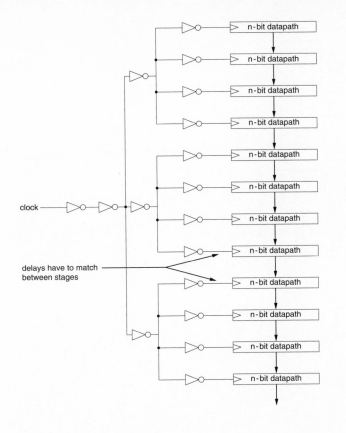

FIGURE 5.82 Clock-tree layout

5.6 I/O Structures

Of all the CMOS circuit structures that will be covered in this text, I/O structures require the most amount of circuit-design expertise in association with detailed process knowledge. Thus it is probably inappropriate for a system designer to contemplate I/O pad design. Rather, well-characterized library functions should be used for whatever process is being used. The following section will summarize some basic design options for I/O pads.

5.6.1 Overall Organization

Pad size is defined usually by the minimum size to which a bond wire can be attached. This is usually of the order of 100 to 150μ square. The spacing of pads is defined by the minimum pitch at which bonding machines can operate. This tends to be in the 150–200μ range (Fig. 5.83a). Extremely high pad counts may be achieved by interdigitating pads as shown in Fig. 5.83(b). Pads are usually designed to be "core-limited" or "pad-limited." In the

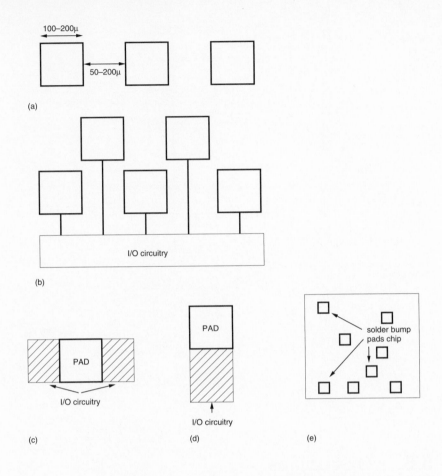

FIGURE 5.83 I/O pad options: (a) pad spacing; (b) interdigitated pads; (c) core limited pads; (d) pad limited pads; (e) solder bump I/O

former, shown in Fig. 5.83(c), the internal core of the chip determines the size of the chip, so thin pads are required. The I/O circuitry is placed on either side of the pad. A pad-limited version is shown in Fig. 5.83(d). Here the I/O circuitry is placed toward the center of the chip. Finally, Fig. 5.83(e) shows an option that is available in some processes where I/O pads may be placed anywhere on the chip. This technology works by plating the pads with solder bumps and then inverting the chips and reflow-bonding them to a substrate. Figure 5.84 illustrates some of these concepts in more detail. A variety of placement of components is shown. Power- and ground-bus widths may be calculated from a worst-case estimate of the power dissipation of a die and from a consideration of providing good supply voltages. Multiple power and ground pads may be used to reduce noise. Some designers advocate placing the lowest circuit voltage (V_{SS}) as the outermost track. With these points in mind, a frame generation program may be easily constructed. This takes a simple description of the pad ordering and produces a

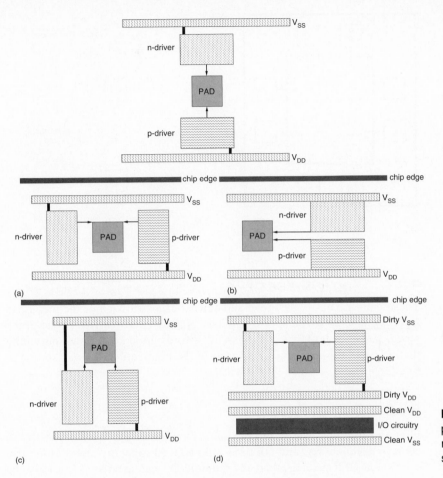

FIGURE 5.84 More detailed pad layouts showing various relationships of power busses, pad, and transistors

finished pad frame. A typical description might be as follows:

```
LEFT:
    INPUT   A;
    INPUT   B;
TOP:
    VDD     VDD;
    INPUT   C;
RIGHT:
    OUTPUT  Z;
    OUTPUT  Y;
BOTTOM:
    OUTPUT  W;
    VSS     VSS;
```

The resulting I/O frame is shown in Fig. 5.85(a).

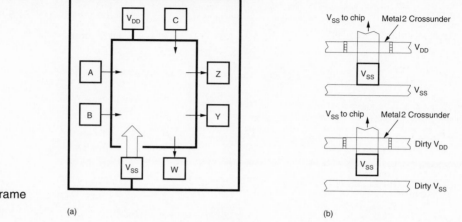

FIGURE 5.85 I/O frame generation example

(a)

(b)

5.6.2 V_{DD} and V_{SS} Pads

These pads are easily designed and consist of a sandwich of the metal pad layers connected to the appropriate bus. A nonplanarity arises at one of the power pads. A two-level metal process affords good crossovers, providing that a large number of vias are used in the connection. This is shown in Fig. 5.85(b).

5.6.3 Output Pads

First and foremost, an output pad must have sufficient drive capability to achieve adequate rise and fall times into a given capacitive load. If the pad drives non-CMOS loads, then any required DC characteristics must also be met. In this discussion we will concentrate on pads to drive CMOS loads. Given a load capacitance and target rise and fall time, the output transistor sizes may be calculated from the equations derived in Chapter 4. One then generally needs buffering to present a lower load to the internal circuitry. As previously discussed, an inter-stage ratio of between 2 and 10 is optimal for speed. Generally, in a pad, an n-stage (n is even) inverter circuit is used to result in a noninverting output stage.

Because large transistors typically are used and I/O currents are high, the susceptibility to latch-up is highest in I/O structures. In particular, latchup will occur when transients rise above V_{DD} or below V_{SS}. These conditions are most likely to occur at I/O pads due to the interface to external circuitry. Hence, the layout guidelines given in Chapter 3 should be used. This means separating n- and p-transistors and using the appropriate guard rings tied to the supply rails. If possible the I/O output transistors (i.e., those whose drains connect directly to external circuitry) should be doubly guard-ringed. This means that an n-transistor should be encircled with a p^+ connection connected to V_{SS} and an n^+ in an n-well connected to V_{DD}. The p output transistor should be encircled with a p^+ ring connected to V_{DD} and an n^+ connected to the substrate and V_{SS}. The rings should be continous in diffusion and strapped with metal where possible. Polysilicon can not be used as a crossover because it breaks the continuity of the diffusion rings. In addition to the double guard rings, if possible dummy collectors consisting of p^+ connections to V_{SS} and n^+ in n-well connections to V_{DD} should be placed between the I/O transistors and any internal circuitry. The dummy collectors and guard rings serve to reduce the stray carriers injected into the substrate when the drain diodes are forward-biased. The I/O transistors should have their sources connected to the "dirty" V_{SS} and V_{DD} connections of the chip; that is, connections from the power supplies that solely feed the I/O transistors. Where possible, separate internal V_{DD} and V_{SS} supply connections should be made to internal circuitry. The dirty and clean V_{SS} signals should be single-point connected (at the bond pad). All V_{DD} and V_{SS} connections should be ohmically connected in metal. The I/O transistors should be constructed from parallelled smaller transistors. In nonsilicided processes this alleviates any RC delay down long gate lines. In addition, it allows parallel metal connections to be made to the I/O transistor to avoid metal migration problems. The I/O transistors often have gates longer than normal to improve the avalanche breakdown characteristics. In an output pad or bidirectional pad these transistors form the output driver transistors.

When driving TTL loads with CMOS gates, the different switching thresholds have to be considered. The V_{IL} of a TTL gate is 0.4 volts; the V_{OL} of a CMOS gate is 0 volts. Thus we have no problem in this respect. The V_{IH} for a TTL gate is 2.4 volts. The V_{OH} for a CMOS gate is 5 volts (for a 5-volt supply), and hence there is no problem here. In the low state, the CMOS buffer must be capable of "sinking" 1.6 mA for a standard TTL load with a V_{OL} of <.4 volts. For typical driver transistors, this is usually not a problem.

5.6.4   Input Pads

In an input buffer, the first stage is connected directly to external circuitry. This means that the gate of an n- and/or p-transistor may experience voltages beyond the normal operating range for the CMOS process. The gate connection of an MOS transistor has a very high input resistance (10^{12} to 10^{13} ohms). The voltage at which the oxide punctures and breaks down is about 40–100 volts. The voltage that can build up on a gate may be determined from

$$V = \frac{I\Delta t}{C_g},$$

(5.13)

where

V = the gate voltage

I = the charging current

Δt = the time taken to charge the gate

C_g = the gate capacitance.

Thus if $I = 10\ \mu A$, $C_g = .03pF$, and $\Delta t = 1\ \mu s$, the voltage that appears on the gate is approximately 330 volts. Usually a combination of a resistance and diode clamps (electrostatic protection) are used to limit this potentially destructive voltage. A typical circuit is shown in Fig. 5.86. Clamp diodes D_1 and D_2 turn on if the voltage at node X rises above V_{DD} or below V_{SS}. Resistor R is used to limit the peak current that flows in the diodes in the event of an unusual voltage excursion. Values anywhere from 200Ω to $3\ K\Omega$ are used. This resistance, in conjunction with any input capacitance, C, will lead to an RC time constant, which must be considered in high-speed circuits. Preferences on the resistor construction have changed over time. Polysilicon resistors used to be used. Current protection structures tend to use a tub resistor (p-diff in an n-well process). Clamping diodes are formed by using n^+ in substrate and p^+ in n-well diffusions. As with I/O transistors these must be doubly guard-ringed because they are diffusions that can be forward-biased by external over- or under-shoots. A popular alternative is to use the drains of the I/O transistors. In an input-only pad the transistor has its gate tied to V_{SS} while the gate of the p-transistor is tied to V_{DD}. A little series-diffusion resistance in the I/O transistors can improve their breakdown characteristics. Figure 5.87 shows an illustrative input pad that embodies these guidelines.

In an n-well process, all n-device I/O circuitry can be designed. In this case n^+-diffused protection resistors, as well as n "punch-through" devices, may be used. A punch-through device has closely spaced source and drain diffusions but no gate. The device affords protection by "avalanching" at around 50V. No wells need be included in this type of I/O.[74]

The input buffer is normally constructed with gate lengths longer than normal to aid the breakdown characteristics. It is followed by a number of

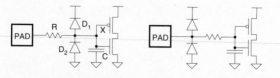

FIGURE 5.86 Input pad electrostatic discharge (ESD) protection

typical input-protection circuits

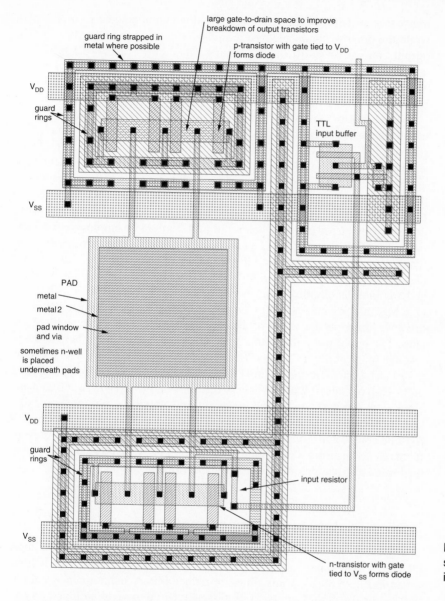

FIGURE 5.87 Input pad symbolic layout showing important features

stages sufficient to drive the internal load. The switching threshold of the input buffer is of importance when being driven by non-CMOS circuitry. For instance, when interfacing TTL logic to CMOS, it is advantageous to place the switching point of the input inverter in the middle of the TTL switching range. For TTL V_{OL} = 0.4 volts and V_{OH} = 2.4 volts. Thus the switching point should be set near 1.4 volts. This is achieved by ratioing the inverter

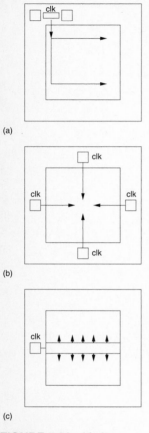

FIGURE 5.88 Various clock driver options: (a) a single driver; (b) a four-sided approach; (c) the "down the center" approach

transistors or using a differential pair with one input connected to the input while the other is connected to a reference voltage. Alternatively, the TTL output can use an additional resistor connected to the 5-volt supply to improve the TTL V_{OH} (the only trouble with this is trying to convince the board designers to do it; i.e., it is fairly impractical). The chip solution to this is to include a resistor inside the pad in the form of a p-transistor tied to V_{DD}.

Clock buffers are a particularly stringent case of input buffers. They usually have very high internal loading and have to provide extremely fast rise and fall times. The chip layout for a moderately sized clock buffer is shown in Fig. 5.88(a). Here the clock buffer is straddled by a power-and-ground pad (or even two power-and-ground pads), which supply power to the clock buffer alone. Because there may be a significant delay between the input clock pad and the on-chip clock, an early clock may be picked off to drive any registered pad (see below) or a PLL might be employed. An alternative clock-driver strategy is shown in Fig. 5.88(b).[75] In this design, four tristate clock drivers are placed in the middle of each pad side. This distributes the clock drive while minimizing the distance between the clock driver and any internal circuitry. This in turn reduces any possible *RC* clock delay. One final clock buffer strategy is shown in Fig. 5.88(c).[76] Here the clock driver is placed in the center of the chip and extends the entire width of the chip. The ouput p-device in this structure is 10 inches long! This strategy resulted in the ability to drive a 200 MHz clock across the chip with less than 0.5 *ns* skew. In addition, the clock is designed to radiate out from the center of the chip in concert with the data so that the relative skew between data and clock is minimized. As evidenced by these examples, clock-buffer design can not be treated as an afterthought. The design of the clock distribution must be considered from the commencement of the chip design. There is little magic to ensuring a good clock-distribution network other than applying the basic electrical engineering theory described in Chapter 4.

5.6.5 Tristate and Bidirectional Pads

The circuit of a tristate buffer is shown in Fig. 5.89(a). By merging an input pad and a tristate pad, a bidirectional pad may be constructed. This is illustrated in Fig. 5.89(b). Many times, to reduce library maintainence a single bidirectional pad that can be discretionarily wired to yield an input, an output, a tristate-output, or bidirectional pad is supplied. To reduce the design costs, most pad libraries use a common power-bus/protection structure that includes dirty power and ground, guard-ringed I/O driver transistors/protection diodes, and a series input resistor. All remaining circuitry is placed either at the side or toward the center of the chip. Figure 5.90 shows a symbolic view of a typical I/O pad. This is also shown in Plate 4.

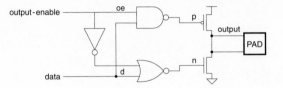

(a)

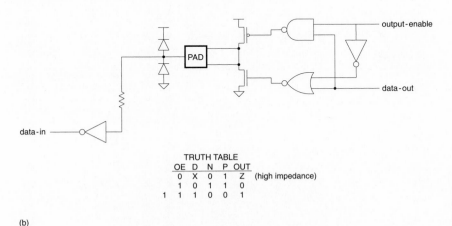

(b)

TRUTH TABLE

OE	D	N	P	OUT	
0	X	0	1	Z	(high impedance)
1	0	1	1	0	
1	1	1	0	0	1

FIGURE 5.89   A tristate pad (a) and a bidirectional pad (b)

5.6.6   Miscellaneous Pads

Many times, pads other than input, output, tristate, or bidirect are required. In this section some of these will be examined.

If a pad has to have a pull-up or pull-down included (for instance to allow the discretionary wiring of a pad), this may be achieved by using a long p- or n-transistor. The required length may be calculated from the desired pull-up/pull-down current. The gates may have to be conditionally turned off to allow for static DC testing. Because these transistors have drains connected to the outside world, these structures should be doubly guard-ringed the same as I/O transistors.

To achieve low setup and hold times for a chip, latches or registers are frequently included with a pad. This alleviates any internal delay that might result if the pad were allowed to drive the input into the chip to an internal storage element. Similarly, to achieve low clock-to-output delays registers might be included in output pads.

Fast-rising output pulses of large amplitude have spectrums well into the UHF range. This can generate interference in radios, cellular telephones, and television sets. In situations where low Radio Frequency Interfence

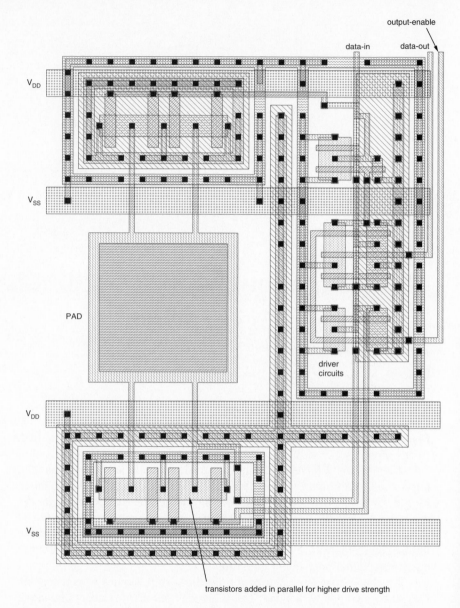

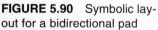

FIGURE 5.90 Symbolic layout for a bidirectional pad

(RFI) is required (i.e., television sets) the basic approach is to reduce the level of the high-order harmonics. A popular approach is to use controlled-slew-rate pads. Here the rise/fall time of the I/O pad is artificially limited to a value that does not impact normal circuit performance. In addition, reducing the I/O swing directly reduces the level of higher order harmonics. Another approach uses a 1-volt signal combined with special pad drivers and receivers to reduce the level of RFI.[77]

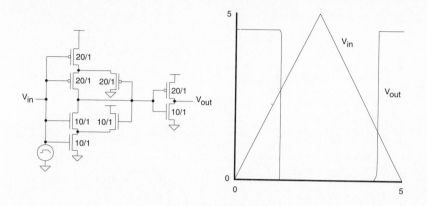

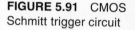

FIGURE 5.91 CMOS Schmitt trigger circuit

Frequently, hysteresis is required on an input pad so that a clean edge is generated by a slowly varying input. A Schmitt trigger may be used for this function. The circuit diagram of a CMOS Schmitt trigger is shown in Fig. 5.91. It works by switching at a different threshold on rising and falling edges.

5.6.7 ECL and Low Voltage Swing Pads

Reducing the voltage swing on pads can also aid in the construction of very fast pads. By using ECL levels, very fast CMOS I/O buffers have been demonstrated.[78,79,80] In one case, specially designed pads actively measure the impedance of the external lines they are driving and automatically match the I/O pad driver to this impedance to reduce reflections.[81]

Figure 5.92 shows the circuitry used in an automatic impedance control CMOS pad that operates at ECL levels.[82] The output driver is composed of a programmable pull-up and pull-down structure comprised of exponentially sized n-transistors. By enabling certain drive transistors, the series impedance driving an external transmission line can be set. The receiver is a differential receiver biased to switch at half the ECL supply. The resistors are diffusion devices. The buffered output of the receiver is fed to the chip and to a set of sample registers that are enabled two inverter delays in time apart. These registers may be used as a discrete time-sampling mechanism to measure the return time of a reflected signal. The output pad is pulsed and cycled through its impedance range, and the return signal time is measured by the sample registers. The correct setting of the output-pad impedance may then be determined by finding the point where the highest derivative in sample-bit position occurs. The pads on a chip are accessed by Boundary Scan techniques to load impedance values, unload the sample register, and drive the outputs. Such a pad occupies 930μ by 150μ in a 0.8μ process. The driver consumes 10 mW + 2 mW/100 MHz, driving a 50-Ω transmission line with a 1-volt supply. A photograph of a portion of a test chip employing these pads is shown in Plate 5.

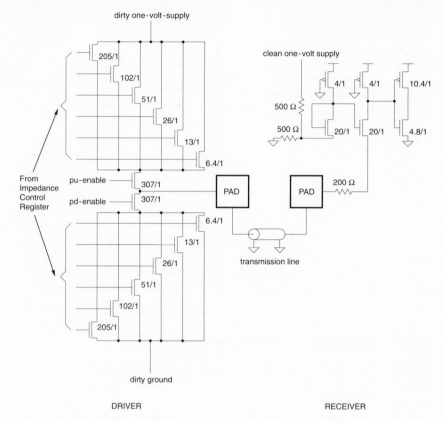

FIGURE 5.92 ECL I/O pad

DRIVER RECEIVER

Another approach to fast low voltage swing chip I/O is shown in Fig. 5.93. This is termed GTL after the inventor, Gunning.[83] Figure 5.93(a) shows two chips utilizing the technique. The bus uses a transmission line with 50Ω termination resistors to a termination voltage of about 1.2 volts. The output driver is an open-drain n pulldown transistor. The input circuit is a differential receiver. Figure 5.93(b) shows an output driver that includes circuitry to limit overshoot, and reduce the turn-off *di/dt*. The $V_{OL(max)}$ is around 0.4V. Figure 5.93(c) shows the input buffer which employs a differential amplifier referenced to an external reference voltage, V_{ref}, which is set to 0.8V. This technique has been used in systems with wide (72 bit) high speed buses.

5.7 Low-power Design

Low-power design with high performance, for battery-operated portable systems, is a strong direction for CMOS system design. In this section we will summarize some of the techniques for achieving low power while maintaining

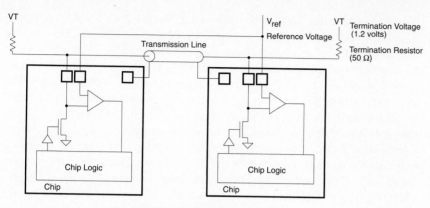

two or more interconnected chips

(a)

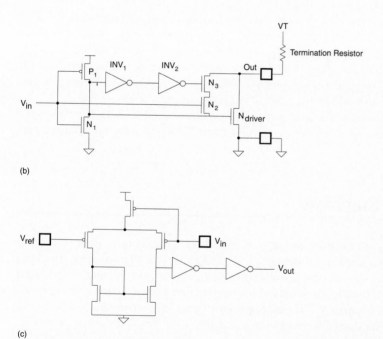

(b)

(c)

FIGURE 5.93 GTL I/O strategy: (a) chip-to-chip connection; (b) output driver; (c) input receiver

performance. This is based on the principles presented so far in this book and on research carried out at the University of California, Berkeley, and elsewhere.[84]

As we have discovered, the power dissipated by a CMOS circuit is the sum of the static power (P_s), the dynamic power (P_d), and the short-circuit power (P_{sc}). P_s may be reduced to that due to leakage, if any circuits that draw DC power such as pseudo-nMOS circuits are eliminated. The dynamic power is dependent on the supply voltage, the stray capacitance, and the frequency of operation. The supply voltage has the greatest effect, so reducing it, in turn, has a large effect on reducing power dissipation. This can be achieved by using an overall lower supply voltage (2.7–3 V is common in battery-operated equipment). This comes at a price in speed. Local supply reductions may be possible where circuitry is not operating at full speed. The stray capacitance may be reduced by using the smallest number of transistors to implement a function. In Chandrakasan et al.,[85] it was found that a static pass gate logic family with reduced threshold voltages was a good performer (see Section 5.4.6). This family retains the advantages of static design while reducing the transistor count substantially. The final parameter that affects speed is the frequency of operation of the circuit. Often, by employing parallelism, it is possible to operate at a lower frequency. While this may increase the area, the overall effect might be lower power dissipation (i.e., the reduction in supply voltage is quadratic, while the speed is inversely proportional to supply voltage).[86,87]

Power-down modes are important for lower power applications. This necessitates techniques for gracefully halting processing and restoring state if necessary when the circuit powers up. An example of one thing to watch is the length of the clock line that remains clocking during power down. It should be stopped as close as possible to the clock pad. A clock line traversing a chip can add appreciably to the power-down current.

In summary, for low-power design one should use the lowest supply voltage and operating frequency consistent with achieving the required performance.

5.8   Summary

In this chapter some alternative CMOS logic circuits have been discussed. The layout and circuit design of CMOS gates was then treated. Clocking strategies were outlined indicating suitable memory elements and logic structures. Finally, the basics of I/O design were covered. Using this material as a base, Chapter 8 will examine some useful subsystems that use a variety of the techniques discussed in this chapter.

5.9   Exercises

In the following exercises, use either the process parameters given at the end of Chapter 2 and hand calculations or an appropriate simulator. Estimate routing capacitance where possible.

1. Design a 32-input NOR gate that is optimized for

   - speed
   - density

   using gate-array complementary logic cells (i.e., NAND, NOR, INV, AOI, OAI), with $W_n = 50\mu$, $L_n = 1\mu$, $W_p = 50\mu$, $L_p = 1\mu$.

2. Redesign the 32-input NOR gate for a standard-cell chip where the increment in transistor size is 5μ but the p-to-n width ratio is fixed at 1. What improvements can you make if the increment is 1μ and n- and p-transistors can assume any width (i.e., a custom approach)?

3. Design a pseudo-nMOS gate that implements the function $F = -(C.(A + B + C) + (D.E.F))$.

4. Using the sea-of-gates base array in Fig. 5.17, design the layout metallization and contacts for the resettable D register shown in Fig. 5.57(a).

5. A chain register employing alternate n- and p-channel pass transistors clocked with a common clock is shown in Fig. 5.72(c). Explain how it works and what are the advantages and limitations of such a register.

6. What limits the V_{DD}-supply voltage level in a given CMOS technology (i.e., how low can you set it and how high can you set it)? Explain what you would expect the effect of these two extremes to be on internal CMOS circuits. Suggest situations where both of these extremes might be of use.

7. Design a pass-transistor network that implements the sum function for an adder:

   $$S = A.B.C + A.-B.-C + -A.-B.C + -A.B.-C$$

8. Design a CVSL gate for the function in the previous exercise.

9. Explain the terms "setup time" and "hold time" in relation to a CMOS D register. If the clock is delayed to a register with regard to the data input, which of these parameters varies and how?

10. Explain how you might estimate and plan the clock-distribution scheme in a chip. Summarize the parameters that are relevant and show how your scheme deals with these.

11. Summarize the approaches you would take to reduce the power dissipation of a CMOS chip that is destined for a palmtop computer.

12. For the Boolean Function Unit shown in Fig. 5.35(d) with the feed-back connection inserted, design the sizes of all transistors in the gate including inverters driving signals P_1–P_4 and the output inverter. Assume that $3\beta_p = \beta_n$ and that the p pull-up is a minimum p-device.

13. Derive the transistor ratios for a pseudo-nMOS NOR gate library for a 3V CMOS process where the V_{OL} noise margin is .2V (assume that $V_{tn} = 0.5$V and $V_{tp} = 0.5$V) using the bias circuit shown in Fig. 5.27.

14. A lower power chip has a clock of 12 MHz. In the power-down mode, the clock driver drives 5 mm of metal1 wire 2μ wide. If the area capacitance of metal is 60 aF/μ^2, what is the power-down dissipation, assuming this is the dominant term? What is the dissipation if the wire is reduced to 50μ?

15. Three current starved inverters are cascaded in a PLL VCO. What kind of inverters would you use to achieve the maximum operating frequency? Explain and/or demonstrate by simulation your proposal.

5.10 References

1. T. Uehara and W. M. van Cleemput, "Optimal layout of CMOS functional arrays," *IEEE Transactions on Computers,* vol. C-30, no. 5, May 1981, pp. 305–311.

2. T. Uehara and W. M. van Cleemput, *op. cit.*

3. Omar Wing, "Interval-graph based circuit layout," *Proceedings, IEEE International Conference on Computer Aided Design,* Santa Clara, Calif.: 1983, pp. 84–85.

4. O. Wing, "Automated gate-matrix layout," *Proc. IEEE International Symposium on Circuits and Systems,* 1982, Rome, Italy, pp. 681–685.

5. Shuo Huang and Omar Wing, "Gate matrix partitioning," *IEEE Transactions on CAD,* vol. 8, no. 7, Jul. 1989, pp. 756–767.

6. Shuo Huang and Omar Wing, "Improved gate matrix layout," *IEEE Transactions on CAD,* vol. 8, no. 8, Aug. 1989, pp. 875–889.

7. Yu Hen Hu and Sao-Jie Chen, "GM Plan: A Gate Matrix Layout Algorithm Based on Artificial Intelligence Planning Techniques," *IEEE Transactions on CAD,* vol. 9, no. 8, Aug. 1990, pp. 836–845.

8. Sung Mo Kang, "A design of CMOS polycells for LSI circuits," *IEEE Transactions on Circuits and Systems,* vol. CAS-28, no. 8, Aug. 1981, pp. 838–843.

9. C. M. Lee, B. R. Chawla, and S. Just, "Automatic generation and characterization of CMOS polycells," *IEEE/ACM Proceedings of the 18th Design Automation Conference,* June 1981, Nashville, Tenn., pp. 220–224.

10. M. A. Brown, M. J. Gasper, J. W. Eddy, and K. D. Kolwicz, "CMOS cell arrays—an alternative to gate arrays," *Proceedings of the Custom Integrated Circuit Conference,* May 1983.

11. M. Shoji, "FET scaling in domino CMOS gates," *IEEE Journal of Solid State Circuits,* vol. SC-20, no. 5, Oct. 1985, pp. 1067–1071.

12. Bernhard Hoppe, Gerd Nevendorf, Doris Schmitt-Landsiedel, and Will Specks, "Optimization of high-speed CMOS logic circuits with analytical models for signal delay, chip area, and dynamic power dissipation," *IEEE Transactions on Computer-Aided Design,* vol. 9, no. 3, Mar. 1990, pp. 236–247.

13. Tohru Furuyama, Yohji Watanabe, Takashi Ohsawa, and Shigeyoshi Watanabe, "A new on-chip voltage converter for submicrometer high-density DRAMs," *IEEE JSSC,* vol. SC-22, no. 3, June 1987, pp. 437–441.

14. Rosalyn B. Ritts, Prasad A. Raje, James D. Plummer, Krishna C. Saraswat, and Kit M. Cham, "Merged BiCMOS logic to extend the CMOS/BiCMOS performance crossover below 2.5-V supply," *IEEE JSSC,* vol.26, no.11, Nov. 1991, pp. 1606–1614.

15. Chih-Liang Chen, "2.5-V bipolar/CMOS circuits for 0.25-μm BiCMOS technology," *IEEE JSSC,* vol. 27, no. 4, Apr. 1992, pp. 485–491.

16. Torkel Arnborg, "Performance predictions of scaled BiCMOS gates using physical simulation," *IEEE JSSC,* vol. 27, no. 5, May 1992, pp. 754–760.

17. Muhammad S. Elrabaa and Mohamed I. Elmasry, "Design and optimization of buffer chains and logic circuits in a BiCMOS environment," *IEEE JSSC,* vol. 27, no. 5, May 1992, pp. 792–801.

18. Samir S. Rofail and Mohamed I. Elmasry, "Analytical and numerical analyses of the delay time of BiCMOS structures," *IEEE JSSC,* vol. 27, no. 5, May 1992, pp. 834–839.

19. Wen Fang, Arthur Brunnschweiler, and Peter Ashburn, "An accurate analytical BiCMOS delay expression and its application to optimizing high-speed BiCMOS circuits," *IEEE JSSC,* vol. 27, no. 2, Feb. 1992, pp. 191–202.

20. Takayasu Sakurai, "A unified theory for mixed CMOS/BiCMOS buffer optimization," *IEEE JSSC,* vol. 27, no. 7, Jul. 1992, pp. 1014–1019.

21. Hyun J. Shin, "Full-swing BiCMOS logic circuits with complementary emitter-follower driver configuration," *IEEE JSSC,* vol. 26, no. 4, Apr. 1991, pp. 578–584.

22. Kenji Sakaue, Yasuro Shobatake, Masahiko Motoyama, Yoshinari Kumaki, Satoru Takatsuka, Shigeru Tanaka, Hiroyuki Hara, Kouji Matsuda, Shuji Kitaoka, Makoti Noda, Youichiro Niitsu, Masayuki Norishima, Hiroshi Momose, Kenji Maeguchi, Manabu Ishibe, Shoichi Shimizu, and Toshikazu Kodama, "A 0.8-mm BiCMOS ATM switch on an 800-Mb/s asynchronous buffered Banyan network," *IEEE JSSC,* vol. 26, no. 8, Aug. 1991, pp. 1133–1144.

23. Hiroyuki Hara, Takayasu Sakurai, Makoto Noda, Tetsu Nagamatsu, Katsuhiro Seta, Hiroshi Momose, Youichirou Niitsu, Hiroyuki Miyakawa, and Yoshinori Watanabe, "A 0.5-μm 2M-Transistor BiPNMOS channelless gate array," *IEEE JSSC,* vol. 26, no. 11, Nov. 1991, pp. 1615–1620.

24. Satoru Aikawa, Yasuhisa Nakamura, and Hitoshi Takanashi, "Multipurpose high-coding-gain 0.8-μm BiCMOS-VLSI's for high-speed multilevel trellis-coded modulation," *IEEE JSSC,* vol. 26, no. 11, Nov. 1991, pp. 1700–1707.

25. Kazuo Yano, Mitsuru Hiraki, Shiji Shukuri, Yasuo Onose, Mitsuru Hirao, Nagatoshi Ohki, Takashi Nishida, Koichi Seki, and Katsuhiro Shimohigashi, "Quasi-complementary BiCMOS for sub-3-V digital circuits," *IEEE JSSC,* vol. 26, no. 11, Nov. 1991, pp. 1708–1719.

26. S. H. K. Embabi, A. Bellaouar, M. I. Elmasry, and R. A. Hadaway, "New full-voltage-swing BiCMOS buffers," *IEEE JSSC,* vol. 26, no. 2, Feb. 1991, pp. 150–153.

27. Chung-Yu Wu, Jinn-Shyan Wang, and Ming-Kai Tsai, "The analysis and design of CMOS multidrain logic and stacked multidrain logic," *IEEE JSSC*, vol. SC-22, no. 1, Feb. 1987, pp. 47–56.

28. Siegfried K. Wiedmann, "Advancements in bipolar VLSI circuits and technologies," *IEEE JSSC*, vol. SC-19, no. 3, June 1984, pp. 282–291.

29. Mark G. Johnson, "A symmetric CMOS NOR gate for high speed applications," *IEEE JSSC*, vol. SC-23, no. 5, Oct. 1988, pp. 1233–1236.

30. Kenneth J. Schultz, Robert J. Francis, and Kenneth C. Smith, "Ganged CMOS: trading standby power for speed," *IEEE JSSC*, vol. SC-25, no. 3, June 1990, pp. 870–873.

31. Kenneth J. Schultz, et al., *op. cit.*

32. Yasoji Susuki, Kaichiro Odagawa, and Toshio Abe, "Clocked CMOS calculator circuitry," *IEEE JSSC*, vol. SC-8, no. 6, Dec. 1973, pp. 462–469.

33. Takayasu Sakurai, Kazutaka Nogami, Masakazu Kakumu, and Tetsuya Iizuka, "Hot-carrier generation in submicrometer VLSI environment," *IEEE JSSC*, vol. SC-21, no. 1, Feb. 1986, pp. 187–191.

34. Damu Radhakrishnan, Sterling R. Whitaker and Gary K. Maki, "Formal design procedures for pass transistor switching circuits," *IEEE JSSC*, vol. SC-20, no. 2, Apr. 1985, pp. 531–536

35. Damu Radhakrishnan, et al., *op. cit.*

36. C. A. Mead and L. Conway, *Introduction to VLSI Systems*, Reading, Mass.: Addison-Wesley, 1980.

37. Guy L. Steele, Jr., "Common Lisp—The Language," Burlington, Mass.: Digital, 1984, pp. 222–223.

38. R. H. Krambeck, Charles M. Lee, and Hung-Fai Stephen Law, "High speed compact circuits with CMOS," *IEEE JSSC*, vol. SC-17, no. 3, June 1982, pp. 614–619.

39. V. Friedman and S. Liu, "Dynamic logic CMOS circuits," *IEEE JSSC*, vol. SC-19, no. 2, Apr. 1984, pp. 263–266.

40. Nelson F. Gonclaves and Hugo J. DeMan, "NORA: a racefree dynamic CMOS technique for pipelined logic structures," *IEEE JSSC*, vol. SC-18, no. 3, June 1983, pp. 261–266.

41. C. M. Lee and E. W. Szeto, "Zipper CMOS," *IEEE Circuits and Systems Magazine*, May 1986, pp. 10–16.

42. L. G. Heller, W. R. Griffin, J. W. Davis, and N. G. Thoma, "Cascade voltage switch logic: a differential CMOS logic family," Proceedings of the IEEE International Solid State Circuits Conference, Feb. 1984, San Francisco, Calif., pp. 16–17.

43. Timothy A. Grotjohn and Bernd Hoefflinger, "Sample-set differential logic (SSDL) for complex high-speed VLSI," *IEEE JSSC*, vol. SC-21, no. 2, Apr. 1986, pp. 367–369.

44. Leo C. M. Pfennings, Wim G. J. Mol, Joseph J. J. Bastiens, and Jan M. F. Van Dijk, "Differential split-level CMOS logic for subnanosecond speeds," *IEEE JSSC*, vol. SC-20, no. 5, Oct. 1985, pp. 1050–1055.

45. Kan M. Chu and David I. Pulfrey, "Design procedures for differential cascode voltage switch circuits," *IEEE JSSC*, vol. SC-21, no. 6, Dec. 1986, pp. 1082–1087.

46. Thomas D. Simon, "A fast static CMOS NOR gate," in *Proceedings of the 1992 Brown/MIT Conference on Advanced Research in VLSI and Parallel Systems* (Thomas Knight and John Savage, eds.) Cambridge, Mass.: MIT Press, pp. 180–192.

47. Morton H. Lewin, *Logic Design and Computer Organization,* Reading, Mass.: Addison-Wesley, 1983, Chapter 3.

48. David Fan, C. Thomas Gray, William Faflow, Thomas Hughes, Wentai Liu, and Ralph K. Cavin, "A CMOS parallel adder using wave pipelining," in *Proceedings of the 1992 Brown/MIT Conference on Advanced Research in VLSI and Parallel Systems* (Thomas Knight and John Savage, eds.) Cambridge, Mass.: MIT Press, pp. 147–164.

49. N. Ohwada, T. Kimura, and M. Doken, "LSIs for digital signal processing," *IEEE JSSC,* vol. SC-14, no. 2, Apr. 1979, pp. 221–239.

50. Shih-Lien Lu and Milos Ercegovac, "A novel CMOS implementation of double-edge-triggered flip-flops," *IEEE JSSC,* vol. 25, no. 4, Aug. 1990, pp. 1008–1010.

51. M. Afghahi and J. Yuan, "Double-edge-triggered D-flip-flops for high-speed CMOS circuits," *IEEE JSSC,* vol. 26, no. 8, Aug. 1991, pp. 1168–1170.

52. H. Jonathan Chao and Cesar A. Johnston, "Behavior analysis of CMOS D flip-flops," *IEEE JSSC,* vol. 24, no. 5, Oct. 1989, pp. 1454–1458.

53. Daniel W. Dobberpuhl, Richard T. Witek, Randy Allmon, Robert Anglin, David Bertucci, Sharon Britton, Linda Chao, Robert A. Conrad, Daniel E. Dever, Bruce Gieseke, Soha M. N. Hassoun, Gregory W. Hoeppner, Kathryn Kuchler, Maureen Ladd, Burton M. Leary, Liam Madden, Edward J. McLellan, Derrick R. Meyer, James Montanaro, Donald A. Priore, Vidya Rajagopalan, Sridhar Samudrala, and Sribalan Santhanam, "A 200-MHz 64-b Dual-Issue CMOS Microprocessor," *IEEE JSSC,* vol. 27, no. 11, Nov. 1992, pp. 1555–1567.

54. F. A. Gardner, "Charge-pump phase-locked loops," *IEEE Transactions on Communications,* vol. COM-28, Nov. 1980, pp. 1849–1858.

55. Kozaburo Kurita, Takashi Hotta, Tetsuo Nakano, and Nouaki Kitamura, "PLL-based BiCMOS on-chip clock generator for very high-speed microprocessor," *IEEE JSSC,* vol. 26, no. 4, Apr. 1991, pp. 585–589.

56. Deog-Kyoon Jeong, Gaetano Borriello, David A. Hoodges, and Randy H. Katz, "Design of PLL-based clock generation circuits," *IEEE JSSC,* vol. SC-22, no. 2, Apr. 1987, pp. 255–261.

57. Ian A. Young, Jeffrey K. Greason, and Keng L. Wong, "A PLL clock generator with 5–110 MHz of lock range for microprocessors," *IEEE JSSC,* vol. 27, no. 11, Nov. 1992, pp. 1599–1607.

58. Mark G. Johnson and Edwin L. Hudson, "A variable delay line PLL for CPU-coprocessor synchronization," *IEEE JSSC,* vol. 23, no. 5, Oct. 1988, pp. 1218–1223.

59. T. J. Chaney and F. U. Rosenberger, "Anomalous behavior of synchronizer and arbiter circuits," *IEEE Transactions on Computers,* vol. C-22, Apr. 1973, pp. 421–422.

60. Fred Rosenberger and Tomas J. Chaney, "Flip-flop resolving time test circuit," *IEEE JSSC,* vol. SC-17, no. 4, Aug. 1982, pp. 731–738.

61. Lance A. Glasser and Daniel W. Dobberpuhl, "The Design and Analysis of VLSI Circuits," Reading, Mass.: Addison-Wesley, 1985, pp. 360–365.

62. F. U. Rosenberger, private communication.

63. Stephen T. Flannagan, "Synchronization Reliability in CMOS Technology," *IEEE JSSC,* vol. SC-20, no. 4, Aug. 1985, pp. 880–882.

64. Harry J. M. Veendrick, "The behavior of flip flops used as synchronizers and prediction of their failure rate," *IEEE JSSC,* vol. SC-15, no. 2, Apr. 1980, pp. 169–176.

65. Jakob H. Hohl, Wendell R. Larsen, and Larry C. Schooley, "Prediction of error probabilities for integrated digital synchronizers," *IEEE JSSC,* vol. SC-19, no. 2, Apr. 1984, pp. 236–244.

66. Lee-Sup Kim and Robert W. Dutton, "Metastability of CMOS latch/flip flop," *IEEE JSSC,* vol. 25, no. 4, Aug. 1990, pp. 942–951.

67. Fred U. Rosenberger and Charles E. Molnar, "Comments on 'Metastability of CMOS latch/flip flop,'" *IEEE JSSC,* vol. 27, no. 1, Jan. 1992, pp. 128–130.

68. Robert W. Dutton, "Reply to Comments on 'Metastability of CMOS latch/flip flop,'" *IEEE JSSC,* vol. 27, no. 1, Jan. 1992, pp. 121–132.

69. Carver Mead and Lynn Conway, *Introduction to VLSI Systems,* Reading, Mass.: Addison-Wesley, 1980, Chapter 7.

70. Nelson F. Goncalves and Hugo J. DeMan, "NORA: a racefree dynamic CMOS technique for pipelined logic structures," *IEEE JSSC,* vol. SC-18, no. 3, June 1983, pp. 261–266.

71. David Renshaw and Choon How Lau, "Race-free clocking of CMOS pipelines using a single global clock," *IEEE JSSC,* vol. SC-25, no. 3, June 1990, pp. 766–769.

72. W. M. Penny and L. Lau, *MOS Integrated Circuits: Theory, Fabrication, Design and Systems Applications of MOS LSI,* New York: Van Nostrand, Reinhold; 1973, Chapter 5.

73. Neil Weste and Kamran Eshraghian, *Principles of CMOS VLSI Design, A Systems Perspective,* Reading, Mass.: Addison-Wesley, 1984, Chapter 5.

74. E. Fujishin, K. Garret, M. P. Louis, R. F. Motta, and M. D. Hartranft, "Optimized ESD protection circuits for high speed MOS/VLSI," *IEEE Proceedings of the Custom Integrated Circuits Conference,* May 1984, pp. 569–573.

75. Darius Tanksalvala, Joel Lamb, Michael Buckley, Bruce Long, Sean Chapin, Jonathon Lotz, Eric Delano, Richard Luebs, Keith Erskine, Scott McMullen, Mark Forsyth, Robert Novak, Tony Gaddis, Doug Quarnstrom, Craig Gleason, Eshan Rashid, Daniel Halperin, Leon Sigal, Harlan Hill, Craig Simpson, David Hollenbeck, John Spencer, Robert Horning, Hoang Tran, Thomas Hotchkiss, Duncan Weir, Donald Kipp, John Wheeler, Patrick Knebel, Jeffrey Yetter, and Charles Kohlhardt, "A 90 MHz RISC CPU designed for sustained performance, *Proceedings of the IEEE Solid State Circuits Conference,* Feb. 1990, San Francisco, Calif., pp. 52–53.

76. Dan Dobberpuhl, et al., "A 200-MHz 64-b Dual-Issue CMOS Microprocessor," *op. cit.*

77. Seigo Suzuki, Kiyoyuki Kawai, and Kunio Muramatsu, "A CMOS chip pair for digital TV," *IEEE JSSC,* vol. SC-22, no. 5, Oct. 1987, pp. 835–840.

78. Tom Knight and Alex Krymm, "A self-terminating low-voltage-swing CMOS output driver," *IEEE JSSC,* vol. 23, no. 2, Apr. 1988, pp. 457–464.

79. Hans-Jurgen Schumacher, Jan Dikken, and Evert Seevinck, "CMOS subnanosecond true-ECL output buffer," *IEEE JSSC,* vol. 25, no. 1, Feb. 1990, pp. 150–154.

80. Michel S. J. Steyaert, Wout Bijker, Pieter Vorenkmap, and Jan Sevenhans, "ECL-CMOS and CMOS-ECL interface in 1.2μm CMOS for 150MHz digital ECL data transmission systems," *IEEE JSSC,* vol. 26, no. 1, Jan. 1991, pp. 18–24.

81. H. B. Bakoglu, *Circuits, Interconnections and Packaging for VLSI,* Reading, Mass.: Addison-Wesley, 1990, Chapter 6.

82. André DeHon, Thomas Knight, and Thomas Simon, "Automatic Impedance Control," to appear in the Proceedings of the IEEE Solid State Circuits Conference, Feb. 1993, San Francisco, CA.

83. Bill Gunning, Leo Yuan, Trung Nguyen, and Tony Wong, "A CMOS Low-Voltage-Swing Transmission-Line Transceiver," *IEEE Proceedings of the International Solid State Circuits Conference,* Feb. 1992, San Francisco, Calif., pp. 58–59.

84. Anatha P. Chandrakasan, Samuel Sheng, and Robert W. Brodersen, "Low-power CMOS digital design," *IEEE JSSC,* vol. 27, no. 4, Apr. 1992, pp. 473–484.

85. Anatha P. Chandrakasan, et al., op. cit.

86. Anatha P. Chandrakasan, et al., *op. cit.*

87. D. Liu and C. Svensson, "Trading speed for low power by choice of supply and threshold voltages," *IEEE JSSC,* vol 28, no. 1, Jan. 1993, pp. 10–17.

88. Jiren Yuan and Christer Svensson, "High-Speed CMOS Circuit Technique," *IEEE JSSC,* vol. 24, no.1, Feb. 1989, pp. 62–70.

Plate Captions

Plate 1   Cross section of a CMOS inverter in an n-well process

Plate 2   nWell CMOS design rules

Plate 3   Symbolic layouts for the CMOS inverter

Plate 4   Symbolic layout for an I/O pad

Plate 5   Chip microphotograph of CMOS ECL level automatic impedance controlled pads

Plate 6   Three level metal standard cell symbolic layout

Plate 7   Combinational adder mask layouts

Plate 8   Mask layout for 6 transistor static RAM

Plate 9   Metal3 standard cell layout for boundary scan tap controller

Plate 10   4-bit Manchester adder symbolic layout

Plate 11   Representative symbolic layouts for filter tap datapath

Plate 12   Chip microphotograph of Ghost Canceller chip

Plate 13   Chip microphotograph of 6-bit flash A/D converter

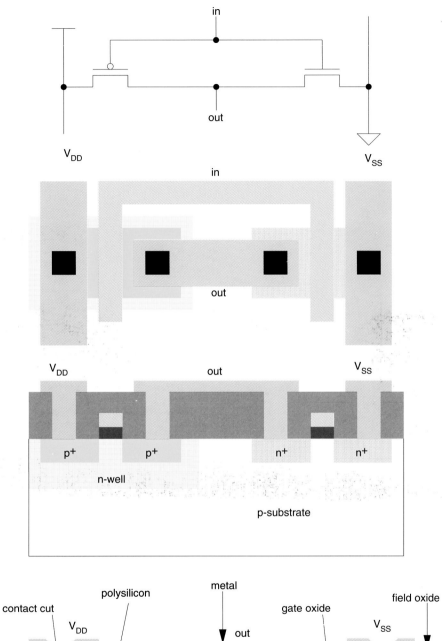

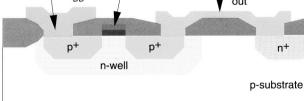

Plate 1

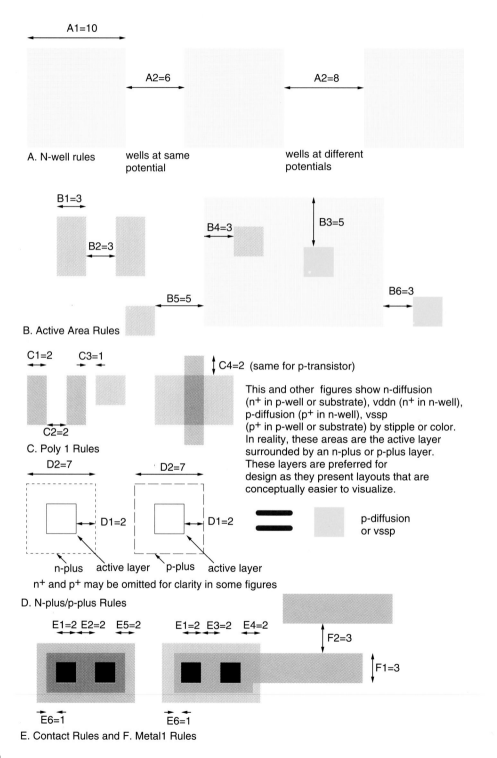

A1=10

A2=6

A2=8

A. N-well rules

wells at same potential

wells at different potentials

B1=3

B2=3

B4=3

B3=5

B5=5

B6=3

B. Active Area Rules

C1=2   C3=1

C2=2

C4=2 (same for p-transistor)

C. Poly 1 Rules

This and other figures show n-diffusion (n⁺ in p-well or substrate), vddn (n⁺ in n-well), p-diffusion (p⁺ in n-well), vssp (p⁺ in p-well or substrate) by stipple or color. In reality, these areas are the active layer surrounded by an n-plus or p-plus layer. These layers are preferred for design as they present layouts that are conceptually easier to visualize.

D2=7   D2=7

D1=2   D1=2

p-diffusion or vssp

n-plus   active layer   p-plus   active layer

n⁺ and p⁺ may be omitted for clarity in some figures

D. N-plus/p-plus Rules

E1=2 E2=2 E5=2   E1=2 E3=2 E4=2

F2=3

F1=3

E6=1   E6=1

E. Contact Rules and F. Metal1 Rules

Plate 2a

G1=2

H1=3

G4=1 G2=3 G3=1

H2=4

G. Via Rules and
H. Metal2 Rules

J1=8

J4=2 I2=3 J3=2
I1=2

J2=5

I. Via2 Rules and
J. Metal3 Rules

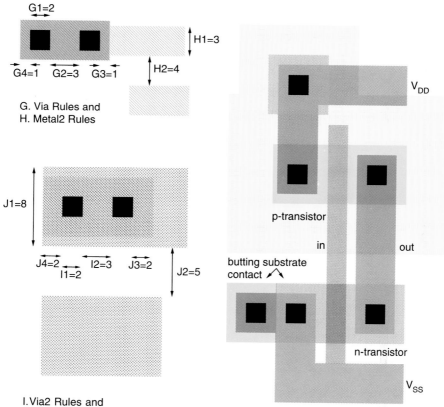

V$_{DD}$

p-transistor

in   out

butting substrate
contact

n-transistor

V$_{SS}$

CMOS n-well inverter designed with Lambda Rules
n$^+$ and p$^+$ layers are omitted

Plate 2b

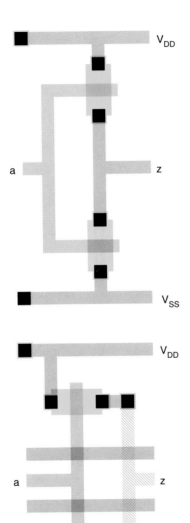
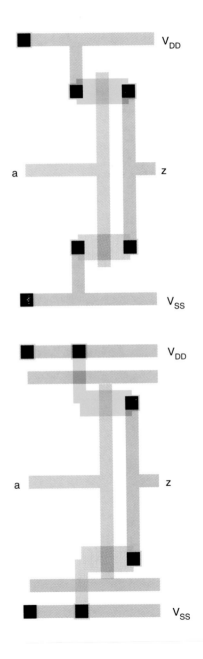

Plate 3a

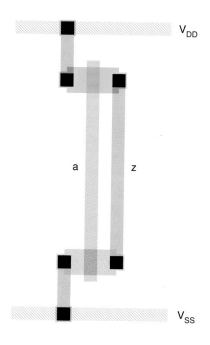

V_{DD}

a z

V_{SS}

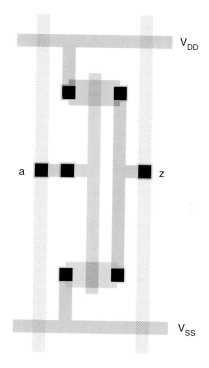

V_{DD}

a z

V_{SS}

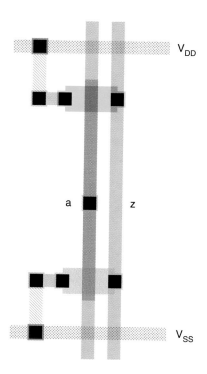

V_{DD}

a z

V_{SS}

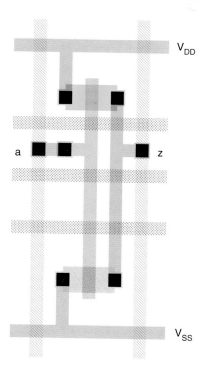

V_{DD}

a z

V_{SS}

Plate 3b

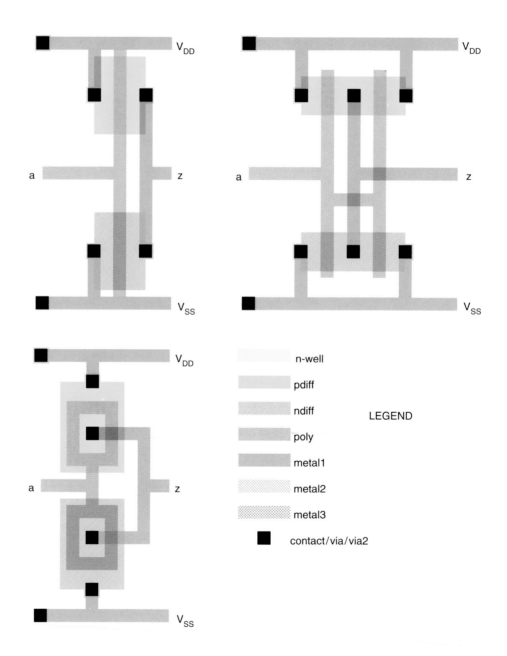

Plate 3c

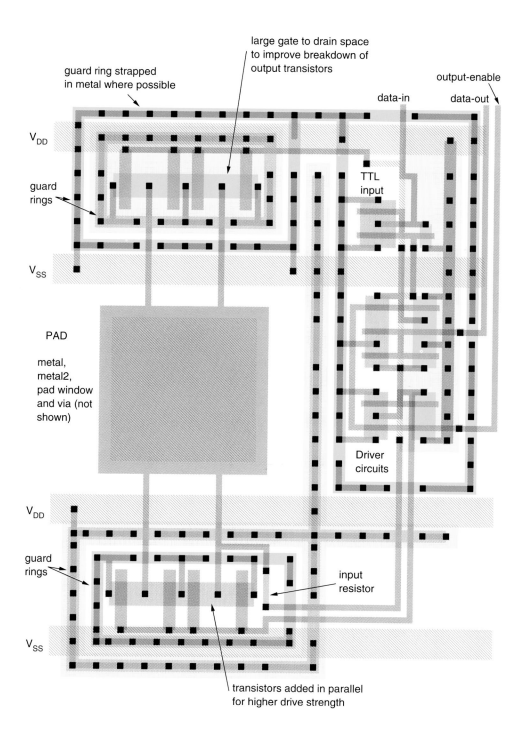

guard ring strapped
in metal where possible

large gate to drain space
to improve breakdown of
output transistors

output-enable

data-in

data-out

V_{DD}

guard
rings

TTL
input

V_{SS}

PAD

metal,
metal2,
pad window
and via (not
shown)

Driver
circuits

V_{DD}

guard
rings

input
resistor

V_{SS}

transistors added in parallel
for higher drive strength

Plate 4

Plate 5

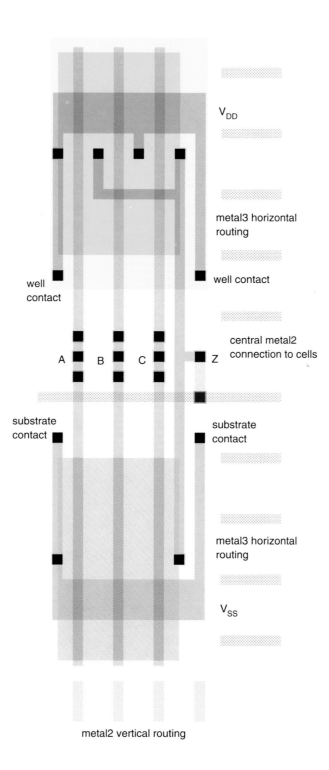

V_{DD}

metal3 horizontal
routing

well contact

well
contact

A   B   C   Z

central metal2
connection to cells

substrate
contact

substrate
contact

metal3 horizontal
routing

V_{SS}

metal2 vertical routing

Plate 6

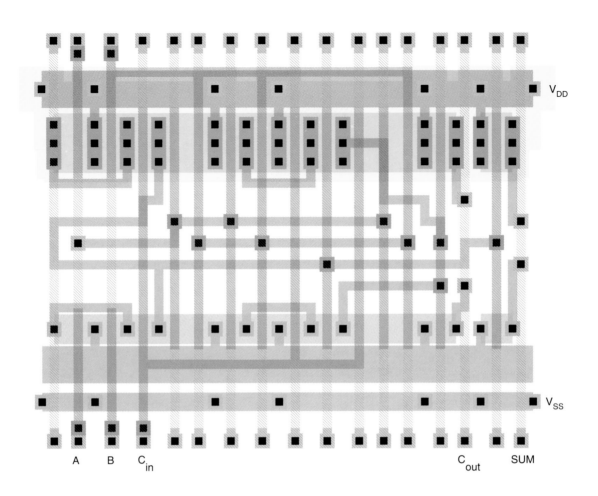

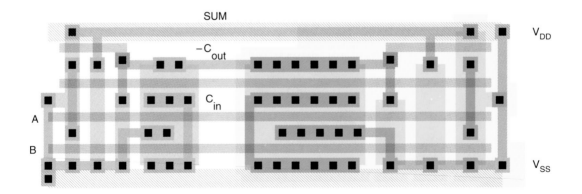

Plate 7

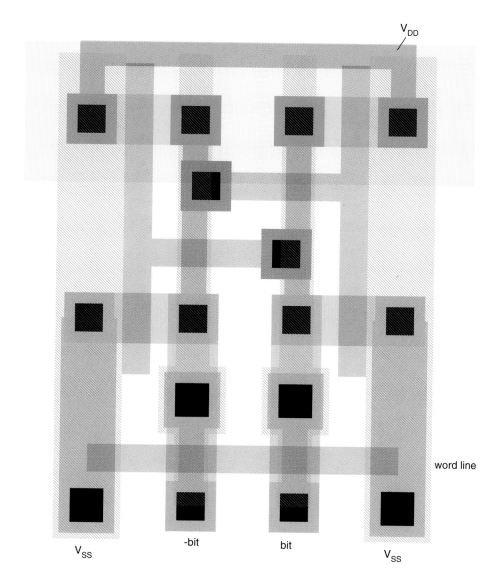

V_DD

word line

V_SS   -bit   bit   V_SS

Plate 8

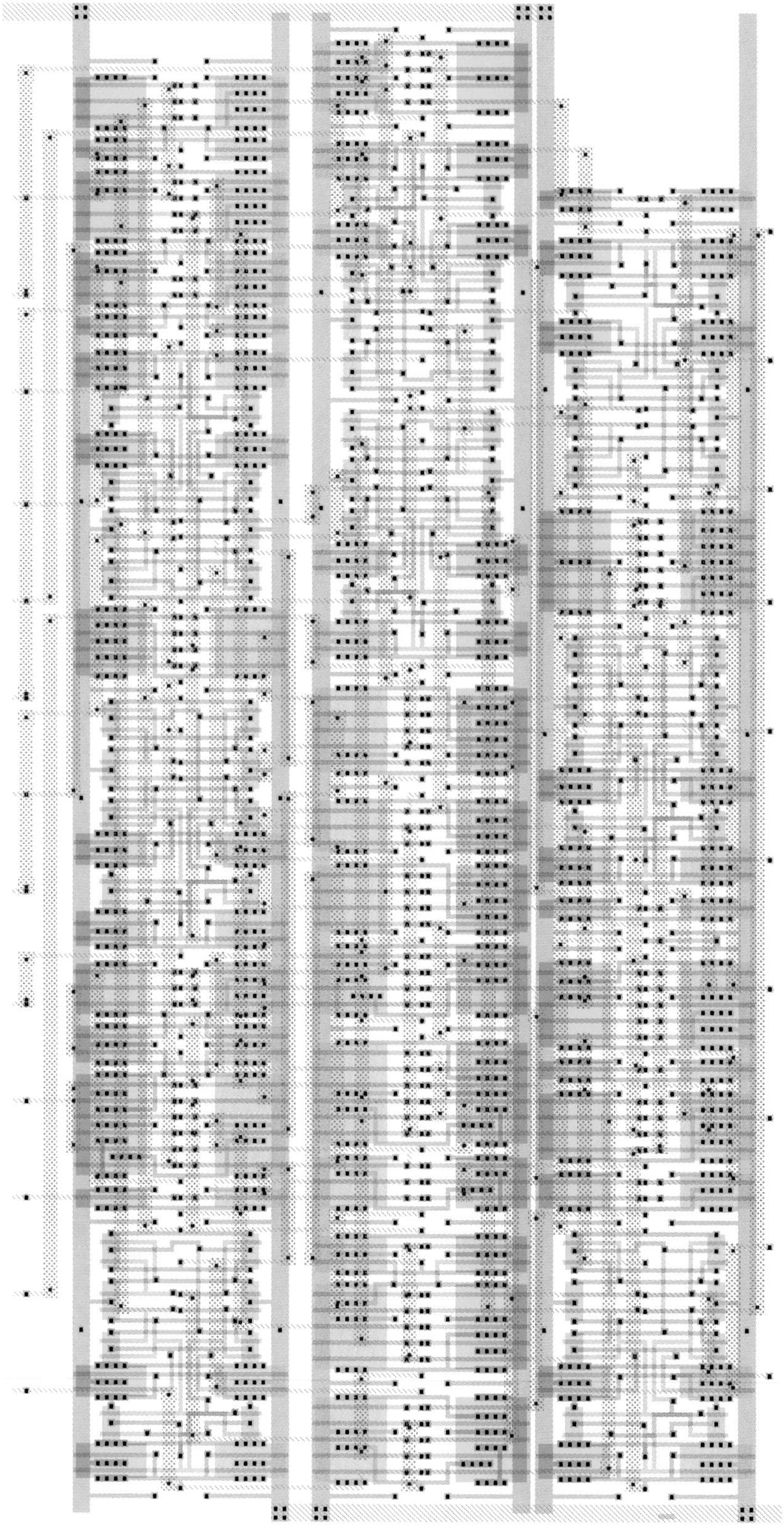

Plate 9

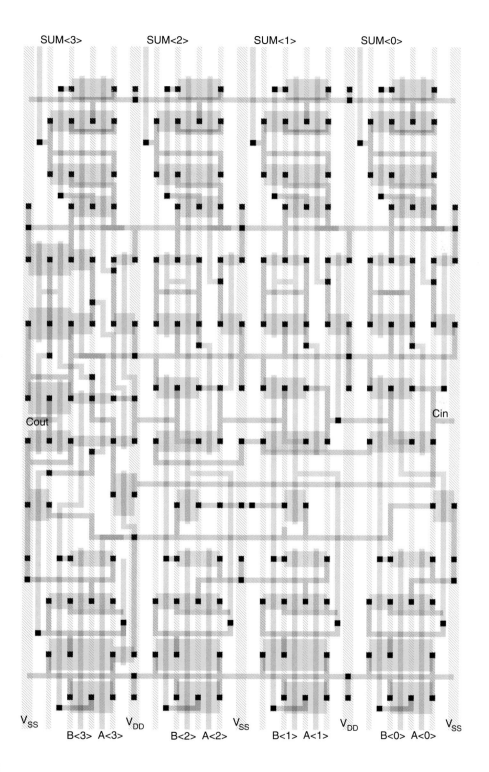

Plate 10

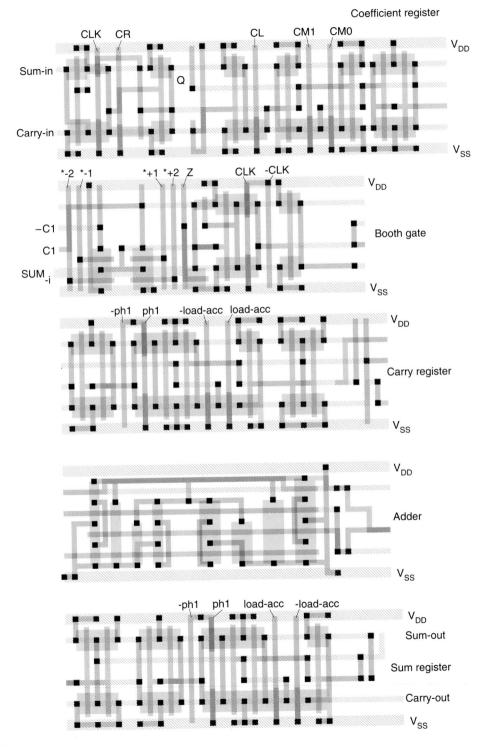

Plate 11

Plate 12

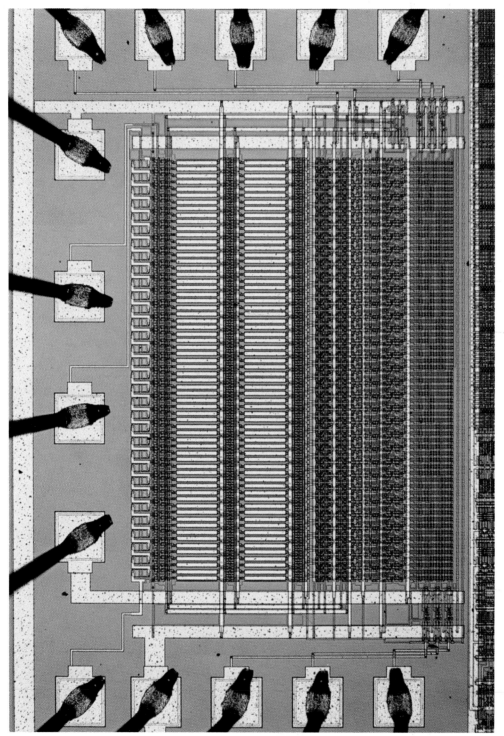

Plate 13

SYSTEMS DESIGN AND DESIGN METHODS

This Part starts with Chapter 6 examining a structured design approach and its application to CMOS system design. This is followed by a discussion of CMOS chip implementation options ranging from Field Programmable Gate Arrays (FPGAs) to full custom layout, which illustrates the trade-offs between implementation, design complexity, and time to market. The chapter then discusses a variety of CMOS design automation options and the various design tools used for CMOS design.

Chapter 7 deals with the important problem of testing CMOS circuits by introducing the reader to the test process, followed by an explanation of test nomenclature. Various methods of designing a testable CMOS circuit are then treated. This treatment centers on a structured approach to testing.

Chapter 8 provides an extensive set of subsystem examples starting with coverage of datapaths. Adders are treated at great depth. The chapter continues with a treatment of the design of memories and concludes by examining various techniques for implementing CMOS control logic.

PART 2

CMOS DESIGN METHODS

6

6.1 Introduction

In Chapter 1 we found that the design description for an integrated circuit may be described in terms of three domains, namely: (1) the behavioral domain, (2) the structural domain, and (3) the physical domain. In each of these domains there are a number of design options that may be selected to solve a particular problem. For instance, at the behavioral level, the freedom to choose, say, a sequential or a parallel algorithm is available. In the structural domain, the decision about which particular logic family, clocking strategy, or circuit style to use is initially unbound. At the physical level, how the circuit is implemented in terms of chips, boards, and cabinets also provides many options to the designer. These domains may be hierarchically divided into levels of design abstraction. Classically these have included the following:

- Architectural or functional level.
- Register-transfer level (RTL).
- Logic level.
- Circuit level.

The relationship between description domains and levels of design abstraction are elegantly shown by the Y-chart[1,2] in Fig. 6.1, which was introduced in Chapter 1. In this diagram, the three radial lines represent the three description domains, namely the behavioral, structural, and physical

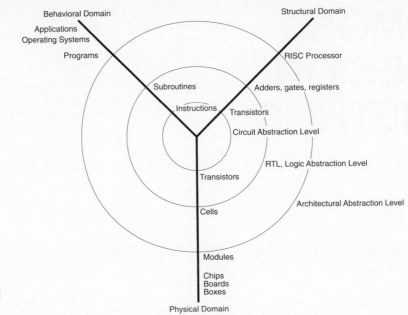

FIGURE 6.1 Y-chart showing description domains and levels of design abstraction

domains. Along each line are enumerated types of objects in that domain. Circles represent levels of similar design abstraction: the architectural, logic, and circuit levels. The particular abstraction levels and design objects may differ slightly, depending on the design method.

In this chapter we will examine the means by which we transform a description in one domain into a description in another domain. We begin by discussing some of the guiding principles that apply to most engineering projects. Then the various design strategies available to the CMOS IC designer are surveyed, ranging from very fast prototyping or small-volume approaches to the more labor-intensive custom design approaches. The CAD tools necessary to achieve the design strategies are then summarized. Finally, we examine the economics of design, which can guide us to the right selection of an implementation strategy.

6.2 Design Strategies

6.2.1 Introduction

The economic viability of an IC is in large part affected by the productivity that can be brought to bear on the design. This in turn depends on the efficiency with which the design may be converted from concept to architecture, to logic and memory, to circuit and hence to a physical layout. A good VLSI design system should provide for consistent descriptions in all three description domains (behavioral, structural, and physical) and at all relevant levels of abstraction (architecture, RTL, logic, circuit). The means by which this is accomplished may be measured in various terms that differ in importance based on the application. These design parameters may be summarized in terms of

- Performance—speed, power, function, flexibility.
- Size of die (hence cost of die).
- Time to design (hence cost of engineering and schedule).
- Ease of test generation and testability (hence cost of engineering and schedule).

Design is a continuous trade-off to achieve adequate results for all of the above parameters. As such, the tools and methodologies used for a particular chip will be a function of these parameters. Certain end results have to be met (i.e., the chip must conform to performance specifications), but other constraints may be a function of economics (i.e., size of die affecting yield) or even subjectivity (i.e., what one designer finds easy, another might find incomprehensible).

Given that the process of designing a system on silicon is complicated, the role of good VLSI-design aids is to reduce this complexity, increase productivity, and assure the designer of a working product. A good method of simplifying the approach to a design is by the use of constraints and abstractions. By using constraints the tool designer has some hope of automating procedures and taking a lot of the "legwork" out of a design. By using abstractions, the designer can collapse details and arrive at a simpler concept with which to deal.

In this chapter we will examine design methodologies that allow a variation in the freedom available in the design strategy. The choice, assuming all styles are equally available, should be entirely economic. According to function, suitable design methods are selected. It may be found that due to inefficiencies in layout, some styles will not be capable of implementing the function. Following these steps, the required die cost is estimated and the quickest means of achieving that die should be chosen. We will focus on structured approaches to design since they offer the best prospects of dealing with large and diverse VLSI problems of the present and future.

6.2.2   Structured Design Strategies

The successful implementation of almost any integrated circuit requires an attention to the details of the engineering design process. Over the years a number of structured design techniques have been developed to deal with both complex hardware and software projects. Not surprisingly the techniques have a great deal of commonality. Rigorous application of these techniques can drastically alter the amount of effort that has to be expended on a given project and also, in all likelihood, the chances of a successful conclusion. Whether under consideration is a small chip designed by a single designer or a large system designed by a team of designers, the basic principles of structured design will improve the prospects of success. In the following sections some of the classical techniques for reducing the complexity of IC design will be summarized.[3,4]

6.2.3 Hierarchy

The use of hierarchy, or "divide and conquer," involves dividing a module into submodules and then repeating this operation on the submodules until the complexity of the submodules is at an appropriately comprehensible level of detail. This parallels the software case where large programs are split into smaller and smaller sections until simple subroutines, with well-defined functions and interfaces, can be written. As we have seen, a design may be expressed in terms of three domains. We can employ a "parallel hierarchy" in each domain to document the design. For instance, an adder may have a subroutine that models the behavior, a gate-connection diagram that specifies the circuit structure, and a piece of layout that specifies the physical nature of the adder. Composing the adder into other structures can proceed in parallel for all three domains, with domain-to-domain comparisons ensuring that the representations are consistent.

At a system level, the use of hierarchy allows one to specify single-designer projects, at which level the schedule is proportional to the number of available personnel.

Example

To illustrate the principle of hierarchy consider the top-level diagram of a raster-graphics vector generator that includes an 8-bit difference engine, shown in Fig. 6.2(a). This engine may be used for a variety of graphics algorithms, including line drawing and linear shading. Operation in the case of drawing a line on a raster display consists of loading the X, Y, count, and direction registers with the initial (X,Y) point, length of the line, and up/down-count control data for the X and Y counters. The difference engine block is loaded with three values—A, B, and C—which are derived from the parameters of the line to be drawn.

A diagram of the difference engine is shown in Fig. 6.2(b). It consists of an A, B, and C register, an adder, and two multiplexers or muxes. The multiplexers, registers, and adder may be decomposed into 1-bit units. The hierarchy is stopped at the level where modules are defined in terms of simulation models and physical layouts. For instance, the adder, multiplexer, and register might be standard cells. Similar decompositions could be completed for the other modules in Fig. 6.2(a).

The hierarchy defined above is a structural hierarchy that reflects functionality, such as the adding, multiplexing, or storing state. An alternative hierarchy for the difference engine is shown in Fig. 6.3, where 8 identical "bit-slices" have been built. Each bit slice has one element of the engine

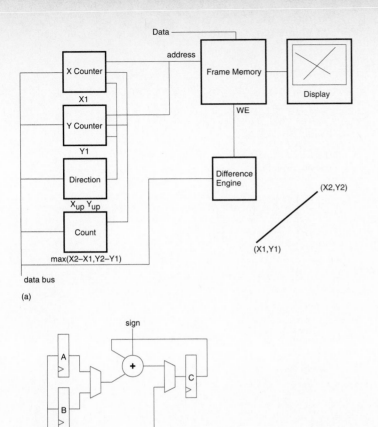

(a)

data bus

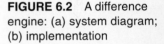

(b)

data bus

8

FIGURE 6.2 A difference engine: (a) system diagram; (b) implementation

shown in Fig. 6.2(b). This is known as a "physical hierarchy" because it might be the decomposition used to build an n-bit difference engine layout. Thus there are at least two "disjoint" hierarchies describing the same structure. The hierarchies "join" at the difference-engine level. Generally, it is good practice to maintain identical hierarchies between the function, structure, and physical aspects of a design because this allows consistent checks between description domains from the lowest level of the hierarchy to the very top levels. Frequently, if the physical hierarchy is designed first without a structural or functional hierarchy, it will be found that the resulting hierarchy is cumbersome. On the other hand structural hierarchies may be defined that do not map well to physical constraints. For instance, consider the floorplan shown in Fig. 6.4 where module *A* has to fit within a certain area constraint. Module *B* has space for some of the contents of Module *A* but, due to the structural hierarchy, the floorplan in Fig. 6.4(a) results. Usually, after a few iterations the physical and structural hierarchies may be reconciled (Fig. 6.4b). Many times the issue is moot because an automatic layout system is able to take the structural hierarchy and create a layout that meets both timing and area requirements.

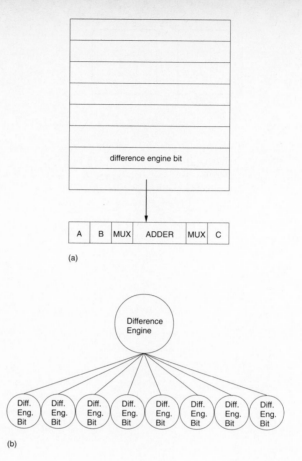

(a)

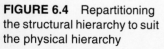

FIGURE 6.3 A physical hierarchy for the difference engine

(b)

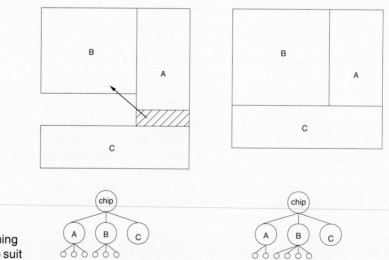

FIGURE 6.4 Repartitioning the structural hierarchy to suit the physical hierarchy

(a)

(b)

386

6.2.4   Regularity

Hierarchy involves dividing a system into a set of submodules. However, hierarchy alone does not necessarily solve the complexity problem. For instance, we could repeatedly divide the hierarchy of a design into different submodules but still end up with a large number of different submodules. With regularity as a guide, the designer attempts to divide the hierarchy into a set of similar building blocks. The use of iteration to form arrays of identical cells is an illustration of the use of regularity in an IC design. However, extended use may be made of regular structures to simplify the design process. For instance, if the designer were constructing a "datapath," the interface between modules (power, ground, clocks, busses) might be common but the internal details of modules may differ according to function. Regularity can exist at all levels of the design hierarchy. At the circuit level, uniformly sized transistors might be used rather than manually optimizing of each device. At the logic-module level, identical gate structures might be employed. At higher levels, one might construct architectures that use a number of identical processor structures. By using regularity in the ways mentioned, a design may be judged correct by construction. Methods for formally proving the correctness of a design may also be aided by regularity.

Regularity allows an improvement in productivity by reusing specific designs in a number of places, thereby reducing the number of different designs that need to be completed.

Example

Continuing the example of the difference engine in Fig. 6.2(b), the multiplexer, adder, and register modules may be defined in terms of identical CMOS inverters and tristate inverters as illustrated in Fig. 6.5. The counters shown in Fig. 6.2(a) might use the same adder, register, and multiplexer used in the difference engine. For every different module that is used (no matter what level), a variety of design checks have to be performed—functional verification, timing verification, layout-connectivity verification, etc. By identifying common operations at a high level, regularization can reduce the number of different modules that need to be designed and verified (i.e., counting = adding 1 = adding). This principle applies at all levels of hierarchy.

6.2.5   Modularity

The tenet of modularity adds to hierarchy and regularity the condition that submodules have well-defined functions and interfaces. If modules are "well-formed," the interaction with other modules may be well-characterized. The notion of "well-formed" may differ from situation to situation, but a good starting point is the criteria placed on a "well-formed" software subroutine. First of all, a well-defined interface is required. In the case of software this is an argument list with typed variables. In the IC case this

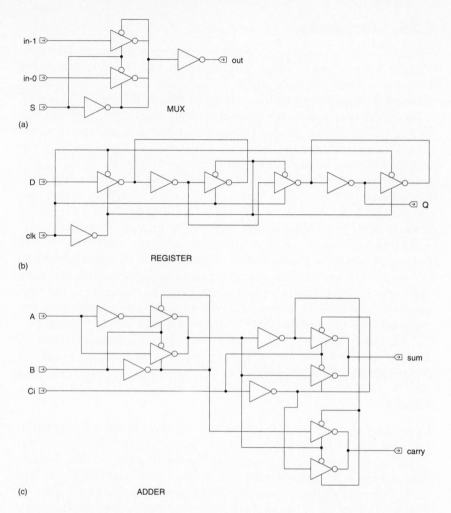

FIGURE 6.5 Regularity at the circuit level by using inverters and tristate buffers: (a) multiplexer; (b) register; (c) adder

corresponds to a well-defined behavioral, structural, and physical interface that indicates the position, name, layer type, size, and signal type of external interconnections, along with logic function and electrical characteristics. For instance, connection points may indicate the power and ground, inputs and outputs to a module. The function must also be defined in an unambiguous manner. Modularity helps the designer to clarify and document an approach to a problem, and also allows a design system to more easily check the attributes of a module as it is constructed. The ability to divide a task into a set of well-defined modules also aids in a team design where each of a number of designers has a portion of a complete chip to design.

In structured programming, proponents advise the use of only three basic constructs. These are concatenation, iteration, and conditional selection. In the IC design world these constructs have parallels. For instance, concatenation is mirrored by cell abutment, where IC cells (in the physical domain) are connected by placing them adjacent to each other and intercell connections are formed on the common boundary. Iteration is handled in the IC case by one- and two-dimensional arrays of identical cells, typified by a memory. The use of conditional selection is typified in a programmable logic array (PLA), the function of which is determined by the location of transistors in an array. When combined with the ability to parametize designs, these three programming notions can greatly aid the designer in modularizing a design. Of course in a Hardware Description Language (HDL) these constructs are used directly.

At a system level, the correct decisions regarding modularity allows one to break up a system with the confidence that when the parts are combined, the whole system will function as specified.

Example

A good example of the use (or ill use) of modularity is the use of transmission gates as inputs to logic modules (especially those using ratioed logic). Normally in CMOS circuits the inputs to logic blocks connect to the gates of MOS transistors. In these cases, the internal behavior of the modules is entirely determined by the arrival time and shape of the input waveform. Consider the case where the inputs are connected to transmission gates that are in turn connected to ratioed circuitry (for instance, consider the situation when the 2-input multiplexer shown in Fig. 6.6(a) is used for the multiplexer blocks in the difference engine example). The internal signal condition is now determined by the source impedance in addition to the input timing. A module-to-module cross-check has to occur to ensure that the driving circuit can adequately drive the mux. This is an example of a poorly modularized circuit. The fix is to use the tristate-inverter-based mux shown in Fig. 6.5(a) or the buffered mux shown in Fig. 6.6(b).

Modules can also be poorly modularized on a temporal basis. Consider a module for the difference engine that uses dynamic CMOS logic but fails to latch or register the inputs. Because external inputs might arrive at various times with respect to the clock, erroneous results might occur unless the timing of each input is individually checked. A modular approach to clocking where all module inputs are registered on entering the module and all outputs are the outputs of registers is the first step in ensuring module-to-module timing consistency.

6.2.6   Locality

By defining well-characterized interfaces for a module, we are effectively stating that the other internals of the module are unimportant to any exterior

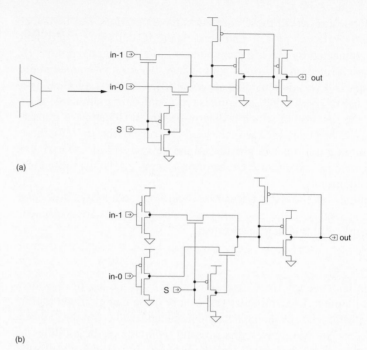

FIGURE 6.6 An example of poor modularity: (a) mux with transmission-gate inputs; (b) a solution—a buffered mux

interface. In this way we are performing a form of "information hiding" that reduces the apparent complexity of that module. In the software world this is paralleled by the reduction of global variables to a minimum (hopefully to zero).

Example:

Increasingly, locality has come to mean "time locality"; that is, modules see a common clock, and hence synchronous-timing methods apply. The first way of ensuring time locality is to pay attention to the clock generation and distribution network (see Chapter 4). Having done this, critical paths, if possible, should be kept within module boundaries. Any global module-to-module signal should have the entire clock cycle to traverse the chip. Repeated chip crossings of critical signals will rapidly lead to inferior timing characteristics. Many times in modern designs, logic is replicated to alleviate cross-chip crossings.

Modules can also be located to minimize the "global wiring" that may be necessary to connect a number of modules in a system. A common imperative in design systems today that applies for both gate-arrays and custom design is use "wires first, then modules"—rather than the more common "place modules, then route them together."

TABLE 6.1 Structured Software and VLSI Hardware Design

	SOFTWARE	HARDWARE
Hierarchy	Subroutines, libraries	Modules
Regularity	Iteration, code sharing, object-oriented procedures	Datapaths, module reuse, regular arrays, gate arrays, standard cells
Modularity	Well-defined subroutine interfaces	Well-defined module interfaces, timing- and loading-data for cells
Locality	Local scoping, no global variables	Local connections through floorplanning, registered inputs and outputs

6.2.7 Summary

There are strong parallels between the methods of design for software systems and for hardware systems. Table 6.1 summarizes some of these parallels for the principles outlined above.

As stated previously the use of HDLs to describe hardware systems in essence merges these two disciplines and the software methods above are used to define hardware. At some level the hardware aspects become relevant as a physical chip is the end product.

6.3 CMOS Chip Design Options

In this section we will examine a range of design options that may be used to implement a CMOS system design. These are arranged in order of "increased design investment," which loosely relates to the time it takes to design the device. The sequence is also somewhat in order of complexity of device that may be implemented.

6.3.1 Programmable Logic

As the investment made in any chip design is significant, designers search for ways in which to amortize the design effort over a large number of devices. This might result from a huge single market for one device or, more probably, multiple smaller markets for a more adaptable device. The larger the unit volume for a part the lower its cost will be to the end user.

Programmability is one way to achieve a wider use for a particular part. This is epitomized by the microprocessor. Often, though, the cost or speed of a microprocessor may not meet system goals and an alternative solution is required. In CMOS, one may divide this spectrum of programmable devices into three areas:

- Chips with programmable logic structures.
- Chips with programmable interconnect.
- Chips with reprogrammable gate arrays.

The CMOS-system designer should be familiar with these options for two reasons:

- First, it allows the designer to competently assess a particular system requirement for an IC and recommend a solution, given the system complexity, the speed-of-operation, cost goals, time-to-market goals, and any other top-level concerns.
- Second, it familiarizes the IC system designer with methods of making any chip design reprogrammable and hence more useful and of wider-spread use.

6.3.2   Programmable Logic Structures

The first broad class of programmable CMOS devices are represented by the programmable logic devices referred to as PALs® (Programmable Array Logic, ®Advanced Micro Devices, Inc.) or PLDs (Programmable Logic Devices).[5,6] Generally, these devices are implemented as AND-OR plane devices as shown in Fig. 6.7. In the design shown a number of inputs feed vertical wires, which are selectively connected to an AND-OR gate. Each AND-OR gate has a variable number of product terms that feed the gate. This gate in turn feeds an I/O cell, which allows registering of the AND-OR signal and the feedback of the registered result into the AND-OR plane. PAL devices come in a large range of sizes with a variable number of inputs, outputs, product terms, and I/O-cell complexity. The 22V10 is an industry-standard device with the following characteristics:

    12 inputs
    10 I/Os
    #product terms 9 10 12 14 16 14 12 10 8
    24 pins

The I/O structure for a 22V10 is shown in the inset in Fig. 6.7. It consists of a register, an output 4:1 mux, a tristate buffer, and a 2:1 input mux. The tristate buffer is used to enable the output. Alternatively, the pin may be used as an input to the array. The 4:1 mux routes the true or complemented ver-

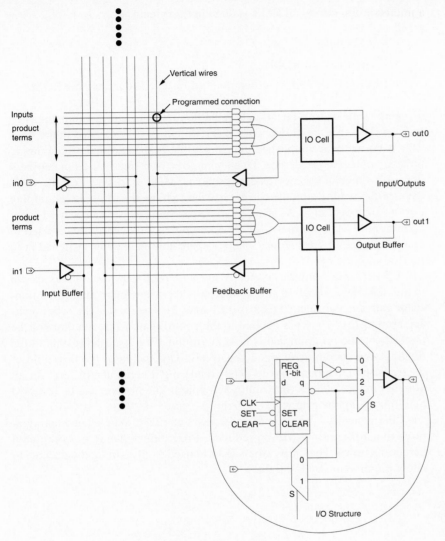

FIGURE 6.7 A typical PAL architecture

sion of the product term or register output to the output. The 2:1 input mux may also select the register output. The register is provided with a global synchronous preset and asynchronous reset.

Typical speeds for a 22V10 in high-speed CMOS are:

- CLK to output—8 *ns*.

- Input to combinational output—15 *ns*.

Typical toggle frequencies with feedback are around 40 MHz.

The programming of PALs is done in three main ways:

- Fusible links.
- UV-erasable EPROM.
- EEPROM (E^2ROM)—Electrically Erasable Programmable ROM.

Fusible links use a metal such as platininum silicide or titanium tungsten to form links that are blown when a certain current is exceeded in the fuse. This is normally accomplished by using a higher than normal programming voltage applied to the device. This technology is normally used in conjunction with a bipolar process (as opposed to a CMOS process) where the small devices can readily sink the current needed to blow the fuses. Programming is a one-time operation. As an alternative to current, a laser can be used to cut aluminum fuses in normal CMOS technologies. Frequently this is used in redundant memory techniques where a spare column may be switched in to replace a failing one.

UV-erasable memories typically use a floating gate structure as shown in Fig. 6.8. Here a floating gate is interposed between the regular MOS transistor gate and the channel (see also Chapter 3). To program the cell, a voltage around 13–14 volts is applied to the control gate while the drain of the transistor to be programmed is held at around 12 volts. This results in the floating gate becoming charged negatively. This increases the threshold of the transistor (to around 7 volts), thus rendering it permanently "off" for all normal circuit voltages (maximum 5–6 volts). The process can be reversed by illuminating the gate with UV light.

"Permanently" means at least 10 years at 125°C. At elevated temperatures the storage time will be reduced. Programming may be completed numerous times. The chips are usually housed in glass-lidded packages to allow illumination by UV light.

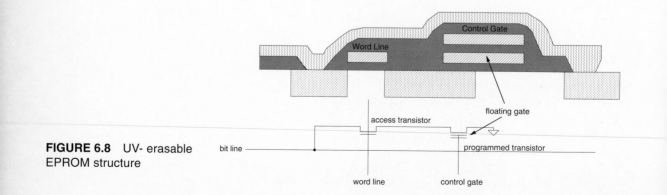

FIGURE 6.8 UV- erasable EPROM structure

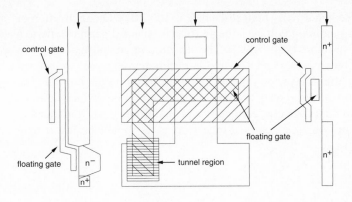

FIGURE 6.9 EEPROM structure (© IEEE 1992)

EEPROM technology allows the electrical programming and erasure of CMOS ROM cells. This type of programming forms the most popular in use today for CMOS and is the one most likely encountered by the IC-system designer in today's foundry processes. A typical structure is shown in Fig. 6.9.[7] Two transistors are typically used in a ROM cell. One is an access transistor, while the other is the programmed transistor. A two-poly sandwich is again used in the programmed transistor with the control gate on the top. A very thin oxide between the floating gate and the drain of the device allows electrons to "tunnel" to or from the floating gate (thus charging the gate oxide) to turn the cell off or on respectively. The series-access transistor allows programming of cells. EEPROM has a testability advantage over fused technologies. Each device can be fully tested before shipment. A range of ROM architectures have been used, including the normal NOR ROM structure[8] and NAND structures.[9]

By way of comparison, in a custom .8 μm CMOS chip a PLA (Programmable Logic Array) of the complexity of the 22V10 (programmed via mask) would be roughly 200μ wide by 500μ tall or .01 mm^2, and would be approximately the same speed or faster in a given technology. On a 100 mm^2 square chip one could fit 5,000 such PLAs (assuming 50% overhead for routing).

6.3.3   Programmable Interconnect

In a PAL the device is programmed by changing the characteristics of the switching element. An alternative would be to program the routing. This has been demonstrated via a number of techniques including Laser Pantography, where a laser lays down paths of metal under computer control. Commercially, programmable routing approaches are represented by products from Actel, QuickLogic, and other companies.

The Actel Field Programmable Gate Arrays[10] are based on an element called a PLICE™ (Programmable Low-Impedance Circuit Element) or antifuse. An antifuse is normally high resistance (>100 MΩ). On application of appropriate programming voltages, the antifuse is changed permanently to a low-resistance structure (200–500Ω). The structure of an antifuse is shown in Fig. 6.10(a). It consists of an ONO (oxide-nitride-oxide) layer sandwiched between a polysilicon layer on top and an n+ diffusion on the bottom. The QuickLogic array is based on a structure called a ViaLink®, which consists of a sandwich of material between metal1 and metal2.[11] This is illustrated in Fig. 6.10(b). The "on" resistance of this structure is somewhat lower than that in Fig. 6.10(a).

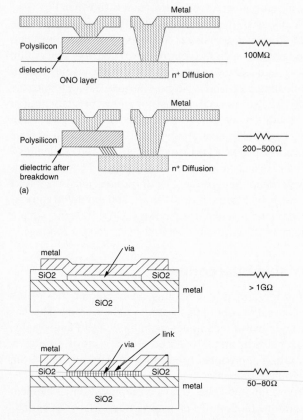

FIGURE 6.10   Programmable interconnect structures: (a) Antifuse®; (b) ViaLink®

One chip architecture that uses the antifuse is shown in Fig. 6.11.[12,13,14] Logic elements are arranged in rows separated by horizontal interconnect. Interconnect permanently connected to the logic elements passes vertically. Both horizontal and vertical segments are segmented into a variety of lengths. Segments may be joined by programming antifuses. Certain special signals such as power and a clock line are routed globally to all logic. The logic elements are surrounded by I/O pads and programming and diagnostic logic. Note the similarity to a gate-array (Section 6.3.5).

A more detailed representation of the interconnect scheme is shown in Fig. 6.12. Pass transistors are used to connect wire segments for the purpose of programming. These may be bypassed by antifuses if the links are required permanently. In the figure transistors N_1, N_2, N_3, N_4, N_9, N_{10}, N_{11}, and N_{12} are the column-access transistors, while transistors N_5, N_6, N_7, and N_8 are the row-access transistors. These are used during programming or may be used for diagnostic purposes to check the state of any signal. In the example shown, the antifuse at the conjunction of N_2's column and N_8's row has been programmed (denoted by a solid dot). This connects the signal in logic module A to the segments that are bolded in the diagram. In addition, the bypass antifuse on N_8 has been programmed, thereby extending the horizontal segment to the next set of logic cells. To program the antifuse at N_2-N_8, all pass transistors in series with N_2 are turned on and the top end is connected to the programming voltage. In addition, all transistors in series with N_8 are turned on and the end connected to the ground supply. When the programming sequence is applied, the antifuse so selected is "blown." Similar addressing techniques allow for the sampling of signals for testing or debugging. The sequencing of the antifuse blowing is carefully determined to ensure that all fuses can be blown.

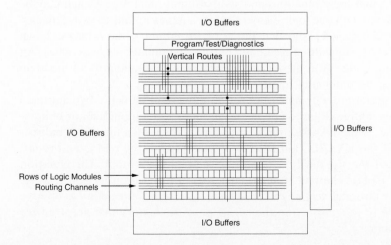

FIGURE 6.11   Actel FPGA chip architecture

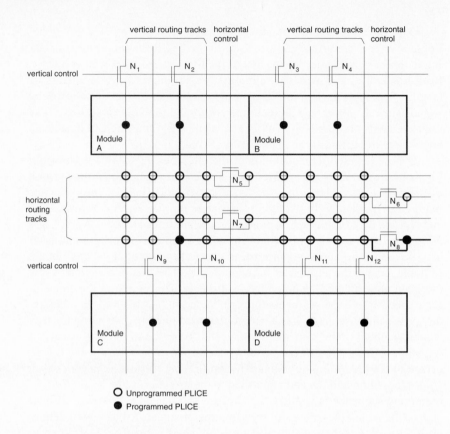

FIGURE 6.12   Actel interconnect example

○ Unprogrammed PLICE
● Programmed PLICE

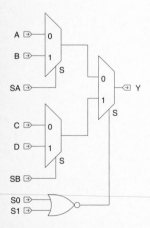

FIGURE 6.13   Actel
logic cell

The structure of the Actel logic element is shown in Fig. 6.13. It consists of three 2-input muxes and a NOR gate. This structure can implement all 2- and 3-input logic functions and some 4-input functions. A latch may be implemented with one logic element, while a register requires two elements. The QuickLogic cell is shown in Fig. 6.14. In addition to the structure shown in Fig. 6.13, it includes a resettable register and numerous logic gates. An interesting trade-off in these types of arrays is the granularity of the logic cell versus the amount of routing.

The Actel programmable I/O pad is shown in Fig. 6.15. Two antifuses allow the configuration to operate as an input pad, output pad, or bidirectional pad. If the *ENABLE* pin is not programmed, then the pad is bidirectional. If the *ENABLE* antifuse to V_{DD} is blown, the pad is an output, whereas if the V_{SS} antifuse is blown, the pad is an input. The isolation devices isolate the pad if necessary during programming and testing. (A highly desirable feature of the Actel architecture is the ability to observe any node in the chip using the series pass-transistors that are used for programming.)

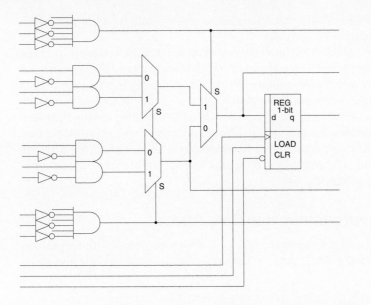

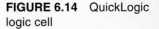

FIGURE 6.14 QuickLogic logic cell

At the time of writing, these arrays could implement 550 logic modules and 70 I/O modules. The speed of a particular circuit depends on the logic element speed and the delay through antifuse elements in any routing. A single logic module exhibits a delay from 7 *ns* to 14 *ns* (5V and 25°C) depending on fan-out in a 2 μm technology. Long route delays through many antifuses can range from 15 *ns* to 35 *ns*. With smaller technologies the logic module delays would decrease while the routing delays might decrease

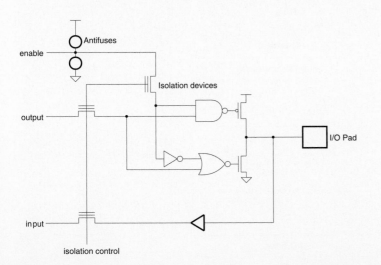

FIGURE 6.15 Actel I/O pad

somewhat. More drastic reductions in the routing delays would come with lower "on-resistance" antifuses.

If a 32-bit adder were implemented in an Actel array, 160 logic modules would be needed, and it would add in approximately 65 *ns*. Thus roughly 3.5 32-bit adders would fit in a single FPGA chip. Currently in a 1μ custom chip 7 mm on a side (6 mm-by-6 mm active area), we could fit 1300 adders (each 560μ by 50μ) running twice as fast, or 600 adders running ten times as fast. Bear in mind that if we only wanted two 32-bit adders and a bit of logic running at 10 MHz, we could get in in an afternoon and for about \$5–10 if we used an FPGA; compared with 6 months and \$200,000 that an application-specific chip would require.

6.3.4 Reprogrammable Gate Arrays

A further class of programmable device is the programmable (or reprogrammable) gate array. These may be further categorized into ad-hoc and structured arrays.

6.3.4.1 The XILINX Programmable Gate Array

An example of an ad-hoc array is a set of products from the XILINX company.[15] The architecture of the XC3000 series is depicted in Fig. 6.16. An array of Configurable Logic Blocks (CLBs) is embedded within a set of horizontal and vertical channels that contain routing that can be personalized to interconnect CLBs. The configuration of the interconnect is achieved by turning on n-channel pass transistors. The state that determines a given interconnect pattern is held in static RAM cells distributed across the chip close

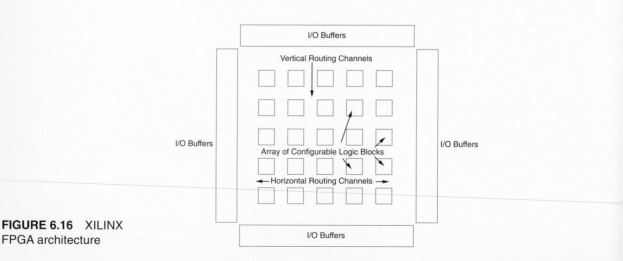

FIGURE 6.16 XILINX FPGA architecture

to the controlled elements. The CLBs and routing channels are surrounded by a set of programmable I/Os.

In detail, the structure of a CLB is shown in Fig. 6.17. It consists of two registers, a number of muxes, and a combinatorial function unit. The latter can generate two functions of four variables, any function of five variables, or a selection between two functions of four variables. The function bit and each mux is controlled by a number of RAM state bits. More recent CLBs feature enhanced table lookup function generators which can be used to build logic functions or used as register storage. Inbuilt support for carry chains means that datapaths can be conveniently built (XC4000 series). Each input and output on a CLB has a particular local interconnect pattern (called direct interconnect by XILINX), which allows most local interconnection between adjacent CLBs to take place. At the junction of the horizontal and vertical routing channels (where the general-purpose interconnect runs),

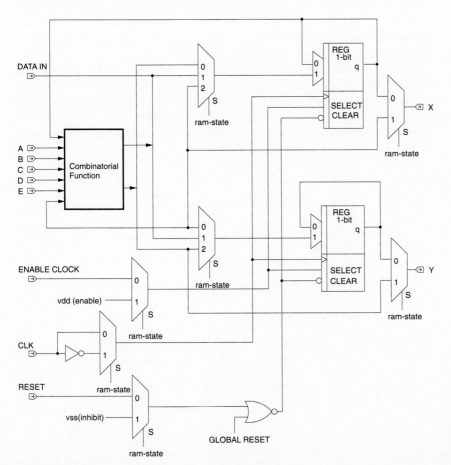

FIGURE 6.17 XILINX Configurable Logic Block® (CLB)

programmable switching matrices are employed to redirect routes. Fig. 6.18 shows a typical CLB surrounded by switching matrices. The switching matrices perform crossbar switching of the global interconnect, which runs both vertically and horizontally. Programmable Interconnect Points or PIPs interconnect the global routing to CLBs. Both PIPs and the switching matrices are implemented as n-channel pass gates controlled by 1-bit RAM cells. Extra special long-distance interconnect is used to route important timing signals with low skew.

Assuming one has a board design finished, design proceeds by mapping the logic design to the CLBs and thence to one or more programmable gate-arrays. Software then "places and routes" the CLBs by loading the internal state RAM with the codes needed to program the I/Os, the CLBs, and the routing. The design is then ready to be tested or used.

Currently, the largest array holds ≈500 CLBs and has approximately 100K bits of state RAM (this will increase with time as processes shrink). In common with the Actel approach, timing is dependent on the basic CLB speed and a routing delay term. Users seem to be able to achieve system-clock rates that are 30–50% of the speed grade. Thus with 250 MHz parts an 80 MHz clock frequency is feasible.

A 32-bit adder would require approximately 62 CLBs, enabling roughly 8 to be implemented on the largest CLB available in 1993. The speed would be approximately 20–50 MHz. Thus the reprogrammable arrays implemented in a more advanced (but standard) process and with probably larger die sizes (hence cost) are roughly of the same complexity as

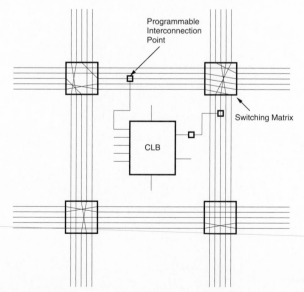

FIGURE 6.18 XILINX crossbar connect and CLB local connect example

the programmable example given in the previous section implemented in a less dense process.

While the XILINX arrays are stand-alone programmable gate arrays, the ideas may be of use to the IC-system designer who wishes to embed some reprogrammable logic within a larger system. In addition, the IC designer may find that prototyping a design in such an array might aid in system debug of a chip function. A significant advantage of the reprogrammable gate array is the ability to redesign the internals of a chip by changing software. This can be of considerable advantage in a product that has to undergo field updates.

6.3.4.2 Algotronix

An example of a regular programmable array is the CAL1024 (Configurable Array Logic) from Algotronix.[16,17] This architecture contains 1024 identical logic cells arranged in a 32-by-32 matrix. At the boundary of the chip, 128 programmable I/O pins allow cascading the chips in even larger arrays. The cell interconnect is shown in Fig. 6.19. Each cell is connected to the East, South, West, and North neighbor. In addition two global-interconnect signals connect to each cell. These are used to supply a low-skew signal to all cells for clocking. Each cell also receives row select lines and bit lines that are used to program RAM bits within the logic cells that dynamically customize the logic cell.

The cell design is shown in Fig. 6.20. It consists of four "through" multiplexers to route single-bit signals entering from the North, South, East, and West. In addition two multiplexers route a selection of signals to a function

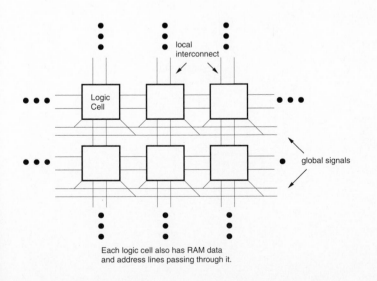

Each logic cell also has RAM data
and address lines passing through it.

FIGURE 6.19 Algotronix
FPGA chip architecture

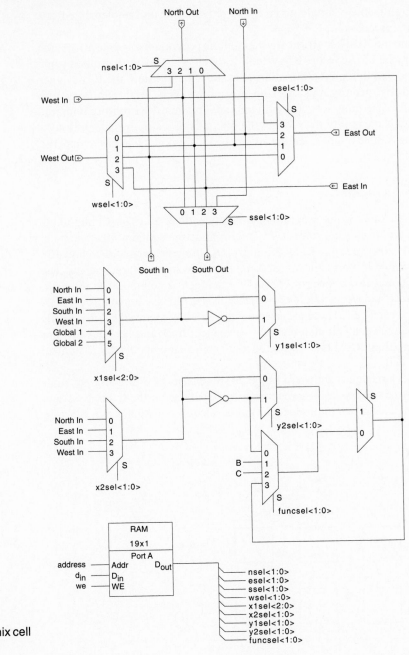

FIGURE 6.20   Algotronix cell design

TABLE 6.2   CAL Logic Cell Functions

NUMBER	FUNCTION	NUMBER	FUNCTION
0	ZERO	8	–X1.X2
1	ONE	9	–X1.–X2
2	X1	10	X1+X2
3	–X1	11	X1+–X2
4	X2	12	–X1+X2
5	–X2	13	xnor(X1,X2)
6	X1.X2	14	–X1+–X2
7	X1.–X2	15	xor(X1,X2)
16	D Clk Latch	17	–D Clk Latch
18	D –Clk Latch	19	–D –Clk Latch

unit. These signals include the signals entering on the orthogonal edges of the cell, two global "clock" signals, and the output of the function block for feedback situations (latches). The muxes are controlled (as in the XILINX array) by small 5-transistor static RAM cells. The functions that the logic cell can implement are detailed in Table 6.2.

The I/O pads are very interesting. The trick is to use only one pin for I/O into and out of the array but have the communicating chips automatically deal with two pins that are outputs. The pads achieve this by using a ternary (three-level) logic scheme to sense when two outputs are driving each other via a contention circuit. This is then used with an XOR gate, as shown in Fig. 6.21, to deduce the correct input value.

Design is similar to both the XILINX and Actel approaches, where substantially automatic techniques can place and route a CAL chip. Unlike both other approaches, however, all routing (save the global clock lines) must pass through cells to get from one point to another. Thus, in the worst case, a signal may have to travel through 64 cells. Although implemented with fast transmission gates, this still can result in a substantial delay. For instance the through routing delay is in the range of 0.5 *ns*–2 *ns*, resulting in a delay of 32 *ns*–128 *ns*. However, an intriguing option with this type of array is that the programming can be changed almost in real time. Thus one can think of

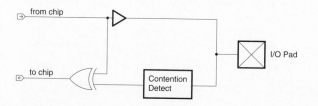

FIGURE 6.21   Algotronix I/O circuit

actually having a computer program that "executes" on such an array many times faster than conventional machines.[18]

From a complexity viewpoint, a single-bit adder would take 4 cells. Thus a 32-bit ripple-carry adder would take 128 cells, and 8 adders could fit on a 1024-cell chip (1.5 μm CMOS). The speed would be in the 500 *ns* range. In all of these array architectures serial arithmetic may be preferable to parallel arithmetic because low-delay connections can be made between adjacent cells.

6.3.4.3 Concurrent Logic

The CLi6000 series is another example of a regular array style FPGA.[19] Current designs have between 1000 and 3136 cells, with prospects of up to 10,000 cells per chip in the next few years. As an example the CLi6005 consists of a 7-by-7 array of superblocks. Each superblock has an array of 8-by-8 logic cells. Each logic cell connects to the four nearest neighbors and to a local and express bus (Fig. 6.22). The cell structure is shown in Fig. 6.23.

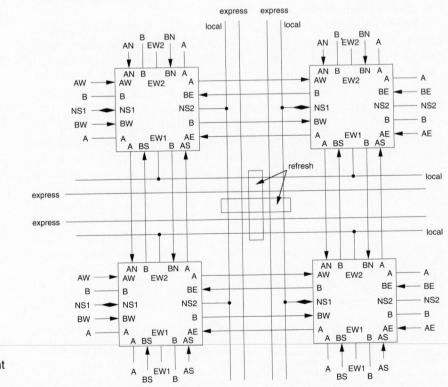

FIGURE 6.22 Concurrent logic array details

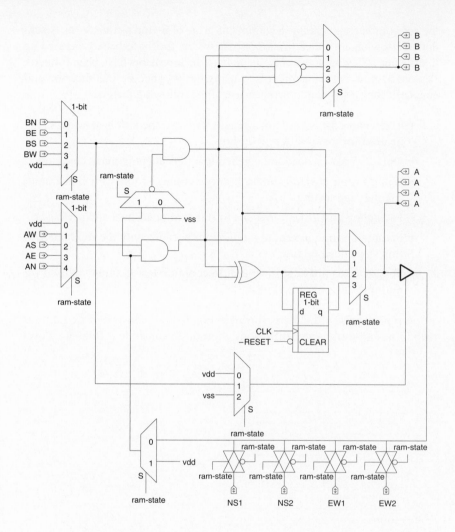

FIGURE 6.23   Concurrent Logic array logic cell

Compared with the Algotronix cell it has considerably more functionality within a cell. A resettable register, XOR, and an AND gate are included. Thus, for instance, a single-counter bit can be implemented in a single cell.

6.3.5   Sea-of-Gate and Gate Array Design

Programming interconnect on chips is a method of reducing the design cost of an integrated circuit. For small-volume chips this can have a direct impact on the part price. The most popular style in use for the implementation of general logic functions is the Sea-of-Gates (SOG) or Gate Array structure, in which the core of the chip contains a continuous array of n- and p-transistors. A vendor stocks what are called master or base wafers that have been processed up to the stage of laying down polysilicon (i.e., the transistors have been formed). Personalization is then achieved by using design-specific metalization and contacts. The cost is kept down because of the following factors:

- The wafer cost is kept low because large numbers of base wafers may be used for many different designs.
- Only 2–5 masks need to be generated, thus keeping mask costs low.
- Design time is small due to highly automated tools for placement, routing, and testing.
- Packaging cost is kept low due to standard bond-outs and packages.
- Processing time is kept to a minimum because only the top metalization steps need be run.
- Testing costs are kept low because common test fixtures are used for multiple designs.

A typical SOG structure is shown in Fig. 6.24. It consists of a continous strip of n- and p-transistor diffusions adjacent to substrate diffusions. Poly-

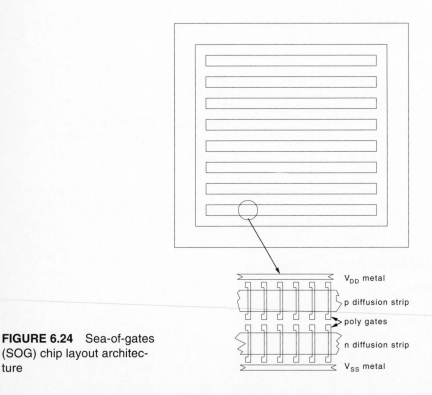

FIGURE 6.24   Sea-of-gates (SOG) chip layout architecture

silicon crossing the n and p diffusions forms a continuous horizontal array of transistors. These rows are repeated vertically. The core of an SOG chip so constructed is surrounded by an array of I/O cells that can also be programmed by metalization. Routing channels are formed by routing over the top of unused transistors. Gate arrays, which predate SOG structures, used fixed-height routing tracks. Wiring between active logic rows in an SOG chip occurs over the top of unused transistors, while in a Gate Array the routing is constrained to a routing channel. Fig. 6.25 shows a collection of gates wired together illustrating the routing over the top of the transistor rows. The necessity to pick a number for the Gate Array routing track density thereby constraining the number of horizontal routes gave way eventually to the continuous-array SOG approach.

A number of design decisions have to be made when designing the base array.[20] These include the following:

- The overall size of the core array.
- The macro structure of the strips:
  - How many n rows and p rows there are per horizontal strip, and how they are routed.
- The micro architecture:
  - The size and ratio of the n- and p-transistors.
  - The number, direction, and layer of routing tracks.
  - The method of logic-gate isolation.
  - The personalization method.

Usually, the core-array sizes vary from small to large die sizes. When a given system is being planned, the actual density of transistors is mapped to an equivalent raw gate number and then to an effective usable gate level that reflects the cost of routing and placement overheads. For instance, if the column pitch of the array is 10μ and the row pitch is 100μ, then an 8 mm-by-

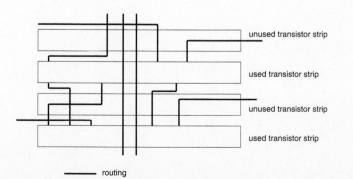

unused transistor strip

used transistor strip

unused transistor strip

used transistor strip

—— routing

FIGURE 6.25   SOG gates wired together showing routes over unused transistors

8 mm core chip would contain $800 \times 80 = 64{,}000$ transistor pairs. This corresponds roughly to 16,000 2-input NAND gates. At a 40% usability index (the number of gates that may be used/the total number of gates), this means we can expect to use 6400 gates on this base core.

Most SOG structures have a single row of n- and p-transistors. Some designers have found it advantageous to use a double row of n-transistors and a double row of p-transistors to aid in the implementation of memories and dynamic logic.[21] Other designers, wishing to implement analog circuits, choose arrays of transistors that are suitable for those applications.[22]

Most designs choose equally sized transistors, presumably because unequal rise and fall times tend to even out. The absolute size of transistors is a trade-off between drive capability, fan-in loading, and the array density required. The size of the transistors also affects the granularity of routing tracks.

Typical examples of geometrically isolated and gate-isolated designs are shown in Fig. 6.26 and Fig. 6.27. In Fig. 6.26 a geometrically isolated design typical of early gate-arrays is shown in which three n-p pairs are coupled to form a cell.[23] N- and p-transistors are equally sized. The polysilicon gates are commoned. The "dog-bone" poly connections on the transistor gates allow for routing. Substrate connections are placed below the n-transistor strips and above the p-transistor strips. A typical SOG design is

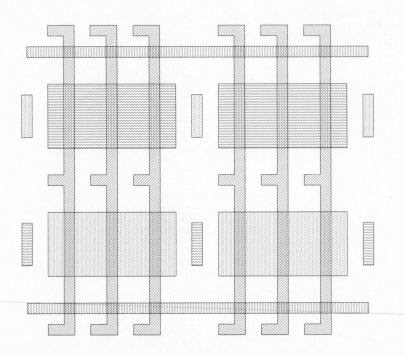

FIGURE 6.26 Geometry isolated SOG or Gate-array base cell

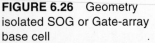

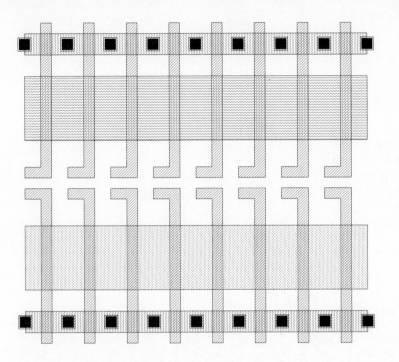

FIGURE 6.27   Transistor- or gate-isolated SOG cell

shown in Fig. 6.27. The key point about the SOG structure is that the transistors at the end of a gate serve to isolate adjacent gates. This is achieved by tying the gate of the n-isolation transistor to V_{SS} and the gate of the p-isolation transistor to V_{DD}. Where adjacent gates share a V_{SS}- or V_{DD}-connected transistor, the isolation transistor is not required. Substrate and well connections run under power busses at the bottom and top of the cell.

The personalization may be completed in a number of ways. For instance, possible methods are:

- Single-layer metal.
- Single-layer metal and contacts.
- Double-layer metal and contacts and vias.
- Triple-layer metal, vias, and contacts.

The personalization of the arrays shown in Fig. 6.26 and 6.27 for a 3-input NAND gate and a D latch are shown in Fig. 6.28. Both cells are arranged to have metal1 running horizontally and metal2 vertically. Note how for the NAND gate the geometrically isolated gate is smaller but for a more complex structure the transistor isolated array is much smaller.

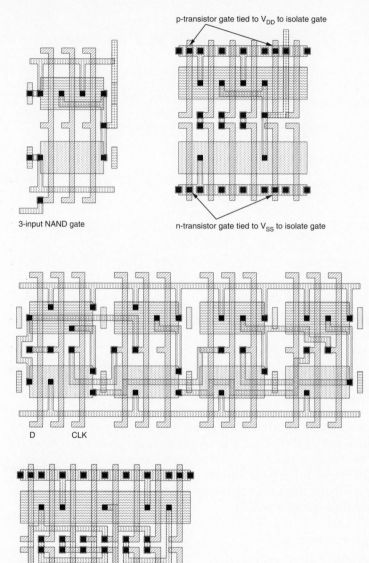

p-transistor gate tied to V_{DD} to isolate gate

3-input NAND gate

n-transistor gate tied to V_{SS} to isolate gate

D   CLK

CLK   D   Q

D latch

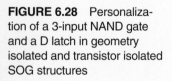

FIGURE 6.28 Personalization of a 3-input NAND gate and a D latch in geometry isolated and transistor isolated SOG structures

The routing style of most SOG orients the macrocells (i.e., NANDs, NORs, REGISTERs) along the rows as illustrated in Fig. 6.27, with routing running horizontally between rows of macrocells. In an alternative strategy, the cells are grouped in columns[24] with routing tracks running vertically between columns of macrocells on the chip. The latter approach has been shown to provide 1.08 to 1.31 higher gate density than the row-based approach.

6.3.6 Standard-cell Design

Whereas gate-array architectures standardize at the chip geometry level, it is possible to standardize at the logic or function level. That is, a specific design for each logic gate in a library can be created. This is the basis for what is termed standard-cell or cell-based design. Library cells are normally created for the following classes of circuits:

- SSI logic (nand, nor, xor, aoi, oai, inverters, buffers, registers).
- MSI logic (decoders, encoders, parity trees, adders, comparators).
- Datapath (alus, adders, register files, shifters, bus extractors, and inserters).
- Memories (RAM, ROM, CAM).
- System-level blocks (multipliers, microcontrollers, UARTs, RISC cores).

A design is captured using the standard cells available in a library via schematic or HDL. The layout is then normally automatically placed and routed by CAD software. For SSI and MSI blocks, the layout style is usually identifiable as rows of constant or near-constant height blocks separated by rows of routing. As the complete layout is being done, optimization of the height of routing channels may be completed by good placement. Most manufacturers have extended the SSI/MSI standard-cell technique to the design of datapaths and other higher-level functions such as microprocessors and their peripherals. Another fundamental component of a standard cell system is a selection of memories. Often these are available as a set of parameterizable modules based on word width, number of words, and number of read- and write-ports.

Compared to gate-arrays, standard-cell designs provide a density advantage at the cost of increased prototype costs and possibly increased design complexity. However, where manufacturers have implemented sizable circuit blocks, the productivity of a standard-cell approach might in fact be better that that of a gate-array because the function does not need to be designed.

6.3.6.1 A Typical Standard-cell Library

The LSI Logic standard cell library[25] is representative of a large number of libraries that are available. Frequently the SSI logic blocks come in a density-optimized version and a speed-optimized version. Figure 6.29 shows possible layouts for a low-power and normal-power 2-input NAND gate. Metal power busses run horizontally at the top and bottom of the cells. Connections to the cells are available at the top and bottom of the cells. As these hypothetical cells are implemented on a salicided process, connections to the inputs of the gates are made in polysilicon. The density-optimized versions use minimum-sized transistors to achieve the smallest-height standard-cell while the speed-optimized versions use large transistors to provide good driving capability. A summary of typically available cell types is summarized in Table 6.3.

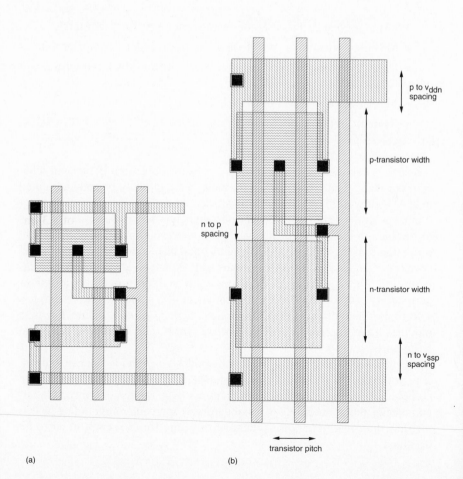

FIGURE 6.29 Typical standard-cell structures showing low-power and regular-power cells

(a)

(b)

TABLE 6.3 Typical SSI Standard-cell Library Summary

GATE TYPE	VARIATIONS	OPTIONS
inverter/buffer/tristate buffers		High-, Normal-, Low-power
nand/and	2–8 inputs	High-, Normal-, Low-power
nor/or	2–8 inputs	High-, Normal-, Low-power
xor	2–3 inputs	High-, Normal-, Low-power
xnor		High-, Normal-, Low-power
aoi (and-or-invert)		High-, Normal-, Low-power
oai (or-and-invert)		High-, Normal-, Low-power
multiplexers	2–8 inputs, inverting/ noninverting	High-, Normal-, Low-power
schmitt trigger	inverting/noninverting	High-, Normal-, Low-power
adder/half-adder	normal, fast	High-, Normal-, Low-power
latches	D, asych/synch clear/ set, scan	High-, Normal-, Low-power
registers	D, JK, asych/synch clear/set, scan	High-, Normal-, Low-power
I/O pads	in, out, tristate, bi-direct, boundary scan, limited slew rate, crystal oscillator	Various current options 1–16 mA

In addition various parameterizable macro cells such as register files, FIFOs, RAMs and ROMs may be provided.

Wide varieties in standard-cell topologies exist. An example of a 3μm library may be found in *CMOS3 Cell Library* by Dennis V. Heinbruch.[26] These cells are very intricate and designed to minimize parasitics and to maximize performance within a given area. Such libraries take a long time to create. Another approach is to abstract the geometry of the cells to allow rapid redeployment in a new technology. Other variations are shown in Fig. 5.15. Where no salicide is available, the polysilicon gates might be strapped in metal2, to eliminate any *RC* delays that might occur in routes that pass through a number of rows of cells.

Figure 6.30 (also Plate 6) shows a three-level-metal standard-cell strategy used at TLW (for a 3-input NAND gate). In this example, internal cell connections are completed in metal1. Connections to gates occur at the center of the cell with a double-via structure from poly to metal2. Metal3 runs horizontally and metal2 runs vertically; thus the cells may be completely covered with routing. With this kind of a cell combined with good automatic placement, very good densities can be achieved. With a library where the

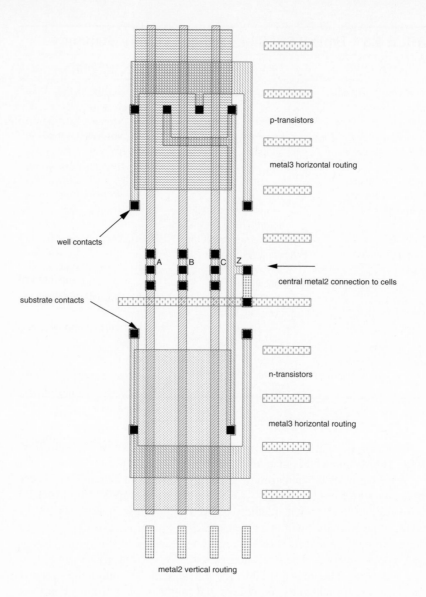

p-transistors

metal3 horizontal routing

well contacts

A B C Z

central metal2 connection to cells

substrate contacts

n-transistors

metal3 horizontal routing

FIGURE 6.30 Three-level-metal standard cell

metal2 vertical routing

size of the transistors may be parameterized and some generator support for regular-array structures such as datapaths and memories is available, the density can rival that done by hand. This means that other than for special analog, memory, or I/O blocks, all layout can be compiled with little impact on die cost but a big impact on productivity.

6.3.7 Full-custom Mask Design

Full-custom design is the name given to the technique where the function and layout of practically every transistor is optimized. Traditionally, this is how most commercial designs have been done from the beginning of IC-design history. Many times, nonconventional circuit forms or clocking methods will be used in an effort to decrease size or increase speed. Design involves detailed manipulation of the geometric layout ("polygon pushing") and detailed circuit simulation of every circuit structure. As a historical point, it is interesting to note that even in the mid-1970s custom design was long on geometry and short on any kind of verification (due to lack of compute cycles). Design entry might have included cutting your own mask from Rubylith®, entering the geometry via a text editor (having drawn it by hand), or digitizing the same hand-drawn layout with a large digitizer. If you were lucky you may have seen the layout briefly on a monochrome storage display. Frequently, using a library cell consisted of deft use of a pair of scissors, some tape, and an old layout plot. Design-verification tools consisted of a large room, knee pads, and coloring pencils to color layout plots. Of course, circuits consisted of $10s$ to $100s$ of transistors.

In these times, for digital CMOS circuits, companies rarely use full custom design due to the high labor content and low productivity. Exceptions to this include the design of memory and commodity parts such as FPGAs and the design of quasi-analog components such as phase-locked loops. In addition, large mega-cells such as RISC microprocessors may be custom-designed for speed and cost reasons.

In 1989, Fey[27] found that for full-custom designs the productivity ranged between 6 and 17 transistors per day for logic transistors and 60 to 230 transistors per day for ROM transistors.

6.3.8 Symbolic Layout

Because a major component of custom design is the physical layout of new modules, IC-designers sought methods of reducing the time of entry of the physical layout. In the early 1970s MOS logic designers frequently used manually drawn shorthand notations for layout structures (the author included). As computers became more prevalent, this practice led to *symbolic* layout systems. These systems attempt to abstract the layout in some manner in order to reduce the complexity of the task, thus increasing productivity. Over the last 15 years a number of strategies have evolved.

6.3.8.1 Coarse-grid Symbolic Layout

The idea behind coarse-grid symbolic layout involves dividing the chip surface into a uniformly spaced grid in both the X and Y directions. The grid size

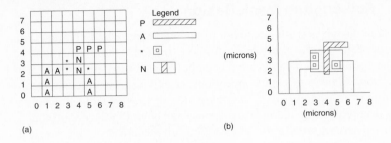

FIGURE 6.31   Coarse-grid symbolic layout

(a)

(b)

represents the minimum feature or placement tolerance that is desired in a given process, and is usually selected by close consultation between design-tool developers and semiconductor-process engineers. For each combination of mask layers that exist at a grid location, a symbol is defined. Figure 6.31 shows a typical symbol set and layout. Given a particular design system, these symbols are then placed on the grid to construct the desired circuit, much the same way as you would tile a floor. Symbol sets could be defined as characters or graphical symbols, which was invaluable in the early days of color displays because character-only color displays were a lot cheaper than color graphics displays.

American Microsystems International (AMI)[28,29] and Rockwell International[30] pioneered the use of character-based symbolic layout.

This style of symbolic layout provided for first-generation symbolic layout with low-cost design entry. In general these systems have been supplanted by more modern approaches.

6.3.8.2   Gate-matrix Layout

A character-based symbolic layout style was developed at Bell Labs[31] specifically for custom CMOS circuitry. It improved on coarse-grid symbolic layout by providing a regular layout style where a matrix of intersecting transistor diffusion rows and polysilicon columns is employed. The intersection of a row and a column is a potential transistor site (poly crossing diffusion). A related style is featured in Piguet et al.[32]

The evolution of this technique from a standard-cell viewpoint is shown in Fig. 6.32. Figure 6.32(a) shows a circuit implemented in terms of standard cells (four 2-input NANDs and one inverter). Note that intercell connections are in metal. Rather than running these connections in metal, we can run vertical polysilicon columns corresponding to each gate signal. The transistors may then be placed on the polysilicon signals and interconnected, as shown in Fig. 6.32(b). Note that vertical columns may be either polysilicon or diffusion. Horizontal rows are transistors and/or metal routing tracks. Metal may also run vertically. A character-symbolic layout for the layout may be

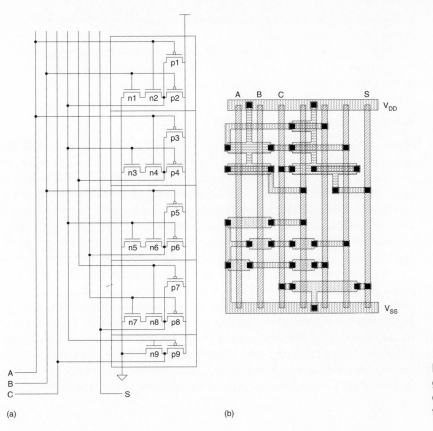

FIGURE 6.32 Evolution of gate-matrix layout: (a) standard cell layout (in schematic form); (b) gate-matrix layout

created (in fact this is how the layouts were first captured). The following rules summarize the gate-matrix technique:

1. Polysilicon runs only in one direction and is of constant width and pitch.
2. Diffusion wires (of constant width) may run vertically between polysilicon columns.
3. Metal may run horizontally and vertically. Any pitch departures from minimum (e.g., power rails) are manually specified.
4. Transistors can only exist on polysilicon columns.

To convert from a character symbolic description to mask artwork, the character matrix is examined and the symbols are expanded to their equivalent mask entities. Operations such as merging horizontal dashes into one metal wire and merging adjacent devices are performed during this phase. Obviously very simple software and limited computer resources are needed to capture designs in this manner.

In common with coarse-grid symbolic layout, gate-matrix-symbolic layout systems have been largely replaced, but the style of layout is still of interest for small- to medium-sized modules.

6.3.8.3   Sticks Layout and Compaction

A popular method of symbolic design is termed "sticks" layout. Here the designer draws a freehand sketch of a layout, using colored lines to represent the various process layers such as diffusion, metal, and polysilicon. Where polysilicon crosses diffusion, transistors are created and where metal wires join diffusion or polysilicon, contacts are formed. Alternatively, specific primitives such as transistors are drawn and interconnected with lines representing conductors. Following this rapid capture of the rough topology, a spacing program or compactor determines the correct spacing between all wires, transistors, and contacts created. The most popular compactor is what is termed a graph compactor. The compactor creates a directed-constraint graph. The nodes of the graph are the primitives, and the branches are used to connect groups that have potential design-rule violations. The weights of the branches are the minimum separations necessary between two nodes. An example of the mapping of a symbolic circuit to a graph is depicted in Fig. 6.33. If there is no spacing necessary between two groups, an edge will not be created between the two groups. Once the graph has been established, the critical path (i.e., the

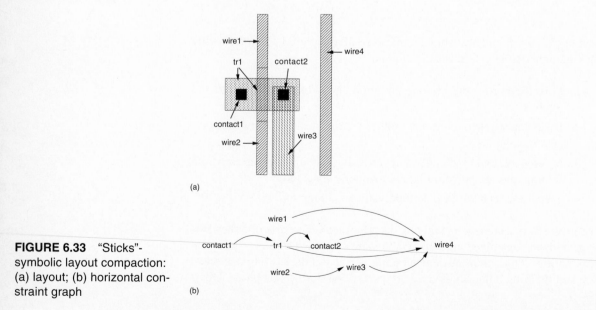

FIGURE 6.33   "Sticks"-symbolic layout compaction: (a) layout; (b) horizontal constraint graph

path with the greatest spacing requirement) through the graph can be determined. The nodes in the path can then be placed sequentially. For any given node, there may be a number of paths to it. The critical path to a given node will determine its minimum placement consistent with all design rules. X and Y passes through the graph are completed to compress the layout.

During the 1980s a large body of CAD research was devoted to "sticks" symbolic layout systems.[33–43] These systems and their commercial derivatives have met with varied success and acceptance.

6.3.8.4   *Virtual-grid Symbolic Layout*

Virtual-grid symbolic layout[44] is a symbolic layout method that draws on the experience gained in coarse-grid symbolic systems, gate-matrix, "sticks"-type systems, and other approaches, such as ICSYS,[45] developed at the University of Edinburgh and Caltech. In essence, the system approaches design at the layout level by manipulating circuit elements such as transistors and wires as opposed to any form of geometric mask description. These elements are placed on a grid to facilitate easy design capture and use of simplified tools, with the final geometric spacing between grid lines determined by the density and interference of circuit elements on neighboring grid locations. This leads to the notion of a *virtual grid*.

The concept is best illustrated by a simple example, as shown in Fig. 6.34(a). Three vertical wires are shown centered on a grid. The result of using a fixed grid of 10 units and a wire width and separation of 10 units leads to the mask description shown in Fig. 6.34(b). By using a grid in which the spacing varies according to topology, the mask description in Fig. 6.34(c) is constructed. The end result for the designer is that placement on the grid may be done without regard to any design rules. In addition to eliminating design rules, the grid is also used to define circuit connectivity in a manner similar to that employed in schematic capture systems. Here, the notion of a "coordinode," as introduced by Buchanan, is used to capture physical location, structural connectivity, and behavioral state. As its name suggests, a coordinode has the properties of a coordinate, namely some xy position that will eventually map to the silicon surface. In addition, it may possess the properties of a node in a circuit, such as voltage or simulation state. Structurally, a coordinode defines the nodes in the network being designed. In the virtual-grid context, a coordinode is mapped to a discrete set of grid points rather than a quasi-continuous set of xy coordinates. The grid coordinates form the lines of action in a circuit, defining the essential communication paths in and through a circuit. Local geometric perturbations are handled by software skilled in the art of manipulating geometry.

MULGA is an example of an integrated design system based on these principles still in use.[46,47] The NS system is a VLSI system that has been in use since 1984, which employs the same principles.[48] Further systems have

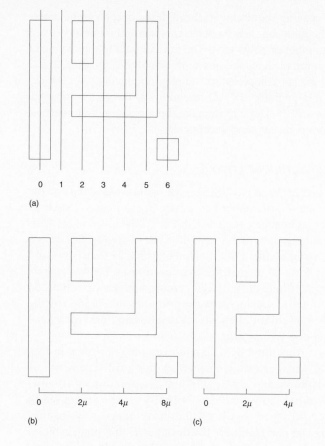

FIGURE 6.34 Virtual-grid symbolic layout

reported the use of similar techniques.[49,50] A virtual-grid-circuit capture system yields the following benefits:

- Design-rule-free topology capture.
- Rapid design capture through the use of point interconnect.
- Fast grid-based algorithms for connectivity audit, compaction, and other design processes.
- Ability to allow simplified parametized cells with automatic geometry generation.
- Hierarchical module assembly.
- Natural target for higher-level silicon compilers (geometry free).

The use of the virtual grid allows a very simple compaction strategy to be used. However, recent virtual-grid systems also use graph based compaction.

In terms of productivity, based on roughly 20 chip designs, by using known circuits and just altering the geometry, productivities similar to standard-cell designs are seen while densities comparable with hand layout result.

Symbolic-layout systems are an acquired taste. Many people (such as the author) could not imagine designing without them at the layout level, while many others design quite happily without them. Recently, there has been somewhat of a renaissance of symbolic layout when designers with fixed-geometry designs are confronted with the effort of porting an old CMOS design to a completely new process.

6.3.9   Process Migration—Retargetting Designs

Of concern to the system designer are methods of retaining the investment in engineering large systems where ever-improving process densities mean that the system cost can be reduced over time by incorporating the five ASICs designed in a given year onto one chip in two years, and onto one-fourth of a chip in four years. The following approaches are possible:

- Recast the architecture.
- Recast the logic design.
- Recast the layout in the new process.

The first option requires a complete reengineering of the problem with the associated system and production test vectors. This frequently provides the best technical solution with a varied saving on previous engineering. In the software domain, this is equivalent to rewriting the program to suit a faster computer with, say, a parallel architecture.

The second option allows a design implemented in today's standard cells to be fairly painlessly translated to the next generation standard cells. Timing analysis still has to be completed, but the process can reuse previous engineering efforts. While this process works well for logic, specialized VLSI structures such as memories may be a problem. Some ASIC vendors counter this by providing n-port memory compilers that work in the target processes. The counterpart in the software world is the use of a new compiler to produce code for a new fast processor.

Finally for the "dusty deck" problem, researchers have turned their attention to the problem of migrating mask designs implemented in old technologies to newer smaller processes. Some success has been recorded with systems that extract symbols from old geometry and then recompact the symbols with the new design rules of the target process.[51]

In all cases, the design effort would still contain a large proportion of simulation and timing analysis, which usually constitutes the major portion of the design effort in today's designs.

6.4   Design Methods

When starting a design project, the designer has a number of options with regard to the specification level of the chip. Usually the designer starts at the behavioral level and progresses to the RTL level, then to the logic level, and then possibly to the structural level, and finally to the layout level. Depending on the complexity of the design, tools exist to synthesize a chip layout from any of these levels of specification.

6.4.1   Behavioral Synthesis

At the behavioral level, the operation of the system is captured without having to specify the implementation. For instance, the pipelining required to meet a certain speed may not be specified. This is the level that provides the fastest emulation of the system and the one that is best used to debug the operation of the complete system. Obviously, this level is technology-independent.

For the synthesis of complex behavioral descriptions including signal-processing architectures, a rich research literature is available. Researchers have had success with high-level synthesis by building systems to synthesize constrained architectures. Good examples are the Cathedral series of silicon compilers, Cathedral I, which concentrated on bit-serial digital filters[52]; Cathedral II, which compiled collections of communicating sequential DSP processors[53]; and Cathedral III, which was aimed at video-signal-processing architectures.[54] Another example is the LAGER compiler for signal processing architectures.[55] These targeted systems are sometimes called Silicon Compilers, because they take a design from the behavioral (code) to the mask level (silicon).

In principle, a behavioral compiler must perform the following operations:

- Decide upon and assign resources based on area and timing requirements.
- Insert pipeline registers to achieve timing constraints.
- Create microcode and/or control logic.

For instance, consider the following behavioral code fragment:

```
a = a + b*c;
```

This specifies a multiply-accumulate step. Depending on the required speed and word size, this may be implemented as anything from a bit-serial multiplier to a fully parallel Booth-encoded Wallace tree multiplier (see Chapter 8).

The vector-drawing architecture shown in Fig. 6.2 is derived from the equation for a straight line,

```
Y = a*x + b;
```

It is not intuitively obvious how the structure in Fig. 6.2 might evolve from the equation above. It took a clever human and difference-equation mathematics to produce the implementation shown.

Behavioral-synthesis systems currently provide very good silicon implementations for narrow (but very useful) classes of problems, and will continue to gain ground as they become more generalized and commercially available.

6.4.2   RTL Synthesis

RTL-synthesis programs take an RTL description and convert it to a set of registers and combinational logic. At this stage of the design process, the architecture has been captured. One research system pioneering this approach to design (and aimed at CMOS) is the Yorktown Silicon Compiler System.[56] There are also a number of commercial systems now available.

Commonly, RTL descriptions are captured using a Hardware Description Language (HDL). In general, RTL HDLs have to capture the following attributes of a design:

- Control flow using if-then-else and case statements.
- Iteration.
- Hierarchy.
- Word widths, bit vectors, and bit fields.
- Sequential versus parallel operations.
- Register specification and allocation.
- Arithmetic, logic, and comparison operations.

An RTL compiler is responsible for converting a description in an HDL into a set of registers and combinational logic. Logic optimization is then used to improve the logic to meet timing or area constraints (Section 6.4.3).

As examples of available commercial systems, some of the transformations that allow RTL descriptions to be synthesized will be given in terms of the VHDL language[57–59] and the Synopsy VHDL Compiler®.[60] Consider the following (sketchy) VHDL description of the difference engine shown in Fig. 6.2(b).

```
package types is
type OP_CODE is (NOP, LOADA, LOADB, LOADF, RUN);
```

```
attribute OP_CODE_ENCODING of OP_CODE:
    type is "000 001 010 011 100";
end types;
```

The **type** section defines a user-defined type called OP_CODE, which will be used to control the difference engine. It states that the 3-bit field has five operation codes to load the three registers, to run the difference engine, or to do nothing (NOP). An optional encoding has been assigned via the attribute keyword. Thus one operation that the HDL synthesizer does is to assign values to unspecified type fields.

```
entity DIFF_ENGINE is
port(
    DATA : in BIT_VECTOR (0 to 7);
    OP : in OP_CODE;
    CLOCK : in BIT;
    SIGN : out BIT;
    );
end DIFF_ENGINE;
```

The **entity** section defines the name of the design (DIFF_ENGINE) and denotes a port interface to the module. For instance, in this case DATA is defined to be an 8-bit bit-vector that is an input.

```
architecture DIFF_ENGINE_1 of DIFF_ENGINE is
signal A,B,F : BIT_VECTOR (0 to 7);
begin
    process
    begin
        wait until (not CLOCK'stable and CLOCK = '1');
            case OP is
                when LOADA => A <= DATA;
                when LOADB => B <= DATA;
                when LOADF => F <= DATA;
                when RUN =>
                    if (SIGN = '1') then F <= F+A
                    else F <= F+B
                    end if;
                when NOP => F <= F
            end case;
        SIGN <= (F<0);
    end process;
end DIFF_ENGINE_1
```

In the above RTL description, a number of statements are illustrated. First, the signal statement defines some internal signals that have local scope within the module. The **process** statement indicates a section of code to be implemented sequentially. The **wait** statement indicates the presence of clocked registers and specifies that the registers are triggered on the rising edge of CLOCK. The **case** operator indicates a multiplexer, as does the if operator. Finally, the "+" operator indicates an addition while the "<" operator indicates a comparison operator. When combined with an appropriate substrate (simulation, logic library, operator definition), the above description may be compiled into a set of logic gates and registers.

In the case of state-machines, RTL compilers need to provide for automatic state-assignment[61] and minimization.

6.4.3 Logic Optimization

Logic optimization programs take logic descriptions as generated by an RTL synthesis (with the registers stripped out) or which are generated directly at the logic level and optimize the network of gates that are required to implement the function specified by the logic description for a given logic library. The registers are then reunited with the optimized logic, and the physical layout for the system may be implemented using largely automatic techniques. The methods for this are well understood[62] and there are a number of very successful commercial systems.

A typical flow through a typical logic synthesis system is shown in Fig. 6.35. The design is commenced with a logic description. This may be in the form of Boolean equations or a schematic netlist of logic gates. The objec-

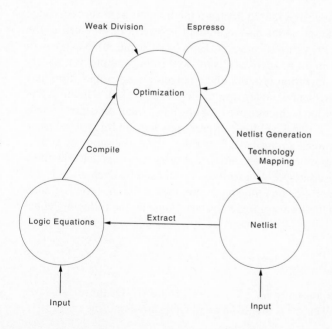

FIGURE 6.35 Logic-optimization flow

tive of a logic optimization scheme is to manipulate the logic to meet speed or area constraints or a combination of both goals. Generally, logic-optimization systems divide the problem into two stages:

- A technology-independent phase in which the logic is optimized according to algebraic and/or Boolean techniques.
- A technology-mapping phase, which translates the technology-independent description derived in the previous step to specific library standard cells, FPGA elements, or other implementable logic gates.

The technology-independent aspect of logic optimization uses a large body of algorithms that operate on logic networks, using both Boolean and algebraic techniques. Most often logic-optimization systems provide a means to read logic networks, manipulate them, perform a technology mapping, and save the resulting structure to be used by an automatic layout program or some other tool. A typical flow through an optimization script is as follows[63]:

- Network organization.
- Two-level minimization.
- Algebraic decomposition.
- Iterative improvement.

Having read the design in, the first step might be to perform tasks such as eliminating constant nodes and redundant inverters or converting the logic to a two-level PLA sum-of-products form (see Chapter 8). Next, a two-level minimization might be invoked. Espresso[64] is an example of a widely used two-level minimization program. Next, algebraic decomposition may be used that introduces new nodes into the logic network in a manner that minimizes the cost. One technique used is known as "weak division."[65] This is used to decompose two-level logic expressions into multiple-level logic expressions. It operates by repeatedly "dividing" the expressions by subexpressions that appear more than once in the set of expressions that constitute the design. The most suitable subexpression is chosen by evaluating a cost function that may be based on reducing the number of literals (area) or other functions related to the levels of logic (speed). Consider the following equations:

```
f1 = aef + bef + ceg
f2 = aeg + bg + def
```

The common subexpressions are ef, ae, eg, and a+b. Of these, ef saves the most literals. When ef is divided into all subexpressions, the result is

```
f1 = (a+b)t1 + ceg
```

```
f2 = aeg + bg + dt1
t1 = ef
```

After this pass the literal `eg` might be chosen, yielding

```
f1 = (a+b)t1 + ct2
f2 = at2 + bg + dt1
t1 =ef
t2 = eg
```

Finally, algorithms are used to iteratively improve the logic structure. This may employ the algebraic techniques of extraction, factoring, and substitution in addition to decomposition.

Following the technology-independent step, a technology mapper is then used to optimize the gates for a particular technology.[66] Two kinds of optimizers are in popular use. The first consists of a rule base consisting of rules in the form

if (antecedent) *then* (precedent).[67]

They are used to map over small sections of circuitry to choose suitable logic gates for an implementation. Figure 6.36 illustrates some typical rules. Figure 6.36(a) eliminates cascaded inverters, while Fig. 6.36(b) converts NOR and INVERT logic gates into an OAI gate. Other rules might bias gate selection toward faster gates such as NAND gates. Another approach is termed Directed-Acyclic-Graph(DAG) covering.[68] In this approach, what is called a *base-function set* is chosen. This might be a 2-input NAND gate and an inverter. All logic gates in the target library are then described in terms of the base-function elements. These are known as *pattern gates*. The logic network is optimized using the base-function set. This creates a *subject graph.* Graph optimization techniques are then used to find an optimized set of target gates. Figure 6.37(a) shows a base function set. Figure 6.37(c) shows examples of pattern graphs. For the 4-input NAND gate shown in Fig. 6.37(c), two possible pattern graphs are shown. Figure 6.37(d) shows a subject graph in which a particular mapping has been identified. Finally, Fig. 6.37(e) shows the resultant mapped logic implementation.

The MIS[69] and BOLD[70] systems are examples of research-based tools that provide logic minimization. These systems provide an environment that contains a number of minimization techniques. These are used to construct minimization scripts that can be adapted for varying styles of logic minimi-

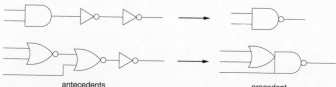

antecedents   precedent

FIGURE 6.36 Rule-based technology mapping

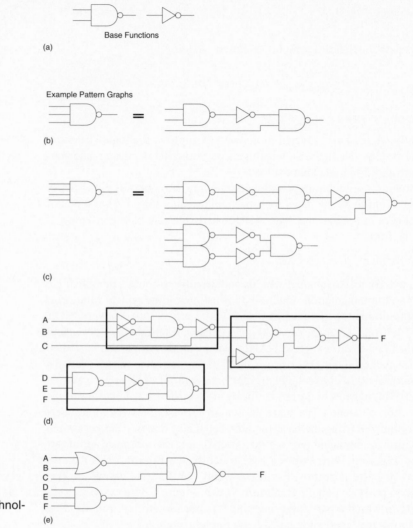

FIGURE 6.37   DAG technology mapping

zation. Figure 6.38 shows the typical inputs to MIS that might consist of a minimization script, a set of equations, and a logic-library definition. The output is a netlist implementing the equations in terms of the logic library. An example of the use of this program is presented in Chapter 8. The EDIF language is frequently used as a common netlist format between design systems.[71]

Apart from increasing design productivity, logic synthesis systems are very useful for transforming between technologies. For instance, a designer might synthesize a circuit in terms of multiple FPGAs, and construct a prototype. This might be used to verify the operation of the circuit under real-world conditions and then a single-chip version may be compiled using a gate-array library and the original logic description.

6.4.4 Structural-to-Layout Synthesis

Once a network of logic gates and registers is available, these may be automatically converted to a layout. Software for this task is very well developed, having been refined over the last 15 to 20 years. Gate arrays and standard-cell designs use this approach. There are two main phases that are required: placement and routing.

6.4.4.1 Placement

Placement is the task of placing modules adjacent to each other to minimize area or cycle time (timing-directed placement). Two main automated algorithms have been developed. The Min-cut algorithm[72] takes the blocks at the top level of the chip or module to be placed and finds two approximately equal area-groupings of subblocks with the minimum number of signal interconnections. These two blocks are then placed in the top and bottom half of a conceptual final layout. This process is repeated for these two halves, splitting the conceptual layout into quarters and so on until the leaf cells are reached. This algorithm is very fast and gives good placements. Another popular technique in which the movement of modules is likened to thermal annealing is also used.[73] Modules are initially allowed to move randomly, and the "temperature" of the layout is evaluated by applying some measure such as routing area or timing. As the layout "cools" the routing and/or timing improves. For each proposed subblock movement, the resulting temperature is calculated. If it is higher than the current temperature, the move is not completed. To avoid local minima, the "melt" is reheated and then recooled according to an "annealing schedule." This process is used in the TimberWolf program that was developed at the University of California, Berkeley[74] and refined at Yale University.

6.4.4.2 Routing

A router takes a module placement and a list of connections and connects the modules with wires. This technology is very mature. Types of routers include channel routers, switchbox routers, and maze routers. Channel routers are typified by the YACR2 router[75] and the Greedy router.[76] These routers route rectangular channels. Switchbox routers can route more complex channel shapes than channel routers. Maze routers[77] can route just about any configuration but have comparatively long running times. They are usually reserved for really tough routing problems.

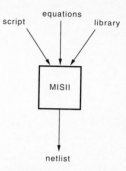

FIGURE 6.38 Logic optimization using MISII

A global router[78] is a special router that works during a placement algorithm to try to plan where routes will travel when the layout is finally placed.

6.4.4.3 An Automatic Placement Example

The standard-cell (constant height cells) placement part of TimberWolf takes as inputs the following:

- The .cel file.
- The .blk file.
- The .par file.
- The .net file.

The .cel file describes the connectivity of standard-cells port locations and signal names. A partial example is shown below:

```
cell 2 INVERTER-2
left -36 right 36 bottom -225 top 225
pin name Z_top signal NEXT_STATE<2> 18 225
equiv name Z_bottom 18 -225
pin name A_top signal [169] -18 225
equiv name A_bottom -18 -225
.
pad 0 %%%pad_TOP_0 orient 3
padside T
left -1 right 1 bottom -1 top 1
pin name top_%io-port signal STATE<3> 0 1
.
pad 1 %%%pad_TOP_1 orient 3
padside T
left -1 right 1 bottom -1 top 1
pin name top_%io-port signal STATE<2> 0 1
```

The first `cell` statement denotes an inverter, `INVERTER-2` with output `NEXT_STATE<2>` and input `[169]`. The bounding box of the inverter is specified by the second line. Each successive line specifies a port and its location. Ports that feed through cells may also be specified. Finally, the ordering and location of I/O pads are specified by the `pad` statements.

The .blk file contains information pertaining to the structure of each row in the layout. An example follows:

```
block height 450 class 1
block height 450 class 1 mirror
block height 450 class 1
```

This defines a three-row layout with the middle row mirrored in *Y*.

The .par file contains various global parameters to be applied to the layout. The following is an example:

```
rowSep 1.1111112
addFeeds
feedThruWidth 36
implicit.feed.thru.range 0.25
do.global.route
do.global.route.cell.swaps
```

For instance, this specifies to add feedthroughs if necessary.

Finally the .net file specifies information about the nets to be routed. The following is an example:

```
allnets HVweights 2.5 1.0
```

This specifies that all nets are to be routed with an equal weighting of 2.5 for horizontal and 1 for vertical routes.

TimberWolf returns the following files

- The .pl1 and .pl2 files which describe the placement of modules.
- The .pin file, which describes the segment list of routes.
- The .twf file.
- The .out file, which is a summary of the program execution.
- The .sv2 and .sav files, which allow restart of the program.

For example, a portion of .pl2 file example appears as follows:

```
.
D-REG-MUX-8 0 502  612 952  2 1
.
INVERTER-2 1188 1452  1260 1902  1 2
.
D-REG-MUX-31 1332 502  1944 952  2 1
.
```

A portion of a .pin file appears as follows:

```
.
[212] 3 NOR2-1 A_bottom 702 502 1 1 0
.
STATE<0> 12 AOI21-35 B_top 774 2852 4 -1 0
.
```

Using this information, a channel router may be called to complete the channel routes specified by TimberWolf. With the addition of power feeds on the two ends of the layout, the standard cells and routing blocks may be placed to create the final layout. A typical standard-cell layout in outline form is shown in Fig. 6.39 and Plate 9.

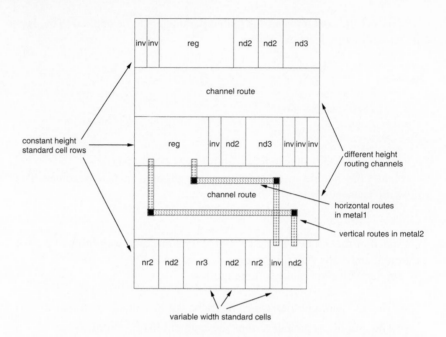

FIGURE 6.39 A typical standard-cell layout style

A large number of mature research, proprietary, and commercial place-and-route systems are available.

6.4.5 Layout Synthesis

The layout of regular structures such as RAMs, ROMs, PLAs, register files, multipliers, and general datapaths may be synthesized by software generators. These programs take a number of parameters as input and automatically create a custom physical layout. Some systems create actual mask layout tuned to a particular process, while others create symbolic layouts that may be compacted to suit a particular technology.[79]

The following is an example of a virtual-grid symbolic description of an inverter from the NS design system. It is specified in Common Lisp.

```
(defaspect-generator ("USER:INVERTER" :VIRTUAL-GRID)
                        (w ratio pw)
;;; transistors
  (part N-CHANNEL-MOSFET :origin (pt 1 3) :WIDTH w)
  (part P-CHANNEL-MOSFET :origin (pt 1 7)
                          :WIDTH (* ratio w))
;;; contacts
  (part VG-TERMINAL :x 0 :y 3
                        :CONNECTED-LAYERS '(N-DIFF METAL))
```

```
    (part VG-TERMINAL :x 2 :y 3
                         :CONNECTED-LAYERS '(N-DIFF METAL))
    (part VG-TERMINAL :x 3 :y 5
                         :CONNECTED-LAYERS '(METAL POLY))
    (part VG-TERMINAL :x 0 :y 7
                         :CONNECTED-LAYERS '(P-DIFF METAL))
    (part VG-TERMINAL :x 2 :y 7
                         :CONNECTED-LAYERS '(P-DIFF METAL))
;;; wires
    (part VG-LOG :from (pt 0 10) :to (pt 4 10)
                         :LAYER 'METAL :WIDTH pw)
    (part VG-LOG :from (pt 0 0) :to (pt 4 0)
                         :LAYER 'METAL :WIDTH pw)
    (part VG-LOG :from (pt 0 0) :to (pt 0 3) :LAYER 'METAL)
    (part VG-LOG :from (pt 0 7) :to (pt 0 10) :LAYER 'METAL)
    (part VG-LOG :from (pt 2 5) :to (pt 2 3) :LAYER 'METAL)
    (part VG-LOG :from (pt 1 0) :to (pt 1 10) :LAYER 'POLY)
    (part VG-LOG :from (pt 3 0) :to (pt 3 10) :LAYER 'POLY)
    (part VG-LOG :from (pt 2 7) :to (pt 2 5) :LAYER 'METAL)
    )
```

The first two statements specify transistors. The next four statements specify inter-layer contacts (the type is specified by the list of layers following the :CONNECTED-LAYERS keyword). The final eight statements specify wires with position keywords (:from :to), a size keyword (:WIDTH), and the :LAYER keyword, specifying the layer on which the wire is routed.

In this example, the width of the n- and p-transistors has been specified in terms of the variables w and ratio. The keyword :WIDTH passes this to the transistor generator. In addition, the power bus width has been specified in terms of pw. Figure 6.40 illustrates a few instances of this layout generator.

At a higher level, the following is the top-level call to a PLA generator (see Chapter 8):

```
(defaspect-generator ("PLA:PLA" :VIRTUAL-GRID) (pla-
                        filename)
  (let* ((inputs (get-number-of-inputs pla-filename))
         (outputs (get-number-of-outputs pla-filename))
         (cells
          (list
           `("PLA-EDGE" :inputs ,inputs :outputs ,outputs
             :edge :bottom)
           `("PLA-MIDDLE" :pla-file ,pla-filename)
           `("PLA-EDGE"   :inputs ,inputs :outputs ,outputs
             :edge :top))))
    (vertically-abut cells)
    (import-all-ports)))
```

This hierarchically calculates some parameters (inputs,outputs) from the file specifying the PLA and calls some other generators (PLA-EDGE and

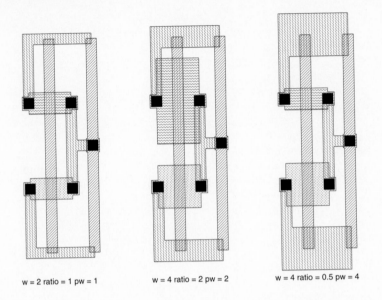

w = 2 ratio = 1 pw = 1　　　w = 4 ratio = 2 pw = 2　　　w = 4 ratio = 0.5 pw = 4

FIGURE 6.40 Various instantiations of a CMOS inverter generated by a generator

PLA-MIDDLE) and then vertically abuts these cells. Following this, the ports are imported from the lower level of the design. As an example of the next level down, the generator that performs the PLA-MIDDLE function is reproduced below:

```
(defaspect-generator ("PLA:PLA-MIDDLE" :VIRTUAL-GRID)
                         (pla-filename)
  (let ((cells (list
              `("PLA-LEFT" :pla-file ,pla-filename)
              `("AND-PLANE" :pla-file ,pla-filename)
              `("AND-OR-JOIN" :pla-file ,pla-filename)
              `("OR-PLANE" :pla-file ,pla-filename)
              `("PLA-RIGHT" :pla-file ,pla-filename))))
    (horizontally-abut cells)
    (import-all-ports)))
```

This horizontally abuts the PLA-LEFT, AND-PLANE, AND-OR-JOIN, OR-PLANE, and PLA-RIGHT cells. Finally at the AND and OR plane level, the generator places transistors according to the PLA personality matrix.

It may be seen that with the combination of symbolic layout, a powerful language and a good CAD substrate, powerful layout generators may be created with minimum effort.

As opposed to the creation of random logic, which the previous section illustrated, layout generators are used for regular arrays or places where a simple algorithm can specify the layout.

6.5 Design-capture Tools

6.5.1 HDL Design

The behavior and/or structure of a system may be captured in a Hardware Description Language. There are a wide variety of proprietary, commercial, and public domain languages including those specifically designated hardware description languages (HDLs), such as VHDL, ELLA, Verilog®, and modified high-level languages, such as C, Pascal, and Lisp. Languages like VHDL allow for the capture of both structure and behavior. For example, Chapter 1 used an example of a structural design coded in Verilog®, while this chapter used an RTL VHDL description.

The popular standard HDLs differ from high-level languages by catering for hardware notions such as bit vectors, signals, and time within the native language. This is reflected in the syntax of the language and the underlying runtime operating support, which includes compilers, debuggers, and simulators. In common with high-level languages, HDLs usually provide all of the elements of modern computer languages—structure, parametization, conditionals, looping, and hierarchy.

6.5.2 Schematic Design

The traditional method of capturing a digital system design is via an interactive schematic editor. Actually, preferences have cycled from textual netlists (when graphics hardware was expensive) to interactive graphic editors to textual HDLs. Many design systems allow a free mix of code and diagrams so that designers can choose. In general, diagrams are more quickly understood ("a picture is worth a thousand words"), but HDLs are more easily modified.

Schematic editors provide a means to draw and connect components. A collection of components may be collected into a module for which an icon may be defined. The icon is a diagram that stands for the collection of components within the module. The shape might suggest the function of the module, while the I/O connections of the module are represented by stubs with signal names. This icon may then be used in another module, and so on, hierarchically, throughout the design. Figure 6.41 shows a typical schematic for a module and its schematic icon.

Primarily, schematic editors are menu-based graphic editors with operations such as:

- Creating, selecting, and deleting parts by pointing or area inclusion.
- Changing the graphic view by panning, zooming, or other means.

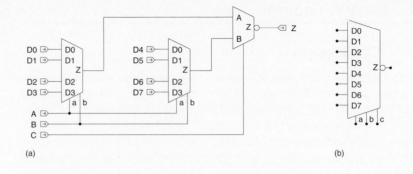

FIGURE 6.41 Structural views of a circuit: (a) schematic; (b) schematic icon

To a basic graphic editor, operations are added that pertain to the electrical nature of the schematic, such as:

- Selecting an electrical node and interrogating it for state, connections, capacitance, etc.
- Running an attached simulator or other electrical network-based tools.

6.5.3 Layout Design

Layout too can be captured via code (in the case of generators) or interactive graphics editors. However, to maintain one's sanity, a good color editor is a strong requirement if substantial layout editing is to be performed. Layout editors, like schematic editors, are based on drawing editors (for instance see Rubin[80]). Differences occur in the way color is treated and sometimes in the way detail is thresholded (although in advanced design systems one editor is usually used for all diagram editing[81]). Because there is usually a large amount of data present, various means of turning off detail are required to alleviate long redraw times. A layout editor might interface to a Design Rule Checking program to allow interactive checking of DRC errors, and to a layout-extraction program to examine circuit-connectivity issues.

6.5.4 Floorplanning

Floorplanning[82,83] is the exercise of arranging blocks of layout within a chip to minimize area or maximize speed. The latter is increasingly the main reason for performing this activity. Floorplan editors provide graphical feedback about the size and placement of modules without showing internal layout details. In addition, the editors show connectivity information between modules in the form of "rat's-nest" wiring diagrams, where the connected ports of modules are connected by straight lines. These kinds of diagrams indicate the relative density of wiring and whether, for instance, ports line up between adjacent modules. Figure 6.42 shows a simple example.

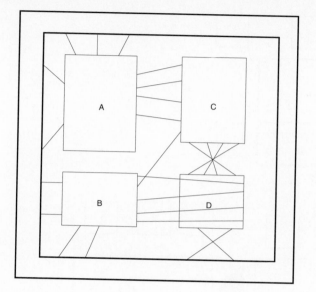

FIGURE 6.42 A floorplan example

This shows that module D should be flipped around the Y axis to improve the routing. Some editors provide shaded color displays of routing density that allows designers to re-place and "rip-up-and-reroute" congested areas of the chip. While floorplanning may be done automatically, many times a much better job can be done manually. Usually, the task is not that time consuming, given the right interactive tools and a knack for doing it.

6.5.5   Chip Composition

Similar to structural synthesis, chip composition, or "block-place-and-route," is the term that is applied to wiring the top level modules in a design. At this point a good placement of modules is assumed. The task consists mainly of routing modules together and then placing a pad ring around the completed chip core. Usually there is a routing strategy that is followed. For instance, the technique of binary composition,[84] has been widely used. Here modules are combined alternately in horizontal and vertical strips from the bottom up until the complete chip is routed. Figure 6.43 illustrates an example that shows the progression of steps of grouping modules and adding routing channels. Figure 6.43(a) shows the unrouted, relatively placed modules. Figure 6.43(b) shows a horizontal composition where, for instance, A and B are routed together by routing cell AB. A vertical composition step is shown in Fig. 6.43(c), where all modules are now connected. Routing block DEF routes module D and the composed module (E,EF,F). The advantage of this

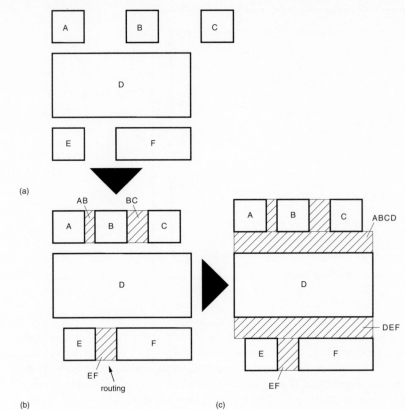

FIGURE 6.43 Binary composition for chip routing

approach is that only channel routes need to be routed, which is a well-solved problem.

6.6 Design Verification Tools

Figure 6.44 shows a conventional flow through a set of design tools to produce a working CMOS chip from a functional specification. Depending on the tools, some steps may be automatic and hidden from the designer but usually are performed by some agent. The design process is commenced with a clock-cycle-accurate functional specification (say, in a high-level language such as C). This is used to verify that the system performs as required. This is translated to a structural RTL or logic description. If done manually, the functionality of the two descriptions has to be proved isomorphic. This is done by applying a stimulus to the functional description and to a logic simulation of the RTL description and comparing the outputs of both forms on a

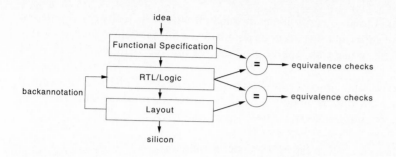

FIGURE 6.44   Design flow through typical CMOS VLSI tools

clock-cycle by clock-cycle basis. RTL simulations may be done with the actual clock timing by estimating the layout loading capacitances. Once the functional equivalence has been satisfied, the structural description is transformed into a physical form (i.e., a layout). Again, this might be automatic for a gate array or standard-cell layout or might be done manually. The problem now is to prove that the layout is a faithful reproduction of the structure of the RTL description (i.e., all signals are routed correctly). In addition, we have to prove that the functionality is still maintained in the temporal domain (timing). This is done by extracting the parasitic routing capacitances introduced by the physical layout and applying them to the RTL simulation model.

Each one of these steps requires a certain set of verification tools. In this section we summarize these tools.

6.6.1   Simulation

Probably the software tool that designers most frequently encounter is some form of simulator that is used to predict and verify the performance of a given circuit. Simulators come in a wide variety depending on the accuracy and speed of simulation required.[85]

6.6.1.1   Circuit-level Simulation

The most detailed and accurate simulation technique is referred to as Circuit Analysis. As the name suggests the simulators operate at the circuit level. Circuit-analysis programs are typified by the SPICE program developed at the University of California at Berkeley[86] and ASTAP developed at IBM.[87] Commercially available versions are typified by the HSPICE program.[88] The basis for this type of program is the solution of the matrix equations relating the circuit voltages, currents, and resistances (or conductances). This type of simulator is characterized by high accuracy but long simulation times. Simulation time is typically proportional to N^m, where N is the number of nonlinear devices in the circuit, and m is between 1 to 2. This type of program is used to verify in detail small circuits or to verify the simulation results of faster but less accurate simulators, such as timing simulators. It is unrealistic to use this type of program for the verification of large VLSI chips.

Circuit simulators used to verify performance of CMOS circuits should not be assumed to accurately predict the performance of designs. There are three basic sources of error. These are as follows:

- Inaccuracies in the MOS model parameters.
- The use of an inappropriate MOS model.
- Inaccuracies in parasitic capacitances and resistances.

Usually, contemporary circuit simulators related to SPICE provide different levels of modeling (specified by the LEVEL parameter). Simple models are optimized for speed of processing, while more complex models are used for more accurate simulation. If possible one should verify actual transistors from known process corners against the DC characteristics predicted by the simulator. A good practice is to include test transistors of both polarities with various widths with the lengths used in the design (usually the minimum and slightly longer for I/O transistors). Because processes are shrunk, the models used by a simulator may no longer be able to accurately predict the performance of the transistors.

Assuming that one has verified the DC performance of the transistors, the AC performance can now be in question. A significant source of error in predicting performance can be the parasitic capacitances that are applied to nodes in the circuit description used in the circuit-analysis program. The gate capacitance is part of the MOS model and should be subtracted from the total capacitance predicted by a layout-extraction program. One should check how drain and source capacitances are added to transistors—often they are added as diodes or as part of the MOS model. In this case they should not be added as stray capacitance on the node. A good practice is to create a check layout with known areas and peripheries on each layer and then check the SPICE deck produced by any extraction program. The bottom line is to be aware of the means of process calibration (i.e., that someone is responsible for it).

6.6.1.2   Timing Simulation

It is possible to simplify the general circuit analysis approach used above to allow simple nonmatrix calculations to be employed to solve for circuit behavior. This usually involves making some approximations about the circuit. Typical of an early simulator using this approach is the MOTIS simulator.[89] More recent examples may be found in White and Sangiovanni-

Vincentelli.[90] The accuracy of such simulators is less than that of SPICE-type simulators, but the execution time is almost two orders of magnitude less.

Implementations may use MOS-model equations to calculate device currents or may use table look-up methods. Calibration of any simulator is advisable, using the techniques described above for circuit simulators. Usually the relative accuracy of such simulators is good; that is, inherently high speed circuits will demonstrate better performance than slower circuits. However, sometimes the absolute accuracy may not be as good as a circuit simulator, especially if no real silicon has been used to check simulation results.

Absolute accuracy is somewhat of a red herring anyway, because process variation, temperature variation, and allowable supply-voltage excursions may vary by a range of three or four to one. The designer is usually trying to predict the slowest the circuit will operate. It is unwise to do this with no margin unless the whole design system is known to accurately predict worst-case performance. Designers generally allow a 10–20% margin in assessing speeds.

6.6.1.3 Logic-level Simulation

Many simulators have evolved to deal with simulation at the logic level. They use primitive models such as NOT, AND, OR, NAND, and NOR gates. Some operate in a "unit delay" mode, where every gate is assumed to have a delay of one time-unit. This type of simulator can be highly optimized for execution speed. Alternatively, timing parameters may be assigned to the logic models based on prior circuit simulation and known circuit parasitics. Because all logic circuits are rarely active simultaneously, logic events may be scheduled on a queue. This means that the state of the network is evaluated on an event-driven basis, rather than on a timing-substep basis, as are most of the implementations of the two previous simulators.

Timing is normally specified in terms of an inertial delay and a load dependent delay for the appropriate edge transitions, as follows:

$$T_{gate} = T_{intrinsic} + C_{load} \times T_{load}, \tag{6.1}$$

where

T_{gate} = the delay of the gate

$T_{intrinsic}$ = the intrinsic gate delay (no load)

C_{load} = the actual load in some units (i.e., *pF* or # normalized gates)

T_{load} = the delay per load in some units (i.e., *ns/pF* or *ns/#normalized gates*)

(A normalized gate might be the minimum gate load of the smallest inverter in a standard-cell library—all other gate inputs would be characterized in terms of this unit.)

Logic simulators with such timing information are quite accurate for CMOS logic configurations or other circuits where the function has been well characterized at the gate level. Nowadays this characterization can be done automatically by running scripts that perform the circuit simulations and extract the relevant data. Where this capability is not available, a considerable manual simulation effort is required to create a new standard-cell library in a new process.

Logic simulators are adequate for well-characterized CMOS circuits that have regular logic counterparts. They are relatively fast and are thus suitable for large circuits. This has been also aided by hardware engines that compute the simulation algorithm. Early logic simulators were not suitable for circuits with transistors used as transmission gates, such as transmission gate multiplexers, memories, or pass-gate logic. However, recent logic simulators do deal with transistor circuits in a limited manner.

6.6.1.4 Switch-level Simulation

Switch-level simulators merge logic-simulator techniques with some circuit-simulation techniques by modeling transistors as switches. RSIM[91] is an example of a switch-level simulator with timing. CMOS gates are modeled as either pull-up or pull-down structures, for which RSIM dynamically calculates a resistance to power or ground. This resistance is used with the output capacitance of the gate to predict rise or fall times.

Switch-level simulators alleviate the need for circuit analysis calibration of CMOS gates, but do have some accuracy limitations when evaluating transmission-gate circuits (they are usually overly pessimistic). In addition, some circuit structures present pathological topology cases, which confuse the simulation algorithms (the "tiny XOR" gate used in the transmission-gate adder in Chapter 8 is an example).

If you design at the transistor level, a switch-level simulator provides a first line of defense as far as simulation. One should probably back up any simulations with a reduced set of simulations using a timing simulator.

6.6.1.5 Mixed-mode Simulators

There now exist very good commercial simulators that merge the good points of functional simulation, logic simulation, switch simulation, timing simulation, and circuit simulation. Each circuit block can be simulated in the appropriate mode. For instance a standard-cell logic block might be simulated at the logic level, a memory might be simulated at the functional level, and a phase-locked loop might be simulated at the circuit level. In this way only those circuits requiring detailed simulation expend expensive compute cycles.

6.6.1.6   Summary

A good simulator is crucial to modern CMOS design. The style of simulator determines the level to which one can safely design. With a logic simulator, one can accurately model well characterized gates and functional blocks. A timing simulator allows design down to the transistor level for most digital circuits and some limited-accuracy analog circuits. Finally, a circuit-analysis program provides enough accuracy (when calibrated to a process) for the most complicated analog circuitry. The simulation times and therefore the amount of circuitry that may be simulated with each kind of simulator varies widely. Logic simulators (particularly unit-delay) are of use at the system level. Timing simulators are useful for modules into the 100–100K transistor range and have been used for 1M+ transistor circuits for a few hundred vectors. Circuit simulators are useful for 10 to 1,000–transistor complexities for short simulation periods. Modern mixed-mode simulators allow a trade-off in simulation accuracy and time of simulation.

6.6.2   Timing Verifiers

Classically, designers simulated with unit-delay simulators to verify functionality. Then they ran simulations with delays to check for timing problems. The detection of such problems is pattern dependent. In other words, if the critical timing vector is not exercised, the critical path will not be found. A timing verifier takes a different approach to temporal verification. Here, the delays through all paths in a circuit are evaluated in a pattern-independent manner and the user is provided with information about these delays. CMOS verifiers in common with simulators may work at the gate or the transistor level.[92] The circuit to be analyzed is first statically examined to determine the direction of signal flow in all transistors. This is necessary to evaluate only those delays that will be critical in actual circuit operation. A recent example of this type of analyzer is the Pearl program.[93] Each transistor is examined and the direction of signal flow is calculated using nine rules. These rules may be determined from:

- Circuit-design methodology rules.
- Electrical rules.
- User-supplied rules.

The Pearl program calculates an RC delay for each node using RSIM. These are then evaluated in a breadth-first manner. Delay paths are qualified by appropriate clocks.

A timing analyzer implemented at the transistor level can provide a designer with rapid feedback about critical paths. Combined with a switch-level simulator for rapid global functional simulation, a timing simulator for detailed module verification, and a circuit-analysis program for critical-path evaluation, the timing analyzer completes a set of powerful verification tools. Timing analyzers implemented at the gate level allow the same quality of design down to the gate level, which is sufficient for a wide range of CMOS systems.

Pitfalls of timing analyzers are false paths and sneak paths. False paths can occur because the timing analyzer does not know how the circuit is used. For instance, that a bus is only used to read or write during a cycle and not for both. False paths are dealt with by blocking them as they are recognized. Sneak paths are paths that for some reason the timing analyzer does not recognize. These can occur in complicated clocking schemes that may be beyond aging timing analyzers. For this reason it may be prudent to timing simulate circuits as a backup unless you are confident that your timing analyzer catches everything. (Many have believed the latter point, only to be ushered back to reality by the outcome of the silicon.)

6.6.3   Network Isomorphism

An electrical network may be represented by a graph where the vertices of the graph are devices such as MOS transistors, bipolar transistors, diodes, resistors, and capacitors. The arcs are the connections between devices. These are the electrical nodes in the circuit. This graph may be in turn represented by some data structure that may be accessed by a variety of software routines interested in the electrical connectivity properties of the circuit. Two electrical circuits are identical if the graphs representing them are isomorphic; that is, each graph has the same number of devices and for every device in one circuit there is a matching device in the other circuit. The matching devices have identical properties such as:

- Transistor width and length.
- Resistance value.
- The number of connections on each terminal (i.e., gate, drain, source).

Each node in one circuit has a matching node in the other circuit. They have identical properties such as:

- The same number of source and drains attached to them.
- And the same number of gates (MOS gate).

Network isomorphism is used to prove that two networks are equivalent and therefore should function equivalently. This is used most often to prove that a layout is equivalent to a network extracted from a schematic schematic or HDL structural netlist. Other uses include proving that two schematics or two layouts are equivalent.

The process of comparing two networks is commonly called "netlist comparison," "network isomorphism," or LVS (layout versus schematic).

Electrical networks may use subnetworks as devices. For instance, in a chip layout standard-cell blocks may be represented by bounding boxes (for a vendor's proprietary library). The layout extract operation then extracts only the routing. This is compared with to the network obtained by expanding the structural description down to the level of gates (but not transistors). Frequently the notion of "logical equivalency" is used. This allows a layout-design system to swap the order of signals on series transistors in logic gates with respect to the structural specification for layout convenience. While this is fine for logic circuits, some problems can occur if it is used in high-performance and mixed-signal circuits. Consider an analog-bias circuit consisting of series transistors where the order of the transistors dictates the behavior—these cannot be swapped.

6.6.4   Netlist Comparison

If a schematic or circuit description is entered to define an IC, at some stage a physical layout is generated. This may be completed automatically, as in the case of a gate array or place-and-route standard-cell system. Alternatively, the physical layout may have a manual component. Ideally, the signal names between parallel representations would be the same, allowing easy comparison between desired and actual circuit by matching node names and the number and type of components connected to each node in the schematic and the layout. In reality, signal names are often omitted from internal nodes in a circuit (especially in layouts) and only applied to I/O ports. Thus there is the problem of comparing two graphs that are labeled in a limited manner. Programs that verify the equivalence or lack thereof of two unnamed circuit graphs are thus needed.

Typical of a program that performs this function is GEMINI.[94] Signatures are calculated for each transistor in the test and reference circuit. Signatures include:

- Fan-in.
- Fan-out.
- Transistor type.
- Bound nets connected to the transistor.

Test and reference circuits are then repeatedly checked to correlate transistors. Discrepancies are either indicated interactively or by a listing of the matched and unmatched nodes.

6.6.5 Layout Extraction

Layout extractors examine the interrelationship of mask layers to infer the existence of transistors and other components. They are related to design rule checkers (Section 6.6.7). Various approaches have been implemented to approach this problem.[95,96] Commonly, parasitic capacitances and resistances are reported in addition to transistor connectivity. Algorithms commonly use geometric-shape intersections to recognize active devices (see Chapter 3). The need for such tools by the system designer will decrease as higher-level design techniques provide "correct-by-construction" modules. However, some form of layout extraction is usually done to create data for the back annotation step described in the next section.

6.6.6 Back-Annotation

Once a layout has been constructed and there is isomorphism between the schematic network and the layout network, one can correlate extracted capacitances from the layout with the schematic and perform simulation or timing analysis to verify performance. This is done by moving the capacitance that appears on a layout node to the corresponding schematic node while accounting for existing capacitance on the schematic node. For instance, the schematic may already have the source-drain and gate load due to the gates connected to the node and only the routing capacitance is required to be added. This operation is known as "back-annotation" (Fig. 6.44).

6.6.7 Design-rule Verification

If mask design is completed manually (and even automatically), it is necessary to verify that the layout conforms to the geometric design rules. This is achieved with a design-rule checker. Many variations exist, but typical approaches are found in Szymanski and Van Wyk,[97] Baker and Terman,[98] and Baird[99] (see also Chapter 3).

Hierarchical design-rule checkers are necessary for large circuits.[100] These design-rule checkers use the hierarchical nature of a design to reduce the number of cells that have to be individually checked.

6.6.8 Pattern Generation

Pattern generation is the last step in the sequence that starts at the architecture for a chip and ends with a database suitable for manufacture. It is the operation of creating the data that is used for maskmaking. Over the years, the format of this data has changed as the methods of generating masks have changed. Originally, the data drove flatbed plotters that cut Rubylith® (a red plastic "mask" layer backed by a clear Mylar® plastic backing). Nowadays, most semiconductor operations use electron-beam–generated masks (i.e., generated by exposing a resist-coated metal film with a focused electron beam). These machines expose the masks in a raster-scan style similar to a television.

A common format is the Electron Beam Exposure System, EBES format.[101] Data is composed of rectangles, parallelograms, or trapezoids. Given a layout captured in a design system, the following steps must be completed to create an EBES file:

- Combine layers to form required mask (i.e., all n^+ and vddn regions for an nplus mask).

- Size-resulting data (i.e., shrink or bloat to account for processing effects such as under-etching or sideways diffusion).

- Canonicalize resulting geometry in terms of base figures (i.e., rectangles).

- Sort the resulting shapes in scanline order.

- Determine polarity of mask (i.e., dark field or light field).

- Output data in suitable format.

Because this is the last step in the design process and because it is hard to detect defects on the masks, manufacturers frequently pattern two or more die patterns on a single-mask reticle and then use differencing techniques to detect differences between pairs of like die patterns to detect mask defects.

6.7 Design Economics

It is important for the IC designer to be able to predict the cost and the time to design a particular IC or sets of ICs. This can guide the choice of an implementation strategy. This section will summarize a simplified approach to estimate these values.

In this study we will concentrate on the cost of a single IC, although one should consider the overall system when making such decisions. System level issues such as packaging and power dissipation may affect the cost of an IC.[102]

The selling price of an integrated circuit may be given by

$$S_{total} = \frac{C_{total}}{1 - m},$$

(6.2)

where

C_{total} = the manufacturing cost of a single IC to the vendor

m = the desired profit margin.

The margin has to be selected to ensure a profit after fixed costs including overhead (G&A), and the cost of sales (marketing and sales costs) have been subtracted out.

The costs to produce an integrated circuit are generally divided into the following:

- Nonrecurring costs (NREs).
- Recurring costs.
- Fixed Costs.

6.7.1 Nonrecurring Engineering Costs (NREs)

The nonrecurring costs are those costs that are spent once during the design of an integrated circuit. They include

- The engineering design cost.
- The prototype manufacturing cost.

These costs are amortized over the total number of ICs sold. F_{total}, the total nonrecurring cost is given by

$$F_{total} = E_{total} + P_{total},$$
(6.3)

where

E_{total} = the engineering cost

P_{total} = the prototype manufacturing cost.

Normally the recurring costs are viewed as an investment for which there is a required rate of return. For instance, if $100K is invested in NRE for a chip then $1M might have to be generated as profit for a rate of return of 10.

6.7.1.1 Engineering Costs

The costs of designing the IC (E_{total}) hopefully happen only once during the chip design process. The costs include:

- Personnel costs.
- Support costs.

The personnel costs might include the labor for:

- Architectural design.
- Logic capture.
- Simulation for functionality.
- Layout of modules and chip.
- Timing verification.
- DRC and tapeout procedures.
- Test generation.

The support costs, amortized over the life of the equipment for the length of the design project, include:

- Computer costs.
- CAD program costs.
- Education or reeducation costs.

6.7.1.2 Prototype Manufacturing Costs

These costs (P_{total}) are the fixed cost to get the first ICs from the vendor. They include:

- The mask cost.
- Test fixture costs.
- Package tooling.

The photo-mask cost is proportional to the number of steps used in the process. Mask costs increase as the process dimensions are reduced, so while newer, smaller processes generally have increased mask costs, masks on the metalization layers can be less expensive than the lower layers. A mask can currently cost between $500 and $1500.

A test fixture consists of a printed wiring board-probe assembly to probe individual die at the wafer level and the interface to the tester. Costs range from $1000 to $5000 depending on the complexity of the interface electronics.

If a custom package is required, it may have to be designed and manufactured (tooled). The time and expense of tooling a package depends on the sophistication of the package. Where possible, standard packages should be used.

6.7.2 Recurring Costs

Once the development cost of an IC has been determined, the IC manufacturer will arrive at a price for the specific IC. This includes a recurring cost; that is, one that recurs every time an IC is sold.

The IC manufacturer will determine a part price for an IC based on the cost to produce that IC and a profit margin. The margin can fall as the revenue increases. An expression for the cost to process an IC follows.

The total cost is

$$C_{total} = C_{process} + C_{package} + C_{test},$$ (6.4)

where

$C_{package}$ = package cost

C_{test} = test cost—the cost to test an IC is usually proportional to the number of vectors and the time to test.

$$C_{process} = \frac{W + P}{N Y_w Y_{pa} Y_{ft}},$$ (6.5)

where

W = wafer cost

P = processing cost

N = gross die per wafer (the number of complete die on a wafer)

Y_w = die yield per wafer

Y_{pa} = packaging yield

Y_{ft} = final test yield.

The wafer yield, Y_w, was dealt with in Chapter 4. The packaging yield is the percentage of successfully diced, bonded and packaged parts. The final test yield is the percentage of packaged parts that pass a final packaged part test sequence.

6.7.3 Fixed Costs

Once a chip has been designed and put into manufacture, the cost to support that chip from an engineering viewpoint may have a few sources. In order for the part to be effectively used, Data Sheets describing the characteristics of the IC have to be written. A data sheet is probably always required, even for application specific ICs that are not sold outside the company that developed them. From time to time Application Notes describing how to use the IC may be needed. In addition specific application support may have to be provided to help particular users. This is particularly true for ASICs, where the designer usually becomes the walking, talking data sheet and application note. Another ongoing task may be failure analysis if the part is in high volume and you desire to increase the yield.

Finally there is what is called "the cost of sales," which is the marketing, sales force, and overhead costs associated with selling each IC. In a captive situation this might be zero.

6.7.4 Schedule

At the outset of a system-design project involving newly designed ICs it is important to be able to estimate the design cost and design time for that system. Estimating the cost can guide the designer as to the method by which the ICs will be designed. Estimating the time is essential to be able to select a strategy by which the ICs will be available in the right timescale and at the right price. This second task (estimating schedules) is usually the least well specified and requires some experience to accurately predict design timescales.

If we assume that for a given IC size $C_{process}$ is constant, the variables left in determining the return on investment of an IC are, E_{total}, the engineering design cost, P_{total} the prototype-manufacturing cost. P_{total} depends on the way in which the IC is implemented. We examined a variety of strategies for the design of CMOS systems in Section 6.2. The fixed costs of prototyping P_{total} are relatively constant, given an implementation technology. The engineering costs depend on the complexity of the chip and the design strategy. For this reason, it is important to be able to estimate a schedule for the design of an IC and then manage the available resources to bring the project to a successful conclusion.

Studies on schedule management for ICs have been carried out by analyzing many IC design projects implemented in a variety of ways.[104] These show that schedule is only a function of personpower, that is, the number of people working on the project. The study showed that below 30 weeks, schedule is proportional to personpower, while beyond about 30 weeks, the schedules become proportional to the cube root of the personpower. Fey and Paraskevopoulos[102] suggest a number of methods for increasing productivity, thereby improving schedules. They include the following:

- Using a high-productivity design method.
- Improving the productivity of a given technique.
- Decreasing the complexity of the design task by partitioning.

A range of various design methods were examined in Section 6.3 which form the basis for achieving some of these goals.

As a final point one should note that increasing the personpower is a poor way of improving a schedule and likely will have the opposite effect.

6.7.5   Personpower

In order to estimate the schedule, one must have some idea of the amount of effort required to complete the design. As we have seen, typical IC projects will involve the following tasks:

- Architectural design.
- Logic capture.
- Simulation for functionality.
- Layout of modules and chip.
- Timing verification.
- DRC and tapeout procedures (ERC, LVS, MEBES).
- Test generation.

If we take each of these activities and apply a productivity figure for a given complexity of design, we may have the basis for the manpower, or person-power, required to complete the project.

Fey has completed productivity studies for custom-chip designs[103, 104] and gate-array designs.[105]

6.7.6   An Example—Gate-array Productivity

Fey's productivity models for gate arrays lead to the following empirically determined equation:

$$P = 16.2\ mG^{0.6}\ 0.5 < G < 25, \tag{6.6}$$

where

P = the productivity in gates/person-day

G = the number of gates in thousands

$m = (0.61^{I})(0.86^{U})(0.64^{R})(1.17^{D}),$

where

$$I = \text{the adjusted I/O} = \frac{(inputs + outputs + bidirects)^{0.5}}{K} \tag{6.7}$$

K = the number of gates

R = the complexity (from 1 (lowest) to 5)

U = the number of gates used = max(0, %gates used – 90%)

D = the design experience = number of previous designs completed by designer

As with automobile advertising, these formulae are for comparison purposes only; your mileage may vary. Other variables that were studied included the number of test vectors per gate, the quality of the design specification and the year of the design.

By normalizing the productivity, we obtain

$$P_N = \frac{P}{m} = 16.2G^{0.6} \quad 0.5 < G < 25 \qquad (6.8)$$

where P_N is the normalized productivity in gates per person-day.

The personpower (M) may be calculated from the productivity by

$$M = \frac{200G}{P_B} \quad 0.5 < G < 25, \qquad (6.9)$$

where

$$P_B = 17.1G^{0.61} \text{ (Eq. 6.8 fitted to experimental data),}$$

Thus

$$M = 11.7G^{0.39} \qquad (6.10)$$

By evaluating M one may then estimate the schedule (T), using the following equation:

$$T = M \qquad M < 29 \qquad (6.11)$$

$$T = 9.1M^{0.34} \quad M \geq 29 \qquad (6.12)$$

Thus for a 1000-gate design these equations would suggest that it takes 9 weeks while a 10K-gate design would take 29 weeks. These equations are included as an analytical guide for estimating schedules. They represent the result of one piece of research work aimed at quantifying design productivity. Nothing replaces experience when it comes to estimating the real thing (and even then that does not always help!).

The general outcome of this research suggests keeping design schedules below 6–7 months. Increasing productivity can increase the number of gates that can be designed in this time. With HDL based synthesis, this should reach 50–100K gates. Other design methods should lead to higher productivities.

6.8   Data Sheets

A data sheet for an IC describes what it does and outlines the specifications for making the IC work in a system. These specifications would include power-supply voltages, currents, input setup times, output-delay times, and clock-cycle times. Also included would be pin loadings and package and pinning details. While commercially produced chips are accompanied by data sheets (and this is a good place to look for examples), chips produced in small volumes internally in an organization may often be introduced into the world without the advantage of a data sheet.

A good habit to acquire is that of compiling a data sheet for any chip you might design. A data sheet is the interface between the chip designer and the board-level designer. In particular, it is good practice and mandatory in industry to compile the data sheet for the chip and give it to the ultimate customer before it is fabricated. This prevents many undesirable scenarios that can arise when perfectly designed chip meets perfectly designed system and creates product nightmare. In this section an outline of a typical data sheet will be reviewed by way of example.

6.8.1   The Summary

A summary of the chip includes the following details:

- The designation and descriptive name of the chip (i.e., ABC1478—FIR Filter Chip).
- A short description of what the chip does.
- A features list (optional for an internal product—but good for your ego).
- A very high level block diagram of the chip function.

This serves to orient the user to the chip and the function it performs.

6.8.2   Pinout

The pinout section should contain a description of the following pin attributes:

- The name of the pin.

- The type of the pin (i.e., whether input, output, tristate, digital, analog, etc.).
- A brief description of the pin function.
- The package pin number.

This documents the external interface of the chip.

6.8.3   Description of Operation

This section should outline the operation of the chip as far as the user of the chip is interested. Programming options, data formats, and control options should be summarized.

6.8.4   DC Specifications

The absolute maximum ratings should be stated for the following:

- Supply voltage.
- Pin voltages.
- Junction temperature.

The style of each I/O (i.e., TTL, CMOS, ECL) should be summarized and the following DC specifications should be given over the operating range (temperature and voltage, i.e., mins and maxes):

- The V_{IL} and V_{IH} for each input.
- The V_{OL} and V_{OH} for each output (at a given drive level).
- The input loading for each input.
- The output drive capability of each output.
- Quiescent current.
- Leakage current.
- Power-down current (if applicable).
- Any other relevant voltages and currents.

This section communicates the power dissipation and required voltages for the chip to correctly operate.

6.8.5   AC Specifications

The following timing specifications should be presented:

- Setup and hold times on all inputs (slowest and fastest).

- Clock (and all other relevant inputs) to output delay times (slowest and fastest).
- Other critical timing, such as minimum pulse widths,.

This data should be tabulated in table form and supported by a timing diagram where necessary. This is probably the most important section and an area where data provided ahead of the chip fabrication will aid the board designer. Designs are frequently snagged, for instance, when chip designers assume infinitely fast external memories and do not allow enough time between outputs changing and the next rising edge of the clock.

6.8.6 Package Diagram

A diagram of the package with the pin names attached should be supplied.

6.9 Summary

This chapter has covered a broad spectrum of design issues that may be encountered when designing CMOS chips. The structured design strategies that were introduced early in the chapter are useful for any kind of CMOS-chip design method. A range of implementation options was given to give the reader an appreciation for the wide spectrum of solutions that are available today. In addition a summary of the design styles was given. Increasingly, the level of design is being pushed upward as logic synthesizers are refined, compilers are optimized, and knowledge is captured from libraries of reusable components. You as a designer must keep abreast of such techniques to ensure that you can bring to bear a productivity that results in timely, cost-effective, and reliable silicon that may be shipped after the first manufacturing run.

6.10 Exercises

1. Explain how you would assess the required design-method for a function that has to be performed by a single chip. Draw a decision chart that shows the various questions that have to be answered, and the resulting actions.

2. Explain the following terms with respect to CMOS-chip design: hierarchy, regularity, modularity, and locality. Give an example of each.

3. Summarize the differences between a SOG chip and a standard-cell chip. What benefits does each implementation style have?

6.11 References

1. Daniel D. Gajski, *Silicon Compilation,* Reading, Mass.: Addison-Wesley, 1988.
2. D. D. Gajski and R. H. Kuhn, "New VLSI tools," *IEEE Computer,* vol. 16, no. 12, 1983, pp. 11–14.
3. Carver Mead and Lynn Conway, *Introduction to VLSI Systems,* Reading, Mass.: Addison-Wesley, 1980.
4. Irene Buchanan, "Modelling and verification in structured integrated circuit design," Ph.D. thesis, Dept. of Computer Science, University of Edinburgh, Scotland, 1980.
5. *PAL Device Data Book,* Sunnyvale, Calif.: Advanced Micro Devices Inc., 1988.
6. *GAL Data Book,* Hillsboro, Ore.: Lattice Semiconductor Corp., 1990.
7. Clinton Kuo, Mark Weidner, Thomas Toms, Henry Choe, Ko-Min Chang, Ann Harwood, Joseph Jelemensky and Philip Smith, "A 512-kb Flash EEPROM Embedded in a 32-b Microcontroller," *IEEE Journal of Solid State Circuits,* vol. 27, no. 4, Apr. 1992, pp. 574–582.
8. Takaaki Nozaki, Toshiaki Tanaka, Yoshiro Kijiya, Eita Kinoshita, Tatsuo Tsuchiya, and Yutaka Hayashi, "A 1-Mb EEPROM with MONOS memory cell for semiconductor disk application," *IEEE JSSC,* vol. 26, no. 4, Apr. 1991, pp. 497–501.
9. Masaki Momodomi, Tomoharu Tanaka, Yoshihisa Iwata, Yoshiyuki Tanaka, Hideko Oodaira, Yasuo Itoh, Riichiro Shirota, Kazunori Ohuchi and Fujio Masuoka, "A 4-Mb NAND EEPROM with tight programmable V_t distribution," *IEEE JSSC,* vol. 26, no. 4, Apr. 1991, pp. 492–496.
10. *ACT Family Field Programmable Gate Array DATABOOK,* San Jose, Calif.: Actel Corporation, 1990.
11. "Very-High-Speed FPGAs," *pASIC 1 Family Data Book,* Santa Clara, Calif.: QuickLogic Corporation, 1992.
12. Esmat Hamdy, John McCollum, Shih-ou Chen, Steve Chiang, Shafy Eltoukhy, Jim Chang, Ted Speers and Amr Mohsen, "Dielectric-based antifuse for logic and memory ICs," *Proceedings of the International Electron Devices Meeting,* 1988, pp. 786–789, Washington, D.C.
13. Abbas El Gamal, Jonathan Greene, Justin Reyneri, Eric Rogoyski, Khaled A. El-Ayat, and Amr Mohsen, "An architecture for electrically configurable gate arrays," *IEEE JSSC,* vol. 24, no. 2, Apr. 1989, pp. 394–398.
14. Khaled A. El-Ayat, Abbas El Gamal, Richard Guo, John Chang, Ricky K. H. Mak, Frederick Chiu, Esmat Z. Hamdy, John McCollum, and Amr Mohsen, "A CMOS electrically configurable gate array," *IEEE JSSC,* vol. 24, no. 3, Jun. 1989, pp. 752–762.
15. *The Programmable Gate Array Data Book,* San Jose, Calif.: XILINX, Inc., 1990.
16. Thomas Andrew Kean, "Configurable logic: a dynamically programmable cellular architecture and its VLSI implementation," Ph.D. thesis, Department of Computer Science, University of Edinburgh, Scotland, 1989.

17. *CA11024 Datasheet,* Edinburgh, Scotland: Algotrinix Ltd., 1990.

18. J. P. Gray and T. A. Kean, "Configurable hardware: a new paradigm for computation," *Proceedings of the 1989 Decennial Caltech Conference, Pasadena, CA,* Cambridge, Mass.: MIT Press, pp. 1-17.

19. *CLi6000 Series Field-Programmable Gate Arrays,* (data sheet), Sunnyvale, Calif.: Concurrent Logic Inc., 1992.

20. Michiel A. Beunder, Juergen P. Kernhof, and Bernd Hoefflinger, "The CMOS gate forest: an efficient and flexible high-performance ASIC design environment," *IEEE JSSC,* vol. 23, no. 2, Apr. 1988, pp. 387–399.

21. Harry J. M. Veendrick, Dré A. J. M. Van Den Elshout, Dick W. Harberts, and Teus Brand, "An efficient and flexible architecture for high-density gate arrays," *IEEE JSSC,* vol. 25, no. 5, Oct. 1990, pp. 1153–1157.

22. Philippe Duchene and Michel J. Declercq, "A highly flexible sea-of-gates structure for digital and analog applications," *IEEE JSSC,* vol. 24, no. 3, Jun. 1989, pp. 576–584.

23. *1.5 Micron Compacted Array Technology Databook,* Milpitas, Calif.: LSI Logic Corp., 1987.

24. Masatomi Okabe, Yoshihiro Okuno, Takahiko Arakawa, Ichiro Tomioka, Takio Ohno, Tomoyoshu Noda, Masahiro Hatanaka, and Yoichi Kuramitsu, "A 400K-transistor CMOS sea-of-gates array with continuous track allocation," *IEEE JSSC,* vol. 24, no. 5, Oct. 1989, pp. 1280–1286.

25. *1.0-Micron Cell-Based Products Databook,* Milpitas, Calif.: LSI Logic, Feb. 1991.

26. Dennis V. Heinbruch, *CMOS3 Cell Library,* Reading, Mass.: Addison-Wesley, 1988.

27. Curt F. Fey and Demetris E. Paraskevopoulos, "Studies in LSI technology economics IV: models for gate array design productivity," *IEEE JSSC,* vol. SC-24, no. 4, Aug. 1989, pp. 1085–1091.

28. D. Gibson and S. Nance, "SLIC—symbolic layout of integrated circuits," *IEEE Proceedings of the 13th Design Automation Conference,* Jun. 1976, pp. 434–440.

29. D. Clary, R. Kirk, and S. Sapiro," SIDS—a symbolic interactive design system," *IEEE/ACM Proceedings of the 17th Design Automation Conference,* Jun. 1980, Minneapolis, Minnesota, pp. 292–295.

30. R. P. Larson, "Versatile mask generation techniques for customer microelectronic devices," *IEEE/ACM Proceedings of the 15th Design Automation Conference*, Jun. 1978, Las Vegas, Nev., pp. 193–198.

31. A. D. Lopez and H-F. S. Law, "A defense gate matrix layout style for MOS LSI," *IEEE JSSC,* vol. SC-15, no. 4, Aug. 1980, pp. 736–740.

32. C. Piguet, J. Zahnd, A. Stauffer, and M. Bertarionne, "A metal-oriented layout for CMOS logic," *IEEE JSSC*, vol. SC-19, no. 3, Jun. 1984, pp. 425–436.

33. J. Williams, "STICKS—a graphical compiler for high level LSI design," *Proceedings of the National Computer Conference,* May 1978, pp. 289–295.

34. A. Dunlop, "SLIM—the translation of symbolic layouts into mask data," *IEEE/ACM Proceedings of the 17th Design Automation Conference,* Jun. 1980, Minneapolis, Minnesota, pp. 595–602.

35. M. Y. Hsueh and D. O. Pederson, "Computer-aided layout of LSI building blocks," *IEEE Proceedings of the 1979 International Symposium on Circuits and Systems,* Jul. 1979, Tokyo, Japan, pp. 474–477.

36. G. Kedem and H. Watanabe, "Graph optimization techniques for IC layout and compaction," *IEEE/ACM Proceedings of the 20th Design Automation Conference,* Jun. 1983, Miami Beach, Fla., pp. 113–120.

37. R. C. Mosteller, "Rest—a leaf cell design system," *Proceedings of IFIP VLSI '81* (J. Gray, ed.), Edinburgh 1981, pp. 163–172.

38. Werner Bonath and Manfred Glesner, "Process-independent 2D-compaction in a symbolic design environment," *Proceedings of IFIP VLSI '89* (G. Musgrave and V. Lauther, eds.), Munich, 1989, pp. 433–443.

39. David G. Boyer, "Symbolic layout compaction review," *IEEE/ACM Proceedings of the 25th Design Automation Conference,* Jun. 1988, Anaheim, Calif., pp. 383–389.

40. W. H. Crocker, C. Y. Lo, and R. Varadarahan, "MACS: a module assembly and compaction system," *Proceedings of the IEEE International Conference on Computer Design,* Nov. 1985, Santa Clara, Calif., pp. 205–208.

41. J. L. Burns and A. R. Newton, "SPARCS: a new constraint-based IC layout symbolic spacer," *Proc. IEEE Custom Integrated Circuits Conference,* May 1986, Rochester, N.Y., pp. 534–539.

42. H. Shin, A. Sangiovanni-Vincentelli, and C. Sequin, "Two-dimensional module compactor based on zone-refining," *Proceedings of the IEEE International Conference on Computer design,* Oct. 1987, Port Chester, N.Y., pp. 201–204.

43. Johan K. J. Van Ginderdeuren, Hugo J. De Man, Bart J. S. De Loore, Hilbradb Vanden Winjingaert, Atoine DeLaruelle, and Guy R. J. Van Den Audenaerde, "A high-quality digital audio filter set designed by silicon compiler CATHEDRAL-I," *IEEE JSSC,* vol. SC-21, no. 6, Dec. 1986, pp. 1067–1075.

44. N. Weste, "Virtual grid symbolic layout," *IEEE/ACM Proc. of the 18th Design Automation Conference*, Nashville, Tenn., Jun. 1981, pp. 225–233.

45. I. Buchanan, "Modelling and verification in structured integrated circuit design," Ph.D. thesis, Dept. of Computer Science, University of Edinburgh, Scotland, 1980.

46. N. Weste, "MULGA—an interactive symbolic layout system for the design of integrated circuits," *Bell System Technical Journal*, vol. 60, no. 6, Jul.-Aug. 1981, pp. 823–858.

47. N. Weste and B. Ackland, "A pragmatic approach to topological symbolic IC design," *IFIP Proc. VLSI '81* (J. Gray, ed.), Edinburgh, Scotland, August 1981, pp. 117–129.

48. James J. Cherry, "CAD programming in an object oriented programming environment," *VLSI CAD Tools and Applications,* (Wolfgang Fichtner and Martin Morf, eds.), Norwell, Mass.: Kluwer Academic Publishers, 1987, Chapter 9.

49. K. Ramachandran, R. R. Cordell, D. F. Daly, D. N. Deutsch and A. F. Kwan, "SYMCELL—a symbolic standard cell system," *IEEE JSSC,* vol. 26, no. 3, Mar. 1991, pp. 449–452.

50. M. C. Revett and P. A. Ivey, "ASTRA—a CAD system to support a structured approach to IC design," *IFIP VLSI '83,* (F. Anceav and E. J. Aas, eds.), 1983, pp. 413–422.

51. Bill Lin and A. Richard Newton, "A circuit disassembly technique for synthesizing symbolic layouts from mask descriptions," *IEEE Transactions on CAD,* vol. 9, no. 9, Sept. 1990, pp. 959–969.

52. Rajeev Jain, Francky Catthoor, Jan VanHoof, Bart J. S. De Loore, Gert Goossens, Nelson F. Goncalvez, Luc. J. M. Claesen, Johan K. J. Van Gindereuren, Joos

VanDeWalle, and Hugo J. De Man, "Custom design of a VLSI PCM-FDM transmultiplexer from system specifications to circuit layout using a computer-aided design system," *IEEE JSSC,* vol. SC-21, no. 1, Feb. 1986, pp. 73–85; and Johan K. J. Van Ginderdeuren et al., *op. cit.*

53. J. Rabaey, H. De Man, J. Vanhoof, G. Goossens, and F. Catthor, "Cathedral II: a synthesis system for multiprocessor DSP systems," *Silicon Compilation* (Daniel D. Gajski, ed.), Reading, Mass.: Addison-Wesley, 1988.

54. F. Catthoor and H. De Man, "Application-specific architectural methodologies for high-throughput digital signal and image processing," *IEEE Transactions on Acoustics, Speech and Signal Processing,* vol. 38, Feb. 1990, pp. 339–349.

55. C. Bernard Shung, Rajeev Jain, Ken Rimey, Edward Wang, Mani B. Srivastava, Brian C. Richards, Erik Lettang, S. Khalid Azim, Lars Thon, Paul N. Hilfinger, Jan M. Rabaey, and Robert W. Brodersen, "An integrated CAD system for algorithm-specific IC design," *IEEE Transactions on CAD,* vol. 10, no. 4, Apr. 1991, pp. 447–463.

56. R. K. Brayton, R. Campansano, G. De Micheli, R. H. J. M. Otten, J. van Eijndhoven, "The Yorktown Silicon Compiler system," *Silicon Compilation* (Daniel D. Gajski, ed.), Reading, Mass.: Addison-Wesley, 1988.

57. David R. Coelho, "The VHDL handbook," Norwell, Mass.: Kluwer Academic Press, 1989.

58. Jean-Michel Bergé, Alain Fonkova, Serge Maginot, and Jacques Roilland, "VHDL designer's reference," Norwell, Mass.: Kluwer Academic Press, 1992.

59. Douglas L. Perry, "VHDL," Summitt, Penn.: McGraw-Hill, 1992.

60. Steve Carlson, *Introduction to HDL-Based Design Using VHDL,* Mountain View, Calif.: Synopsys, 1991.

61. Srinivas Devadas, Hi-Keung Ma, Richard Newton, and A. Sangiovanni-Vincentelli, "MUSTANG: state assignment of finite state machines targeting multilevel logic implementations," *IEEE Transactions on CAD,* vol. 27, no. 12, Dec. 1988, pp. 1290–1300.

62. R. K. Brayton, G. D. Hachtel, and A. L. Sangiovanni-Vincentelli, "Multilevel logic synthesis," *Proceedings of the IEEE,* vol. 78, no. 2, Feb. 1990, pp. 264–300.

63. Richard Rudell, "Logic Synthesis," Custom Integrated Circuits Conference '91 Educational Session; Session III: Design Automation (lecture notes).

64. R. K. Brayton, G. D. Hachtel, C. T. McMullen, and A. Sangiovanni-Vincentelli, *ESPRESSO-IIC: Logic Minimization Algorithms for VLSI Synthesis, The Netherlands*, Norwell, Mass.: Kluwer Academic, 1984.

65. R. K. Brayton and C. McMullen, "The decomposition and factorization of Boolean expressions," *Proc. IEEE Int. Symposium on Circuits and Systems,* 1982, pp. 49–54.

66. Kurt Keutzer, "DAGON: technology binding and local optimization by DAG matching," *IEEE Proc. 24th DAC,* 1987, pp. 341–347.

67. David Gregory, Karen Bartlett, Aart De Geus, and Gary Hachtel, "Socrates: a system for automatically synthesizing and optimizing combinational logic," *Proceedings of the 23rd DAC,* Jun.-Jul., 1986, pp. 79–85.

68. Kurt Keutzer, op. cit.

69. R. Brayton, E. Detjens, S. Krishna, T. Ma, P. McGeer, L. Pei, N. Phillips, R. Rudell, R. Segal, A Wang, R. Yung, and A. Sangiovanni-Vincentelli, "Multiple-level logic optimization system," *Proc. IEEE ICCAD 1986,* pp. 356–359; and R. Brayton, R. Rudell, A. Sangiovanni-Vincentelli, and A. R. Wang, "MIS: a multi-

ple-level logic optimization system," *IEEE Transactions on CAD of Integrated Circuits and Systems,* vol. CAD-6, no. 6, Nov. 1987, pp. 1062–1081.

70. D. Bostick, G. D. Hachtel, R. Jacoby, M. R. Lightner, P. Moceyunas, C. R. Morrison, and D. Ravenscroft, "The Boulder Optimal Logic Design system," *Proc. Int. Conf. on Computer-Aided Design,* Nov. 1987, pp. 62–65.

71. *EDIF Electronic Design Interchange Format Version 2 0 0* (Paul Stanford and Paul Mancuso, eds.), Washington, D.C.: Electronic Industries Association, 1989.

72. Ulrich Lauther, "A min-cut placement algorithm for general cell assemblies based on a graph," *Proceedings of the 16th Design Automation Conference,* 1979, pp. 1–10.

73. S. Kirkpatrick, C. Gelatt, and M. Vecchi, "Optimization by simulated annealing," *Science,* vol. 220, no. 4598, May 13, 1983, pp. 671–680.

74. C. Sechen and A. Sangiovanni-Vincentelli, "TimberWolf3.2: a new standard cell placement and global routing package," *Proceedings of the 23rd Design Automation Conference,* Las Vegas, Nev., 1986, pp. 432–439.

75. A. Sangiovanni-Vincentelli, A. Santamauro, and J. Reed, "A new gridless channel router: Yet Another Channel Router the second (YACRII)," *Proceedings of the International Conference on Computer-Aided Design,* 1984, pp. 72–75.

76. Ronald L. Rivest and Charles Fiducia, "A 'greedy' channel router," *Proceedings of the 19th Design Automation Conference,* 1982, pp. 418–424.

77. J. Soukup, "Fast maze router," *Proceedings of the 15th Design Automation Conference,* Las Vegas, Nev., 1978, pp. 100–102.

78. J. Soukup, "Global router," *Proceedings of the 16th Design Automation Conference,* Las Vegas, Nev., 1978, pp. 481–484.

79. Hung-fai Steven Law, Graham Wood, and Mindy Lam, "An Intelligent Composition Tool for Regular and Semiregular VLSI Structures," *Silicon Compilation* (Daniel D. Gajski, ed.), Reading, Mass.: Addison-Wesley, 1988.

80. Steven M. Rubin, *Computer Aids for VLSI Design,* Reading, Mass.: Addison-Wesley, 1987, Chapter 11.

81. James J. Cherry, *op. cit.*

82. William R. Heller, G. Sorkin, and Klim Maling, "The planar package planner for system designers," *Proceedings of the 19th Design Automation Conference,* Jun. 1982, pp. 253–260.

83. Ralph H. J. M. Otten, "Automatic floorplan design," *Proceedings of the 19th Design Automation Conference,* Jun. 1982, pp. 261–267.

84. C. L. Wardle, C. R. Watson, C. A. Wilson, J. C. Mudge, and B. J. Nelson, "A declarative design approach for combining macrocells by directed placement and constructive routing," *Proceedings of the 21st Design Automation Conference,* 1984, pp. 594–601.

85. Christopher J. Terman, "Simulation tools for VLSI," *VLSI CAD Tools and Applications* (Wolfgang Fichtner and Martin Morf, eds.), Norwell, Mass.: Kluwer Academic, 1987, Chapter 3.

86. L. W. Nagel, "SPICE2: a computer program to simulate semiconductor circuits," *Memo ERL-M520,* Dept. Electrical Engineering and Computer Science, University of California at Berkeley, May 9, 1975.

87. W. T. Weeks, A. J. Jimenez, G. W. Mahoney, D. Mehta, H. Qasemzadeh, and T. R. Scott, "Algorithms for ATSAP—a network analysis program," *IEEE Transactions on Circuit Theory,* vol. CT-20, Nov. 1973, pp. 628–634.

88. *HSPICE User's Manual H9001,* Campbell, Calif.: Meta-Software, 1990.

89. B. R. Chawla, H. K. Gummel, and P. Kozak, "MOTIS—an MOS timing simulator," *IEEE Transactions on Circuits and Systems,* vol. 22, no. 12, Dec. 1975, pp. 901–910.

90. J. White and A. Sangiovanni-Vincentelli, *Relaxation Techniques for the Simulation of VLSI Circuits,* Hingham, Mass.: Kluwer Academic, 1987.

91. C. Terman, "Timing simulation for large digital MOS circuits," in *Advances in Computer-Aided Engineering Design,* Volume 1 (A. Sangiovanni-Vincentelli, ed.), JAI Press, 1984, pp. 1–91.

92. Thomas G. Szymanski, "LEADOUT: a static timing analyzer for MOS circuits," *IEEE International Conference on Computer-Aided Design,* Santa Clara, Calif., Nov. 1986, pp. 130–133.

93. James J. Cherry, "Pearl: a CMOS timing analyzer," *IEEE/ACM Proceedings of the 25th Design Automation Conference,* Anaheim, Calif., 1988, pp. 148–153, and "Pearl User's Guide," Parsec, Inc., Palo Alto, Calif. 1992.

94. C. Ebeling and O. Zajicek, "Validating VLSI circuit layout by wirelist comparison," *Proceedings of IEEE Int. Conf. on CAD,* Sept. 1983, pp. 172–173.

95. M. Hofmann and V. Lauther, "HEX: an instruction driven approach to feature extraction," *Proceedings of the 20th Design Automation Conference,* Jun. 1983, pp. 331–336.

96. C. M. Baker and C. J. Terman, "Tools for verifying integrated circuit designs," *Lambda Magazine (VLSI Design),* 4th quarter 1980, pp. 22–30.

97. T. G. Szymanski and C. J. Van Wyk, "Space efficient algorithms for VLSI artwork analysis," *Proceedings of the 20th Design Automation Conference,* June 1983, pp. 734–739.

98. C. M. Baker and C. J. Terman, *op. cit.*

99. H. S. Baird, "Fast algorithms for LSI artwork analysis," *Proceedings of the 14th Design Automation Conference,* 1977, pp. 303–311.

100. T. Whitney, "A Hierarchial Design Analysis Front End," *VLSI '81,* 1981, pp. 217–225.

101. Steven M. Rubin, *op. cit.,* Appendix E.

102. Curt F. Fey and Demetris E. Paraskevopoulos, "Studies in LSI technology economics II: a comparison of product costs using MSI, gate arrays, standard cells, and full custom VLSI," *IEEE JSSC,* vol. SC-21, no. 2, Apr. 1986, pp. 297–303.

103. Demetris E. Paraskevopoulos and Curt F. Fey, "Studies in LSI technology economics III: design schedules for application-specific integrated circuits," *IEEE JSSC,* vol. SC-22, no. 2, Apr. 1987, pp. 223–229.

104. Curt F. Fey, "Custom LSI/VLSI chip design productivity," *IEEE JSSC,* vol. SC-20, no. 2, Apr. 1985, pp. 555–561.

105. Curt F. Fey and Demetris E. Paraskevopoulos, "Studies in LSI technology economics IV: models for gate array design productivity," *IEEE JSSC,* vol. SC-24, no. 4, Aug. 1989, pp. 1085–1091.

CMOS
TEST
METHODS

7

7.1 The Need for Testing

While in real estate the refrain is "Location! Location! Location!," the comparable advice in IC design should be "Testing! Testing! Testing!" While most problems in VLSI design have been reduced to algorithms in readily available software, the responsibilities for the various levels of testing and testing methodology can be a significant burden on the designer.

In Chapter 4 we noted that the yield of a particular IC was the number of good die divided by the total number of die per wafer. Due to the complexity of the manufacturing process not all die on a wafer correctly operate. Small imperfections in starting material, processing steps, or in photomasking may result in bridged connections or missing features. It is the aim of a test procedure to determine which die are good and should be used in end systems.

Testing a die (chip) can occur:

- at the wafer level.
- at the packaged-chip level.
- at the board level.
- at the system level.
- in the field.

By detecting a malfunctioning chip at an earlier level, the manufacturing cost may be kept low. For instance, the approximate cost to a company of detecting a fault at the above levels is[1]:

- wafer $0.01–$.1
- packaged-chip $0.10–$1
- board $1–$10
- system $10–$100
- field $100–$1000.

Obviously, if faults can be detected at the wafer level, the cost of manufacturing is kept the lowest. In some circumstances, the cost to develop adequate tests at the wafer level, mixed signal requirements or speed considerations may require that further testing be done at the packaged-chip level or the board level. A component vendor can only test at the wafer or chip level. Special systems, such as satellite-borne electronics, might be tested exhaustively at the system level.

Tests may fall into two main categories. The first set of tests verifies that the chip performs its intended function; that is, that it performs a digital filtering function, acts as a microprocessor, or communicates using a particular protocol. In other words, these tests assert that all the gates in the chip, acting in concert, achieve a desired function. These tests are usually used early in the design cycle to verify the functionality of the circuit. These will be called *functionality tests* in this book. They may be lumped into the verification activity. The second set of tests verifies that every gate and register in the chip functions correctly. These tests are used after the chip is manufactured to verify that the silicon is intact. They will be called *manufacturing tests* in this book. In many cases these two sets of tests may be one and the same, although the natural flow of design usually has a designer considering function before manufacturing concerns.

It is interesting to note that of most first-time failures of silicon, it is the functionality of the design that is to blame; that is, the chip does exactly what the simulator said it would but for some reason (almost always human error) that function is not what the rest of the system expects.

7.1.1   Functionality Tests

Functionality tests are usually the first tests a designer might construct as part of the design process. Does this adder add? Does this counter count? Does this state-machine yield the right outputs at the right clock cycles?

For most systems, functionality tests involve proving that the circuit is functionally equivalent to some specification. That specification might be a verbal description, a plain-language textual specification, a description in some high-level computer language such as C, FORTRAN, Pascal, or Lisp or in a hardware-description language such as VHDL, ELLA, or Verilog®, or simply a table of inputs and required outputs. Functional equivalence involves running a simulator at some level on the two descriptions of the chip (say, one at the gate level and one at a functional level) and ensuring for all inputs applied that the outputs are equivalent at some convenient check-points in time. The most detailed check might be on a cycle-by-cycle basis.

Functional equivalence may be carried out at various levels of the design hierarchy. If the description is in a behavioral language (such as the last two categories mentioned), the behavior at a system level may be verifiable. For instance, in the case of a microprocessor, the operating system might be booted and key programs might be run for the behavioral description. However, this might be impractical (due to long simulation times) for a gate-level model and extremely impractical for a transistor-level model. The way out of this impasse is to use the hierarchy inherent within a system to verify chips and modules within chips. That, combined with well-defined modular interfaces, goes a long way in increasing the likelihood that a system composed of many VLSI chips will be first-time functional. Remember too, at the lowest levels of the hierarchy, timing tests must be run to validate that a particular function such as addition is achieved at a given clock frequency.

There is no good theory on how to ensure that good functional tests be written. The best advice is to simulate the chip or system as closely as possible to the way it will be used in the real world. Often this is impractical due to slow simulation times and very long verification sequences. One approach is to move up the simulation hierarchy as modules become verified at lower levels. For instance, the gate-level adder and register modules in a video filter might be replaced by functional models and then the filter itself might be replaced by a functional model. At each level, small tests are written to verify the equivalence between the new higher-level functional model and the lower-level gate or functional level. At the top level, the filter functional-model may be surrounded with a software environment that models the real-world use of the filter. For instance, a carefully selected subsample of a video frame might be fed to the filter and the output of the functional model compared with what is expected theoretically. The video output might also be observed on a video frame buffer to check that it looks correct (by no means an exhaustive test, but a confidence builder). Finally, if enough time is available, all or part of the functional test may be appplied to the gate level and even the transistor level if transistor primitives have been used. One approach that is becoming more popular and feasible is to model chips as collections of reprogrammable gate arrays (see Chapter 6). Commercial hardware is available to aid this activity.

Remember the following statement, culled from many years of IC-design experience, whenever you are tempted to give test work short shrift:

"If you don't test it, it won't work! (Guaranteed.)"

7.1.2 Manufacturing Tests

Whereas functionality tests seek to verify the function of a chip as a whole, manufacturing tests are used to verify that every gate operates as expected. The need to do this arises from a number of manufacturing defects that might occur during chip fabrication or during accelerated life testing (where the chip is stressed by over-voltage and over-temperature operation). Typical defects include:

- layer-to-layer shorts (i.e., metal to metal).
- discontinous wires (i.e., metal thins when crossing vertical topology jumps).
- thin-oxide shorts to substrate or well.

These in turn lead to particular circuit maladies, including:

- nodes shorted to power or ground.
- nodes shorted to each other.[2]
- inputs floating/outputs disconnected.

Tests are required to verify that each gate and register is operational and has not been compromised by a manufacturing defect. Tests are normally carried out at the wafer level to cull out bad die, and then on the packaged parts. The length of the tests at the wafer level might be shortened to reduce test time based on experience with the test sequence.

Apart from the verification of internal gates, I/O integrity is also tested through completing the following tests:

- I/O-level test (i.e., checking the noise margin for TTL, ECL, or CMOS I/O pads).
- Speed test.
- I_{DD} test.

The last of these tests checks the leakage if the circuit is composed of complementary logic. Any value markedly above the expected value for a given wafer normally indicates an internal shorting failure (or very bad leakage). Wafer tests may be done at high speed or low speed (1 MHz) due to possible power and ground bounce effects that may be present in older testers.

In general, manufacturing-test generation assumes that the circuit/chip functions correctly, and ways of exercising all gate inputs and of monitoring all gate outputs are required.

To illustrate the difference between a functional test and a manufacturing test, consider the testing of a microprocessor at a functional level, which might be the first concern of the designer (to see whether the microprocessor worked as a whole). To test any instruction, a sequence of instructions that use that instruction might be used (i.e., does the ADD instruction add?). While this might prove that the control logic that yields that instruction is intact, it does not, for instance, prove that the instruction works for all possible addresses and data. At this level of test it is assumed that the adders, muxes, gates, and registers in the microprocessor datapaths operate correctly.

Tests that exercise all bits in the datapaths have to be written to verify the chip at the manufacturing level. These tests might include a test to check that registers can store a 1 and a 0 and a test that exercises each bit in any adder and ensures that the carry chain is not broken (i.e., does the adder add for all inputs?). The inputs have to be chosen carefully to check for all possible manufacturing defects. The manufacturing tests may be the only tests applied to a microprocessor prior to its being placed in a socket and booted.

7.1.3   A Walk Through the Test Process

As a designer you may be responsible for part or all of the tests that are written to test a particular chip (often called the stimulus). "Written" might include a number of methods of test specification from applying waveforms or logic values manually to a simulator to, more probably, writing a program in a high-level language to apply stimuli to a description of the circuit. When the stimulus is applied to a circuit via a simulator, the output of the simulator may be dumped to a file (often called an *activity file*). If this output is filtered so that only the chip inputs and chip outputs are retained and further filtered so that only the quiescent signal values are kept after an input or inputs change, then the resulting file may be used to generate a "test program."

Depending on whether you are testing a wafer or a packaged part, a probe card or "device-under-test" (DUT) board would be needed to connect the tester outputs and inputs to the die I/O pads or chip package pins. Probe cards are normally constructed by experts, while DUT boards are well within the capabilities of the electronic hobbyist.

The next requirement for a chip tester is the existence of this "test program." This is a file with a format of inputs and outputs that suit the chip tester that is to be used to test your chip. A simple format is shown below for the case of a single-bit adder:

```
      III    OO
             SC
             UA
             MR
              R
      ABC    Y
  0   000    00
```

```
1    001   10
2    010   10
3    011   01
4    100   10
5    101   01
6    110   01
7    111   11
```

The first line designates the signal directions and shows three inputs and two outputs. The next five lines designate the signal names. Thereafter, each line designates a new test vector. The first column is the test vector number. The next three columns are the binary value of the inputs, and the next two columns are the expected output values. Each line represents a certain length clock cycle that is asserted by the tester. Signals normally change soon after an internal clock running at the tester period. Clock generation may be carried out in two different ways. First, the clock can be regarded as any other signal, in which case it takes two tester cycles to complete a single clock cycle—one for the clock low and one for the clock high. Alternatively, a timing generator may be used, which allows the clock rising edge (for instance) to be placed anywhere in the tester cycle. So, for instance, if the inputs are changed at the start of the tester cycle, the clock might be programmed to rise at the middle of the cycle.

Sundry other setup files are normally required by the tester. Normally a mapping file is required that maps a given input or output in the test program to a physical connection (pin) in the tester. This pin may be programmed to be an input, output, tristate, bidirect, or, in some cases, a multiplexed data pin. Each pin on the tester is driven by a function memory that is used either to assert a value or to check a value at a DUT pin. In addition, various control memories may be present to control the drive on the tester pin (i.e., to control a tristate pin) or to mask data from the chip (i.e., to ignore certain pins at certain times). These memories have finite length, so sometimes with older testers more than one vector load has to be used to test a part. This normally slows testing because the reload procedure may be slow. Modern testers seldom suffer from this problem.

The clock speed is specified (by specifying a test cycle time, T_C), as are supply-voltage levels and pins on the tester and probe card or DUT board. The time at which outputs are asserted or inputs are sampled is also specified on a pin-by-pin basis (T_S). The format of the test data may usually be chosen from Non Return to Zero (NRZ), Return To Zero (RTZ), or other formats, such as Surround By Zero (SBZ). For instance, an RTZ output would transition (if the pin were driven high) at T_S and return to zero at T_C.

The probe card or DUT board is connected to the tester. The test program is compiled and downloaded into the tester, and the tests are applied to the circuit. The tester samples the chip outputs and compares the values with those provided by the test program. If there are any differences, the chip is marked as faulty (with an ink dot) and the miscomparing vectors may be displayed for reference. In the case of a probe card, the card is raised, moved to the next die on the wafer, and lowered, and the test procedure is repeated. In the case of a DUT board with automatic part handling, the tested part is binned into a good or bad bin and a new part is fed to the DUT board, and the test is repeated. In most cases these procedures take a few seconds for each part tested.

The ability to vary the voltage and timing on a per-pin basis with a tester allows a process known as "schmooing" to be carried out. For instance, one might vary the V_{DD} voltage from 3 to 6 volts on a 5-volt part while varying the tester cycle time. This yields a graph that shows the speed sensitivity of the part with respect to voltage. Another "schmoo" test that is frequently exercised is to skew the timing on inputs with respect to the chip clock to look for setup and hold variations.

7.2   Manufacturing Test Principles

A critical factor in all LSI and VLSI design is the need to incorporate methods of testing circuits. This task should proceed concurrently with any architectural considerations and not be left until fabricated parts are available (which is a recurring temptation to designers).

Figure 7.1(a) shows a combinational circuit with n-inputs. To test this circuit exhaustively a sequence of 2^n inputs (or test vectors) must be applied and observed to fully exercise the circuit. This combinational circuit is converted to a sequential circuit with addition of m-storage registers, as shown in Fig. 7.1(b). The state of the circuit is determined by the inputs and the pre-

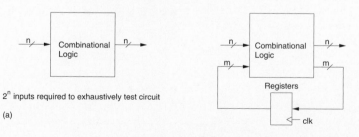

2^n inputs required to exhaustively test circuit

(a)

2^{n+m} inputs required to exhaustively test circuit

For n = 25 m = 50, 1μS/test, the test time is over 1 billion years (Williams)

(b)

FIGURE 7.1   The combinational explosion in test vectors

vious state. A minimum of $2^{(n+m)}$ test vectors must be applied to exhaustively test the circuit. To quote Williams[3]:

> With LSI, this may be a network with n = 25 and m = 50, or 2^{75} patterns, which is approximately 3.8×10^{22}. Assuming one had the patterns and applied them at an application rate of 1 μs per pattern, the test time would be over a billion years (10^9).

Clearly, this is an important area of design that has to be well understood.

7.2.1 Fault Models

7.2.1.1 Stuck-At Faults

In order to deal with the existence of good and bad parts it is necessary to propose a "fault model," that is, a model for how faults occur and their impact on circuits. The most popular model is called the "Stuck-At" model. With this model, a faulty gate input is modeled as a "stuck at zero" (Stuck-At-0, S-A-0, SA0) or "stuck at one" (Stuck-At-1, S-A-1, SA1). This model dates from board-level designs where this was determined to be an adequate set of models for modeling faults. Figure 7.2 illustrates how an S-A-0 or S-A-1 fault might occur. These faults most frequently occur due to thin-oxide shorts (the n-transistor gate to V_{SS} or the p-transistor gate to V_{DD}) or metal-to-metal shorts.

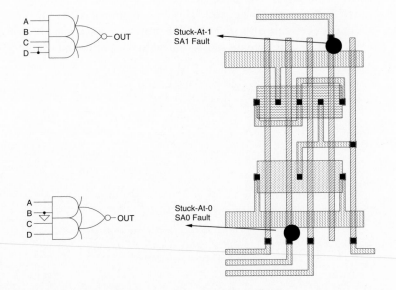

FIGURE 7.2 CMOS stuck-at faults

7.2.1.2   *Short-Circuit and Open-Circuit Faults*

Other models include "stuck-open"[4] or "shorted" models. Two shorted faults are shown in Fig. 7.3. Considering the faults shown in Fig. 7.3, the short *S1* is modeled by an S-A-0 fault at input *A,* while short *S2* modifies the function of the gate. What becomes evident is that to ensure the most accurate modeling, faults should be modeled at the transistor level, because it is only at this level that the complete circuit structure is known. For instance, in the case of a simple NAND gate, the intermediate node in the series n-pair is "hidden" by the schematic. What this implies is that test generation must be done in such a way as to take account of possible shorts and open circuits at the switch level.[5] Although the switch level may be the most appropriate level, expediency dictates that most existing systems rely on Boolean logic representations of circuits and S-A-0 and S-A-1 fault modeling.

A particular problem that arises with CMOS is that it is possible for a fault to convert a combinational circuit into a sequential circuit. This is illustrated for the case of a 2-input NOR gate in which one of the transistors is rendered ineffective (stuck open or stuck closed) in Fig. 7.4. This might be due to a missing source, drain, or gate connection. If one of the n-transistors (*A* connected to gate) is stuck open, then the function displayed by the gate will be

$$F = (not\ (A + B)) + (A\ .\ (not\ B)\ .\ F_n),$$

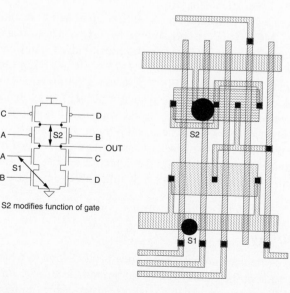

FIGURE 7.3   CMOS bridging faults

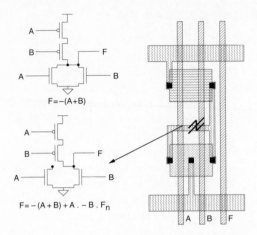

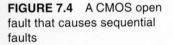

FIGURE 7.4 A CMOS open fault that causes sequential faults

$$F = -(A + B)$$

$$F = -(A + B) + A . - B . F_n$$

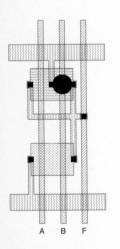

FIGURE 7.5 A defect that causes static I_{DD} current

where F_n is the previous state of the gate. Similarly if the B n-transistor drain connection is missing, the function is

$$F = (not (A + B)) + ((not A) . B . F_n).$$

If either p-transistor is open, the node would be arbitrarily charged (i.e., it might be high due to some weird charging sequence) until one of the n-transistors discharged the node. Thereafter it would remain at zero, bar charge-leakage effects. This problem has caused researchers to search for new methods of test generation to detect such behavior.[6]

Currently debate ranges over whether an SA0/SA1 approach to testing is adequate for testing CMOS. It is also possible to have switches (transistors) exhibit a "stuck-open" or "stuck-closed" state. Stuck-closed states can be detected by observing the static V_{DD} current (I_{DD}) while applying test vectors. Consider the gate fault shown in Fig. 7.5, where a p-transistor in a 2-input NAND gate is shorted. This could physically occur if stray metal overlapped the source and drain connections or if the source and drain diffusions shorted. If we apply test vector 11 to the A and B input and measure the static I_{DD} current, we will notice that it rises to some value determined by the β ratios of the n- and p-transistors. While the debate continues and test cycles are at a premium, the SA0/SA1 model will suffice for some time to come.

7.2.2 Observability

The observability of a particular internal circuit node is the degree to which one can observe that node at the outputs of an integrated circuit (i.e., the pins). This measure is of importance when a designer/tester desires to measure the output of a gate within a larger circuit to check that it operates correctly. Given a limited number of nodes that may be directly observed, it is the aim of well-designed chips to have easily observed gate outputs, and the adoption of some basic test design techniques can aid tremendously in this respect. Ideally, one should be able to observe directly or with moderate indirection (i.e., one may have to wait a few cycles) every gate output within an integrated circuit. While at one time this aim was hindered by limited gate-count processes and a lack of design methodology, current design practices and processes allow one to approach this ideal. Section 7.3 examines a range of methods for increasing observability.

7.2.3   Controllability

The controllability of an internal circuit node within a chip is a measure of the ease of setting the node to a 1 or 0 state. This measure is of importance when assessing the degree of difficulty of testing a particular signal within a circuit. An easily controllable node would be directly settable via an input pad. A node with little controllability might require many hundreds or thousands of cycles to get it to the right state. Often one finds it impossible to generate a test sequence to set a number of poorly controllable nodes into the right state. It should be the aim of a well-designed circuit to have all nodes easily controllable. In common with observability, the adoption of some simple design for test techniques can aid tremendously in this respect.

7.2.4   Fault Coverage

A measure of goodness of a test program is the amount of fault coverage it achieves; that is, for the vectors applied, what percentage of the chip's internal nodes were checked. Conceptually, the way in which the fault coverage is calculated is as follows. Each circuit node is taken in sequence and held to 0 (S-A-0), and the circuit is simulated, comparing the chip outputs with a known "good machine"—a circuit with no nodes artificially set to 0 (or 1). When a discrepancy is detected between the "faulty machine" and the good machine, the fault is marked as detected and the simulation is stopped. This is repeated for setting the node to 1 (S-A-1). In turn, every node is stuck at 1 and 0, sequentially. The total number of nodes that, when set to 0 or 1, do result in the detection of the fault, divided by the total number of nodes in the circuit, is called the percentage-fault coverage.

The above method of fault analysis is called sequential fault grading. While this might be practical for small circuits, or by using hardware simulation accelerators on medium circuits, the time to complete the fault grading may be very long. On average KN cycles (assuming that, on average, N/2 cycles are needed to detect each fault) need to be simulated, where K is the number of nodes in the circuit and N is the length of the test sequence. For $K = 1000$ and $N = 12,000$, 12 million cycles are required. At 1 ms per cycle, this yields 12,000 seconds or 3 hrs 20 minutes. For $K = 100,000$ and

$N = 360,000$, 3.6×10^9 cycles are required. At 1 s per cycle, 1040 years would be required to do sequential fault grading.

To overcome these long simulation times many ingenious techniques have been invented to deal with fault simulation.

7.2.5   Automatic Test Pattern Generation (ATPG)

Historically in the IC industry, designers designed circuits, layout drafts-people completed the layout, and the test engineer wrote the tests. In many ways, the test engineers were the Sherlock Holmes of the industry, reverse engineering circuits and devising tests that would test the circuits in an adequate manner. For the longest time, test engineers implored circuit designers to include extra circuitry to ease the burden of test generation. Happily, as processes have increased in density and chips have increased in complexity, the inclusion of test circuitry has become less of an overhead for both the designer and the manager worried about the cost of the die. In addition, as tools have improved, more of the burden for generating tests has fallen on the circuit/logic designer. To deal with this burden, methods for automatically generating tests have been invented. Collectively these are known as ATPG, for Automatic Test Pattern Generation. This section summarizes one approach to ATPG to provide background for the reader. In practice, one may find that ATPG is of great use in the generation of test vectors or that for a variety of reasons it is not applicable.

Historically, most ATPG approaches have been based on simulation. A five-valued logic[7] form is commonly used to implement test generation algorithms (more advanced algorithms use up to 10 level logic). This consists of the states 1, 0, D, $\overline{D}$, and X. 0 and 1 represent logical zero and logical one respectively. X represents the unknown or DON'T-CARE state. D represents a logic 1 in a good machine and a logic 0 in a faulty machine while $\overline{D}$ represents a logic 0 in a good machine and a logic 1 in a faulty machine. The truth tables for inverters, AND, and OR gates are shown in Tables 7.1, 7.2, and 7.3.

TABLE 7.1   Inverter
Z = NOT A

A	Z
0	1
1	0
X	X
D	$\overline{D}$
$\overline{D}$	D

TABLE 7.2   2-input AND gate Z = A AND B

A	B	0	1	X	D	$\overline{D}$
0		0	0	0	0	0
1		0	1	X	D	$\overline{D}$
X		0	X	X	X	X
D		0	D	X	D	0
$\overline{D}$		0	D	X	0	D

We can examine the use of this five-valued logic by considering the circuit shown in Fig. 7.6 where an S-A-0 fault is to be detected at node h. We will alternatively call a circuit a *machine,* which is customary in test nomenclature. Thus node h would have value D. There are two objectives. The first is to propagate the D on node h to one or more *primary outputs* (POs). A primary output is a directly observable signal, such as a pad or, as we shall learn later, a scan output. This path to the primary output (or outputs) is called the *sensitized path.* The second objective is to set node h to state D via a set of *primary inputs* (PIs). A primary input is one that can be directly set via a pad or some other means. The gate driving node h is the *Gate Under Test* or GUT. From node h we backtrack to the primary inputs (a, b, c, d, e) to find the necessary input vector required to set node h to a 1. Because the gate driving node h is an AND gate from the above definition (a D is a 1 in a good machine), both inputs (f, g) have to be set to 1 to set h to 1. Proceeding further toward the inputs, to assert node f as a 1, both nodes a and b have to be set to a 1. Because node g is driven by an OR gate, either node c or node d need to be set to a 1 to assert node g. Thus a vector $\{a,b,c,d\}$ of $\{1,1,1,0\}$ or $\{1,1,0,1\}$ is required to control node h. To observe that node g has been set to a D, input node e has to be set to a 1. Thus the resultant test vector is

TABLE 7.3   2-input OR gate Z = A OR B

A	B	0	1	X	D	$\overline{D}$
0		0	1	X	D	$\overline{D}$
1		1	1	1	1	1
X		X	1	X	X	X
D		D	1	X	D	1
$\overline{D}$		$\overline{D}$	1	X	1	D

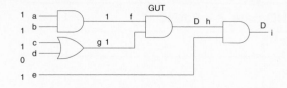

FIGURE 7.6 The *D* algorithm—sensitization step

$\{a,b,c,d,e\} = \{1,1,0,1,1\}$ or $\{1,1,1,0,1\}$. If we are checking for an S-A-1 fault at node *h*, we must be able to set it to 0. By similar reasoning to that for the S-A-0 case the test vector would be $\{a,b,c,d,e\} = \{0,1,X,X,1\}$ or $\{1,0,X,X,1\}$ or $\{0,0,X.X,1\}$ or $\{1,1,0,0,1\}$. Similarly, for other nodes a summary of the vectors is as in Table 7.4.

The next step is to collapse the vectors into the least set that covers all nodes. A possible set is $\{1,1,0,1,1\}$, $\{0,0,1,0,1\}$, $\{1,1,0,0,1\}$.

The reason for using a five-valued logic is shown in Fig. 7.7. Here an additional AND gate and INVERT gate have been added to the circuit. We can see that a fault at node *h* is essentially unobservable (due here to the non-sensical logic). This circuit suffers from what is called reconvergent fan-out.

The usual basis for manual generation of tests by test engineers and many current automatic test-pattern generation programs is the *D*-algorithm (DALG).[8] PODEM[9] and PODEM-X[10] are improved algorithms that are more efficient than the original DALG and in addition treat error-correcting circuits composed of XOR gates with reconvergent fan-out. Another ATPG algorithm is called FAN[11] and an improved efficiency algorithm[12] dealing with tristate drivers called ZALG has been developed.[13] Other work has concentrated on dealing at a module level rather that a gate level.[14] In basis, these algorithms start by propagating the *D* value on an internal node to a primary output. This is called the *D-propagation* phase. The selection of which gates to pass through to the output is guided by observability indexes assigned to gates. At any particular gate input, the gate with the highest observability is selected. Once the *D* value is observable at a primary output, the next step is to determine the primary input values that are required to

TABLE 7.4   Node-vector Summary of *D* Algorithm (Fig. 7.6)

NODE	TEST	VECTOR $\{a,b,c,d,e\}$
h	S-A-0	$\{1,1,0,1,1\}$, $\{1,1,1,0,1\}$
h	S-A-1	$\{0,1,X,X,1\}$, $\{1,0,X,X,1\}$, $\{0,0,X,X,1\}$, $\{1,1,0,0,1\}$
f	S-A-0	$\{1,1,0,1,1\}$, $\{1,1,1,0,1\}$
f	S-A-1	$\{0,0,0,1,1\}$, $\{0,0,1,0,1\}$
g	S-A-0	$\{1,1,0,1,1\}$, $\{1,1,1,0,1\}$, $\{1,1,1,1,1\}$
g	S-A-1	$\{1,1,0,0,1\}$

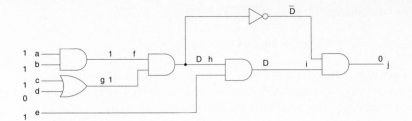

FIGURE 7.7   Reconvergent fan-out with D notation

enable the fault to be observed and tested. This proceeds by backtracking from the faulted signal and sensitized path-enables toward the primary inputs. The selection of which path to proceed along toward the inputs is aided by controllability indices assigned to nodes. This is known as the *backtrace* step.

Controllabilities and observabilities can be assigned statically (that is, without regard to the logic state of the network) or dynamically[15] (that is, according to the current state of the network). The SCOAP[16] algorithm is one method of assigning controllabilities and observabilities. In the SCOAP system the following six testability measures (TMs) are defined for each circuit node:

- $CC0(n)$—combinatorial 0 controllability of node n (i.e., the extent to which a combinatorial node can be forced to a zero).
- $CC1(n)$—combinatorial 1 controllability of node n.
- $CO(n)$—combinatorial observability of node n.
- $SC0(n)$—sequential 0 controllability of node n.
- $SC1(n)$—sequential 1 controllability of node n.
- $SO(n)$—sequential observability of node n.

FIGURE 7.8
NAND gate

The combinatorial measures are applied to the ouputs of logic gates, while the sequential measures apply to registers and other "sequential" modules. As an example, for the AND gate shown in Fig. 7.8 the $CC1$ value is

$$CC1(z) = CC1(a) + CC1(b) + 1.$$

That is, the 1-controllability of the output of the AND gate is the sum of the 1-controllabilities of each input because each input has to be set to 1 to set the output to 1. The 1 is added at the end because the AND gate represents one stage of combinatorial logic. The sequential 1-controllability is given by

$$SC1(z) = SC1(a) + SC1(b).$$

The combinatorial 0-controllability is given by

$$CC0(z) = min[CC0(a), CC0(b)] + 1.$$

This arises due to the fact that either a 0 on a or b forces a 0 at the output. Therefore the easiest controllable input may be used (the lowest combinatorial controllability). The sequential controllability is given by

$$SC0(z) = min[SC0(a), SC0(b)].$$

The combinatorial observability of a is given by

$$CO(a) = CO(z) + CC1(b) + 1;$$

that is, the observability of z added to the combinatorial 1-controllability of b. This occurs because b has to be forced to a 1 to make a observable. The sequential observability of a is given by

$$SO(a) = SO(z) + SO(b).$$

Similar equations may be derived for other gate types. The SCOAP algorithm proceeds by first calculating the circuit controllabilities by propagating controllabilites from the logic inputs. Following this, the observabilities are propagated from the logic outputs. Figure 7.9(a) shows a logic circuit with the 1-controllabities annotated. Figure 7.9(b) shows the observabilities.

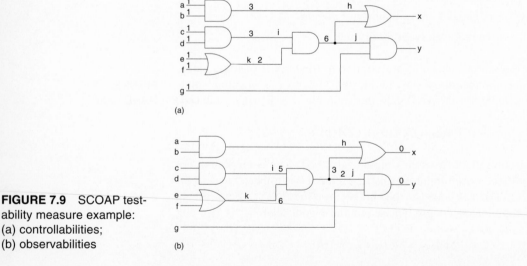

(a)

(b)

FIGURE 7.9 SCOAP testability measure example: (a) controllabilities; (b) observabilities

In cases of multiple fan-out, the minimum observability measure is used. The presence of high controllability numbers indicates a node that is difficult to control, while the presence of high observability numbers indicates nodes that are difficult to observe. As mentioned above, the testability measures are used to guide the selection of paths in the D-propagation and backtrace phase of the D-algorithm–based ATPG procedures.

Other testability measures, such as COP[17] and LEVEL,[18] are also used. COP testability measures are probablistic in nature.

More recently authors have proposed the use of massively parallel methods for ATPG.[19] Methods have also been developed that model faults as changes to a Boolean network. Equivalence checking is used to prove that the two networks are not equivalent. These methods, when combined with random-fault generation and fault simulation, have demonstrated a great deal of success.[20]

7.2.6 Fault Grading and Fault Simulation

Fault grading consists of two steps. First, the node to be faulted is selected. Normally global nodes such as reset lines and clock lines are excluded because faulting them can lead to unnecessary simulation (i.e., if the reset or clock line is stuck, then not much is going to happen in the circuit). A simulation is run with no faults inserted, and the results of this simulation (that is, the primary output responses for each input test vector) are saved. Following this process, in principle, each node or line to be faulted is set to 0 and then 1 and the test vector set is applied. If, and when, a discrepancy is detected between the faulted circuit response and the good circuit response, the fault is said to be detected and the simulation is stopped, and the process is repeated for the next node to be faulted. If the number of nodes to be faulted is K, and the average number of test vectors is N, the number of simulation cycles, S_K, is approximately given by

$$S_K = 2\frac{N}{2}K + N$$
$$= K(N+1) \approx KN. \tag{7.1}$$

This serial fault simulation process is therefore running K sets of the test vector set. With a small vector set, simple circuit, or very fast simulator, this approach is feasible. However, for large test sets and circuits, it is highly impractical.

To deal with this problem, a number of ideas have been developed to increase the speed of fault simulation.

Parallel Simulation is one method for speeding up simulation of multiple machines. In this method m words in an n-bit computer are used to encode the state of n "machines" for a 2^m-state simulator. Two n-bit words may be used to encode n machines for a three-state simulator. More computer words may be used to encode simulators with more states. Moreover this principle has been extended to special-purpose hardware where the computer word length could be optimized to deal with substantially more circuits in parallel. Now if M circuits can be simulated in parallel, then

$$S_K = \frac{KN}{M}.$$ (7.2)

Concurrent Simulation[21] is currently the most popular method for software-based fault simulators. The technique uses a nonfaulted version of the circuit to create a "good" machine model. Each fault creates a new faulty machine that is simulated in parallel with the good machine. Thus $N + 1$ simulations may have to be completed, where N is the number of faults. Concurrent simulators rely on a number of heuristics to reduce the amount of simulation. For instance, when a difference is noted between a faulted machine and a good machine at an externally observable point (i.e., the pads), the faulty machine is dropped from the simulation queue and the fault is "detected." If the bad machine has an X or Z compared to a 1 or 0 for the good machine, the fault is a "possible detect." Obviously, the more externally observable nodes a circuit has, the quicker bad machines get dropped from the simulation. Normally, only the good machine state is stored, with each node listing the fault machines that differ with the good machine. The different state is often small, which implies that there is a small amount of extra simulation to be done. In other words, most simulation for a faulty machine is exactly the same as the good machine. This is what concurrent simulation exploits. Fault collapsing occurs when two different faults result in the same faulty machine. This is noted, and one of the faulty machines may be dropped. Some machines perform static fault collapsing prior to simulation. For instance, an SA0 fault on the input of an inverter is the same as an SA1 fault at the output of the same inverter. With some fault simulators it is possible to create a fault dictionary. This is a cross reference that maps an observed fault to a set of possible internal faults. It is of use when the tester wishes to track down the actual internal failure (such as to perform yield improvement) rather than just cull the part.

Apart from software-based simulations, hardware-fault simulation accelerators that can provide a speedup over software-based simulators are also available.

7.2.7   Delay Fault Testing

The fault models we have dealt with to this point have neglected timing. Failures that occur in CMOS could leave the functionality of the circuit untouched, but affect the timing. For instance, consider the layout shown in

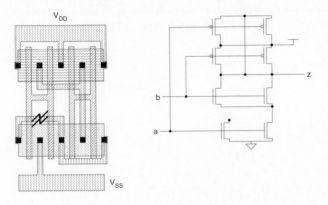

FIGURE 7.10   An example of a delay fault

Fig. 7.10 for a high-power NAND gate composed of paralleled n- and p-transistors. If the link illustrated was opened, the gate would still function, but with increased pull-down time. In addition, the fault now becomes sequential because the detection of the fault depends on the previous state of the gate and the simulation clock speed.

7.2.8   Statistical Fault Analysis

Conventional fault analysis can consume large CPU resources and take a long time. An alternative to this is what is called statistical fault analysis (STAFAN).[22] This method of fault analysis relies on estimating the probability that a fault will be detected. In summary, a fault free simulation is performed on a circuit in which some extra statistics are gathered by a modified simulator on a per-input vector basis. These are as follows:

- Zero-counter—The 0 count on each gate input when a $1 \rightarrow 0$ change of the output is detected.
- One-counter—The 1 count on each gate input when a $0 \rightarrow 1$ change of the output is detected.
- Sensitization-counter—incremented if the input change causes the output to be sensitized.
- Loop-counter—used to detect and deal with feedback.

The one-controllability of line l is given by

$$C1(l) = one\text{-}count/N,$$

where N is the number of vectors.

The zero-controllability is given by

$$C0(l) = zero\text{-}count/N.$$

TABLE 7.5   Statistical Fault Analysis 1 And 0 Observabilities

GATE TYPE	$B1(l)$	$B0(l)$
AND	$B1(m) \cdot \dfrac{C1(m)}{C1(l)}$	$B0(m) \cdot \dfrac{S(l) - C1(m)}{C0(l)}$
OR	$B1(m) \cdot \dfrac{S(l) - C0(m)}{C1(l)}$	$B0(m) \cdot \dfrac{C0(m)}{C0(l)}$
NAND	$B0(m) \cdot \dfrac{C0(m)}{C1(l)}$	$B1(m) \dfrac{S(l) - C1(m)}{C0(l)}$
NOR	$B0(m) \dfrac{S(l) - C1(m)}{C1(l)}$	$B1(m) \cdot \dfrac{C1(m)}{C0(l)}$
NOT	$B0(m)$	$B1(m)$

The one-level sensitization probability is

$$S(l) = sensitization\text{-}count/N.$$

The observabilities are calculated by propagating from gate outputs to gate inputs. For common gates, Jain and Agrawal derive the one-observabilities ($B1$) and zero-observabilities ($B0$) for common gates, as shown in Table 7.5.

Methods also exist to deal with fan-out where two observabilities must be combined. Once these observability and controllability measures have been determined, the probability of fault detection may be calculated as follows:

$$D1(l) = B0(l) . C0(l),$$

where $D1(l)$ is the probability of detection that line l is SA1.

$$D0(l) = B1(l) . C1(l),$$

where $D0(l)$ is the probability of detection that line l is SA0.

From these values the fault coverage of the circuit may be calculated. The results of using this technique follow very closely the results generated by conventional fault simulation.

7.2.9   Fault Sampling

Another approach to fault analysis is known as fault sampling. This is used in circuits where it is impossible to fault every node in the circuit. Nodes are randomly selected and faulted. The resulting fault-detection rate may be statistically inferred from the number of faults that are detected in the fault set and the size of the set. As with all probabalistic methods it is important that the randomly selected faults be unbiased. Although this approach does not yield a specific level of fault coverage, it will determine whether the fault coverage exceeds a desired level. The level of confidence may be increased by increasing the number of samples.

7.3   Design Strategies for Test

7.3.1   Design for Testability

The key to designing circuits that are testable are the two concepts that we have introduced called controllability and observability. Restated, controllability is the ability to set (to 1) and reset (to 0) every node internal to the circuit. Observability is the ability to observe either directly or indirectly the state of any node in the circuit.

We will first cover three main approaches to what is commonly called *Design for Testability*. These may be categorized as:

- ad-hoc testing.
- scan-based approaches.
- self-test and built-in testing.

Following this we will look at the application of these techniques to particular types of circuits. In this treatment we will look at:

- random logic (multilevel standard-cell, two-level PLA).
- regular logic arrays (datapaths).
- memories (RAM, ROM).

7.3.2   Ad-Hoc Testing

Ad-hoc test techniques, as their name suggests, are collections of ideas aimed at reducing the combinational explosion of testing. Common techniques involve:

- partitioning large sequential circuits.
- adding test points.
- adding multiplexers.
- providing for easy state reset.

Long counters are good examples of circuits that can be tested by ad-hoc techniques. For instance imagine you have designed an 8-bit counter and want to test it. Figure 7.11(a) shows a naive implementation in which the counter only has a *RESET* and a *CLOCK* input, with the terminal count (*TC*) being observable. The designer probably thought that a reset and 256 clock cycles, followed by the observation of *TC*, would be adequate for testing purposes. Apart from the nonobservability of the count value ($Q<7:0>$), the

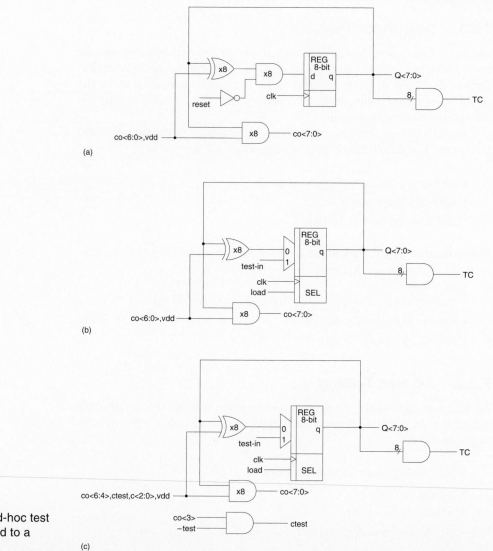

FIGURE 7.11   Ad-hoc test techniques applied to a counter

main problem is the number of cycles required to test a single counter. Possible ad-hoc test techniques are shown in Fig. 7.11(b) and Fig. 7.11(c). In Fig. 7.11(b), a parallel-load feature is added to the counter. This enables the counter to be preloaded with appropriate values to check the carry propagation within the counter. Another technique is to reduce the length of each counter to, say, 4 bits, as shown in Fig. 7.11(c). This is achieved by having the test signal block the carry propagate at every 4-bit boundary. With this method 16 vectors exhaustively can test each 4-bit section. The carry propagate between 4-bit sections may be tested with a few additional vectors.

Another technique classified in this category is the use of the bus in a bus-oriented system for test purposes. This is shown on Fig. 7.12(a) for a very simple accumulator. Each register has been made loadable from the bus and capable of being driven onto the bus. Here the internal logic values that exist on a data bus are enabled onto the bus for testing purposes. A more general scheme is illustrated in Fig. 7.12(b), where the normally inaccessible inputs are set and the outputs are observed via the bus.

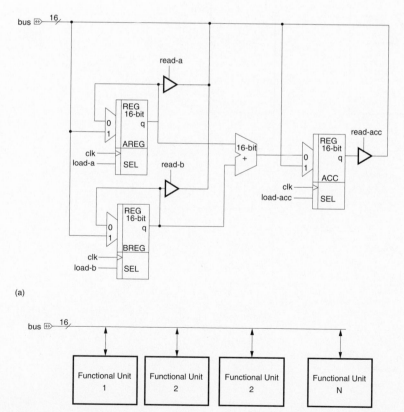

(a)

(b)

FIGURE 7.12 Bus-oriented test techniques

Frequently, multiplexers may be used to provide alternative signal paths during testing. In CMOS, transmission gate multiplexers provide low area and speed overhead. Figure 7.13(a) shows a scheme called a Design for Autonomous Test[23], which uses multiplexers. Figure 7.13(b) shows the circuit configured for normal use, while Fig. 7.13(c) shows the circuit configured to test module A.

Any design should always have a method of resetting the internal state of the chip within a single cycle or at most a few cycles. Apart from making testing easier, this also makes simulation faster because a few cycles are required to initialize the chip.

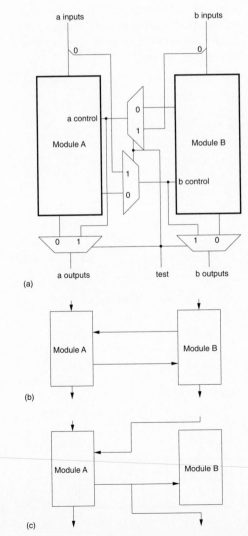

FIGURE 7.13 Multiplexer based testing

In general, ad-hoc testing techniques represent a bag of tricks developed over the years by designers to avoid the overhead of a systematic approach to testing, which will be described in the next section. While these general approaches are still quite valid, process densities and chip complexities necessitate a structured approach to testing.

7.3.3   Scan-Based Test Techniques

A collection of approaches have evolved for testing that lead to a structured approach to testability. The approaches stem from the basic tenets of controllability and observability outlined earlier in this chapter.

7.3.3.1   *Level Sensitive Scan Design (LSSD)*

A popular approach is called Level Sensitive Scan Design, or the LSSD approach, introduced by IBM.[24,25,26] This is based on two tenets. First, that the circuit is level sensitive. According to Williams[27],

> A logic system is *level-sensitive* if, and only if, the steady state response to any allowed input state change is independent of the circuit and wire delays within the system. Also, if an input state change involves the changing of more than one input signal, then the response must be independent of the order in which they change. Steady state response is the final value of all logic gate outputs after all change activity has terminated.

The second principle of LSSD is that each register may be converted to a serial shift register.

The basic building block in LSSD is the Shift Register Latch, or SRL. A block-level implementation of a polarity-hold SRL is shown in Fig. 7.14(a). It consists of two latches, L_1 and L_2. L_1 has a serial data port, I, and an enable, A. It also has a data port, D, and an enable, C. When A is high, the value of L_1 (T_1) is set by the value of I, while when C is high, L_1 is set by D. A and C can not be simultaneously high. When signal B in L_2 is high, T_1 is passed to T_2. A gate-level implementation of the SRL is shown in Figs. 7.14(b) and 7.14(c). In normal operation, the D input is the normal input to the register, while the T_2 signal is the output. L_1 is the master while L_2 is the slave. SRLs may be connected in series by using the T_2 output and the I input of successive latches. During normal system operation, A is held low and C and B may be thought of as a two-phase nonoverlapping clock. When data is to be loaded into the SRLs or dumped out of the SRLs, A and B are used as a two-phase shift clock.

Figure 7.15(a) shows a typical LSSD scan system. An expanded view is shown in Fig. 7.15(b). The first rank of SRLs have inputs driven from a preceding stage and have outputs $QA1$, $QA2$, and $QA3$. These outputs feed a block of combinational logic. The output of this logic block feeds a second

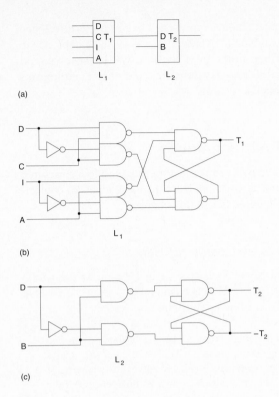

FIGURE 7.14   A shift register latch

rank of SRLs with outputs $QB1$, $QB2$, and $QB3$. Figure 7.15(c) shows a typical clocking sequence. Initially the *shift-clk* and $c2$ are clocked three times to shift data into the first rank of SRLs ($QA1$–3). $c1$ is asserted, and then $c2$ is asserted, clocking the output of the logic block into the second rank of SRLs ($QB1$–3). *shift-clk* and $c2$ are then clocked three times to shift $QB1$, $QB2$, and $QB3$ out via the serial-data-out line. Testing proceeds in this manner of serially clocking the data through the SRLs to the right point in the circuit, running a single "system" clock cycle and serially clocking the data out for observation. In this scheme, every input to the combinational block may be controlled and every output may be observed. In addition, running a serial sequence of 1's and 0's (such as 110010) through the SRLs can test them.

Test generation for this type of test architecture may be highly automated. ATPG techniques may be used for the combinational blocks, and as mentioned, the SRLs are easily tested. The prime disadvantage is the complexity of the SRLs (i.e., impacting density and speed).

7.3.3.2   Serial Scan

Level Sensitive Scan went to great pains to provide a hazard-free latching scheme. Faster clock speeds and design for smaller overhead in the registers

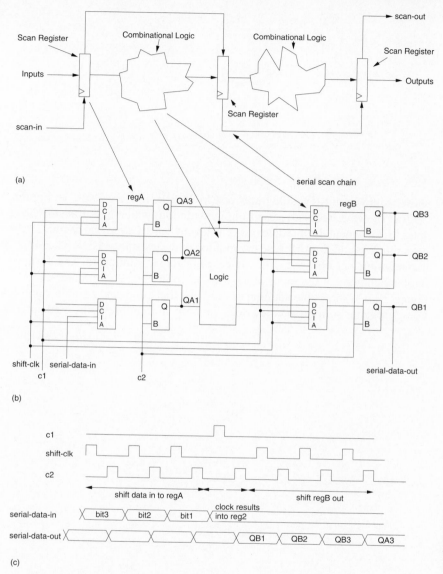

(a)

(b)

(c)

FIGURE 7.15 An LSSD scan chain: (a) basic architecture; (b) example circuit; (c) example timing

has led to simplifications in the SRL that give up a little on the hazard front but retain the scan principles mentioned above. (The hazard is moved inside the register, which with careful design can be guaranteed to be race free for a particular process and environmental characteristics.)

A schematic for a commonly used CMOS edge-sensitive scan-register is shown in Fig. 7.16. A MUX is added before the master latch in a conventional *D* register. *TE* is the Test Enable pin, and *TI* is the Test Input pin. When *TE* is enabled, *TI* is clocked into the register by the rising edge of *CLK*. Figure 7.17

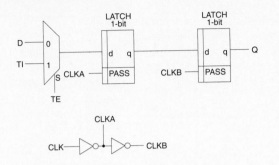

FIGURE 7.16   A typical CMOS scan-register

shows some circuit-level diagrams of CMOS SRL implementations. Figure 7.17(a) shows a frequently used implementation, which uses transmission gates to implement the multiplexers. The layout density overhead for this latch is minimal. In addition, because the addition of the testability MUX places two transmission gates in series, the increase in delay is minimized. Two further implementations of the input MUX are shown in Figs. 7.17(b) and 7.17(c). Figure 7.17(b) shows the addition of only two transistors and a single control line. A register so implemented does have the normal problems associated

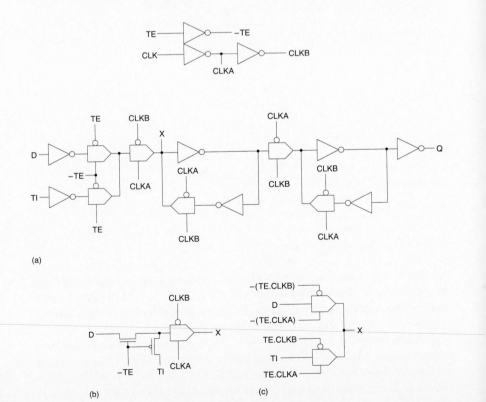

FIGURE 7.17   Various CMOS scan-latch options

with used single-polarity transmission gates (see Chapter 5). Alternatively, the clocks may be gated, as shown in Fig. 7.17(c). While this minimizes transistors, it may lead to unacceptable hold-time constraints on the register. Because the signals applied to the master latch are delayed with respect to the main clock, the data has to be held for a longer time at the input.

7.3.3.3   Partial Serial Scan

Quite often in a design, one may not find it area- and speed-efficient to implement scan registers in every location where a register is used. This occurs, for instance, in signal-processing circuits where many pipeline registers might be used to achieve high speed. If these are in the data-flow section of the chip, then one can think of the logic that has to be tested as the logic with the pipeline registers removed. In this case only the input and output registers need be made scannable.[28] This technique of testing is known as partial scan, and depends on the designer making decisions about which registers need to be made scannable.

Consider the design shown in Fig. 7.18 (from Gupta et al.[29]). In a full-scan test strategy all registers would have to be scannable. A partial-scan design is shown in Fig. 7.18(a) where only two registers have been made scannable ($R6$ and $R3$). In addition, these registers have the ability to hold their state dependent on a HOLD control. The part of the circuit that is being tested and monitored by the scan registers (known as the *kernel*) is shown in Fig. 7.18(b). It may be proven that, by holding the vectors at the input of the kernel for three clock cycles, the kernel may be represented by the combinational-equivalent circuit shown in Fig. 7.18(c). This circuit may be used by an ATPG program to generate test vectors.

7.3.3.4   Parallel scan

One can imagine that serial-scan chains can become quite long, and the loading and unloading sequence can dominate testing time. An extension of serial scan is called random-access or parallel scan.[30]

The basic idea is shown in Fig. 7.19. Each register in the design is arranged on an imaginary (or real) grid where registers on common rows receive common data lines and registers in common columns receive common read- and write-control signals. In the figure, an array of 2-by-2 registers is shown. The D and Q signals of the registers are connected to the normal circuit connections. Any register output may be observed by enabling the appropriate column read line and setting the appropriate address on an output data multiplexer. Similarly, data may be written to any register.

Figure 7.20 shows a D-register implementation called a Cross-Controlled Latch.[31] It consists of a normal CMOS master-slave edge-triggered register augmented by two small n-transistors, N_1 and N_2. When *-test-write-enable* is high, *Probe[j]* is high, and *clk* is low, the value of node Y (D) may

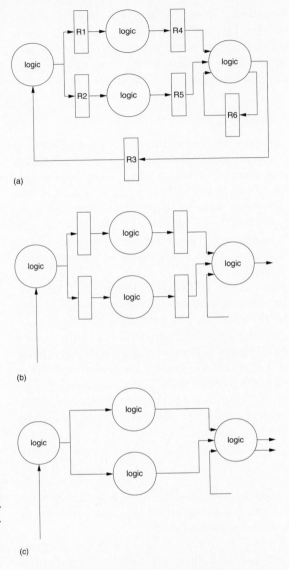

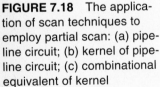

FIGURE 7.18 The application of scan techniques to employ partial scan: (a) pipeline circuit; (b) kernel of pipeline circuit; (c) combinational equivalent of kernel

be sensed on *Sense[i]* via transistor N_2. When *-test-write-enable* is low, *Probe[j]* is high, and *clk* is high, the value on *Sense[i]* can be driven onto node *Y*. This is seen immediately at the output of the register. The net effect on the register-timing parameters of the extra transistors is to slightly increase the minimum clock-pulse width. The area impact for an ASIC-based register is around 3%.

The large number of observable outputs (one for every register in the design) are compressed using signature analysis (see Section 7.3.4.1). The large number of observable outputs leads to very efficient concurrent-fault simulation.

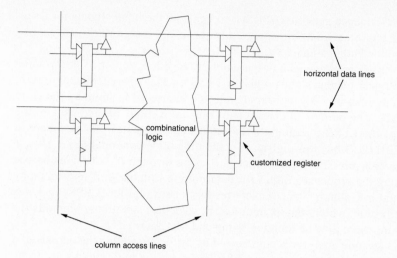

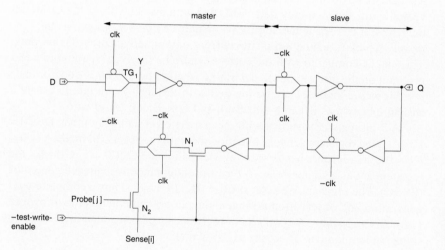

FIGURE 7.19 Parallel scan—basic structure

FIGURE 7.20 Parallel scan register (a cross-controlled latch)

7.3.4 Self-Test Techniques

Self-test and built-in test techniques, as their names suggest, rely on augmenting circuits to allow them to perform operations on themselves that prove correct operation.

7.3.4.1 Signature Analysis and BILBO

One method of incorporating a built-in test module is to use signature analysis[32,33] or cyclic-redundancy checking. This involves the use of a pseudo–

random sequence generator (PRSG) to generate the input signals for a section of combinational circuitry and then using a signature analyzer to observe the output signals.

A PRSG implements a polynomial of some length N. It is constructed from a linear feedback shift register (LFSR), which is constructed, in turn, from a number of 1-bit registers connected in a serial fashion, as shown in Fig. 7.21. The outputs of certain shift bits are XORed and fed back to the input of the LFSR to calculate the required polynomial. For instance, in Fig. 7.21, the 3-bit shift register is computing the polynomial $f(x) = 1 + x + x^3$. For an n-bit LFSR, the output will cycle through $2^n - 1$ states before repeating the sequence. Tables for determining suitable shift registers may be found in Golumb.[34] A complete feedback shift register (CFSR) includes the zero state, which may be required in some test situations. Methods for designing these may be found in Wang and McCluskey.[35]

A signature analyzer is constructed by cyclically adding the outputs of a circuit to a shift register or an LFSR if successive logic blocks are to be tested in a like manner. A typical circuit is shown in Fig. 7.22(a). As each test vector is run, the incoming data is XORed with the contents of the LFSR. At the end of a test sequence, the LFSR contains a number, known as the *syndrome,* which is a function of the current output and all previous outputs. This can be compared with the correct syndrome (derived by running a test program on the good logic) to determine whether the circuit is good or bad.

Signature analysis can be merged with the scan technique to create a structure known as BILBO—for Built-In Logic Block Observation.[36]

A 3-bit register is shown with the associated circuitry. In mode D ($C0 = C1 = 1$), the registers act as conventional parallel registers. In mode A ($C0 = C1 = 0$), the registers act as scan registers. In mode C ($C0 = 1$, $C1 = 0$), the registers act as a signature analyzer or pseudo–random sequence generator (PRSG). The registers are reset if $C0 = 0$ and $C1 = 1$. Thus a complete test-generation and observation arrangement can be implemented, as shown in Fig. 7.22(b). In this case two sets of registers have been added in addition to some random logic to effect the test structure.

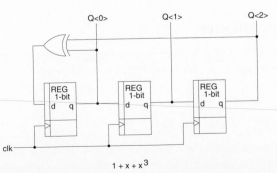

FIGURE 7.21 Pseudo–random sequence generator (PRSG)

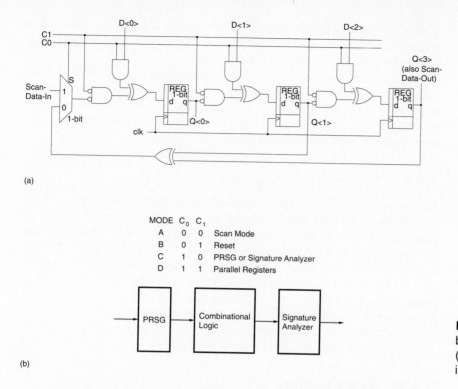

FIGURE 7.22 Built-in logic block observation (BILBO): (a) individual register; (b) use in a system

A chip set for FFT applications was designed with local testing based on pseudo–random pattern generation and signature analysis.[37] With a 28-bit pattern generator and a 17-bit signature at 10 MHz it took 26 seconds to test the part.

7.3.4.2   Memory Self-Test

Embedding self-test circuits for memories in higher-speed circuits not only may be the way of testing the structures at speed but can save on the number of external test vectors that have to be run. A typical read/write memory (RAM) test program for an M-bit address memory might be as follows[38,39]:

```
FOR i=0 to M-1 write(datā)
FOR i=0 to M-1 read(datā) then write(data)
FOR i=0 to M-1 read(data) then write(datā)
FOR i=M-1 to 0 read(datā) then write(data)
FOR i=M-1 to 0 read(data) then write(datā)
```

data is 1 and datā is 0 for a 1-bit memory or a selected set of patterns for an *n*-bit word. For an 8-bit memory data, might be x00, x55, x33, and x0F. An address counter, some multiplexers, and a simple-state machine result in a fairly low overhead self-test structure for read/write memories. Oshawa et al.[40] describe a 4-Mbit RAM with self-test. The self-test consists of 256K cycles that input a checkerboard pattern to test for cell-to-cell interference.

This is followed by 256K cycles in which the data is read out. Then a complemented checkerboard is written and read. A total of 1 million cycles provide a test sufficient for system maintenance.

ROM memories may be tested by placing a signature analyzer at the output of the ROM and incorporating a test mode that cycles through the contents of the ROM. A significant advantage of all self-test methods is that testing may be completed when the part is in the field. With care, self-test may even be performed during normal system operation.

7.3.4.3   *Iterative Logic Array Testing*

Arrays of logic[41,42] present an interesting problem to the test architect because the replication can be used to advantage in reducing the number of tests. In addition, by augmenting the logic extremely high fault coverage rates are possible. An iterative logic array (ILA) is a collection of identical logic modules (such as an n-bit adder). An ILA is C-testable if it can be tested with a constant number of input vectors independent of the iteration count. An ILA is I-testable if a particular fault that occurs in any module as a result of an applied input vector is identical for all modules in the ILA. Assuming that only one module is faulty, the detection of a fault may be made by using an equality test on the ILA outputs.

7.3.5   IDDQ Testing

An increasingly popular method of testing for bridging faults is called IDDQ (V_{DD} supply current Quiescent) or current-supply monitoring.[43,44] This relies on the fact that when a complementary CMOS logic gate is not switching, it draws no DC current (except for leakage). When a bridging fault occurs, for some combination of input conditions a measurable DC I_{DD} will flow. Testing consists of applying the normal vectors, allowing the signals to settle, and then measuring I_{DD}. To be effective any circuits that draw DC power such as pseudo-nMOS gates or analog circuits have to be disabled. Because many circuits now require SLEEP modes to reduce power, this may not be a substantial additional overhead.

Because current measuring is slow, the tests must be run slower than normal, thus increasing the test time. However, this technique gives a form of indirect massive observability at little circuit overhead.

7.4   Chip-Level Test Techniques

In this chapter we have discussed the principles behind testing ICs, and covered some techniques aimed at making testing easier. In the past the design process was frequently divided between a designer who designed the circuit and a test engineer who designed the test to apply to that circuit. The advent of the ASIC, small design teams, the desire for reliable ICs, and rapid times to market have all forced the "test problem" earlier in the design cycle. In fact, the designer who is only thinking about what functionality has to be implemented and not about how to test the circuit will quite likely cause product deadlines to be slipped and in extreme cases products to be stillborn. In this section we will examine some practical methods of incorporating test requirements into a design. This discussion is structured around the main types of circuit structure that will be encountered in a digital CMOS chip.

7.4.1   Regular Logic Arrays

Partial serial scan or parallel scan is probably the best approach for structures such as datapaths. One approach that has been used in a Lisp microprocessor is shown in Fig. 7.23.[45] Here the input busses may be driven by a serially loaded register. These in turn may be used to load the internal datapath registers. The datapath registers may be sourced onto a bus, and this bus may be loaded into a register that may be serially accessed. All of the control signals to the datapath are also made scannable.

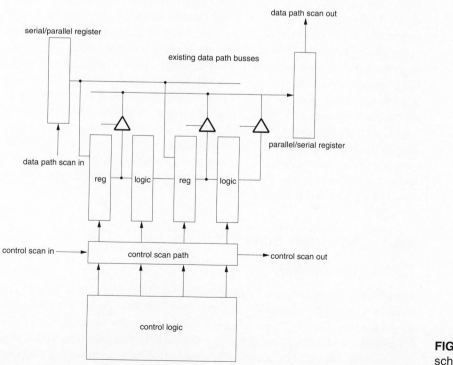

FIGURE 7.23 Datapath test scheme

7.4.2 Memories

Memories may use the self-testing techniques mentioned in Section 7.3.4.2. Alternatively, the provision of multiplexers on data inputs and addresses and convenient external access to data outputs enables the testing of embedded memories. It is a mistake to have memories indirectly accessible (i.e., data is written by passing through logic, data is observed after passing through logic, addresses can not be conveniently sequenced). Because memories have to be tested exhaustively, any overhead on writing and reading the memories can substantially increase the test time and, probably more significantly, turn the testing task into an effort in inscrutability.

7.4.3 Random Logic

Random logic is probably best tested via full serial scan or parallel scan.

7.5 System-Level Test Techniques

Up to this point we have concentrated on the methods of testing individual chips. Traditionally at the board level, "bed-of-nails" testers have been used to test boards. In this type of a tester, the board under test is lowered onto a set of test points (nails) that probe points of interest on the board. These may be sensed (the observable points) and driven (the controllable points) to test the complete board. At the chassis level, software programs are frequently used to test a complete board set. For instance, when a computer boots, it might run a memory test on the installed memory to detect possible faults.

The increasing complexity of boards and the movement to technologies like Multichip Modules (MCMs) and surface-mount technologies (with an absence of through-board vias) resulted in system designers agreeing on a unified scan-based methodology for testing chips at the board (and system level). This is called Boundary Scan.

7.5.1 Boundary Scan

7.5.1.1 Introduction

The IEEE 1149 Boundary Scan architecture[46] is shown in Fig. 7.24. In essence it provides a standardized serial scan path through the I/O pins of an IC. At the board level, ICs obeying the standard may be connected in a variety of series and parallel combinations to enable testing of a complete board or, possibly, collection of boards. The description here is a precis of the pub-

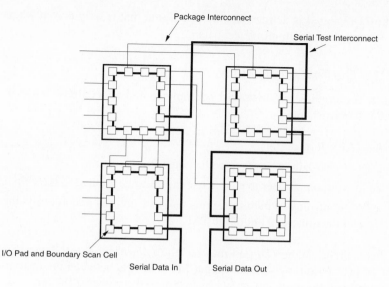

FIGURE 7.24 Boundary scan architecture

lished standard. The standard allows for the following types of tests to be run in a unified testing framework:

- Connectivity tests between components.
- Sampling and setting chip I/Os.
- Distribution and collection of self-test or built-in-test results.

7.5.1.2 The Test Access Port (TAP)

The Test Access Port (or TAP) is a definition of the interface that needs to be included in an IC to make it capable of being included in a Boundary-Scan architecture. The port has four or five single-bit connections, as follows:

- *TCK* (The Test Clock Input)—used to clock tests into and out of chips.
- *TMS* (The Test Mode Select)—used to control test operations.
- *TDI* (The Test Data Input)—used to input test data to a chip.
- *TDO* (The Test Data Output) used to output test data from a chip.

It also has an optional signal

- *TRST** (The Test Reset Signal) used to asynchronously reset the TAP controller; also used if a power-up reset signal is not available in the chip being tested.

The *TDO* signal is defined as a tristate signal that is only driven when the TAP controller is outputting test data.

7.5.1.3 The Test Architecture

The basic test architecture that must be implemented on a chip is shown in Fig. 7.25. It consists of:

- the TAP interface pins.
- a set of test-data registers to collect data from the chip.
- an instruction register to enable test inputs to be applied to the chip.
- a TAP controller, which interprets test instructions and controls the flow of data into and out of the TAP.

Data that is input via the *TDI* port may be fed to one or more test data registers or an instruction register. An output MUX selects between the instruction register and the data registers to be output to the tristate *TDO* pin.

7.5.1.4 The TAP Controller

The TAP controller is a 16-state FSM that proceeds from state to state based on the *TCK* and *TMS* signals. It provides signals that control the test data registers, and the instruction register. These include serial-shift clocks and update clocks.

The state diagram is shown in Fig. 7.26. The state adjacent to each state transition is that of the *TMS* signal at the rising edge of *TCK*.

The reader is referred to the standard for complete descriptions of these states. It is probably best to understand them by examining a typical test sequence. Starting initially in the Test-Logic-Reset state, a low on TMS tran-

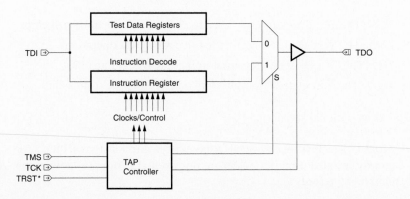

FIGURE 7.25 TAP architecture

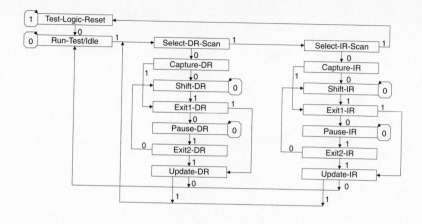

FIGURE 7.26 TAP controller state diagram

sitions the FSM to the Run-Test/Idle mode. Holding *TMS* high for the next three *TCK* cycles places the FSM in the Select-DR-Scan, Select-IR-Scan, and finally Capture-IR mode. In this mode two bits are input to the *TDI* port and shifted into the instruction register. Asserting *TMS* for a cycle allows the instruction register to pause while serially loading to allow tests to be carried out. Asserting *TMS* for two cycles, allows the FSM to enter the Exit2-IR mode on exit from the Pause-IR state and then to enter the Update-IR mode where the Instruction Register is updated with the new IR value. Similar sequencing is used to load the data registers.

A CMOS implementation of the Tap Controller based on that in the standard is shown in Fig. 8.89.

7.5.1.5 The Instruction Register (IR)

The instruction register has to be at least two bits long, and logic detecting the state of the instruction register has to decode at least three instructions, which are as follows:

- BYPASS—This instruction is represented by an IR having all zeroes in it. It is used to bypass any serial-data registers in a chip with a 1-bit register. This allows specific chips to be tested in a serial-scan chain without having to shift through the accumulated SR stages in all the chips.

- EXTEST—This instruction allows for the testing of off-chip circuitry and is represented by all ones in the IR.

- SAMPLE/PRELOAD—This instruction places the boundary-scan registers (i.e., at the chips' I/O pins) in the DR chain, and samples or preloads the chips I/Os.

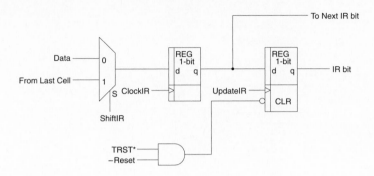

FIGURE 7.27   Instruction-register bit implementation

In addition to these instructions, the following are also recommended:

- INTEST—This instruction allows for single-step testing of internal circuitry via the boundary-scan registers.
- RUNBIST—This instruction is used to run internal self-testing procedures within a chip.

Further instructions may be defined as needed to provide other testing functions.

A typical IR bit is shown in Fig. 7.27.

7.5.1.6   Test-Data Registers (DRs)

The test-data registers are used to set the inputs of modules to be tested, and to collect the results of running tests. The simplest data-register configuration would be a boundary-scan register (passing through all I/O pads) and a bypass register (1-bit long). Figure 7.28 shows a generalized view of the data registers where one internal data register has been added. A multiplexer under the control of the Tap controller selects which particular data register is routed to the *TDO* pin.

7.5.1.7   Boundary Scan Registers

The boundary scan register is a special case of a data register. It allows circuit-board interconnections to be tested, external components tested, and the state of chip digital I/Os to be sampled. Apart from the bypass register, it is the only data register required in a Boundary Scan compliant part.

A single structure (in addition to the existing I/O circuitry) can be used for all I/O pad types, depending on the connections made to the cell. It consists of two multiplexers and two edge-triggered registers. Figure 7.29(a) shows this cell used as an input pad. Two register bits allow the serial shifting of data through the boundary-scan chain and the local storage of a data

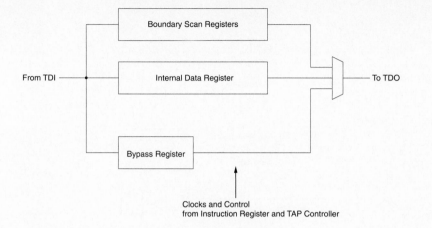

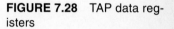

FIGURE 7.28   TAP data registers

bit. This data bit may be directed to internal circuitry in the INTEST or RUNBIST modes (*Mode* = 1). When *Mode* = 0, the cell is in EXTEST or SAMPLE/PRELOAD mode. A further multiplexer under the control of *ShiftDR* controls the serial/parallel nature of the cell. The signal *ClockDR* and *UpdateDR* generated by the Tap Controller load the serial and parallel register, respectively.

An output cell is shown in Fig. 7.29(b). When *Mode* = 1, the cell is in EXTEST, INTEST, or RUNBIST modes, communicating the internal data to

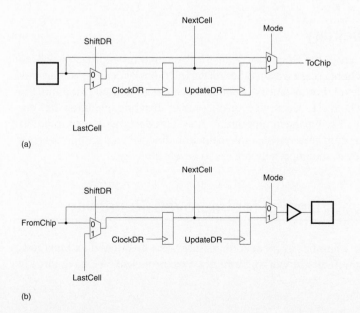

FIGURE 7.29   Boundary scan (a) input and (b) output cells

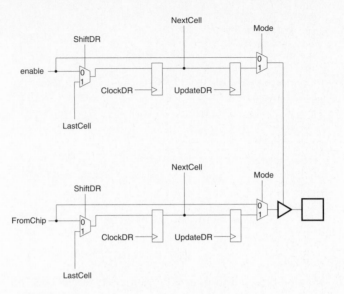

FIGURE 7.30 Boundary scan tristate cell

the output pad. When *Mode* = 0, the cell is in the SAMPLE/PRELOAD mode.

Two output cells may be combined to form a tristate boundary-scan cell, as shown in Fig. 7.30. The output signal and tristate-enable each have their own muxes and registers. The *Mode* control is the same for the output-cell example.

Finally, a bidirectional pin combines an input and tristate cell, as shown in Fig. 7.31.

7.5.2 Summary

At the system level, the Boundary Scan–Test Access Port approach has been summarized. There are, however, other related methods of dealing with testing at the system level. For instance, a boundary-scan method used in a multichip workstation, which uses a central controller rather than implementing the controller in each chip, has been reported.[47] A system designer has to trade off aspects, such as chip area versus implementation time, when deciding on a test strategy. However, the important thing is to *have* a strategy.

7.6 Layout Design for Improved Testability

In this chapter a number of models for failure were postulated and methods for detecting the faults in working circuits were proposed. We have already

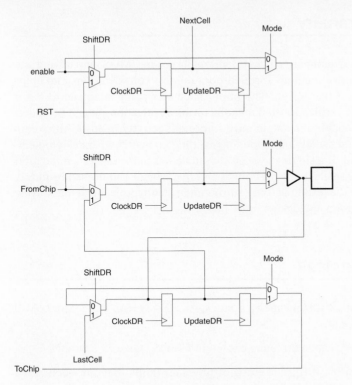

FIGURE 7.31   Boundary scan bidirectional cell

discussed a circuit-design technique to detect stuck-open faults in CMOS in Section 7.3.3.2. An interesting question arises. Can we construct the physical layouts to reduce the likelihood of such failures? This question has interested many researchers, and there is a body of literature that discusses the possible answers.[48–52]

In order to predict layout styles that improve testability, a designer has to have some idea of the nature and frequency of defects for a particular process. The types of defects that commonly occur may be divided into those that short together conductors and those that create open circuits. Shorts are possible intralayer in all layers used for connections, i.e., diffusion, polysilicon, metal1, metal2, and metal3, if used. The gate oxide may also short to the substrate or to either the source or drain. The source and drain regions may also short. Similarly for open circuits, all conducting layers might have open circuits. In addition, contacts may be misaligned, missing, or badly etched, leading to interlayer opens.

For open circuits, the ideas proposed in the literature to increase the immunity to open-circuit faults usually involve incorporating connection redundancy.

7.7 Summary

This chapter has summarized the important issues in CMOS chip testing and has provided some methods for incorporating test considerations into chips from the start of the design. The importance of writing adequate tests for both functional verification and manufacturing verification can not be understated. It is probably the single most important activity in any CMOS chip design cycle and usually takes the longest time no matter what design methodology is used. If a single message should be left in the reader's mind after reading this chapter, it should be that a chip designer should be absolutely rigorous about the testing activity surrounding a chip project and that testing should rank first in any design trade-offs.

7.8 Exercises

1. Explain what is meant by a Stuck-At-1(SA1) fault and a Stuck-At-0 (SA0) fault.

2. How are sequential faults caused in CMOS? Give an example.

3. Explain the different kinds of physical faults that can occur on a CMOS chip, and relate them to typical circuit failures.

4. Explain the terms controllability, observability, and fault coverage.

5. Explain how serial-scan testing is implemented.

6. Explain how a pseudo–random sequence generator (PRSG) may be used to test a 16-bit datapath. How would the outputs be collected and checked?

7. Design a block diagram of a test generator for an $8 \times 4K$ static RAM.

7.9 References

1. T. W. Williams, "Design for testability," in *Computer Design Aids for VLSI Circuits* (P. Antogneti, D. O. Pederson, and H de Man, eds.), NATO AIS Series, The Netherlands: Martinus Nijhoff Publishers, 1986, pp. 359–416.

2. G. S. Visweswaran, Akhtar-uz-zaman M. Ali, Parang K. Lala, and Carlos R. P. Hartmann, "The effects of transistor source-to-gate bridging faults in complex CMOS gates," *IEEE Journal of Solid State Circuits,* vol. 26, no. 6, Jun. 1991, pp. 893–896.

3. Thomas W. Williams and Kenneth P. Parker, "Design for testability—a survey," *Proceedings of the IEEE,* vol. 71, no. 1, Jan. 1983, pp. 98–112.

4. Anura P. Jayasumana, Yashwant K. Malaiya, and Rochit Rajsuman, "Design of CMOS circuits for stuck-open fault testability," *IEEE JSSC,* vol. 26, no. 1, Jan. 1991, pp. 58–61.

5. J. Galiay, Y. Crouzet, and M. Verginiault, "Physical versus logical fault models MOS LSI circuits: impact on their testability," *IEEE Transactions on Computers,* vol. C-29, no. 6, Jun. 1980, pp. 527–531.

6. Y. M. El-ziq and R. J. Cloutier, "Functional level test generation for stuck-open faults in CMOS VLSI," *Digest of Papers, IEEE International Test Conference,* Oct. 1981, pp. 536–546.

7. J. P. Roth, "Diagnosis of automata failures: a calculus and a method," *IBM Journal of Research and Development,* vol. 10, Jul. 1966, pp. 278–291.

8. *Ibid.*

9. Prabhakar Goel, "An implicit enumeration algorithm to generate tests for combinational logic circuits," *IEEE Transactions on Computers,* vol. c-30, no. 3, Mar. 1981, pp. 215–222.

10. Prabhakar Goel and Barry C. Rosales, "PODEM-X—an automatic test generation system for VLSI logic structures," *IEEE Proceedings of the 18th Design Automation Conference,* Jun. 1981, pp. 260–268.

11. H. Fuijiwara and T. Shimono, "On the acceleration of test generation algorithms," *IEEE Transactions on Computers,* vol. C-32, no. 12, Dec. 1983, pp. 1137–1144.

12. Michael H. Schulz, Erwin Trischler, and Thomas M. Sarfert, "SOCRATES: a highly efficient automatic test pattern generation system," *IEEE Transactions on CAD,* vol. 7, no. 1, Jan. 1988, pp. 126–137.

13. Noriyoshi Itazaki and Kozo Kinoshita, "Test pattern generation for circuits with tri-state modules by Z-algorithm," *IEEE Transactions on CAD,* vol. 8, no. 12, Dec. 1989, pp. 1327–1333.

14. John D. Calhoun and Franc Brglez, "A framework and method for hierarchical test generation," *IEEE Transactions on CAD,* vol. 11, no. 1, Jan. 1992, pp. 45–67.

15. Andre Ivanov and Vinod K. Agarwal, "Dynamic testability measures for ATPG," *IEEE Transactions on CAD,* vol. 7, no. 5, May 1988, pp. 598–608.

16. Lawrence H. Goldstein and Evelyn L. Thigpen, "SCOAP: Sandia controllability/observability analysis program," *Proceedings of the 17th Design Automation Conference,* Jun. 1980, pp. 190–196.

17. F. Brglez, P. Pownall, and R. Hum, "Application of testability analysis: from ATPG to critical delay path tracing," *Proceedings 1984 Test Conference,* Oct. 1984, pp. 705–712.

18. A. Lioy and M. Mezzalama, "On parameters affecting ATPG performance," *Proc. CompEuro 1987,* May 1987, pp. 394–397.

19. Srimat T. Chakradhar, Michael L. Bushnell, and Vishwani D. Agrawal, "Toward massively parallel automatic test generation," *IEEE Transactions on CAD,* vol. 9, no. 9, Sept. 1990, pp. 981–994.

20. Michael H. Schulz et al., *op. cit.*

21. E. G. Ulrich and T. Baker, "The concurrent simulation of nearly identical digital networks," *IEEE/ACM Proceedings of the 10th Design Automation Conference,* Jun. 1973, pp. 145–150.

22. Sunil K. Jain and Vishwani D. Agrawal, "STAFAN: an alternative to fault simulation," *Proceedings of the ACM IEEE 21st Design Automation Conference,* June 1984, Albuquerque, N.M., pp. 18–23.

23. E. J. McCluskey and S. Bozorgui-Nesbat, "Design for autonomous test," *IEEE Transactions on Computers,* vol. C-30, no. 11, Nov. 1981, pp. 866–875.

24. E. B. Eichelberger and T. W. Williams, "A logic design structure for LSI testing," *IEEE/ACM Proceedings of the 14th Design Automation Conference,* June 1977, New Orleans, Louisiana, pp. 462–468.

25. E. B. Eichelberger and T. W. Williams, "A logic design structure for LSI testing," *Journal of Design Automation and Fault Tolerant Computing,* vol. 2, no. 2, May 1978, pp. 165–178.

26. S. DasGupta, E. B. Eichelberger, and T. W. Williams, "LSI chip design for testability," *Digest of Technical Papers, IEEE International Solid State Circuits Conference,* San Francisco, Feb. 1978, pp. 216–217.

27. T. W. Williams, *op. cit.*

28. Rajesh Gupta, Rajiv Gupta, and Melvin A. Breuer, "An efficient implementation of the BALLAST partial scan architecture," *IFIP Proceedings of the International VLSI '89 Conference,* Aug. 1990, Munich, pp. 133–142.

29. *Ibid.*

30. H. Ando, "Testing VLSI with random access scan," *IEEE/ACM Digest of Papers COMPCON 80,* Feb. 1980, pp. 50–52.

31. Susheel J. Chandra, Tom Ferry, Tushar Gheewala, and Kerry Pierce, "ATPG based on a novel grid-addressable latch element," *IEEE/ACM Proceedings of the 28th IEEE Design Automation Conference,* June 1991, San Francisco, Calif., pp. 282–286.

32. R. A. Frowerk, "Signature analysis—a new digital field service method," *Hewlett Packard Journal,* May 1977, pp. 2–8.

33. H. J. Nadig, "Signature analysis—concepts, examples and guidelines," *Hewlett Packard Journal,* May 1977, pp. 15–21.

34. S. W. Golumb, *Shift Register Sequences,* Revised Edition, Laguna Hills, Calif.: Aegean Park Press, 1982.

35. Laung-Terng Wang and Edward J. McCluskey, "Complete feedback shift register design for built-in self test," *Proceedings of 1986 IEEE International Conference on Computer-Aided Design (ICCAD-86),* Nov. 1986, Santa Clara, Calif., pp. 56–59.

36. B. Koenemann, J. Mucha, and G. Zwiehoff, "Built-in logic block observation techniques," *Digest 1979 IEEE Test Conference,* 79CH1509-9C, Oct. 1979, pp. 37–41.

37. John Fox, Giuseppe Surace, and Paul A. Thomas, "A self-testing 2-μm CMOS chip set for FFT applications," *IEEE JSSC,* vol. SC-22, no. 1, Feb. 1987, pp. 15–19.

38. Ravindra Nair, Staish M. Satte, and Jacob A. Abraham, "Efficient algorithms for testing semiconductor random-access memories," *IEEE Transactions on Computers,* vol. C-27, no. 6, June 1978, pp. 572–576.

39. Rob Dekker, Frans Beenker, and Loek Thijssen, "A realistic fault model and test algorithms for static random access memories," *IEEE Transactions on CAD,* vol. 9, no. 6, June 1990, pp. 567–572.

40. Takashi Oshawa, Tohru Furuyama, Yohji Watanabe, Hiroto Tanaka, Natsuki Kushiyama, Kenji Tsuchida, Yohsei Nagahama, Satoshi Yamano, Takeshi Tanaka, Satoshi Shinozaki, and Kenji Natori, "A 60ns 4-Mbit CMOS DRAM

with built-in self-test function," *IEEE JSSC,* vol. SC-22, no. 5, Oct. 1987, pp. 663–668.

41. W. H. Kautz, "Testing for faults in cellular logic arrays," *Proceedings of the 8th Annual Symposium on Switching and Automation Theory,* 1967, pp. 161–174.

42. Thirumalai Sridar and John P. Hayes, "Design of easily testable bit-sliced systems," *IEEE Transactions on Computers,* vol. C-30, no. 11, Nov. 1981, pp. 842–854.

43. John M. Acken, "Testing for bridging faults (shorts) in CMOS circuits," *Proceedings of the 20th IEEE/ACM Design Automation Conference,* June 1983, Miami Beach, Fla., pp. 717–718.

44. Kuen-Jong Lee and Melvin A. Breuer, "Design and test rules for CMOS circuits to facilitate IDDQ testing of bridging faults," *IEEE Transactions on CAD,* vol. 11, no. 5, May 1992, pp. 659–670.

45. Patrick Bosshart and Thirumalai Sridhar, "Test methodology for a 32-bit processor chip," *IEEE Digest of Technical Papers,* ICCAD-86, Nov. 1986, pp. 12–14.

46. IEEE Standard 1149.1: "IEEE standard test access port and boundary-scan architecture," New York: IEEE Standards Board.

47. Bulent I. Dervisoglu, "Application of scan hardware and software for debug and diagnostics in a workstation environment," *IEEE Transactions on CAD,* vol. 9, no. 6, June 1990, pp. 612–620.

48. J. Galiay, Y. Crouzet, and M. Vergniault, "Physical versus logical fault models MOS LSI circuits: impact on their testability," *IEEE Transactions on Computers,* vol. C-29, no. 6, pp. 527–531.

49. Wojciech Malay, "Realistic fault modeling for VLSI testing," *IEEE/ACM Proceedings of the 24th IEEE Design Automation Conference,* Miami Beach, Fla., 1987, pp. 173–180.

50. Siegmar Koeppe, "Optimal layout to avoid CMOS stuck-open faults," *IEEE/ACM Proceedings of the 24th Design Automation Conference,* Miami Beach, Florida, 1987, pp. 829–835.

51. Jose Joao H. T. de Sousa, Fernando M. Goncalves, and J. Paulo Teixeira, "Physical design of testable CMOS digital integrated circuits," *IEEE JSSC,* vol. 26, no. 7, July 1991, pp. 1064–1072.

52. Marc E. Levitt and Jacob A. Abraham, "Physical design of testable VLSI: techniques and experiments," *IEEE JSSC,* vol. 25, no. 2, April 1990, pp. 474–481.

SUBSYSTEM DESIGN

8.1 Introduction

Most digital functions can be divided into the following categories:

- datapath operators.
- memory elements.
- control structures.
- I/O cells.

CMOS system design consists of partitioning the system to be designed into components that may be categorized into the above groups. Once those groupings have been determined, CMOS subsystems that implement those functions are designed. Many options exist that trade speed, density, programmability, ease of design, and many other variables. In this chapter we present a number of subsystems built with the circuits developed in Chapter 5. These subsystems may be used to build systems (chips, chip sets, or boards) of considerable complexity.

8.2 Datapath Operators

Datapath operators form an important subclass of VLSI circuit design that benefit from the structured design principles of hierarchy, regularity, modular-

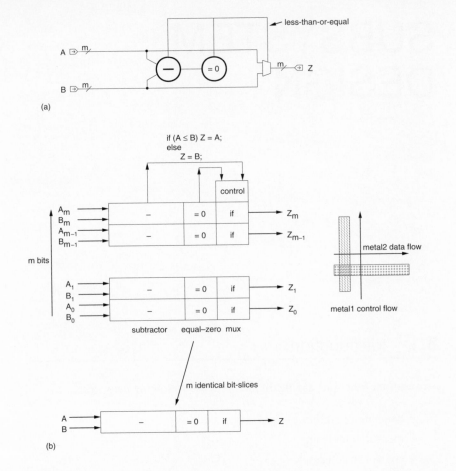

FIGURE 8.1 Datapath Example

ity, and locality. This arises because n-bit data is generally processed, which naturally leads to the ability to use n identical circuits to implement the function. In addition, data operations may generally be sequenced in time or space, which leads to the notion of physically placing linked data operators adjacent to each other. Generally, data may be arranged to flow in one direction, while any control signals are introduced in an orthogonal direction to the dataflow. This mirrors the physical reality of a CMOS chip, which usually has at least two good routing layers (i.e., metal1 and metal2, or metal2 and metal3).

Consider the magnitude comparator shown in Fig. 8.1(a). This may be implemented by the layout shown in Fig. 8.1(b), where data operators are arranged horizontally and data bits are arranged vertically. Data is relayed from operator to operator by horizontal wires (say, in metal2), while control information is routed vertically (say, in metal1). Datapaths allow optimization of the area of the layout by incorporating the regular routing strategy into the operator cell design. Usually, the data routing may be passed over the active circuitry, while the control signals are passed over or through the cells. Little area more than the basic area that the transistors take to implement a function is consumed. This efficiency is hard to achieve in random logic. The VLSI designer can exploit the regularity of datapaths by having to design one "bit-slice" of the design, which is a horizontal slice through the structure, shown in Fig. 8.1(b).

The rest of this section is devoted to describing a variety of data-processing elements that can be cast as datapaths.

8.2.1 Addition/Subtraction

Addition forms the basis for many processing operations from counting to multiplication to filtering. As a result, adder circuits that add two binary numbers are of great interest to digital system designers. A wide variety of adder implementations are available to serve different speed/density requirements. The truth table for a binary full adder was introduced in Chapter 1 and is reproduced in Table 8.1, along with some functions that will be of use during the discussion of adders.

A and B are the adder inputs, C is the carry input, SUM is the sum output, and $CARRY$ is the carry output. The generate signal, G $(A.B)$, occurs when a carry output ($CARRY$) is internally generated within the adder. When the propagate signal, P $(A + B)$, is true, the carry-in signal (C) is passed to the carry output ($CARRY$) when C is true. (In some adders $A \oplus B$ is used as the P term because it may be reused to generate the sum term.)

8.2.1.1 Single-Bit Adders

Probably the simplest approach to designing an adder is to implement gates to yield the required majority logic functions. From the truth table these are:

$$SUM = ABC + A\overline{B}\,\overline{C} + \overline{A}\,\overline{B}C + \overline{A}B\overline{C}, \qquad \textbf{(8.1)}$$

TABLE 8.1   Adder Truth Table

C	A	B	A.B(G)	A + B(P)	A ⊕ B	SUM	CARRY
0	0	0	0	0	0	0	0
0	0	1	0	1	1	1	0
0	1	0	0	1	1	1	0
0	1	1	1	1	0	0	1
1	0	0	0	0	1	1	0
1	0	1	0	1	1	0	1
1	1	0	0	1	1	0	1
1	1	1	1	1	0	1	1

which may be factored as follows:

$$= C(AB + \bar{A}\,\bar{B}) + \bar{C}(A\bar{B} + \bar{A}B)$$

$$= A \oplus B \oplus C \tag{8.2}$$

$$CARRY = AB + AC + BC,$$
which may be factored as follows

$$= AB + C(A + B). \tag{8.3}$$

The gate schematic for the direct implementation of Eqs. (8.2) and (8.3) is shown in Fig. 8.2(a). This implementation uses a 3-input XOR gate. A transistor-level implementation is shown in Fig. 8.2(b). This uses a total of 32 transistors. An implementation that does not use XOR gates is shown in Fig. 8.3(a). This uses an alternative implementation that is achieved by realizing that the *CARRY* term may be reused in the *SUM* term as a common subexpression. In this implementation, shown in Fig. 8.3(b),

$$SUM = ABC + (A + B + C)\,\overline{CARRY}$$

$$= ABC + (A + B + C)\,\overline{(AB + C(A + B))}. \tag{8.4}$$

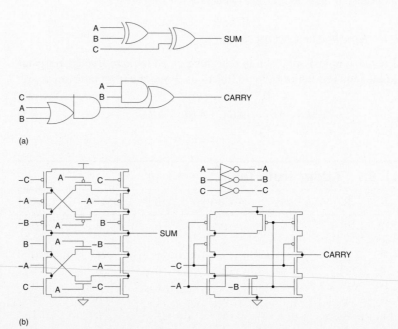

FIGURE 8.2 Single-bit adder schematic (3-input XOR)

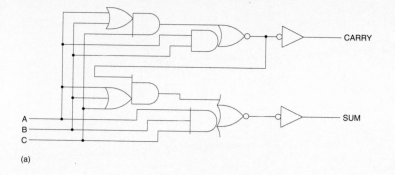

(a)

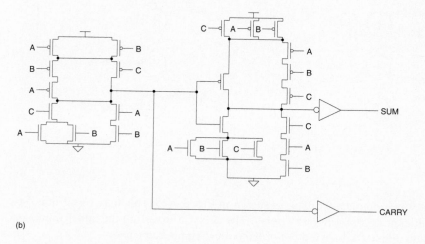

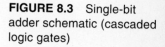

(b)

FIGURE 8.3 Single-bit adder schematic (cascaded logic gates)

The transistor schematic for this implementation is shown in Fig. 8.3(b). It uses 28 transistors.

8.2.1.2   Bit-Parallel Adder

An n-bit adder may be constructed by cascading n 1-bit adders, as shown in Fig. 8.4(a). This is called a Ripple Carry Adder. The inputs are n-bit A and B values. The *CARRY* signal of stage i is fed to the C signal of stage $i + 1$ and the *SUM* signal forms the n-bit output. The nth bit of the *SUM* indicates the sign of the result, while the nth *CARRY* signal indicates whether an overflow condition has occurred. Because the carry-output signal (*CARRY*) is used in the generation of *SUM* in the circuit shown in Fig. 8.3(a), *SUM* will be delayed with respect to *CARRY*. In the case of an n-bit parallel adder, the carry delay has to be minimized, because the delay associated with the adder is $T_n = nT_c$, where T_n is the total add time, n is the number of stages, and T_c is the delay of one carry stage. To optimize the carry delay, the inverter at the output of the carry gate can be omitted. In this case, every other stage oper-

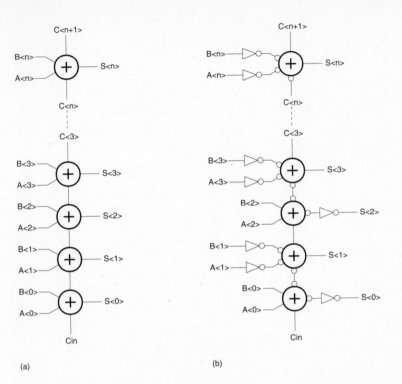

Figure 8.4   Parallel adder implementations

(a)   (b)

ates on complement data, as shown in Fig. 8.4(b). This may result in a significant decrease in carry delay. The delay in inverting the adder inputs or sum outputs is finessed out of the critical-ripple carry path.

An n-bit subtractor may be constructed by inverting one operand to an n-bit adder and adding 1 to the adder via the carry input, as shown in Fig. 8.5(a). An adder/subtractor may be constructed from XOR gates and adders, as shown in Fig. 8.5(b).

The transistor schematic for the adder in Fig. 8.3(a) is shown redrawn in Fig. 8.6(a). The propagate term $(A + B)$ and generate term $(A.B)$ can be clearly seen. To aid in a uniform layout, the p-chain is not the exact dual of the n-chain. It is left to the reader to verify the equivalence. Figure 8.6(b) shows how the transistors in the carry stage might be sized to optimize the delay through the carry stage. Sizing up the transistors in the carry gate while keeping the other transistors small decreases the effective load of these transistors and any parasitic routing capacitance. Using the styles of layout presented so far, two possible mask layouts for the combinational adder are depicted in Fig. 8.7 (also Plate 7). The choice of aspect ratio would depend very much on the environment. In a standard-cell environment, the layout in Fig. 8.7(a) might be appropriate where a single row of n- and p-transistors is used. The routing for the $A, B,$ and C inputs is shown inside the cell although it is quite possible it could be placed outside the cell because external routing

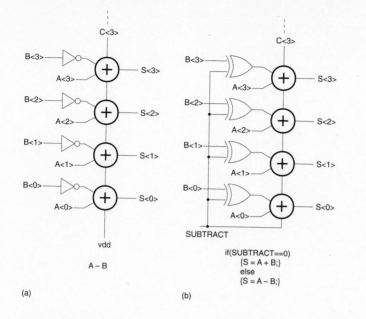

FIGURE 8.5   Arithmetic operators: (a) subtractor; (b) adder/subtractor

tracks have to be assigned to these signals anyway. Figure 8.7(b) shows a layout that might be appropriate for a datapath. Here the transistors are rotated and all of the wiring is completed in polysilicon and metal1. This allows metal2 bus lines to pass over the cell horizontally. In addition, the size of the transistors in the adder may be increased without impacting the bit-pitch (height) of the datapath. The following optimizations may be made to the combinational adder (Fig. 8.6):

1. Arrange the transistors switched by the carry in signal (C) close to the output. This will enable the input signals to settle the gate such that the C transistors are least influenced by body effect.

2. Make all transistors in the sum gate whose gate signals are connected to *CARRY* minimum size. This minimizes the capacitive load on this signal. Keep routing on this signal to a minimum and minimize the use of diffusion as a routing layer.

3. Sizing of series transistors can be determined by simulation. It may or may not pay to increase the size of the series n-transistors and p-transistors. For instance, it may not pay to increase the size of the series transistors connected to A and B in the carry gate in a ripple-carry adder, because these signals will have time to settle in the upper bits of the adder while the carry is rippling. It may be of advantage to increase the size of the C transistors in the carry gate to override the effects of stray capacitance. For a parallel adder, the *SUM*

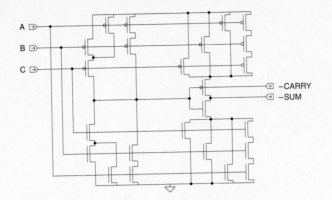

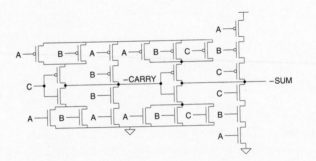

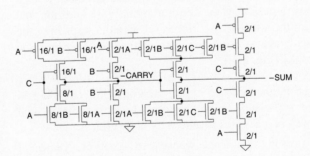

Figure 8.6 Optimized combinational adder schematic

gate transistors may be made minimum size, while for a serial adder the *CARRY* and *SUM* delays would have to be more balanced.

8.2.1.3   Bit-Serial Adders, Carry-save Addition, and Pipelining

Rather than construct a ripple carry adder, a serial adder, shown in Fig. 8.8, may be constructed. This uses a single adder and constructs the *SUM*

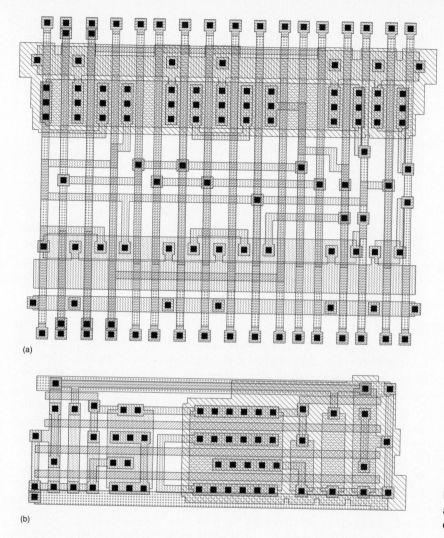

(a)

(b)

FIGURE 8.7 Combinational adder layouts: (a) standard cell; (b) datapath

sequentially. At time *t*, the *SUM* is calculated and the *CARRY* stored in a register. At time *t* + 1, the sum uses *CARRY[t]* to calculate a new *SUM*.

$$CARRY[t+1] = A[t+1].B[t+1] + C[t].(A[t+1] + B[t+1])$$

$$SUM[t+1] = CARRY[t+1].(A[t+1] + B[t+1] + C[t])$$
$$+ A[t+1].B[t+1].C[t] \tag{8.5}$$

The two inputs to the adder are stored in *n*-bit registers. The *SUM* output is stored in an *n*-bit result register. An illustrative add cycle is shown in Fig. 8.8. Addition is commenced by clearing the carry register. Then the

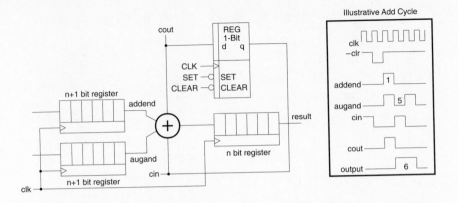

Figure 8.8   Bit-serial adder implementation

operands are serially applied to the inputs of the adder, the least significant bit first. The example shows 1 added to 5 to form 6 at the output. It takes n clock cycles to complete an n-bit add. In a serial adder, equal *SUM* and *CARRY* delays are advantageous, because these delays determine the fastest clock frequency at which the adder can operate.

Bit-serial architectures have been used successfully for a variety of signal-processing applications, especially with technologies in the 2–5μ range.[1] Reasons for using bit-serial architectures include reduced signal routing (1-bit signals instead of n-bit signals), reduced module sizes, and higher-speed operation (one adder and a register rather than an n-bit adder). Multilevel-metal CMOS technologies have largely solved the signal routing problems while more advanced processes have drastically reduced the size and increased the speed of adders and registers to the point where the design problems lie elsewhere (for instance, in correctly completing a large design on schedule). However, the general principle of breaking an n-bit addition into smaller additions may be applicable to current design situations. Apart from bit-serial adders, nibble (4-bit) and byte (8-bit) adders are frequently used. The reason for using reduced-size adders might range from size to power dissipation considerations.

Adders, such as the 1-bit serial adder shown in Fig. 8.8, where both the carry and sum are registered on each cycle, are often called *carry-save adders* (CSAs).[2] This can be extended to an n-bit adder by registering n carries and n sums. The carries are left shifted, with a new carry input introduced to the D of the LSB carry register and the carry output available at the MSB carry register Q. An n-bit CSA would have $2n$ registers. Figure 8.9 illustrates a circuit which uses two 4-bit CSAs, which is representative of a structure that might be used in a digital filter. The inputs $SIN<3:0>$ and $CIN<2:0>$ are added to constant $A<3:0>$ in the left column (first rank) of CSA bits, and then $B<3:0>$ is added in the next column (second rank). Each bit of a CSA has the binary output encoded in the sum and carry of each bit. The carry output of each CSA stage is left shifted to feed the carry input of

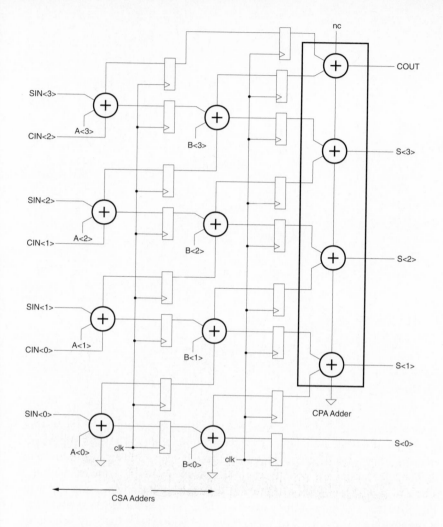

FIGURE 8.9 Carry-save adder (CSA) example

the next rank. The binary output may be extracted by feeding the sum and carry of each bit of the CSA to the inputs of a *carry-propagate adder* (CPA), as illustrated by the ripple-carry adder in Fig. 8.9 (or Fig. 8.4a). Usually a different, faster, architecture is used for the final carry-propagate adder. In applications such as filtering where many additions have to occur and many *n*-bit adders have to be used, use of cascaded carry-save adders reduces the critical path to the sum of the clock to *Q* time of the register, the adder delay, and the setup time into the register. Current CMOS processes allow operation in excess of 200 MHz, and operation above 1 GHz is very close. Figure 8.10 shows a CPA adder structure that can be used for the CPA shown in Fig. 8.9. Registers are used at the input and output of the CPA to ensure that the inputs arrive at the same time as the carry and that the outputs all appear

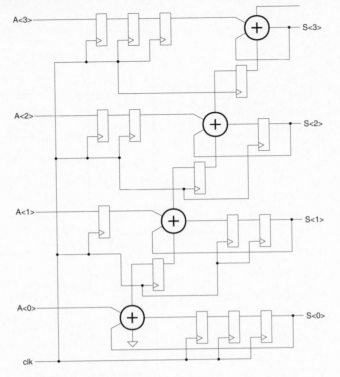

Figure 8.10 Pipelined carry-propagate (CPA) example

at the correct time. Figures 8.9 and 8.10 illustrate two methods of increasing the speed of a basic n-bit adder by the use of *pipelining*. The cost of pipelining is *latency;* that is, the time it takes from when operands are introduced to the data processing element to when outputs are available from the module. The adder in Fig. 8.9 has a latency of two clock cycles, while the adder in Fig. 8.10 has a latency of three cycles. A filter built with k CSAs would have a latency of $k + 3$ cycles. The *throughput* is k adds/cycle. Latency is usually not important in DSP applications, such as filtering, but is important in applications such as microprocessors where for a variety of reasons (including control) an add operation (32 bits or more) has to occur in a single clock cycle. On the other hand, throughput is all important to DSP applications. In the rest of this section some alternative techniques for improving adder speed will be introduced. In so doing, we will discover some classical examples of trading space for time. In other words, by increasing the size of a data element, we can often improve the speed.

8.2.1.4 *Transmission-Gate Adder*

A rather different implementation of an adder uses a novel exclusive-or (XOR) gate. The schematic for this XOR gate is shown in Fig. 8.11. As a

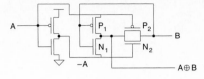

FIGURE 8.11 Transmission-gate XOR (tiny XOR)

point to note, switch-level simulators have problems with this gate. The operation of the gate is explained as follows:

1. When signal A is high, $-A$ is low. Transistor pair P_1 and N_1 thus act as an inverter, with $-B$ appearing at the output. The transmission gate formed by transistor pair P_2 and N_2 is open.

2. When signal A is low, $-A$ is high. The transmission gate (P_2, N_2) is now closed, passing B to the output. The inverter (P_1, N_1) is partially disabled (level reduced B passed to output by P_1, N_1).

Thus this transistor configuration forms a 6-transistor XOR gate. By reversing the connections of A and $-A$, an exclusive-nor (XNOR) gate is constructed.

By using four transmission gates, four inverters, and two XOR gates, an adder may be constructed according to Fig. 8.12.[3] $A \oplus B$ and the complement are formed using the TG XOR gate shown in Fig. 8.11. The *SUM* ($A \oplus B \oplus C$) is formed by a multiplexer controlled by $A \oplus B$ (and complement). Examining the adder truth-table reveals that *CARRY* = C when $A \oplus B$ is true. When $A \oplus B$ is false, *CARRY* = A (or B). This adder has 24 transistors, the same as the combinational adder, but has the advantage of having equal *SUM* and *CARRY* delay times. In addition, the *SUM* and *CARRY* signals are non-inverted. The number of transistors may be reduced if speed is not the ulti-

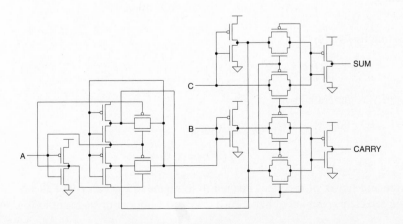

FIGURE 8.12 Transmission-gate adder

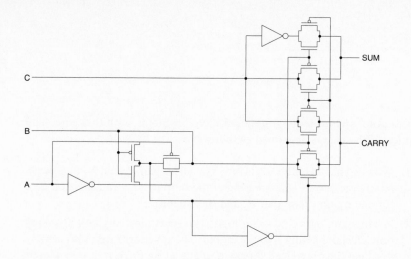

SUM

C

B

CARRY

A

Figure 8.13 Optimized-area TG adder

mate goal. Two transistors may be eliminated by using an inverter on the output of the XOR gate. In addition with some optimization, the output buffers may be eliminated, as shown in Fig. 8.13.[4]

8.2.1.5 Carry-Lookahead Adders

The linear growth of adder carry-delay with the size of the input word for an n-bit adder may be improved by calculating the carries to each stage in parallel. The carry of the ith stage, C_i, may be expressed as

$$C_i = G_i + P_i \cdot C_{i-1}, \qquad (8.6)$$

where

$$G_i = A_i \cdot B_i \qquad generate\ signal$$
$$P_i = A_i + B_i \qquad propagate\ signal.$$

Expanding this yields

$$C_i = G_i + P_i G_{i-1} + P_i P_{i-1} G_{i-2} + \ldots + P_i \ldots P_1 C_0. \qquad (8.7)$$

The sum S_i is generated by

$$S_i = C_{i-1} \oplus A_i \oplus B_i$$

$$\text{or } C_{i-1} \oplus P_i \text{ (if } P_i = A_i \oplus B_i). \qquad (8.8)$$

The size and fan-in of the gates needed to implement this carry-lookahead scheme can clearly get out of hand. As a result, the number of stages of lookahead is usually limited to about four. For four stages of lookahead, the appropriate terms are

$$C_0 = G_0 + P_0 CI$$

$$C_1 = G_1 + P_1 G_0 + P_1 P_0 CI$$

$$C_2 = G_2 + P_2 G_1 + P_2 P_1 G_0 + P_2 P_1 P_0 CI$$

$$C_3 = G_3 + P_3 G_2 + P_3 P_2 G_1 + P_3 P_2 P_1 G_0 + P_3 P_2 P_1 P_0 CI. \qquad \textbf{(8.9)}$$

Figure 8.14 shows a generic carry-lookahead adder. The PG generation and SUM generation circuits surround a carry-generate block. A possible implementation of the carry gate for this kind of carry-lookahead adder for 4 bits is shown in Fig. 8.15. Note that the gates have been partitioned to keep the number of inputs less than or equal to four. This is typical of the type of carry lookahead that would be used in a gate-array or standard-cell design. The circuit and layout are quite irregular. Taking the term of C_3, we note that it may be expressed as

$$C_3 = G_3 + P_3 \cdot (G_2 + P_2 \cdot (G_1 + P_1 \cdot (G_0 + P_0 \cdot CI))). \qquad \textbf{(8.10)}$$

This function may be implemented as a domino CMOS (nMOS) gate, as shown in Fig. 8.16(d). Carry $C_0 - C_2$ are generated similarly. Note that the worst-case delay path in this circuit has six n-transistors in series. A high-speed static version of the carry-lookahead gate for C_3 is shown in Fig. 8.17.[5]

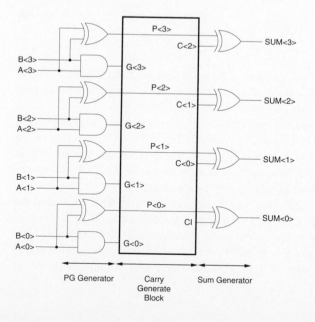

PG Generator   Carry Generate Block   Sum Generator

FIGURE 8.14   Generic carry-lookahead adder (CLA)

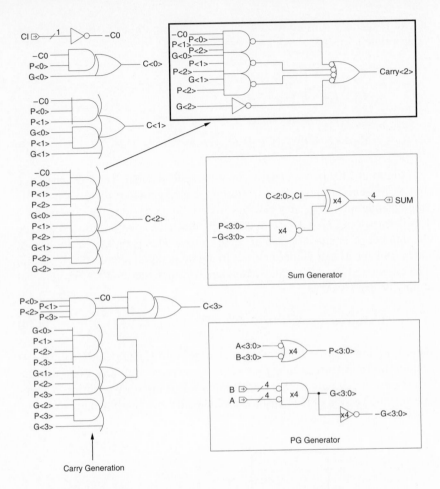

Figure 8.15 4-bit CLA

This uses pseudo-nMOS gates to achieve high-speed static operation. An adder using this stage may be constructed by using a 4-bit adder block with local ripple carry and this gate as the block carry generator.

The Manchester adder stage improves on the carry-lookahead implementation by using a single C_3 gate, as shown in Fig. 8.16(d). A selection of the elemental carry stages is shown in Fig. 8.18. The first, shown in Fig. 8.18(a), is a dynamic stage. Operation proceeds as follows. When *CLK* is low, the output node is precharged by the p pull-up transistor. When *CLK* goes high, the n pull-down transistor turns on. If carry generate ($A.B$) is true, then the output node discharges. If carry propagate ($A + B$) is true, then a previous carry may be coupled to the output node, conditionally discharging it. Note that in this circuit *CARRY* is actually propagated. A static stage is shown in Fig. 8.18(b). This requires P to be generated as $A \oplus B$. A multiplexer-based

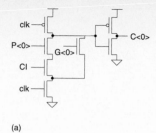

(a)

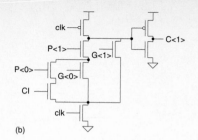

(b)

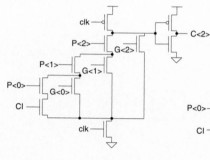

(c)

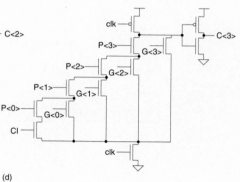

(d)

FIGURE 8.16 Dynamic carry gates

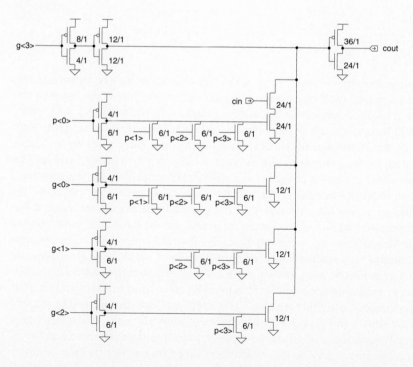

FIGURE 8.17 High-speed carry lookahead logic

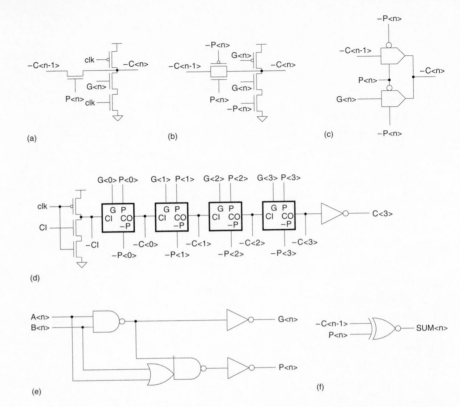

Figure 8.18 Manchester-adder circuits: (a) dynamic stage; (b) static stage; (c) MUX stage; (d) 4-bit section; (e) PG logic; (f) SUM logic

implementation is shown in Fig. 8.18(c). A 4-bit adder may be constructed by cascading four such stages and adding the circuitry to supply the required signals. This is commonly called a Manchester carry adder. Thus a 4-bit adder would be constructed as shown in Fig. 8.18(d). There is some similarity with the domino carry circuit. However, the intermediate carry gates are no longer needed, because the carry values are available in a distributed fashion. The 4-bit adder is chosen to reduce the number of series-propagate transistors, which improves the speed. Note that if all propagate signals are true, and CI is high, six series n-transistors pull the output node low in the case of the dynamic gate while five transistors are in series in the static gate. In addition to four Manchester stages, the adder requires four PG generator blocks, one representative implementation being shown in Fig. 8.18(e). Four SUM generate blocks (an XNOR gate), shown in Fig. 8.18(f), complete the adder. This worst-case propagation time can be improved by bypassing the four stages if all carry-propagate signals are true.[6] The additional circuitry needed to achieve this is shown in Fig. 8.19(a). It consists of an AND gate, which turns on a carry-bypass signal if all carry propagates are true. The optimum number of cascaded stages may be calculated for a given technology by simulation. A final implementation of a 4-bit Manchester adder is shown in Fig. 8.19(b).

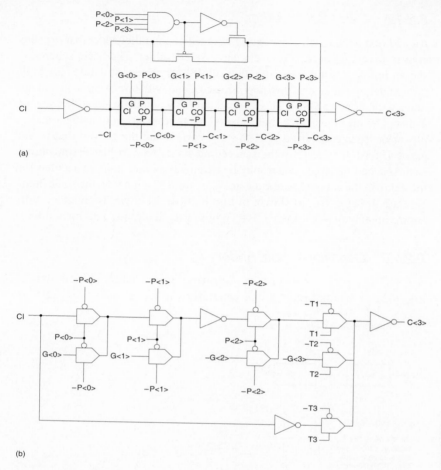

(a)

(b)

FIGURE 8.19 Manchester adder with carry bypass: (a) simple; (b) conflict free

This implementation[7] uses a "conflict-free" bypass circuit, which improves the speed by using a 3-input multiplexer that prevents conflicts at the wired OR node in the adder, shown in Fig. 8.19(a). The control signals T_1, T_2, and T_3 are respectively generated by

$$T_1 = -(P_0 P_1 P_2).P_3$$
$$T_2 = -P_3$$
$$T_3 = P_0 P_1 P_2 P_3.$$

Note that in this version the inverter present on the c_{in} signal has been moved to the center of the carry chain to improve speed (there are now a maximum of two transmission gates in series with an inverter). Very wide, fast adders may be constructed by extending the carry bypass shown in Fig. 8.19(b).[8]

8.2.1.6   Carry-Select Adder

An additional approach to increase the speed of a parallel adder that expends area in favor of speed is to use a carry-select adder. The basic scheme is shown in Fig. 8.20(a).[9] Usually, two ripple-carry-adder structures are built (although any adder structure may be used), one with a zero carry-in and the other with a one carry-in. This is repeated for a certain sized adder, say, of 4-bits. The previous carry then selects the appropriate sum using a multiplexer or tristate adder gates. The stage carries and the previous carry are gated to form the carry for the succeeding stage. As a further optimization, each succeeding ripple adder may be extended by one stage to account for the delay in the carry-lookahead gate. Thus for a 32-bit adder, the stage numbers are 4-4-5-6-7-6, as shown in Fig. 8.20(b). This yields an adder with approximately $(4 + 1 + 1 + 1 + 1 + 1)$, or 9, gate delays for a 32-bit addition.

8.2.1.7   Conditional-Sum Adder

A CMOS implementation of a conditional-sum adder[10] is shown in Fig. 8.21.[11] A conditional block generates C_0, C_1, S_0, and S_1 signals, as

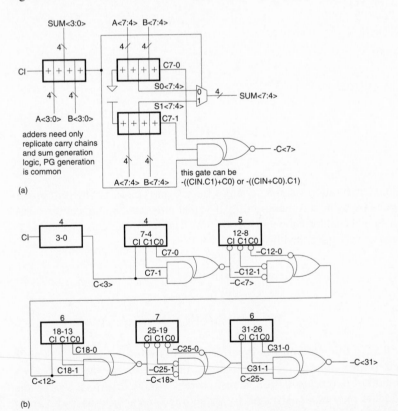

Figure 8.20 Carry-select adder: (a) basic architecture; (b) 32-bit carry-select adder example

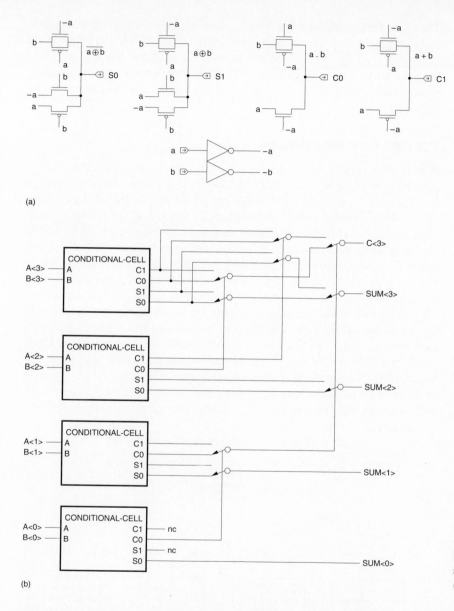

(a)

(b)

FIGURE 8.21 Conditional-sum adder: (a) basic circuits; (b) 4-bit adder example

shown in Fig. 8.21(a). Here pass logic has been used to generate

$$S_0 = -(A \oplus B)$$
$$S_1 = A \oplus B$$
$$C_0 = A \cdot B$$
$$C_1 = A + B.$$

The C_0 and C_1 signals are fed to successive stages, selecting between the S_0 and S_1 signals using transmission gate multiplexers. *SUM*s and *CARRY*s are generated in a tree-like fashion as shown in Fig. 8.21(b). For a 32-bit adder there are six transmission gates in series. If the stray capacitance in the series-transmission gates can be minimized, it is claimed that this adder can be quite fast.

8.2.1.8   Very Wide Adders

Adders with very large word sizes (>32 bits) can be constructed hierarchically by combining smaller "block" adders typically with a word width of 16 bits. Figure 8.22 shows a 64-bit adder composed of four 16-bit blocks. Each 16-bit block outputs a block P and block G signal that are fed to a block-carry generator. This module in turn feeds the carry-in to each 16-bit block.

A single bit of a typical block adder[12] is shown logically in Fig. 8.23(a), while a transmission-gate and inverter implementation is shown in Fig. 8.23(b). This is divided into three sections, which generate the local propagate signal (P), the block propagate (P_{out}), and the block generate (G_{out}) signals and the sum signal (*SUM*). These single bits are cascaded to form a 16-bit adder block. The block propagate and block-generate signals pass through series connections of transmission gates and inverters. These can be accelerated by using bypass techniques similar to that shown in Fig. 8.19(b). Figure 8.24(a) shows a representation of the block-generate chain for a 16-bit adder with bypassing. For example the final transmission gates are controlled by signals S_1, S_2, and S_3. These are generated as follows:

$$S_1 = -T_2.P_{15} \qquad \text{(passes } G\text{<15> to output)}$$

$$S_2 = -T_1.T_2.P_{15} \quad \text{(passes } G\text{<10> to output)}$$

$$S_3 = T_1.T_2.P_{15} \qquad \text{(passes } G\text{<5> to output),}$$

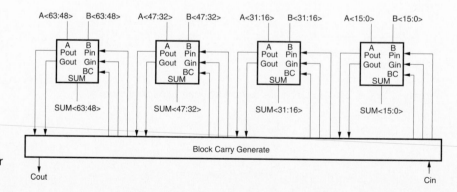

Figure 8.22   A 64-bit adder block diagram

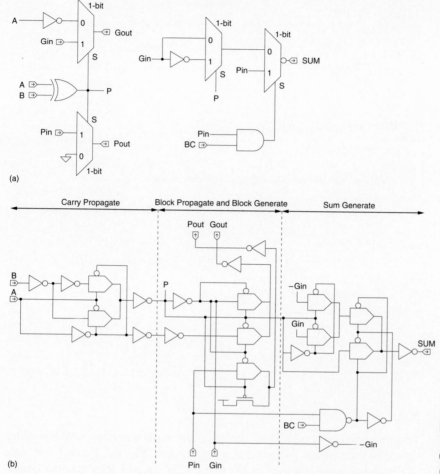

(a)

(b)

FIGURE 8.23 The cells in a 16-bit adder block used in the 64-bit adder: (a) gate diagram; (b) circuit diagram

where

$$T_1 = P_6.P_7.P_8.P_9.P_{10}$$
$$T_2 = P_{11}.P_{12}.P_{13}.P_{14}.$$

The block-carry generator for the 112-bit adder is shown in Fig. 8.24(b). The block generates and propagates from seven 16-bit adders are combined into seven carry-bypass multiplexers. These in turn are bypassed by the transmission gates, reducing the maximum number of series transmission gates from seven to four. The *−BC* signals are the block carries that are fed to each 16-bit adder. The 112-bit adder that Figs. 8.23 and 8.24 are based on yielded a 8.5 *ns* 112-bit adder in a 0.8μ three-level-metal technology.

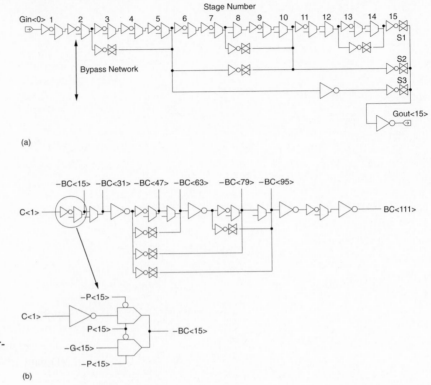

Figure 8.24 Bypass techniques used in a 112-bit adder: (a) 16-bit adder generate bypass logic; (b) 112-bit carry-bypass logic

8.2.1.9 Summary

With the number of adders presented, the natural question is "What adder should be used where?" In general, a ripple adder (Fig. 8.2, Fig. 8.3, Fig. 8.6, or Fig. 8.12) should be used as the first choice because they are small, simple, and relatively fast. The adder shown in Figs. 8.6 and 8.7 is one of the smallest that can be designed and is especially suited to pipelined adders. If a faster nonpipelined adder is required, standard-cell libraries frequently supply the adder shown in Fig. 8.15 or some other 4-bit adder with a look-ahead generator. The carry-select adder is a good choice for a faster n-bit adder because it can easily be assembled from ripple adders and multiplexers and the speed can be improved by adding adders and multiplexers. This ease of construction comes at the expense of area. The Manchester adders are good choices for custom-designed datapaths with word widths from 16 to 32 bits because they are regular, small and fast. The transmission-gate adder (Fig. 8.12 or 8.13) is of use where the *SUM* and *CARRY* propagation times must be similar (e.g., in multiplier arrays). For adders used in floating-point ALUs, the adder outlined in Figs. 8.23 and 8.24 may be suitable.

8.2.2   Parity Generators

A function related to binary addition is parity generation, that is, detecting whether the number of ones in an input word is odd or even. Frequently it is necessary to generate the parity of, say, a 16- or 32-bit word. The function is

$$PARITY = A_0 \oplus A_1 \oplus A_2 \oplus A_3 \ldots \oplus A_n. \tag{8.11}$$

Figure 8.25(a) shows a conventional implementation. A dynamic dual-rail logic version is shown in Fig. 8.25(b). A number of these may be cascaded to perform a 32-bit parity function.[13] A static 4-input XOR that could be used is shown in Fig. 8.25(c).[14] In a data path, Fig. 8.25(a) may be implemented as a linear column with a tree-routing channel connecting the XOR gates.

8.2.3   Comparators

A magnitude comparator is useful to compare the magnitude of two binary numbers. One can build a comparator from an adder and a complementer, as shown in Fig. 8.26. A zero detect (NOR gate) provides the $A = B$ signal while the final carry output provides the $B > A$ signal. Other signals—such as $A < B$ or $A \leq B$—may be generated by logical combinations of these signals. The generation of $B < A$ is shown in Fig. 8.26.

If equality comparison is required, then XNOR gates and an AND gate are all that is required, as shown in Fig. 8.27(a). Rather than a gate implementation, a pass-gate logic structure may be used, as shown in Fig. 8.27(b). Single-polarity transmission gates have been used here as might be appropriate in a low-power circuit, but of course complementary transmission gates may also be used. This structure does not draw any DC current but may be slow for long comparators. The final circuit shown in Fig. 8.27(c) is a merged XNOR/NOR gate using pseudo-nMOS. This gate draws DC current but is very small and very fast.

8.2.4   Zero/One Detectors

Detecting all ones or all zeros on wide words requires large fan-in AND or OR gates. One can build a tree of AND gates, as shown in Fig. 8.28(a). Here alternate NAND and NOR gates have been used. The delay to the output is proportional to $\log N$, where N is the bit width of the word. If the word being checked has a natural skew in the outputs (such as at the output of a ripple adder), the designer might consider mimicking the adder delay in the zero or one detect as shown in Fig. 8.28(b). Here the delay from the last changing output to the zero/one detect is a constant one gate delay. Similar to the comparator example in the last section, a small and fast ONE/ZERO detection

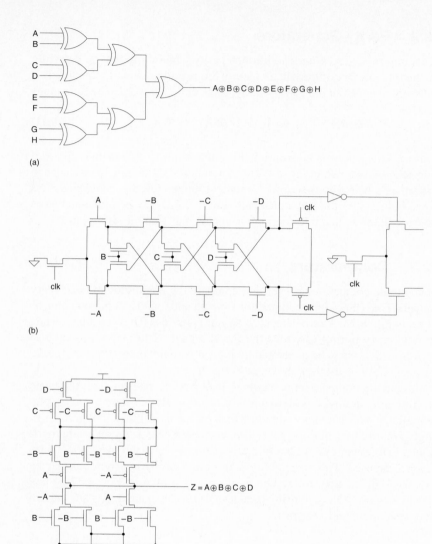

$A \oplus B \oplus C \oplus D \oplus E \oplus F \oplus G \oplus H$

(a)

(b)

Z = A ⊕ B ⊕ C ⊕ D

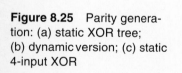

Figure 8.25 Parity generation: (a) static XOR tree;
(b) dynamic version; (c) static
4-input XOR

(c)

circuit for word widths of less that 32 bits is the pseudo-nMOS NOR gate. At
large word widths, self-loading may require the pseudo-nMOS gate to be
split into 8- or 16-bit chunks.

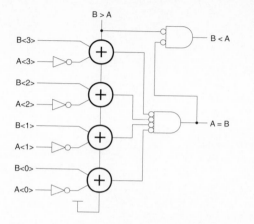

FIGURE 8.26 Comparator using an adder

8.2.5   Binary Counters

Binary counters are used to cycle through a sequence of binary numbers. An asynchronous counter has outputs that change at varying times with respect to the clock edge, whereas a synchronous counter has outputs that change at substantially the same time.

8.2.5.1   Asynchronous Counters

A "ripple-carry" binary counter is shown in Fig. 8.29. This is based on the toggle register introduced in Chapter 5. The T register (which is a single-counter stage) is reproduced in Fig. 8.29(a). This counter stage may be cascaded, as shown in Fig. 8.29(b). Note that the clocking of each stage is carried out by the previous counter stage, and thus the time it takes the last counter stage to settle can be quite large for a long counter chain. This counter is shown mainly for historical and reference purposes and should not be used as shown. Note that it has no reset signal, thus making it extremely difficult to test.

8.2.5.2   Synchronous Counters

A general synchronous up/down counter is shown in Fig. 8.30. It uses an adder and a D register per bit position. The speed that this counter can operate is determined by the ripple-carry time from the LSB to the MSB. This can be improved using any of the carry-lookahead techniques discussed in Section 8.2. If only an incrementer is required, the adder circuit degenerates into a synchronous counter stage, comprising an XOR gate, an AND gate,

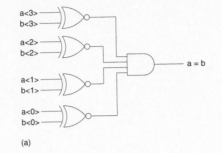

(a)

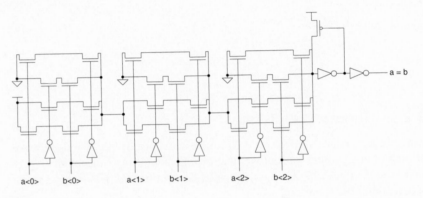

(b)

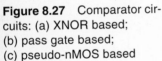

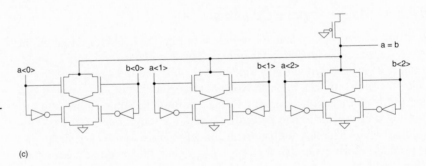

Figure 8.27 Comparator circuits: (a) XNOR based; (b) pass gate based; (c) pseudo-nMOS based

(c)

and a D register, as shown in Fig. 8.31. A multiplexer on the D input of the register allows a value to be loaded into the register for initialization. Remembering that an XOR can be implemented with a multiplexer yields the counter structure shown in Fig. 8.32(a). A reset register allows initialization, while the XOR function is provided by the multiplexer on the register D input. The multiplexer selects between the true and complement values of the register, based on the carry-input value. A more detailed version of the counter cell is shown in Fig. 8.32(b).

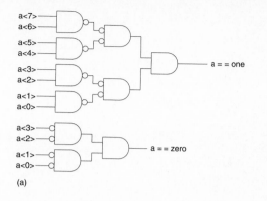

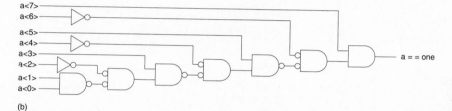

FIGURE 8.28 One- and zero-detect circuits: (a) tree; (b) "ripple"

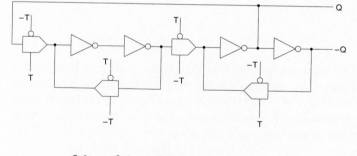

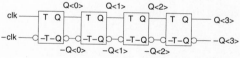

FIGURE 8.29 Asynchronous counter

8.2.6 Boolean Operations—ALUs

Boolean operations are most easily accomplished by using the multiplexer-based circuit shown in Fig. 5.35. This is shown in multiplexer format in Fig. 8.33. An Arithmetic Logic Unit (ALU) requires both arithmetic (add,

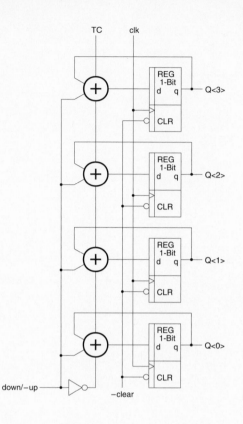

Figure 8.30 Synchronous up/down counter using adders and registers

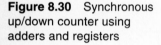

subtract) and Boolean operations. One may either multiplex between an adder and a Boolean unit or merge the Boolean unit into the adder as in the classic TTL 181 ALU.[15] A 1-bit CMOS implementation of the latter circuit that uses a Manchester carry stage is shown in Fig. 8.34. Signal *mode* is false for arithmetic operations and true for Boolean operations. Signals $S<3:0>$ control the operation type. For instance, in Boolean mode for $S<3>=0$, $S<2>=0$, $S<1>=0$, $S<1>=1$, b is passed to the output.

8.2.7 Multiplication

In many digital signal processing operations—such as correlations, convolution, filtering, and frequency analysis—one needs to perform multiplication. Multiplication algorithms will be used to illustrate methods of designing different cells so that they fit into a larger structure. In order to introduce these designs, simple serial and parallel multipliers will be introduced. The appropriate texts should be consulted for more definitive system architectures. The most basic form of multiplication consists of forming the product of two

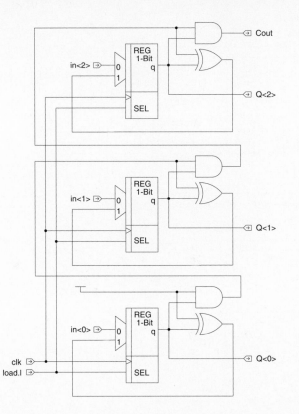

FIGURE 8.31 Incrementer

positive binary numbers. This may be accomplished through the traditional technique of successive additions and shifts in which each addition is conditional on one of the multiplier bits. For example, the multiplication of two positive binary integers, 12_{10} and 5_{10}, may proceed using the shift-and-add method in the following manner:

$$
\begin{array}{lrl}
\text{multiplicand:} & 1100 & : \ 12_{10} \\
\text{multiplier} & \underline{0101} & : \ 5_{10} \\
& 1100 & \\
& 0000 & \\
& 1100 & \\
& \underline{0000} & \\
& 0111100 & : \ 60_{10}
\end{array}
$$

Therefore, the multiplication process may be viewed to consist of the following two steps:

1. Evaluation of partial products.
2. Accumulation of the shifted partial products.

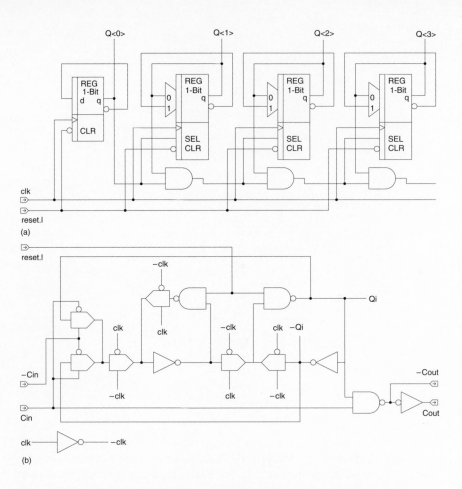

Figure 8.32 Compact synchronous counter

It should be noted that binary multiplication is equivalent to a logical AND operation. Thus evaluation of partial products consists of the logical ANDing of the multiplicand and the relevant multiplier bit. Each column of partial products must then be added and, if necessary, any carry values passed to the next column. There are a number of techniques that may be used to perform multiplication. In general, the choice is based on factors such as speed, throughput, numerical accuracy, and area. As a rule, multipliers may be classified by the format in which data words are accessed, namely:

- serial form.
- serial/parallel form.
- parallel form.

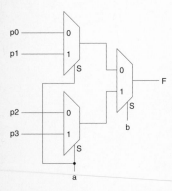

FIGURE 8.33 Boolean logic unit as MUXes

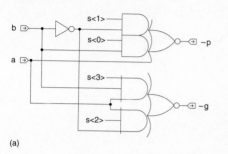

(a)

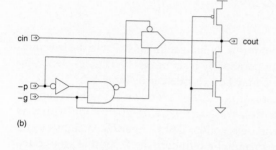

(b)

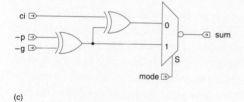

(c)

FIGURE 8.34 181 ALU

8.2.7.1 Array Multiplication

A parallel multiplier is based on the observation that partial products in the multiplication process may be independently computed in parallel. For example, consider the unsigned binary integers X and Y.

$$X = \sum_{i=0}^{m-1} X_i 2^i$$

$$Y = \sum_{j=0}^{n-1} Y_j 2^j$$

The product is found by

$$P = X \times Y = \sum_{i=0}^{m-1} X_i 2^i \cdot \sum_{j=0}^{n-1} Y_j 2^j$$

$$= \sum_{i=0}^{m-1} \sum_{j=0}^{n-1} (X_i Y_j) \, 2^{i+j}$$

$$= \sum_{k=0}^{m+n-1} P_k 2^k.$$

Thus P_k are the partial product terms called summands. There are mn summands, which are produced in parallel by a set of mn AND gates. For 4-bit numbers, the expression above may be expanded as in Table 8.2.

An $n \times n$ multiplier requires $n(n-2)$ full adders, n half adders, and n^2 AND gates. The worst-case delay associated with such a multiplier is $(2n+1)\tau_g$, where τ_g is the worst-case adder delay. Figure 8.35 shows a cell that may be used to construct a parallel multiplier. The X_i term is propagated diagonally from top right to bottom left, while the Y_j term is propagated horizontally. Incoming partial products enter at the top. Incoming *CARRY IN* values enter at the top right of the cell. The bit-wise AND is performed in the cell, and the *SUM* is passed to the next cell below. The *CARRY OUT* is passed to the bottom left of the cell. Figure 8.36 shows the multiplier array with the partial products enumerated. This arrangement may be drawn as a square array, as shown in Fig. 8.37, which is the most convenient for implementation. In this version the degeneracy of the first two rows of the multiplier are shown. The first row of the multiplier adders has been replaced with AND gates while the second row employs half-adders rather than full adders. This optimization might not be done if a completely regular multiplier were required (i.e., one

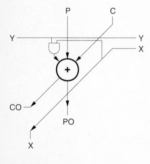

FIGURE 8.35 Array multiplier cell

TABLE 8.2   4-bit Multiplier Partial Products

				X3	X2	X1	X0	Multiplicand
				Y3	Y2	Y1	Y0	Multiplier
				X3Y0	X2Y0	X1Y0	X0Y0	
			X3Y1	X2Y1	X1Y1	X0Y1		
		X3Y2	X2Y2	X1Y2	X0Y2			
	X3Y3	X2Y3	X1Y3	X0Y3				
P7	P6	P5	P4	P3	P2	P1	P0	Product

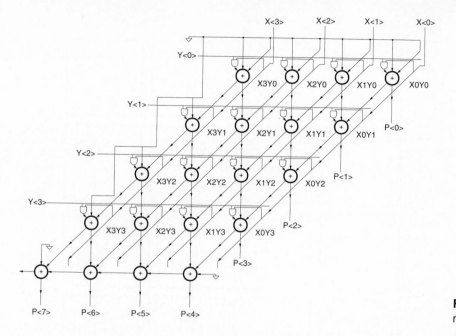

FIGURE 8.36   A 4 × 4 array multiplier

array cell). In this case the appropriate inputs to the first and second row would be connected to ground, as shown in Fig. 8.36.

The cell design for this multiplier is relatively straightforward, with the main attention paid to the adder. An adder with equal carry and sum propagation times is advantageous, because the worst-case multiply time depends on both paths.

8.2.7.2   Radix-n Multiplication

The structure shown in Figs. 8.36 and 8.37 computes the partial-products in a Radix-2 manner, that is by observing one bit of the multiplicand at a time. Higher radix multipliers may be designed to reduce the number of adders and hence the delay required to compute the partial sums. The best known method is called Booth recoding, which is a Radix-4 multiplication scheme.

A Booth-recoded multiplier examines three bits of the multiplicand at a time to determine whether to add zero, 1*, −1*, 2*, or −2* of that rank of the multiplicand. Table 8.3 shows the operation to be performed based on the current two bits of the multiplicand and the previous bit. In addition three control values are shown: *ZERO* zeroes the operand, *NEG* inverts the operand, and *TWO* multiplies the value by 2 (left shift).

Figure 8.38 shows a 16 × 16 Booth-recoded multiplier. Figure 8.38(a) shows the top level schematic and a possible floorplan. The schematic shows the multiplier divided into two parts—one the Booth array and the other a

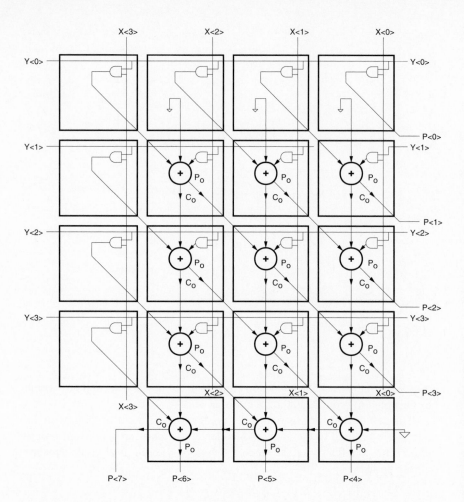

Figure 8.37 A square version of the 4 × 4 array multiplier

TABLE 8.3 Booth-recoding Values

X_{i-1}	X_i	X_{i+1}	OPERATION	NEG	ZERO	TWO
0	0	0	add 0	1	1	0
0	0	1	add 2	0	0	1
0	1	0	sub 1	1	0	0
0	1	1	add 1	0	0	0
1	0	0	sub 1	1	0	0
1	0	1	add 1	0	0	0
1	1	0	sub 2	1	0	1
1	1	1	add 0	0	1	0

carry propagate adder (CPA). The Booth array accepts two 16-bit inputs, *MIER*<15:0> (the multiplier) and *MCAND*<15:0> (the multiplicand) and feeds the CPA. The CPA also accepts a 32-bit input (*SUM-IN*<31:0>), which is used to perform multiple-accumulates. The floorplan divides the layout according to the schematic hierarchy using an array block and a CPA block. As the only 32-bit datapath is the final CPA, this structure is folded to obtain a set of datapaths that are roughly 16 bits high. Figure 8.38(b) shows the schematic and floorplan for the array section of the multiplier. It consists of 8 ranks of adders each 17 bits wide (or tall for the floorplan shown). The first rank (Booth-First-16) degenerates to the schematic shown in Fig. 8.38(c), while the remaining ranks are represented by the schematic in Fig. 8.38(e). Both ranks use a Booth decode cell which is shown in Fig. 8.38(d). This cell observes 3 bits of the multiplier (*MIER*) and produces the control signals *X1*, *X2* and *N*<1:0> which are used in the array adders (Figs. 8.38f) and

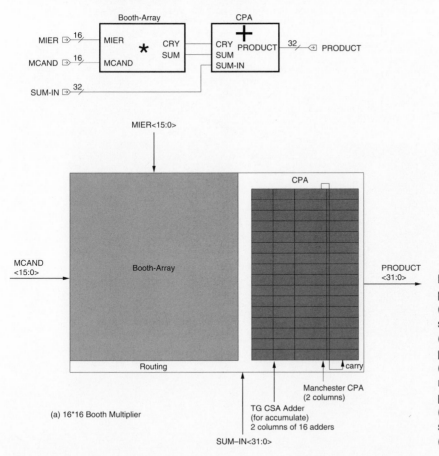

FIGURE 8.38 Radix-2 multiplier (Booth-recoded): (a) 16 × 16 multiplier top level schematic and floorplan; (b) array schematic and floorplan; (c) first rank schematic; (d) Booth decoder; (e) adder rank schematic, rank floorplan, and bit floorplan; (f) Booth gate; (g) array adder schematic and mask layout; (h) final adder

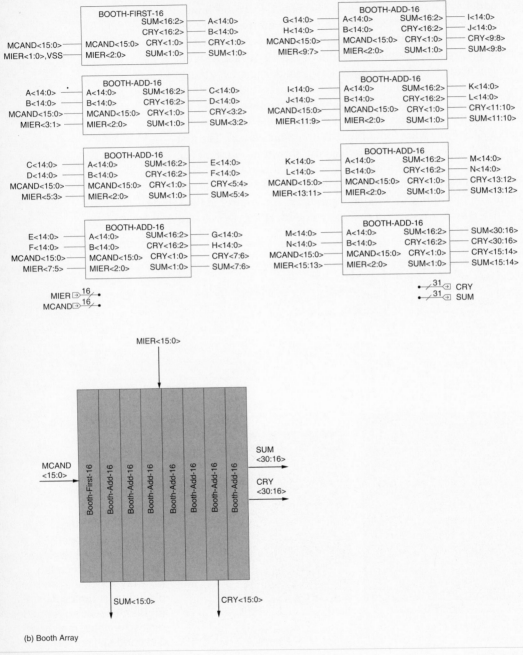

(b) Booth Array

Figure 8.38 *(continued)*

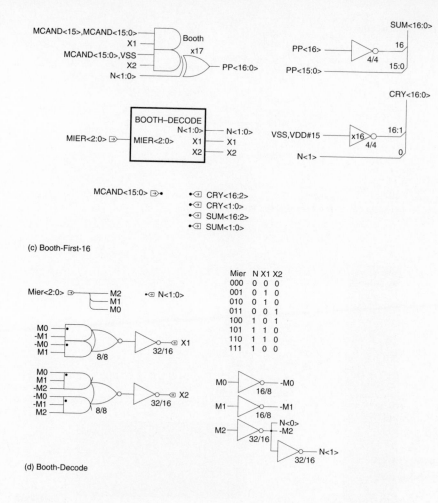

(c) Booth-First-16

Mier	N	X1	X2
000	0	0	0
001	0	1	0
010	0	1	0
011	0	0	1
100	1	0	1
101	1	1	0
110	1	1	0
111	1	0	0

(d) Booth-Decode

FIGURE 8.38   *(continued)*

8.38g). From Fig. 8.38(b) it may be seen that $MIER<1:0>$ and V_{SS} are fed to the first rank (Booth-First-16), $MIER<3:1>$ to the second rank and so on. Each rank "retires" two bits of the partial product sum (SUM) and carry (CRY) so by the last adder rank (lower right of schematic in Fig. 8.38b) 31 SUM, CRY pairs have been produced. These are used by the CPA to produce a 32-bit result. A possible floorplan of the Booth array is shown in Fig. 8.38(b). It consists of the 8 ranks of adder abutted horizontally. The circuit diagram for an adder rank appears in Fig. 8.38(e). It consists of a Booth decode (Fig. 8.38d), 17 Booth gates (Fig. 8.38f), and a 17-bit carry-save adder. The latter circuit consists of a 15-bit CSA for the LSBs and two inverters for the top 2 bits. The floorplan of the adder rank and adder bit is shown in Fig. 8.38(e). The adder rank consists of 15 Booth-Adder modules,

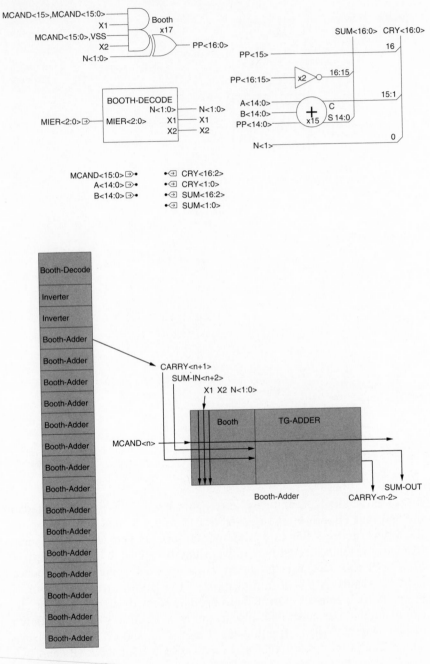

Figure 8.38 *(continued)*   (e) Booth-Add-16

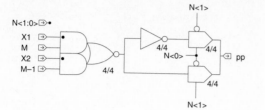

(f) Booth

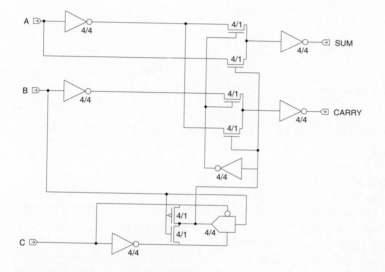

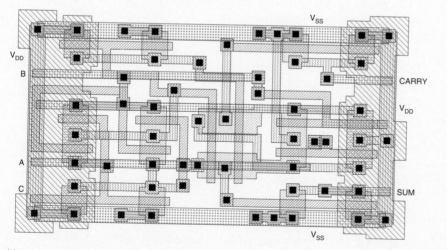

(g)

FIGURE 8.38 *(continued)*

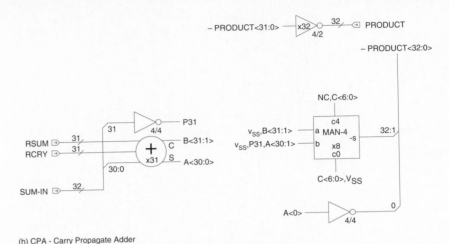

Figure 8.38 *(continued)*   (h) CPA - Carry Propagate Adder

2 inverters and the Booth Decoder vertically abutted. The Booth-Adder consists of a Booth gate and adder stage horizontally abutted. The *X1*, *X2* and *N<1:0>* lines feed vertically from the Booth-Decode logic to the Booth gate in each adder (or inverter). At each rank the *SUM* shifts right by two bit positions while the *CRY* shifts right by one bit position. The adder shown in Fig. 8.38(g) shows one possible adder implementation that has been optimized for size by using the transmission-gate adder with n-pass transistors. A possible mask layout is also shown. Figure 8.38(h) shows the final adder (CPA). It consists of a CSA to add in the *SUM-IN* signal and 32 stages of Manchester adder (8 MAN-4) to produce the final output. Any fast 32-bit CPA could be used here. Of course this multiplier may be made faster by including appropriate pipeline registers.

Figure 8.39 shows some alternative Booth related circuits. Figure 8.39(a) shows an alternative Booth-decoder stage along with a generic multiplier cell (Fig. 8.39b). It is implemented with a multiplexer, an XOR gate, an AND gate, and an adder. This circuit may be highly optimized at the circuit level. Figure 8.39(c) shows one particular implementation for the pre-adder gating that uses n-channel pass transistors.

Radix-8 multiplication carries Radix-4 multiplication one step further by requiring that +1,−1,+2,−2,+3,−3,+4,−4 and 0 times the multiplicand need to be calculated. The *3 is the hard term to calculate, requiring an adder. However, in some circumstances a Radix-8 multiplier might be appropriate.

8.2.7.3   Wallace Tree Multiplication

If Table 8.1, showing the truth table for an adder, is examined, it may be seen that an adder is in effect a "one's counter" that counts the number of 1's on the *A*, *B*, and *C* inputs and encodes them on the *SUM* and *CARRY* outputs. Table 8.4 below summarizes this.

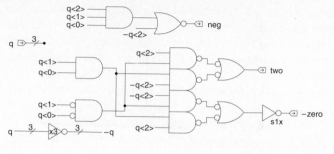

(a)

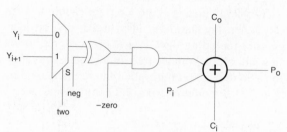

(b)

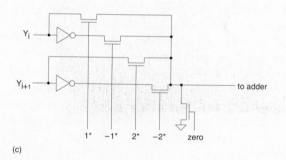

(c)

FIGURE 8.39   Booth-recoded multiplier cells

TABLE 8.4   An Adder as a 1's Counter

ABC	CS	Number of 1's
000	00	0
001	10	1
010	10	1
011	01	2
100	01	1
101	10	2
110	10	2
111	11	3

A 1-bit adder provides a 3:2 compression in the number of bits. The addition of partial products in a column of an array multiplier may be thought of as totaling up the number of 1's in that column, with any carry being passed to the next column to the left. Consider the 6×6 multiplication table shown in Table 8.5.

Considering the product P5, it may be seen that it requires the summation of six partial products and a possible column carry from the summation of P4. Figure 8.40 enumerates the adders required in a multiplier based on this style of addition. The adders have been arranged vertically into ranks that indicate the time at which the adder output becomes available. While this small example shows the general Wallace addition technique, it does not show the real speed advantage of a Wallace tree. In Fig. 8.40 there is an identifiable "array part" and a CPA part, which is at the top right. While this has been shown as a ripple-carry adder, any fast CPA can be used here. The delay through the array addition (not including the CPA) is proportional to log (base3/2) n, where n is the width of the Wallace tree. In a simple array multiplier it is proportional to n. So in a 32-bit multiplier where the maximum number of partial products is 32, the compressions (3:2 compressors) are

$$32 \rightarrow 22 \rightarrow 16 \rightarrow 12 \rightarrow 8 \rightarrow 6 \rightarrow 4 \rightarrow 3 \rightarrow 2.$$

Thus there are 9 adder delays in the array. In an array multiplier (Booth-recoded) there are 16 (note that the Booth recoding may also be used with a Wallace tree adder). To get the total addition time, the final CPA time has to be added to the array propagation times. For a 64-bit multiplier the comparison is 11 for a Wallace tree versus 32 for an array.

Apart from 3:2 compression, 4:2 compression (really 5:3) is often used. An improvement over two cascaded adders may be achieved by using the 4:2 compressor shown in Fig. 8.41. This has three XOR delays in the sum path rather than the four that would be present if two adders were used. A regular

TABLE 8.5   A 6×6 Multiplier

						X5	X4	X3	X2	X1	X0	Multiplicand
						Y5	Y4	Y3	Y2	Y1	Y0	Multiplier
						X5Y0	X4Y0	X3Y0	X2Y0	X1Y0	X0Y0	
					X5Y1	X4Y1	X3Y1	X2Y1	X1Y1	X0Y1		
				X5Y2	X4Y2	X3Y2	X2Y2	X1Y2	X0Y2			
			X5Y3	X4Y3	X3Y3	X2Y3	X1Y3	X0Y3				
		X5Y4	X4Y4	X3Y4	X2Y4	X1Y4	X0Y4					
	X5Y5	X4Y5	X3Y5	X2Y5	X1Y5	X0Y5						
P11	P10	P9	P8	P7	P6	P5	P4	P3	P2	P1	P0	Product

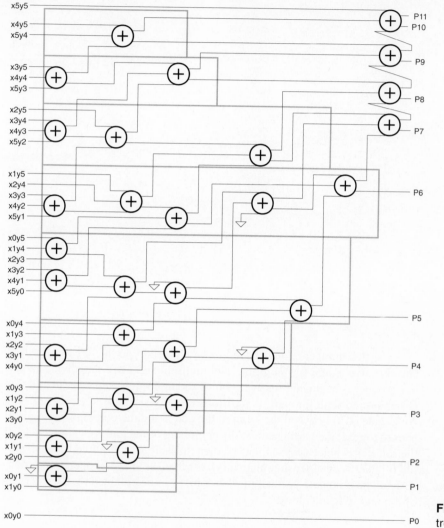

FIGURE 8.40 Wallace adder tree (for 6 × 6 multiplier)

layout for a 54-by-54 bit multiplier using the compressor shown in Fig. 8.41 may be found in Goto et al.[16]

8.2.7.4 Serial Multiplication

Multiplication may be performed serially. The simplest form of serial multiplier, shown in Fig. 8.42, uses the successive addition algorithm and is implemented using a full adder, a logical AND circuit, a delay element (i.e., either static or dynamic flip-flop), and a serial-to-parallel register.

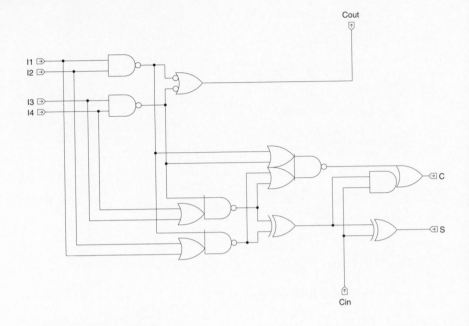

Figure 8.41 A 4:2 (5:3) compressor circuit

The two numbers X and Y are presented serially to the circuit (at different rates to account for multiplier and multiplicand word-lengths). The partial product is evaluated for every bit of the multiplier, and a serial addition is performed with the partial additions already stored in the register. The AND gate ($G2$) between the input to the adder and the output of the register is used to reset the partial sum at the beginning of the multiplication cycle. If the register is made of $N-1$ stages, then the 1-bit shift required for each partial product is obtained automatically. As far as the speed of operation is concerned, the complete product of $M+N$ bits can be obtained in MN intervals of the multiplicand clock.

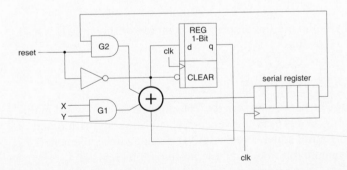

FIGURE 8.42 Serial multiplier

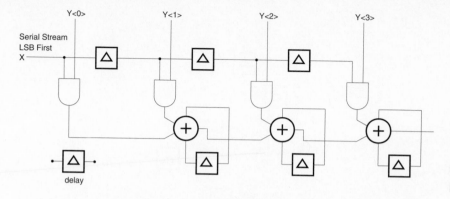

FIGURE 8.43   Serial/parallel multiplier

Using the general approach discussed previously, it is possible to realize a serial/parallel multiplier with a very modular structure that can easily be modified to obtain a pipelined system. The basic implementation is illustrated by Fig. 8.43. In this structure, the multiplication is performed by means of successive additions of columns of the shifted partial products matrix. As left-shifting by one bit in serial systems is obtained by a l-bit delay element, the multiplier is successively shifted and gates the appropriate bit of the multiplicand. The delayed, gated instances of the multiplicand must all be in the same column of the shifted partial-product matrix. They are then added to form the required product bit for the particular column.

This structure requires $M + N$ clock cycles to produce a product. The main limitation is that the maximum frequency is limited by the propagation through the array of adders. The structure of Fig. 8.43 can be pipelined with the introduction of two delay elements in each cell, as shown in Fig. 8.44. If rounding or truncation of the product term to the same word length as the input is tolerated, then the time necessary to produce a product is 2M clock cycles. In this case the multiplier accumulates partial product sums, starting with the least significant partial product. After each addition, the result is an

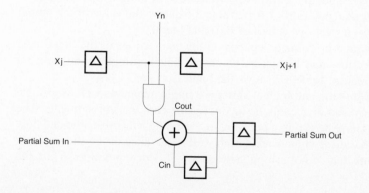

FIGURE 8.44   Pipelined serial/parallel multiplier

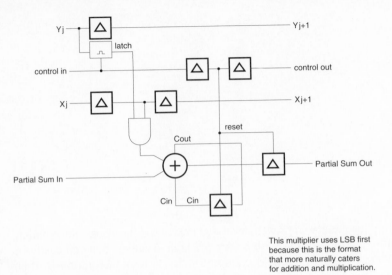

This multiplier uses LSB first
because this is the format
that more naturally caters
for addition and multiplication.

Figure 8.45 Lyon serial multiplier

N-bit number that shortens to N-1 bits before the next partial product is
added. Here, it can be noted that the chip area increases linearly with the
length of the multiplier.

Figure 8.45 shows a schematic of a two-stage serial-multiplier stage
based on the work of Lyon[17], in which the basic solution described so far has
been modified so that both words are in serial form.[18] Multipliers of this type
are frequently useful in FPGAs.

8.2.8 Shifters

Shifters are important elements in many microprocessor designs for arithmetic shifting, logical shifting, and rotation functions. A 4-by-4 barrel
shifter is shown in Fig. 8.46(a), constructed from complementary transmission gates. The input to the shifter is the value to be shifted a (*literal<6:0>*)
and the shift amount (*shift<3:0>*). Table 8.6 shows the value of the output
(*result<3:0>*) for various values of shift and literal.

The function performed depends on the connections of the literal bus.
These connections may be made with an additional multiplexer on the front
of the shift matrix. Table 8.7 shows the functions.

Both arithmetic and logical shifts are implemented as well as rotates.
Figure 8.46(b) shows a symbolic layout for the core transmission gate. The
control lines have been run in polysilicon, assuming either that silicided poly
is used or that these signals are set up well in advance of the literal input.
Other layouts that do not use polysilicon are of course possible. While the

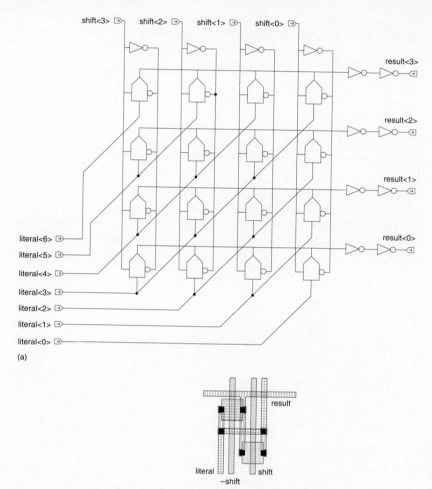

(a)

(b)

FIGURE 8.46 Array shifter using transmission gates: (a) circuit; (b) cell layout

TABLE 8.6   Shifter Operations

SHIFT	RESULT
1	LITERAL<3:0>
2	LITERAL<4:1>
4	LITERAL<5:2>
8	LITERAL<6:3>

TABLE 8.7   Modified Shifter Operations

LITERAL<6:0>	OPERATION
VSS,VSS,VSS,A<3:0>	LOGICAL RIGHT SHIFT
A<3>,A<3>,A<3>,A<3:0>	ARITHMETIC RIGHT SHIFT
A<3:0>,VSS,VSS,VSS	LOGICAL LEFT SHIFT
A<3:0>,A<2:0>	LEFT ROTATE
A<2:0>,A<3:0>	RIGHT ROTATE

circuit shown in Fig. 8.46(a) is fine for transistor level design, it is not really appropriate for a gate-level implementation.

Figure 8.47 shows a shifter that uses multiplexers (which of course can be transmission gates). An implementation for a logical left shift, arithmetic right shift is shown. The shifter is divided into two halves, one of which

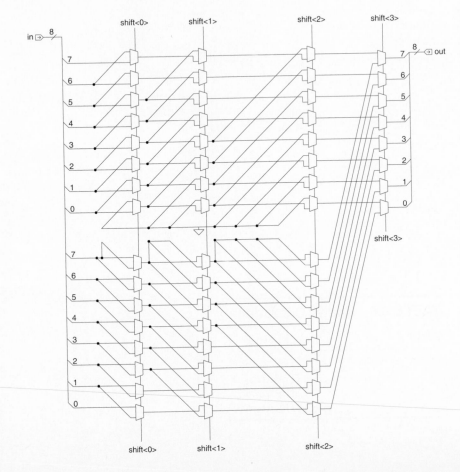

Figure 8.47   Multiplexer-based shifter

shifts right and one of which shifts left. The fill values can be set by appropriate connections at the ends of the shifter ranks. The output of the two shifters is muxed to form a final result. The value of *SHIFT<2:0>* gives the amount of the shift with *SHIFT<3>* = 1 producing a left shift, while *SHIFT<3>* = 0 produces an arithmetic right shift. Left and right rotates may be implemented by wrapping the end connections conditionally to the opposite end bits.

Shifters implemented with transmission gates are notorious for fooling timing analyzers unless the directionality of the pass transistors are somehow communicated to the timing analyzer. The multiplexer shifter may use buffered (inverting, if need be) multiplexers, which can aid in speeding up the long lines in large shifters. The multiplexer version directly takes the shift amount as control, while the array version requires an *n:m* decoder (2:4 for the one shown in Fig. 8.46a). For these reasons the multiplexer version may be favored in CMOS although the version shown in Fig. 8.46 can be compact.

Other shift options are frequently required, for instance, shuffles, bit-reversals, and interchanges. One can either use the complementary transmission gate, static single-pass transistors (usually n-channel). Precharged versions of single-pass transistor shifter circuits are generally cumbersome. Large capacitances can be associated with the intermediate mux nodes and these must all be precharged to prevent charge-sharing problems. The speed of an *n*-bit shifter is proportional to $\log(n)$, so combined with the fast speed of transmission gates, shifting can be a fast operation.

8.3   Memory Elements

Memory elements form critical components in the implementation of CMOS systems. While off-the-shelf memories are limited by the number of I/O pins, the speed of driving into the chip, and large off-chip output nodes, on-chip memories can be engineered to be very fast and to have unique access paths. In general, CMOS ASIC processes will not compete with the density of state-of-the-art DRAM memory, but may be very competitive with high-speed static memories. Memory elements may be divided into the following categories:

- Random access memory.
- Serial access memory.
- Content access memory.

Random access memory at the chip level is classed as memory that has an access time independent of the physical location of the data. This is contrasted with serial-access memories, which have some latency associated

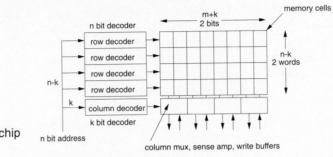

FIGURE 8.48 Memory-chip architecture

with the reading or writing of a particular datum and with content-addressable memories. Within the general classification of random access memory, we can consider read only memory (ROM) or read/write memory (commonly called RAM). ROMs usually have a write time much greater than their read time (programmable ROMs have write times of the order of milliseconds), while RAMs have very similar read and write times. Both types of memory may be further divided into static-load, synchronous, and asynchronous categories. Static-load memories require no clock. Synchronous RAMs or ROMs require a clock edge to enable memory operation. The address to a synchronous memory only needs to be valid for a certain setup time after the clock edge. Asynchronous RAMs recognize address changes and output new data after any such change. Static-load and synchronous memories are easier to design and usually form the best choice for a system-level building block, because they can generally be clocked by the system clock.

The memory cells used in RAMs can further be divided into static structures and dynamic structures. Static cells use some form of latched storage, while dynamic cells use dynamic storage of charge on a capacitor. We will concentrate on static RAMs because they are easier to design and potentially less troublesome than dynamic RAMs. Static RAMs tend to be faster (but much larger) than dynamic RAMs.

A typical memory-chip architecture is shown in Fig. 8.48. Central to the design is a memory array consisting of 2^n by 2^m bits of storage (actually 2^{n-k} by 2^{m+k}). A row (or word) decoder addresses one word of 2^m bits out of 2^{n-k} words. The column (or bit) decoder addresses 2^k of 2^m bits of the accessed row. This column decoder accesses a multiplexer, which routes the addressed data to and from interfaces to the external world.

8.3.1   Read/Write Memory

8.3.1.1   RAM

Figure 8.49 shows one row and one column of a generic RAM architecture with the support circuits required by the RAM cell. The row decoder is a 1 of $n-k$ decoder which may generally be thought of as an AND gate. One of the 2^{n-k}

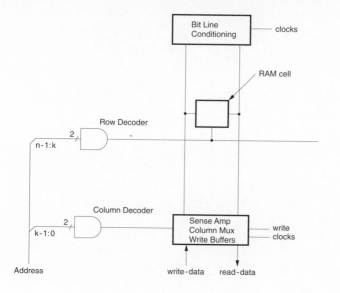

FIGURE 8.49 Generic RAM circuit

row lines is accessed at one time. The bit-line-conditioning circuitry, the ram-cell, the sense amplifiers, column multiplexers, and the write buffers form a tightly coupled circuit that provides for the hazard-free reading and writing of the memory cell. The bit lines are normally run as complementary signals. There are many variations of these circuits to achieve varying density/speed/ noise–margin requirements. We shall look at a variety of schemes for implementing static RAMs. The column decoder is similar to the row decoder but is a 1 of k decoder. k is normally less than n and the decoder drives a multiplexer (rather than a selector). Frequently, the column decoder may be merged with the column multiplexer.

Starting with the RAM cell itself, various circuits are shown in Fig. 8.50.[19] The most commonly used in ASIC memories is the 6-transistor, cross-coupled inverter circuit shown in Fig. 8.50(a). A typical mask-level layout for a 6-transistor circuit is shown in Fig. 8.51 (also Plate 8). The p-transistors may be replaced with high-value polysilicon resistors if the process supports this option (Fig. 8.50b). The value of the resistor has to be such that it prevents leakage from changing any value stored in the RAM cell. Generally the resistors are in the 100's to 1000's of Megaohms. Deleting one of the bit-line pass transistors results in a 5-transistor RAM cell. Writing such a cell has to be considered carefully (see later in this section). A 4-transistor dynamic RAM cell may be achieved by deleting the p loads of the static cell, as shown in Fig. 8.52(a). This cell and the other dynamic cells have to be refreshed to retain the contents of the memory. A 3-transistor cell is shown in Fig. 8.52(b). The cell stores data on the gate of the storage transistor. Separate read and write control lines are used. Multiple read-ports may be added easily, by adding read transistors. In addition, separate or

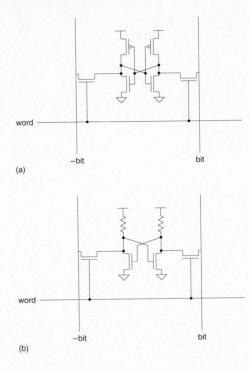

(a)

−bit   bit

word

Figure 8.50 Static RAM cell circuits

(b)

−bit   bit

word

merged read and write data busses may be used. A 1-transistor cell is shown in Fig. 8.52(c).[20] The memory value is again stored on a capacitor. The capacitor can be implemented as a transistor as shown in Figs. 8.52(d) and 8.52(e). Sense amplifiers sense the small change in voltage that results when a particular cell is switched onto the *bit* line. This type of cell (Fig. 8.52c) forms the basis for most high-density DRAMs.[21–28] The cell shown in Fig. 8.52(d) can be implemented in a conventional two metal, single poly process. The dominant problem that arises with this type of memory when used in an ASIC process is the loss of the stored charge due to leakage or stray substrate currents created by surrounding digital logic.

As far as the average CMOS-system design is concerned, the static 6-transistor cell should be used since it involves the least amount of detailed circuit design and process knowledge and is the safest with respect to noise and other effects that may be hard to estimate before silicon is available. In addition, current processes are dense enough to allow large static RAM arrays. As a general system-design principle, large amounts of memory should only be included in a design if the performance of the system is affected. Commercial RAM manufacturers are much better at designing RAMs than the average system designer. If dense memory can be partitioned

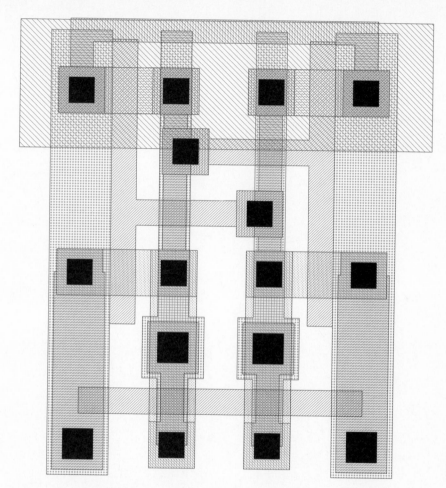

FIGURE 8.51 Mask layout for 6-transistor static RAM

off-chip with no performance degradation or cost impact, then this is a good approach to take.

8.3.1.1.1 Static RAM—read

We will begin our examination of CMOS static RAMs by considering a read operation. Imagine that the bit lines of the circuit shown in Fig. 8.49 are at some value and that the word line is asserted. The one node on the memory cell will attempt to pull the bit line up through the access transistor and the p load. The zero node will attempt to pull the bit line down through the access transistor and the n channel pull-down. As an n-channel transistor is poor at passing a one and the p-channel transistors in the RAM cell are generally small (or in the case of a resistive load, the resistors are very large), design of

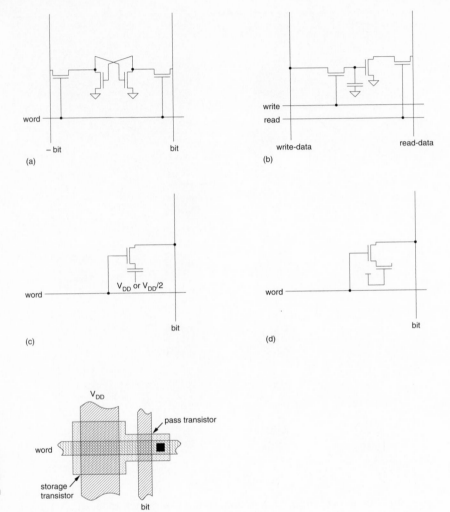

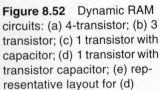

Figure 8.52 Dynamic RAM circuits: (a) 4-transistor; (b) 3 transistor; (c) 1 transistor with capacitor; (d) 1 transistor with transistor capacitor; (e) representative layout for (d)

the RAM circuit concentrates on pulling the bit line from high to low. Thus one method of reading a RAM cell would be to precharge the bit lines high and then enable the word-line decoder. For a given pair of bit lines, one RAM cell will attempt to pull down either the *bit* or *–bit* line depending on the stored data. The bit-line pull-up circuit may use p-channel transistors to precharge each bit line (Fig. 8.53a). In this example, the sense amplifier is an inverter that forms a single-ended sense amplifier. The sense time is roughly

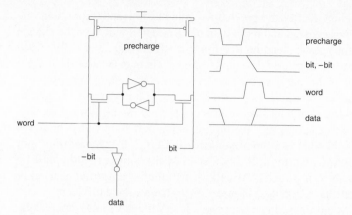

(a)

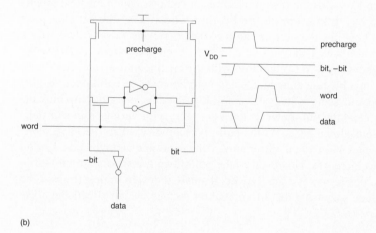

(b)

FIGURE 8.53 RAM read options: (a) V_{DD} precharge; (b) V_{DD}–V_{tn} precharge

the time it takes one RAM cell pull-down and access transistor to reach the inverter threshold. To optimize speed, one might set the inverter threshold above the V_{DD} midpoint, but below an adequate noise margin down from the V_{DD} rail. Alternatively, one can precharge the bit lines with n-channel transistors, which results in the bit lines being precharged to an n threshold down from V_{DD} (Fig. 8.53b). This can dramatically improve the speed of the RAM cell access. In addition, it reduces power dissipation because the bit lines do not change by the supply voltage. The key aspect of the precharged RAM read cycle is the timing relationship between the RAM addresses, the precharge pulse, and the enabling of the row decoder. If the word-line assertion precedes the end of the precharge cycle, the RAM cells on the active word-line will see both bit lines pulled high and the RAM cells may flip state. If the addresses change after the precharge cycle has finished, more than one word line will be accessed and more than one RAM cell will have the chance to pull the bit lines down, leading to erroneous READ data. Normally, RAM designers generate a carefully designed timing chain than ensures the correct temporal relationships between precharge, row access, and sense operations.

A RAM access method that does not require precharge is shown in Fig. 8.54(a). Here n-channel load transistors pull up the bit lines statically. When the word line is asserted, the bit line being pulled down by the RAM cell falls to a value that is a function of the pull-up size, the pass-transistor size, and the RAM inverter pull-down size. At the same time, the pull-up must not be able to flip the RAM cell. A differential amplifier is used to amplify the bit-line difference. Figure 8.54(b) shows the equivalent circuit of the pull-down circuit during a read operation. Voltage V_1 must safely clear the input threshold of the RAM cell inverters. A value of .5–1V is appropriate. Voltage V_2 yields the bit-line difference voltage, which must be amplified to detect a transition on the bit line. The size of the bit-line load determines how fast the bit line can recover (to prevent false writes) after a write operation where the bit line may have been driven to V_{SS}. The sense amplifier is designed in conjunction with the bit-line pull-up and RAM cell to amplify this bit-line change. Design margins must be valid over all process, temperature, and voltage extremes. Figure 8.55 shows the zero bit voltage ($V_{bit(0)}$) and the pull-down voltage ($V_{pulldown}$) for various ratios of pull-up beta to pull-down betas. As the pull-up becomes weaker, the $V_{bit(0)}$ voltage approaches V_{SS} and the differential voltage between a high and a low on the bit lines increases. However, as the pull-down transistors are limited in size by the desire to keep the RAM cell small, a design trade-off has to be made between speed and the differential bit voltage, which affects the noise

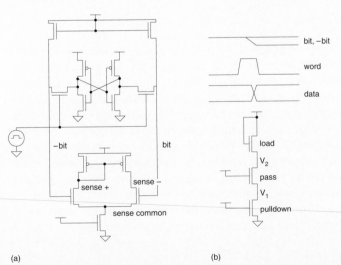

Figure 8.54 RAM read operation model

(a) (b)

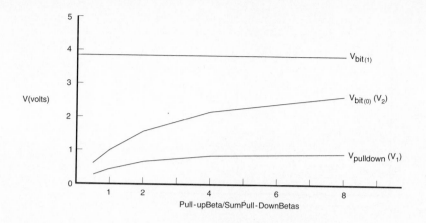

immunity of the cell and the write characteristics. To a first order, the bit-line voltage (V_2) is given by

$$V_2 = (V_{DD} - V_{tn})\left(1 - \sqrt{\frac{1}{1 + \dfrac{\beta_{pullup}}{\beta_{driver\text{-}eff}}}}\right),$$

where β_{pullup} is the gain of the load and $\beta_{driver\text{-}eff}$ is the gain of the combination of the pass and pull-down transistor in series. When the gain of the pull-up is high compared with the pull-down path, the bit-line voltage rises towards $V_{DD} - V_{tn}$. When the gain of the pull-up is very small, the bit-line voltage approaches zero. The pull-down voltage V_1 is a result of resistive divider action between the word-access transistor and the RAM-cell pull-down. While these transistors are in the linear region, V_1 is roughly given by

$$V_1 = V_2 \frac{\beta_{pass}}{\beta_{pass} + \beta_{pulldown}}.$$

The RAM cell and the sense amplifier draw static current, which affects power dissipation. Figure 8.56 shows typical SPICE waveforms for the word line, bit lines, and sense amplifier. In this design the bit line pulls down to about 2 volts, while the bit-line high level is about 4 volts. During access, the RAM cell low value is pulled up to about 1 volt, leaving about 1 volt of margin to the switching point of the RAM cell inverter. The sense amplifier can be seen starting to switch just as the bit lines start diverging. The period between word line deassertion and *bit* nearing-*bit* is the *recovery time* (during which no other word line should be asserted in order to prevent false writes).

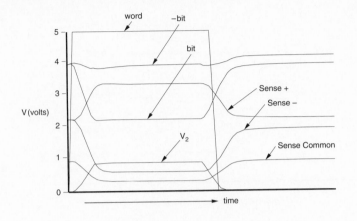

Figure 8.56 Static RAM–read waveforms

Current mode sensing may also be used.[29,30,31] In this technique, the current change in the bit lines is detected using special circuits. The theory is that by using low-impedance circuits, the RC delay inherent in driving the bit lines may be decreased.

8.3.1.1.2 Static RAM—write

The objective of the RAM write operation is to apply voltages to the RAM cell such that it will flip state (a condition we do not desire during the read operation). Figure 8.57(a) shows a straightforward write circuit. In this circuit, the write-enable transistors (N_1,N_2) are enabled to allow the data and complement to move to the bit lines. The word line is then asserted (actually the turn-on order is not important). Either the *bit* or *–bit* line is driven to V_{SS}, while the other bit line is driven to a threshold down from V_{DD}. Figure 8.57(b) shows a more detailed view of the situation. The figure shows a zero stored in the cell. During a WRITE cycle where a one is to be written, node *–Cell* has to be pulled below the RAM-cell inverter threshold and at the same time node *Cell* has to be pulled above the RAM-cell inverter threshold. In the former case, n-transistors N_D (the driver n-transistor), N_1 (the write-access transistor), and N_3 (the word-access transistor) have to pull P_{bit} (the RAM inverter pull-up) below the inverter threshold. In addition N_5 (the bit-line pull-up) has to be pulled low by N_1 and N_D. On the other bit-line side, P_D, N_2 and N_4 have to pull N_{bit} as high as possible. To augment the write operations it may be necessary to use complementary write-access transistors, as shown in Fig. 8.57(c). Correct WRITE operation must be verified over all process, temperature, and voltage extremes. Figure 8.58 shows a plot of the waveforms during a WRITE operation. The SPICE circuit used to model the RAM write operation is shown at the top of the figure. *write-data* and *–write-data* were driven antiphase into the write transistors N_2 and N_6. The cell switches when *–write-data* = 3V and *write-data* = 2V.

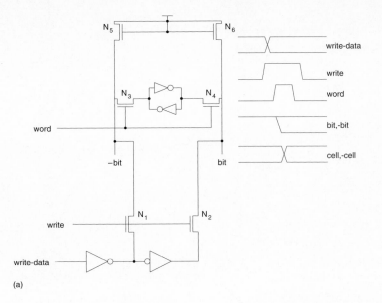

(a)

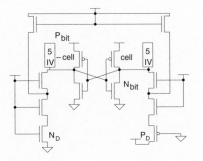

(b)

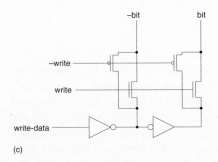

(c)

FIGURE 8.57 Static RAM–write circuits: (a) n-channel pass transistors; (b) circuit model during write; (c) complementary transmission gate version

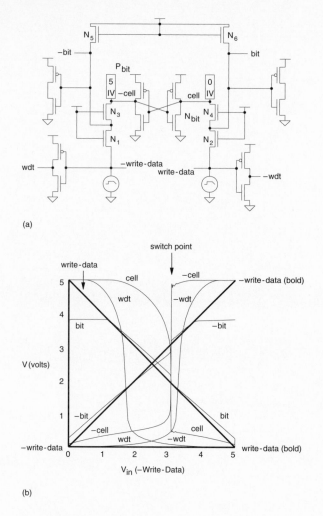

Figure 8.58   Static RAM–write waveforms and circuit model

8.3.1.1.3   Row decoders

The simplest row decoder is an AND gate. Figure 8.59 shows two straight-forward implementations. The first in Fig. 8.59(a) is a static complementary NAND gate followed by an inverter. This structure is useful for up to 5–6 inputs or more if speed is not critical. The NAND transistors are usually made minimum size to reduce the load on the buffered address lines because there are 2^{n-k} ($N_{load} + P_{load}$)'s on each address line. The second implementation, shown in Fig. 8.59(b), uses a pseudo-nMOS NOR gate buffered with two inverters. The NOR gate transistors can be made minimum size, and the inverters can be scaled appropriately to drive the word line. Large fan-in AND gates can also be constructed from smaller NAND and NOR gates, as shown in Fig. 8.59(c). Figure 8.60 shows two possible layout styles (in sym-

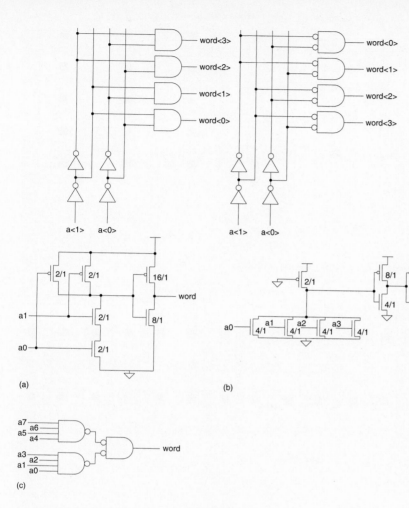

word<3>
word<2>
word<1>
word<0>

word<0>
word<1>
word<2>
word<3>

a<1>   a<0>

a<1>   a<0>

2/1   2/1   16/1

word

a1   2/1   8/1

a0   2/1

(a)

2/1

8/1   32/1

4/1   16/1

a0

a1   a2   a3
4/1   4/1   4/1   4/1

(b)

a7
a6
a5
a4

a3
a2
a1
a0

word

(c)

FIGURE 8.59 Row-decoder circuits: (a) complementary AND gate; (b) pseudo-nMOS gate; (c) cascaded NAND, NOR gates

bolic form) for the row decoders. One passes the address lines over the decode gates, while the other uses a more standard cell style. Choice would depend on the size of the decoder in relation to the size of the RAM cell. Often, speed requirements or size restrict the use of single-level decoding, such as that shown in Fig. 8.59. The alternative is a predecoding scheme, which is illustrated in Fig. 8.61(a). Here the $(n\text{-}k)$ row address lines are split into a p-bit predecode field and a q-bit direct decode field. The q-bit decode field requires a gate per word line, so q is chosen to suit the pitch of the RAM cell. The p-bit predecode field generates 2^p predecode lines (4 in this example), each of which is fed vertically to 2^{n-k}-row decode gates (8 in this example). Figure 8.61(b) shows a possible implementation of a predecode scheme, where the predecode gate is a NAND gate and the word-decode gate is a NOR gate. An additional input ($-clk$) has been included in the NOR gate

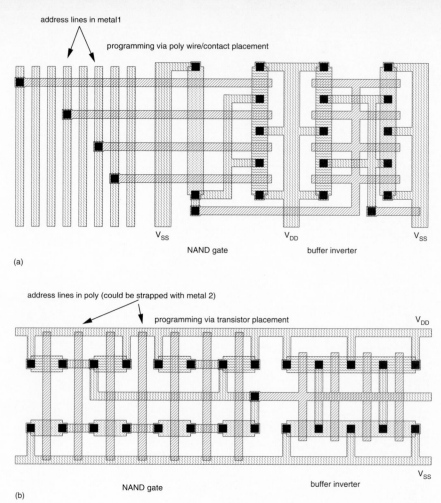

address lines in metal1

programming via poly wire/contact placement

V_{SS} V_{DD} V_{SS}

NAND gate   buffer inverter

(a)

address lines in poly (could be strapped with metal 2)

programming via transistor placement V_{DD}

NAND gate   buffer inverter

V_{SS}

(b)

Figure 8.60 Typical symbolic layouts of row decoders

to allow the enabling of the gate, which is necessary to ensure correct timing of the word signal. A slow rise time and fast fall time on a word-decode gate might be advantageous because it ensures that any RAM cells on a word line transitioning low are isolated before RAM cells on a high-transitioning word line are accessed. Figure 8.61(c) shows a pseudo-nMOS AND row decode gate. Finally, Fig. 8.62 shows a few more row decoder circuits. Figure 8.62(a) shows some obvious ways of building large fan-in AND gates from smaller fan-in gates. Figure 8.62(b) is a pseudo-nMOS decoder that minimizes draw static power. Figure 8.62(c) shows a predecode scheme where the predecode gates power the word-line driver.[32] Figure 8.62(d) shows a domino dynamic AND gate implementation.

8.3.1.1.4 Column decoders

The column decoder is responsible for selecting 2^k out of 2^m bits of the accessed row. A tree decoder is shown in Fig. 8.63. Here the data is routed

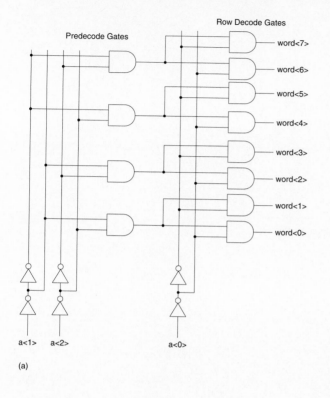

(a)

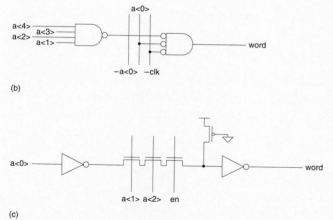

(b)

(c)

FIGURE 8.61 Predecode circuits: (a) basic approach; (b) actual implementation; (c) pseudo-nMOS example

via pass gates enabled by the column-address lines. The address decoding is in essence distributed. Decoders for *bit* and *–bit* lines are shown, although one of these may be omitted for single-ended read operations. The read (and, usually of lesser importance, write) operations are somewhat delayed by the series-transmission gates. However, in comparison with gate delays these

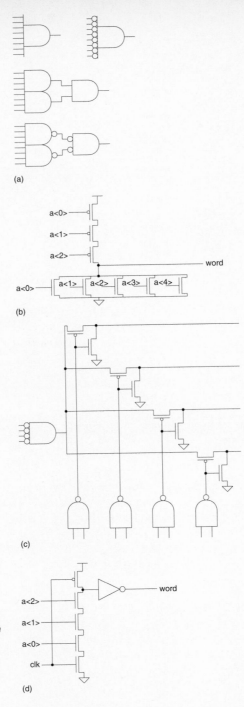

Figure 8.62 Various other row decoder circuits: (a) methods of building large fan-in AND gates; (b) power saving pseudo-nMOS gate; (c) decoder powered; (d) domino

usually are small for a low number of series transistors (2 to 4). Complementary transmission gates may also be used, if required, by either the read operation or write operation.

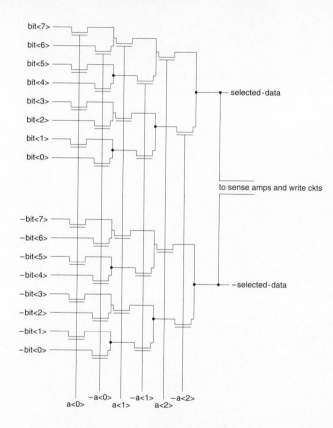

bit<7>
bit<6>
bit<5>
bit<4>
bit<3>
bit<2>
bit<1>
bit<0>

selected-data

to sense amps and write ckts

−bit<7>
−bit<6>
−bit<5>
−bit<4>
−bit<3>
−bit<2>
−bit<1>
−bit<0>

−selected-data

a<0> −a<0> −a<1> −a<2>
       a<1> a<2>

FIGURE 8.63   Tree-style column decoder

If the delay of the series-pass gates was troublesome, the decoder shown in Fig. 8.64 could be used. Here a NAND decoder is employed on a bit-by-bit basis to enable complementary transmission gates (single transistors may be used where possible) onto a common pair of data lines. These are then routed to a sense amplifier and write circuitry.

8.3.1.1.5   Sense amplifiers

Many sense amplifiers have been invented to provide faster sensing, smaller layouts, and lower power-dissipation sensing.[33] The simple inverter sense amplifier provides for low power sensing at the expense of speed. The differential sense amplifier can consume a significant amount of DC power (Fig. 8.54). Alternatively, one can employ clocked sense amplifiers similar to the SSDL gate shown in Fig. 5.40.

8.3.1.1.6   RAM timing budget

The critical path in a static RAM read cycle includes the clock to address delay time, the row address driver time, row decode time, bit-line sense time, and the setup time to any data register. The column decode is usually not in

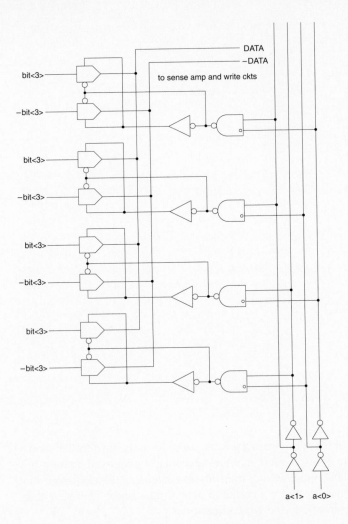

Figure 8.64 Decoded column decoder

the critical path because the decoder is usually smaller and the decoder has the row access time and bit-line sense time to operate. The write operation is usually faster than the read cycle because the bit lines are being actively driven by larger transistors than the memory cell transistors. However, the bit lines may have to be allowed to recover to their quiescent values before any more access cycles take place. In the static load RAM, this speed depends on the size of the static pull-up. Apart from carefully sizing transistors, the RAM speed may be increased by pipelining the row decode signal.

8.3.1.2 Register Files

Register files are generally fast RAMs with multiple read and write ports. Conventional RAM cells may be made multiported by adding pass transistors. Such a configuration is shown in Fig. 8.65(a). A single-write-port, double-read-port memory is shown. For a cross-coupled inverter RAM cell, the write lines generally have to be differential. However, the read lines can be single ended. Figure 8.65(b) shows a modified RAM cell with a single write port and two read ports. This general technique has been used on a 17-port register file that had an overall bandwidth of 1.4 Gigabytes/s.[34]

An alternative register file structure that can be easily changed for a wide variety of read and write ports is shown in Fig. 8.66.[35] Figure 8.66(a) shows a single-write-port, double-read-port cell. The write port is a single-ended implementation where the write pass transistor (N_1) is used to overdrive a weak feedback inverter (N_3, P_3). The threshold of storage inverter (N_2, P_2) is biased towards V_{SS} by increasing the size of N_2 with respect to P_2 to aid in the writing of the cell. The storage inverters drive a buffer inverter (N_4, P_4) which in turn drives two read lines through pass transistors (N_5, N_6). Additional read ports are constructed by adding transistors at the output of inverter (N_4, P_4) and adding additional read-row decode lines. Additional write lines are added by adding transistors that drive inverters (N_2, P_2) and (N_3, P_3). A benefit of this design is that no matter what load appears on the output of the buffer inverter, the state of the memory cell can not be flipped.

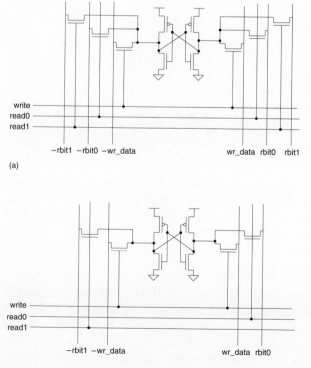

(a)

(b)

FIGURE 8.65 Multiported (2R-1W) RAM cell: (a) fully differential; (b) single-ended read

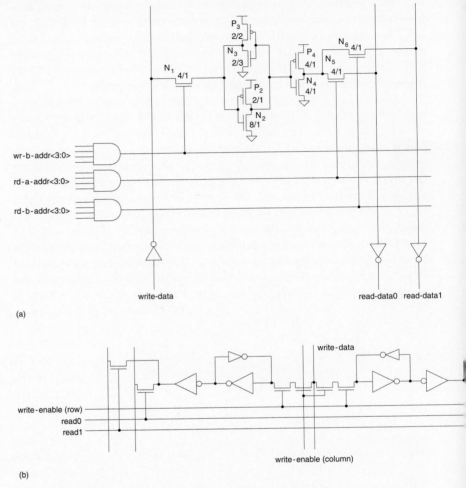

wr-b-addr<3:0>

rd-a-addr<3:0>

rd-b-addr<3:0>

write-data

read-data0 read-data1

(a)

write-data

write-enable (row)
read0
read1

write-enable (column)

(b)

Figure 8.66 Expandable register file cell: (a) row accessed; (b) column accessed and row accessed

The design in Fig. 8.66(a) is used where there is no column multiplexing. The version shown in Fig. 8.66(b) is used where column multiplexing is required. An additional transistor (two in this design for symmetry) is added per column to enable a column for writing.

8.3.1.3 FIFOs, LIFOs, SIPOs

Using the basic RAM memory cell, multiport register cells, or variations of these, a variety of special-purpose memories can be constructed.

A First In First Out (FIFO) memory is useful for buffering data between two asynchronous data streams. Figure 8.67 shows a block diagram that outlines the operation of a FIFO. A stream writes into the FIFO when a *WRITE*

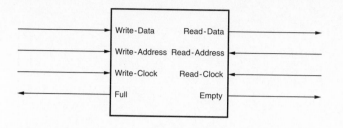

FIGURE 8.67 FIFO interface signals

clock is asserted and observes a *FULL* flag, which is raised when the FIFO can accept no more input data. Another stream reads data when a *READ* clock is asserted until an *EMPTY* flag is asserted. Ideally, the two ports can read and write independently. Due to other system delays and latencies it may be desirable to have *ALMOST-FULL* and *ALMOST-EMPTY* flags so that impending fullness or emptiness can be communicated. The simplest implementation of a FIFO uses a dual port RAM or register file with a read and write counter. An example design of the addressing logic that is useful for synchronous read and write signals is shown in Fig. 8.68. Two counters control the read pointer (*RP*) and the write pointer (*WP*) that are addresses to the dual port memory. A further difference circuit is incremented, in the case of a write, or decremented, in the case of a read. The output of this counter is examined to determine the *EMPTY* and *FULL* flags. In this case a *ZERO* detect determines when the FIFO is *EMPTY*. Alternative implementations of FIFOs may use distributed forms of row decoders, where full and empty bits are propagated by serial shift registers in the word direction of the memory.

A Last In First Out (LIFO) memory, or push-down stack, is of use in such applications as subroutine stacks in microcontrollers. In common with FIFOs, regular RAMs or register files may be used or special distributed row decoders may be designed as the address pointer moves sequentially from row to row. The former usually are more straightforward to design, while the latter may save some space. (See also Section 9.2.4.3.)

A Serial In Parallel Out (SIPO) memory is of use to convert serial data to a parallel form. These memories are often of use in signal-processing applications. An example of the memory cell used in this type of memory is shown in Fig. 8.69. Data is shifted in at a high rate via the complementary clocks *clk* and *−clk,* which should be nonoverlapping to prevent data feedthrough. Data may be read in parallel through access transistor N_1 with an appropriately timed clock pulse (i.e., when the Q data is valid).

8.3.1.4 Serial-Access Memory

Serial-access memories (shift registers) are also of use in signal-processing applications for storage and delaying signals. A serial-access memory may

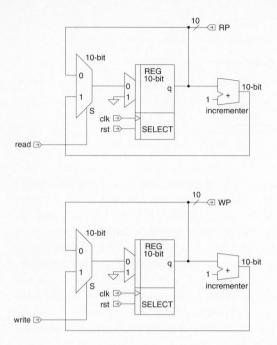

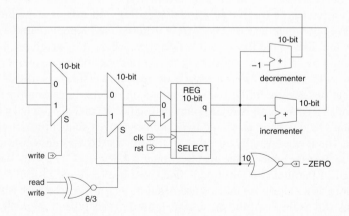

Figure 8.68 FIFO address control design

be simulated by a RAM, and probably for most applications this provides the smallest implementation because the CMOS static RAM cell is a very area-efficient structure. However, the RAM is surrounded by peripheral circuits, such as row and column decoders and sense amplifiers, and in the case of a serial-access memory, a counter. In some circumstances, a dedicated shift-register memory may be appropriate from a density, speed, or floorplanning viewpoint. (See Chapter 9.)

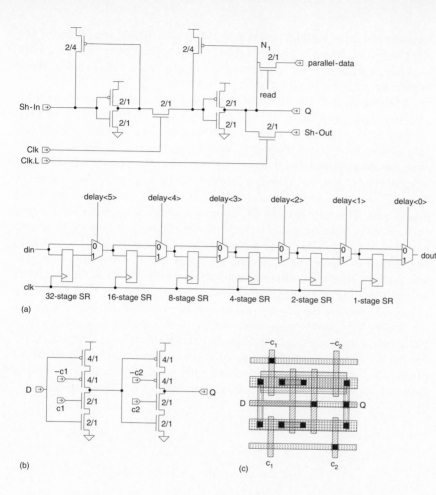

FIGURE 8.69 SIPO cell design

FIGURE 8.70 Tapped delay line: (a) architecture; (b) circuit; (c) symbolic layout

Figure 8.70(a) shows an example of a 64-byte tapped delay line that might be used in a video processing system. Blocks of byte-wide shift registers are delayed by 32, 16, 8, 4, 2, and 1 clock cycles, and multiplexers control the pass-around of the delay blocks to yield the appropriate delay amount. Each memory cell is a shift register, as shown in Fig. 8.70(b). A typical layout for the shift register cell is shown in Fig. 8.70(c). Here the 2-phase clocks are run horizontally between bits of the shift register. The horizontal metal2 power busses are run over the transistors.

8.3.2 Read Only Memory

Read Only Memory cells may be implemented with only one transistor per bit of storage. A ROM is a static memory structure in that the state is retained indefinitely—even without power. A ROM array is usually implemented as a NOR array, as shown in Fig. 8.71. Note that a NAND array may be used if ultra-small ROMs[36,37] are required, but as discussed in Chapter 5, these implementations will be quite slow.

The electrical details of the NOR structure may embody any of the NOR gate structures studied so far, including the pseudo-nMOS NOR and the domino NOR gate. One problem with the domino gate is that the pull-down path passes through two transistors, one the programmed transistor and the other the virtual ground pull-down. This can slow the bit-line transition for large ROMs. A dynamic CMOS alternative to the domino NOR is shown in Fig. 8.72. Here the word lines are forced low while the bit lines are being precharged. This ensures that DC current does not flow. After the bit-line pull-ups have been turned off, the word-line drivers are asserted and one word line is active. The timing chain to ensure this sequence of events has to be carefully designed and simulated. Where DC power dissipation is acceptable and the speed is sufficient, the pseudo-nMOS ROM is the easiest to design, requiring no timing. The DC power dissipation may be significantly reduced by turning the pull-ups on according to the column address decoding. Figure 8.73 shows an example of this where only one bit line in four is being pulled up at any one time. Row decoders for ROMs are similar to those for RAMs except that they are usually very constrained by the ROM bit pitch. This usually means that some form of a predecode structure is required. Column decoders for ROMs are usually simpler than those for

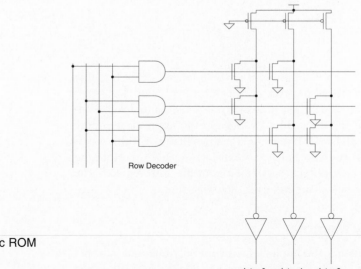

Figure 8.71   Basic ROM architecture

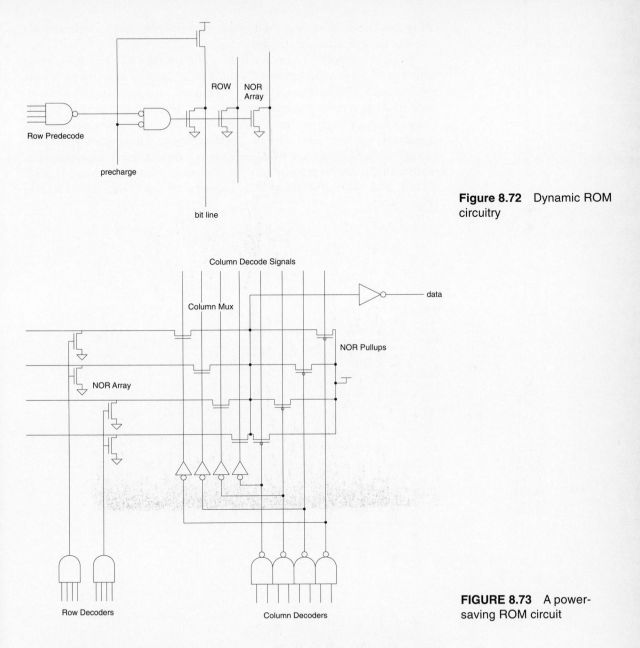

Figure 8.72 Dynamic ROM circuitry

FIGURE 8.73 A power-saving ROM circuit

RAMs because only read operations are employed and single-ended sensing is usually employed.

Mask programmability may be achieved via contact programming, presence or absence of a transistor, or an implant to turn a transistor permanently off or on. Other technology options may be possible, such as electrically erasable random access memories.

Several symbolic layouts for ROM cells are shown in Fig. 8.74, along with a programming techinque. Running word lines in polysilicon is only appropriate for slow speed ROMS or (perhaps) silicided poly. In a microcode ROM in a microprocessor, transistor programming would be preferable, because this would minimize the dynamic power dissipation (less capacitance on word lines). It can also affect speed if the load on word lines can be balanced in a sparse ROM. In a generic circuit that is mask-programmable, metal programming may be desirable. Strapping the poly with metal2 every 4 to 8 ROM sites is appropriate for higher speed ROMs (Fig. 8.74c).

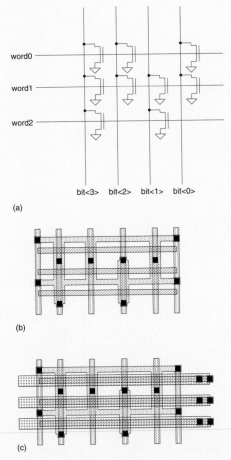

(a)

(b)

(c)

Figure 8.74 ROM layouts: (a) circuit; (b) poly word lines; (c) poly and metal2 strapped word lines

8.3.3 Content-Addressable Memory

A content addressable memory[38,39] is shown in Fig. 8.75(a). The CAM portion examines a data word and compares this data with internally stored data. If any data word internally matches the input data word, the CAM signals that there is a match. These match signals can be passed as word lines to a RAM to enable a specific data word to be output (Fig. 8.75b). This structure may be used as a translation look-aside buffer in the virtual memory lookup in a microprocessor.

A typical CMOS CAM memory cell is shown in Fig. 8.76(a). It consists of a normal static RAM cell with additional transistors N_1 and N_2, which form an XOR gate, and N_3, which is a distributed NOR pull-down. The memory cell may be written and read in the conventional manner. Writes are used to store the match data in the cells, whereas reads are used for testing purposes. A MATCH operation proceeds by placing the data to be matched on the bit lines but not asserting the word line. A 1 appears on the gate of N_3 if the data in the cell is not equal to the data on the bit lines. The drains of N_3 transistors of cells in the same row are commoned, as shown in Fig. 8.76(b). These form a distributed NOR gate, which may be dynamic (with appropriate timing) or pseudo-nMOS (if speed is not critical). Each match line (*match<3:0>*) remains high if the data in the row matches the data placed on the bit lines. These lines may be used to assert the word lines on a RAM.

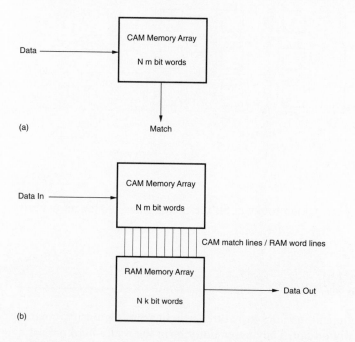

(a)

(b)

FIGURE 8.75 CAM architecture: (a) basic CAM; (b) typical application as translation lookaside buffer

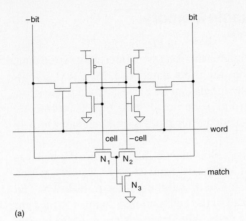

(a)

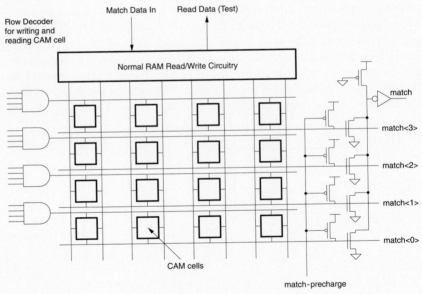

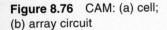

Figure 8.76 CAM: (a) cell; (b) array circuit

(b)

Another NOR gate, which looks at all the match lines, yields an overall match signal.

8.4 Control

While arithmetic and memory structures benefit from regularity, control structures usually do not. They perennially form the really hard part of a design—the part that takes the longest time to design, verify, and test. Usually the control portion of a design is also the last to solidify in the design cycle. Thus it is prudent to look for methods of designing control structures that are highly automated and therefore quick to design. This section begins with a discussion of finite-state machines (FSMs) and then examines various methods for implementing these and other control structures.

8.4.1 Finite-State Machines

A finite-state machine (Fig. 5.43a) provides an organized structure for capturing control sequencing and operation. Diagrammatically, a state machine may be represented by a state-transition diagram (or graph) in which the labelled nodes of the graph represent states and the labelled directed arcs represent transitions between states. A state-transition diagram can be constructed in which the nodes are drawn as circles and the transitions are drawn as lines with arrows. Two basic types of state machines can be designed. A Mealy state machine uses logic to determine the outputs from the inputs and the current state, stored in state registers. A Moore machine determines the outputs from the current state alone. Figure 8.77 shows the two types of machine.

8.4.1.1 FSM Design Procedure

While the design of complex state machines is liable to be machine assisted, small state machines may be designed by hand. As an example, a state machine that might control a toll-booth on a highway will be used. In the *idle* state, the tollbooth, with its gate lowered and its green "proceed" light off, awaits a car. When a car enters the tollbooth, a pressure sensor detects the car and passes a signal to the controller. The controller then awaits the *correct toll* signal and on receiving this, raises the gate and turns the green light on. When the car exits the tollbooth, the controller reenters the *idle* state (green light off and gate down).

The following steps are illustrative of the design of a state machine to perform this function.

1. Draw the state-transition diagram

First the state machine is captured in a state-transition diagram. The inputs to the controller are three signals: a *RESET* (*R* for short) signal, a *CAR-IN-BOOTH* (*A* for short) signal (indicating the car is in the tollbooth), and the *CHANGE-OK* (*C* for short) signal indicating the correct toll has been tendered. There is one output from the controller, the *GREEN-LIGHT* signal used to raise the gate and turn the green light on. The example above may be represented by three states: the *IDLE* state, the *WAIT-FOR-COIN* state and the *WAIT-FOR-CAR-TO-EXIT* state. These are represented as circles in Fig. 8.78(a). Arcs are drawn between states to represent the state transitions. For example, when the controller is in the *IDLE* state it

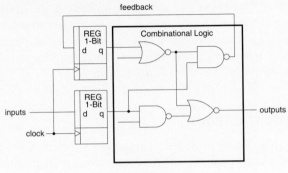

MOORE MACHINE

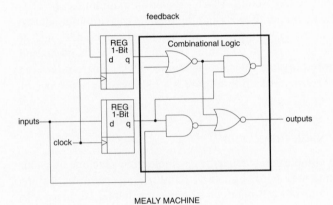

MEALY MACHINE

Figure 8.77 Mealy and Moore state machines

transitions to the *WAIT-FOR-COIN* state when there is a car in the booth (i.e., the input *CAR-IN-BOOTH* or *A*, is true). When *CAR-IN-BOOTH* is false, the *IDLE* state loops to itself (*–A*). Additionally, the *RESET* signal causes the *IDLE* state to be entered.

2. Check the state diagram

There are some simple checks that may be made on the state-transition diagram. These are as follows:

A. Ensure that all states are represented, including the *IDLE* state.

B. Check that the OR of all transitions leaving a state is TRUE. This is a simple method of determining that there is a way out of a state once entered.

C. Verify that the pairwise XOR of all exit transitions is TRUE. This ensures that there are not conflicting conditions that would lead to more than one exit-transition becoming active at any one time.

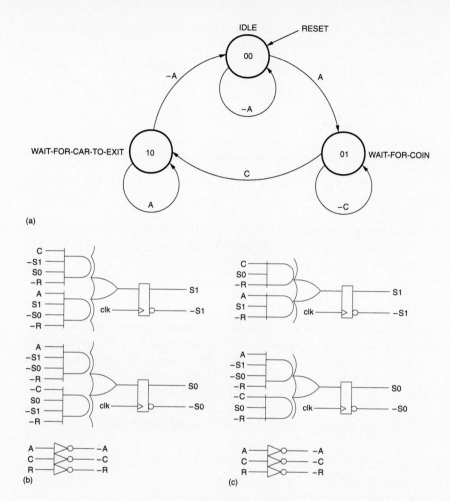

(a)

(b)   (c)

FIGURE 8.78 State-machine transition diagram

D. Insert loops into any state if it is not guaranteed to otherwise change on each cycle. In other words, if a machine enters a state and stays there until some condition occurs, insert the appropriate transition, which is a loop to the state itself. For instance, in the *WAIT-FOR-COIN* state, if the toll is not correct, the controller stays in the *WAIT-FOR-COIN* state.

3. Write the state equations

For each transition, the state equation may be represented as follows:

$$if(state == oldstate \; \& \; condition) \; next\text{-}state = newstate$$

For instance, for the state transition diagram shown in Fig. 8.78, the state equations may be written as

```
next-state IDLE when state==IDLE & !CAR-IN-BOOTH
           OR state==WAIT-FOR-CAR-TO-EXIT & !CAR-IN-BOOTH
next-state WAIT-FOR-COIN when state==IDLE & CAR-IN-BOOTH
           OR state==WAIT-FOR-COIN & !CHANGE-OK
next-state WAIT-FOR-CAR-TO-EXIT
           when state==WAIT-FOR-COIN & CHANGE-OK
           OR state==WAIT-FOR-CAR-TO-EXIT & CAR-IN-BOOTH
```

The first state equation states that the next state is *IDLE* if any of the following conditions are true:

The current state is *IDLE* and the input *CAR-IN-BOOTH* is false.

The current state is *WAIT-FOR-CAR-TO-EXIT* and the input *CAR-IN-BOOTH* is false.

4. Assign the states

The minimum number of state bits that can represent a state machine is $\log_2 K$ where K is the number of states. To ease the problem of state assignment, more states than the minimum may be used. In this example, there are three states, so two state bits (S_1, S_0) are required. Alternatively, unary state assignment would require three bits. The following design guidelines are useful:

A. Assign the *ZERO* state $(S_1, S_0 = 0,0)$ to the most complex state (state *IDLE*).

B. Assign the adjacent states in a Gray code manner (such that they differ by one bit).

C. Otherwise assign states to minimize logic.

Thus for this example an assignment would be:

$$
\begin{aligned}
\text{IDLE} &= 0\ 0 \\
\text{WAIT-FOR-COIN} &= 0\ 1 \\
\text{WAIT-FOR-CAR-TO-EXIT} &= 1\ 0
\end{aligned}
$$

The state assignments are labelled at the center of each state (the bolded circles) in Fig. 8.78. At this point a truth table may be constructed that describes the state machine (see Table 8.8).

Each line of Table 8.8 describes an arc on the state-transition diagram. For instance, the first row shows that the *IDLE* state is entered (00) when *RESET* is set to 1. The second line shows that when in state *IDLE,* the next state is *IDLE* if there is no car in the tollbooth. The *CHANGE-OK* is a don't-

TABLE 8.8 Tollbooth-state table

reset	car-in-booth	change-ok	state<1>	state<0>	state<1>	state<0>
		current state			next state	
1	x	x	x	x	0	0
0	0	x	0	0	0	0
0	1	x	0	0	0	1
0	x	0	0	1	0	1
0	x	1	0	1	1	0
0	1	x	1	0	1	0
0	0	x	1	0	0	0

care in this case. Examining the state bits one by one, logic equations may be written for each bit. For instance,

```
next-state<0> =  !reset & car-in-booth & !state<0> & !state<1>
                 + !reset & !change-ok & state<0> & !state<1>
next-state<1> =  !reset & !change-ok & state<0> & !state<1>
                 + !reset & car-in-booth & !state<0> & !state<1>
```

The logic for any outputs must also be generated. In this example, the output *GREEN-LIGHT* is simply *state<1>*. Notice that the state assignment may have been done in a way that would have necessitated some logic to decode this signal (i.e., state *WAIT-FOR-CAR-TO-EXIT* = 11).

5. Construct the resulting logic and registers

The resulting logic and registers are shown in Fig. 8.78(b). This may be simplified to yield the design in Fig. 8.78(c).

8.4.2   Control Logic Implementation

Control logic in CMOS is constructed in two main ways, with two-level sum-of-products logic and with multilevel logic. Two-level sum-of-products representations have a straightforward geometric implementation in the form of a Programmable Logic Array (PLA). Both two-level and multilevel-logic may be implemented in terms of CMOS logic gates (either static or dynamic).

8.4.2.1   PLA Control Implementation

A programmable logic array (PLA) is a structure that provides a regular structure for implementing combinatorial and sequential logic functions. A

PLA may be used to take inputs and perform some combinatorial function of these inputs to yield outputs, or additionally some of the outputs may be fed back to the inputs via registers, thus forming a finite-state machine. Two-level logic minimization is a well-understood problem. The program Espresso[40] is representative of programs that minimize sum-of-products forms.

A typical PLA uses a two-level sum-of-products AND-OR structure similar to that shown in Fig. 8.79. This implementation also shows clocks to latch inputs and outputs. The basis of a PLA is a sum-of-products form of representation of binary expressions. For example, consider the following expressions that have to be evaluated:

$$z_0 = x_0$$
$$z_1 = x_1 + (-x_0 \cdot -x_1 \cdot -x_2)$$
$$z_2 = -x_1 \cdot -x_2$$
$$z_3 = (-x_0 \cdot -x_1 \cdot x_2) + (-x_0 \cdot x_1 \cdot -x_2)$$

where z_0, z_1, z_2, and z_3 are the four output terms (or sums) and x_0, x_1, and x_2 are the input variables. There are five product terms, namely, x_0, x_1, $-x_0 \cdot -x_1 \cdot x_2$, $-x_1 \cdot x_2$, and $-x_0 \cdot x_1 \cdot -x_2$. Thus these terms would be formed in the AND array of the PLA, as shown in Fig. 8.80. The four outputs are formed by ORing the appropriate product terms. Normally, high-speed PLAs are implemented as two NOR arrays, as shown in Fig. 8.79 (although NAND arrays may be used for slow applications). By using inverting inputs and outputs, the AND-OR structure is maintained.

The electrical design of a CMOS PLA depends on the generic style of PLA. A straightforward physical implementation for a PLA is represented

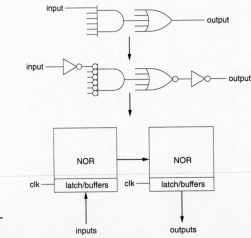

Figure 8.79 PLA Architecture

$$z_0 = x_0$$
$$z_1 = x_1 + (-x_0.-x_1.-x_2)$$
$$z_2 = -x_1.-x_2$$
$$z_3 = (-x_0.-x_1.x_2) + (-x_0.x_1.-x_2)$$

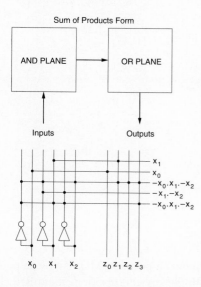

Sum of Products Form

Inputs

Outputs

x_1
x_0
$-x_0.x_1.-x_2$
$-x_1.-x_2$
$-x_0.x_1.-x_2$

x_0 x_1 x_2 $z_0 z_1 z_2 z_3$

FIGURE 8.80 PLA example

by Fig. 8.80. Variations of this involve multiple-sided access (Fig. 8.81) and various folded structures.

A generic floorplan for a "simple" PLA is shown in Fig. 8.82. This has been designed as a set of tiles, designated by letters. In the treatment of various circuit options this naming convention will be used to designate particular cells. Brief descriptions of the cells are as follows:

AN   AND-plane programming cell

OR   OR-plane programming cell

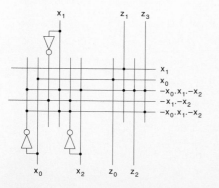

x_1 z_1 z_3

x_1
x_0
$-x_0.x_1.-x_2$
$-x_1.-x_2$
$-x_0.x_1.-x_2$

x_0 x_2 z_0 z_2

FIGURE 8.81 Multisided PLA access

TL	TI	TI	TM	TO	TO	TR
LA	AN	AN	AO	OR	OR	RO
LA	AN	AN	AO	OR	OR	RO
BL	BI	BI	BM	BO	BO	BR

FIGURE 8.82   Generic PLA floorplan

AO   AND-OR communication cell

TI   Top AND-plane input cell

BI   Bottom-AND plane input cell

TO   Top OR-plane output cell

BO   Bottom OR-plane output cell

LA   Left AND-plane cell

RO   Right OR-plane cell

BL   Bottom-left cell

BM   Bottom-middle cell

BR   Bottom-right cell

TL   Top-left cell

TA   Top AND cell

TM   Top-middle cell

TO   Top OR cell

TR   Top-right cell

The most straightforward PLA design uses a pseudo-nMOS NOR gate. Figure 8.83 shows the circuit diagram with the key cell positions identified. Cell AO can either be a layer-change cell or can be used to buffer the AND array outputs. Figure 8.84 shows a PLA for the tollbooth example. Design of

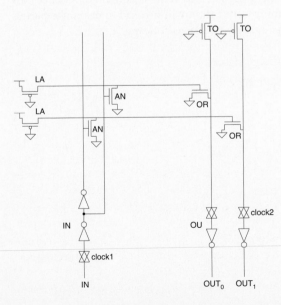

Figure 8.83   Pseudo-nMOS PLA

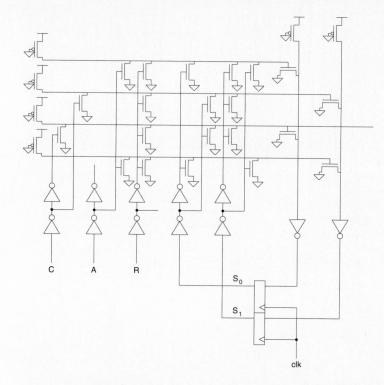

FIGURE 8.84 PLA for toll-booth example

the pseudo-nMOS NOR gates would follow the guidelines given in previous chapters. Advantages of this PLA include simplicity and small size. Disadvantages occur due to the static power dissipation of the NOR gates and possible speed problems (the pull-ups may become slow on large terms). Any convenient register may be used; a static register is shown. This PLA could be fairly independent of the overall system-clocking strategy. Cells TL, TA, BL, BM, TM, TR, RO, and BR are used to route power and clocks as necessary.

By using dynamic CMOS, the circuit shown in Fig. 8.85(a) may be used. Both the AND plane and OR plane have to be supplied with clocks similar to those shown at the bottom of the diagram (Fig. 8.85b). On the rising edge of the clock the input latches store the input data. Following this the AND and OR planes are precharged and then evaluated. When the AND plane outputs are valid, the OR plane may be evaluated. The waveforms in Fig. 8.85(b) may be generated from a multiphase clock or, more probably, in a single-phase clocking scheme by self-timed circuits. Figure 8.86 shows some possible circuits for self timing the PLA operation. The AND precharge may be timed of the rising edge of the clock and the worst case time it takes an AND line to pullup. This may be accomplished by using the circuit shown in Fig. 8.86(a), which uses a dummy AND row. This row has every AND programming transistor inserted (*NPD*) to ensure the load capac-

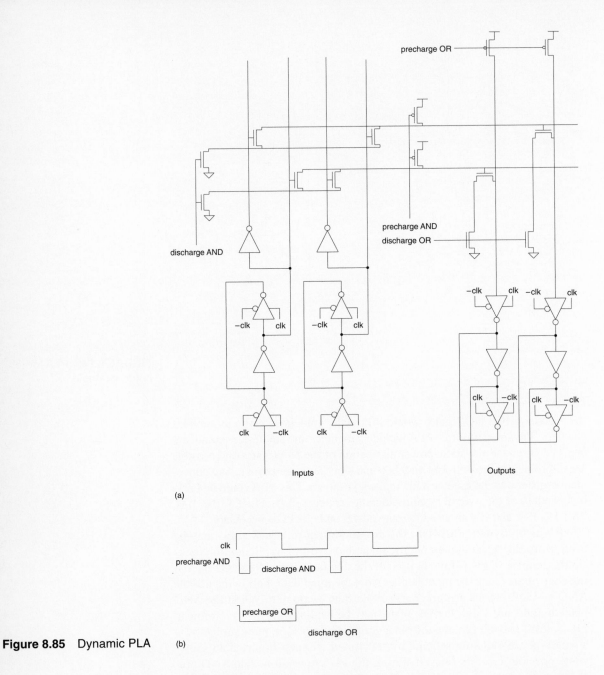

(a)

clk

precharge AND   discharge AND

precharge OR

discharge OR

Figure 8.85   Dynamic PLA   (b)

itance (C_{load}) is a maximum. In addition, the p pull-up (PU) is made smaller than the normal p pull-ups to give some timing margin. An inverter and a NAND gate complete the timing circuit. The OR precharge clock may be self-timed using the circuit shown in Fig. 8.86(b). Here a dummy AND row

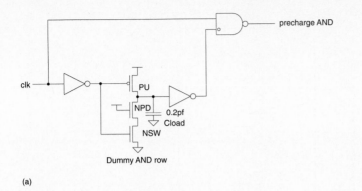

(a)

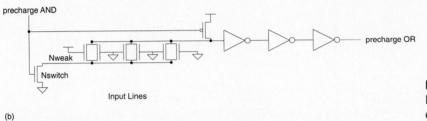

(b)

FIGURE 8.86 Self-timed PLA circuits: (a) AND precharge; (b) OR precharge

is used to determine the worst-case fall time of an AND row. The row is fully populated, with transistors turned off. One smaller-than-normal n pull-down (N_{weak}) in conjunction with a smaller-than-usual ground switch (N_{switch}) is used to pull down the heavily loaded AND line. This is delayed by a few inverters and fed to the OR-place precharge/discharge.

A single-clock PLA that combines a pseudo-nMOS AND plane and a dynamic OR plane is shown in Fig. 8.87.[41] When the clock is high, the OR plane is precharged while all the product terms are forced low by the clocked n-transistor in the AND plane. When the clock transitions low, the product terms conditionally evaluate and then the precharged OR-plane outputs evaluate. The AND plane–transistor ratios are designed according to normal pseudo-nMOS techniques. The inputs must be held constant during the period when the clock is low. This PLA cuts down on the DC dissipation of a fully pseudo-nMOS PLA, while requiring only one clock.

In general PLAs have not found as much acceptance in CMOS as in nMOS technologies. This is due to a number of reasons, some due to CMOS technology and some to the passage of time:

- PLAs have a fixed floorplan and fairly fixed I/O, so extra routing often overshadows any area benefit.
- Dynamic PLAs are cumbersome to design in CMOS, whereas pseudo-nMOS PLAs dissipate DC power.

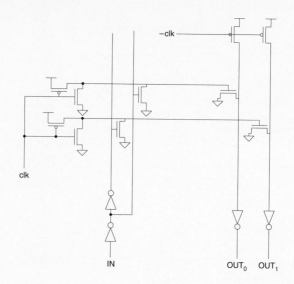

Figure 8.87 Hybrid dynamic pseudo-nMOS PLA

- PLAs are not very compatible with Gate Array technologies, which are popular for ASICs.
- Automatic multilevel logic synthesis has improved dramatically.
- Processes are smaller, hence logic gates are cheaper.
- Large PLAs can be slow.

However, PLAs are compact in themselves and provide a very straightforward way to automate the generation of control logic.

8.4.2.2 ROM Control Implementation

Frequently, control structures may be implemented as a sequenced ROM. A ROM is a special case of a PLA where the AND plane is fully populated. Figure 8.88 shows a simple example of a ROM controller that has a condition-code input and a jump capability implemented by a mux. The ROM has five fields: a next-address field, which provides the next address to the ROM if a branch is not taken; a jump-address field, which is the address taken if the condition code is true; a condition-code select field, which selects which of a number of external conditions to select and the polarity; and an output field, which provides control outputs. Programming consists of writing a microprogram that controls the values of different fields of the ROM. For instance, for the tollbooth example the symbolic microcode might be given as in Table 8.9.

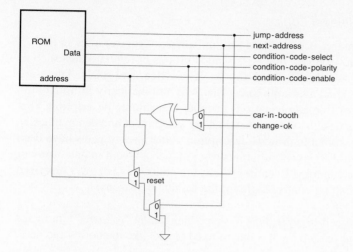

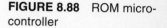

FIGURE 8.88 ROM micro-controller

Here an instruction field has been constructed from the condition-code field as follows:

$$nop = select\ true\ input \qquad 00$$
$$car\text{-}in\text{-}booth = select\ car\text{-}in\text{-}booth\ input\ 01$$
$$change\text{-}ok = select\ change\text{-}ok\ input \quad 10$$

The polarity of the condition controls the XOR gate.

If a metal-programmed ROM is constructed, then a readily changed microsequencer may be constructed. With the addition of a simple datapath, the microsequencer can be extended into a more general microcontroller, which can be used in many low-speed control applications. Megacell libraries frequently contain core microcontrollers that implement standard instruction sets that are supported by a wealth of software.

TABLE 8.9 Symbolic Microcode for the Tollbooth Example

ADDRESS	LABEL	INSTRUCTION	JUMP-ADDRESS	OUTPUT
0	idle:	nop		
1		!car-in-booth	jmp idle	
2	cib:	change-ok	jmp exit	
3		nop	jmp cib	
4	exit:	!car-in-booth	jmp idle	green
5		car-in-booth	jmp exit	green

8.4.2.3   Multilevel Logic

The most commonly used method for implementing control logic in CMOS is to use multilevel logic, that is, cascaded groups of regular gates such as INVERTERS, BUFFERS, NANDs, NORs, XORS, and AOIs. There are many CAD systems available today that will automatically minimize the logic for a set of Boolean equations or other algorithmic description. Furthermore, some design systems can do state assignment and can synthesize state machines from a high-level description. After a set of gates have been generated, automatic layout programs can produce a layout in gate-array or standard-cell technology. Even in full-custom chips, this is now a preferred method of generating control logic for the following reasons:

- Standard-cell logic is fluid in shape and can be "reflowed" into gaps that occur in chip layouts due to fixed blocks like memories and data-paths.
- The designer (or synthesis program) has a large amount of control over speed through basically fast gates, gate sizing, and the ability to trade area for speed.
- The automatic logic-gate layout generation is a mature technology—in double-level metal it might be half as dense as a customized layout, but in triple-level metal the density difference is even less.

Most standard-cell control-logic layout is composed of rows of pre-defined logic and storage cells separated by routing. Programs have been written to automate the generation of control logic from the transistor level. Many of these use the gate-matrix layout style (see Chapter 6). While for small sections of control logic this technique works, there are a number of problems for large sections of control logic. In particular, the layouts get sparse, and internal gate connections completed by long horizontal metal lines tend to produce low-performance gates. Other techniques for custom generating the required logic gates on the fly have included generating dynamic CVSL gate layouts and connecting them in a standard cell style.

8.4.2.4   An Example of Control-Logic Implementation

Figure 8.89 shows the logic schematic for the Boundary Scan–state machine described in Chapter 7. The state-transition diagram appears in Fig. 7.26. A target cycle time of 100 *ns* was desired.

A state-machine description was written that was automatically fed to the MISII[42] logic-synthesis program. This Lisp-based state-machine language description is shown below:

```
(defpal TAP-FSM-AOI-MUX prototype
  (ipin 2 reset)
```

```
(ipin 3 TMS)
(ipin 4 -TCK)
(opin 100 clockir)
(opin 101 updateir)
(rpin 102 shiftir)
(opin 103 clockdr)
(opin 104 updatedr)
(rpin 105 shiftdr)
(rpin 106 enable)
(opin 107 select)
(rpin 108 -reset)
(rpin 20 state<0> h :polarity-fuse)
(rpin 21 state<1> h :polarity-fuse)
(rpin 22 state<2> h :polarity-fuse)
(rpin 23 state<3> h :polarity-fuse)


(setq updateir (and state-update-ir  -TCK))
(setq updatedr (and state-update-dr  -TCK))
(setq clockir (not (and (or state-shift-ir state-capture-ir)
             -TCK)))
(setq clockdr (not (and (or state-shift-dr state-capture-dr)
             -TCK)))
(setq shiftir state-shift-ir :clock -tck )
(setq -reset (not state-test-logic-reset) :clock -tck)
(setq shiftdr state-shift-dr :clock -tck)
(setq enable (or state-shift-ir state-shift-dr) :clock -tck)
(setq select (not (or state-exit2-ir state-exit1-ir
             state-shift-ir
             state-pause-ir state-run-test-idle
                 state-update-ir
             state-capture-ir state-test-logic-reset)))

(macro
  (state-machine
    ;;State transitions
    '(;;Idle—wait for car
          (test-logic-reset
            (:next run-test-idle (not tms))
            (:next test-logic-reset tms)
            (:reset reset))
          (run-test-idle
            (:next select-dr-scan tms)
            (:next run-test-idle (not tms)))
          (select-dr-scan
            (:next select-ir-scan tms)
            (:next capture-dr (not tms)))
          (capture-dr
            (:next exit1-dr tms)
            (:next shift-dr (not tms)))
```

```
(shift-dr
  (:next exit1-dr tms)
  (:next shift-dr (not tms)))
(exit1-dr
  (:next update-dr tms)
  (:next pause-dr (not tms)))
(pause-dr
  (:next exit2-dr tms)
  (:next pause-dr (not tms)))
(exit2-dr
  (:next update-dr tms)
  (:next shift-dr (not tms)))
(update-dr
  (:next select-dr-scan tms)
  (:next run-test-idle (not tms)))
(select-ir-scan
  (:next test-logic-reset tms)
  (:next capture-ir (not tms)))
(capture-ir
  (:next exit1-ir tms)
  (:next shift-ir (not tms)))
(shift-ir
  (:next exit1-ir tms)
  (:next shift-ir (not tms)))
(exit1-ir
  (:next update-ir tms)
  (:next pause-ir (not tms)))
(pause-ir
  (:next exit2-ir tms)
  (:next pause-ir (not tms)))
(exit2-ir
  (:next update-ir tms)
  (:next shift-ir (not tms)))
(update-ir
  (:next select-dr-scan tms)
(:next run-test-idle (not tms)))
  )
;;State number assignments
  '((exit2-dr 0)
  (exit1-dr 1)
  (shift-dr 2)
  (pause-dr 3)
  (select-ir-scan 4)
  (update-dr 5)
  (capture-dr 6)
  (select-dr-scan 7)
  (exit2-ir 8)
```

```
    (exit1-ir 9)
    (shift-ir #xA)
    (pause-ir #xB)
    (run-test-idle #xC)
    (update-ir #xD)
    (capture-ir #xE)
    (test-logic-reset #xF))
;;Base name for state variables (optional)
"state"))
    )
```

This simple language defines the inputs and outputs and then lists the states and their transitions. State assignment is done manually, although in general, software is available to do this task.

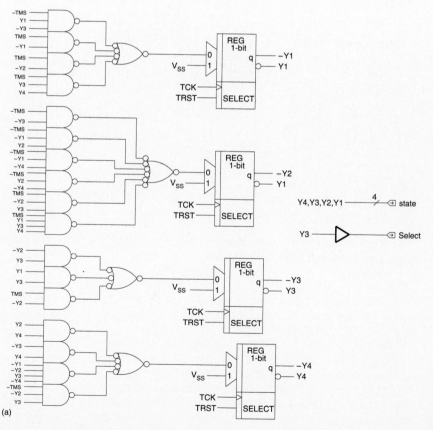

(a)

FIGURE 8.89 Boundary-scan tap controller design

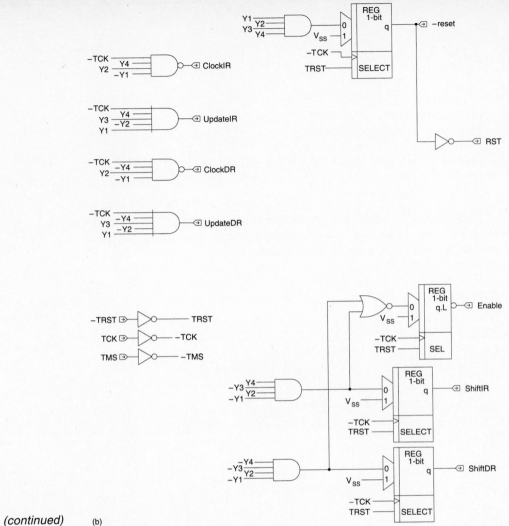

FIGURE 8.89 *(continued)* (b)

When compiled, this state-machine description built the following set of logic equations that were fed to MISII.

```
INORDER = RESET TMS STATE<2> -TCK STATE<0> STATE<1> STATE<3> ;
OUTORDER = NEXT_STATE<3> NEXT_STATE<2> NEXT_STATE<1> NEXT_STATE<0>
NEXT_-RESET SELECT NEXT_ENABLE NEXT_SHIFTDR UPDATEDR CLOCKDR
NEXT_SHIFTIR UPDATEIR CLOCKIR ;
CLOCKIR = !(-TCK * !STATE<0> * STATE<1> * STATE<3>);
UPDATEIR = !(!-TCK + !STATE<0> + STATE<1> + !STATE<2> + !STATE<3>);
NEXT_SHIFTIR = !(STATE<0> + !STATE<1> + STATE<2> + !STATE<3>);
CLOCKDR = !(-TCK * !STATE<0> * STATE<1> * !STATE<3>);
UPDATEDR = !(!-TCK + !STATE<0> + STATE<1> + !STATE<2> + STATE<3>);
NEXT_SHIFTDR = !(STATE<0> + !STATE<1> + STATE<2> + STATE<3>);
NEXT_ENABLE = !(STATE<0> + !STATE<1> + STATE<2>);
```

```
SELECT = !(STATE<3>);
NEXT_-RESET = !(STATE<0> * STATE<1> * STATE<2> * STATE<3>);
     NEXT_STATE<0> = (!STATE<1>*TMS + !STATE<0>*STATE<1>*TMS +
                         STATE<0>*!STATE<2>*!TMS + RESET +
STATE<0>*STATE<1>*STATE<2>*STATE<3>*TMS);
NEXT_STATE<1>=(!STATE<2>*!TMS+STATE<1>*STATE<2>*!STATE<3>*!TMS+
           !STATE<0>*!STATE<1>*STATE<2>*!STATE<3> + RESET +
                  STATE<0>*STATE<1>*STATE<2>*STATE<3>*TMS +
                !STATE<0>*STATE<1>*STATE<2>*STATE<3>*!TMS +
                         STATE<0>*!STATE<1>*STATE<2>*TMS +
!STATE<0>*!STATE<1>*STATE<2>*STATE<3>*TMS);
NEXT_STATE<2>=(!STATE<0>*!STATE<1>*STATE<2>+STATE<0>*STATE<2>+
!STATE<1>*!STATE<2>*TMS + RESET);
          NEXT_STATE<3> = (!STATE<0>*STATE<2>*STATE<3>*!TMS +
              STATE<1>*STATE<3>*TMS + !STATE<2>*STATE<3>*!TMS +
!STATE<0>*!STATE<1>*STATE<2>*!STATE<3> + RESET +
!STATE<1>*!STATE<2>*STATE<3>*TMS TATE<0>*!STATE<1>*STATE<2>*!TMS
STATE<0>*STATE<1>*STATE<2>*STATE<3>*!TMS);
```

This description is read into MISII, converted to sum-of-products form and written out as a PLA description. The following is the expanded input in sum of products form:

```
.i 7
.o 13
.ilb RESET TMS STATE<2> -TCK STATE<0> STATE<1>
STATE<3>
.ob NEXT_STATE<3> NEXT_STATE<2> NEXT_STATE<1> NEXT_-
STATE<0> NEXT_-RESET SELECT NEXT_ENABLE NEXT_SHIFTDR
UPDATEDR CLOCKDR NEXT_SHIFTIR UPDATEIR CLOCKIR
.p 41
1------ 1000000000000
-0----1 1000000000000
--0---1 1000000000000
-----11 1000000000000
--1-000 1000000000000
001-10- 1000000000000
1------ 0100000000000
--1-1-- 0100000000000
-1---0- 0100000000000
--1--0- 0100000000000
1------ 0010000000000
000---- 0010000000000
-00--1- 0010000000000
-0--01- 0010000000000
-11--0- 0010000000000
-0---10 0010000000000
-11-1-1 0010000000000
--1-000 0010000000000
1------ 0001000000000
```

```
-1--0-- 0001000000000
-1---0- 0001000000000
000-1-- 0001000000000
-11-111 0001000000000
--0---- 0000100000000
----0-- 0000100000000
------0 0000100000000
-----0- 0000100000000
------0 0000010000000
--0-01- 0000001000000
--0-010 0000000100000
--11100 0000000010000
---0--- 0000000001000
----1-- 0000000001000
-----0- 0000000001000
------1 0000000001000
--0-011 0000000000100
--11101 0000000000010
---0--- 0000000000001
----1-- 0000000000001
-----0- 0000000000001
------0 0000000000001
.e
```

The initial statements indicate the number of inputs (.i), outputs (.o), and product terms (.p). For each product term (line), the six inputs on the left are coded in terms of zero (0), one (1) or don't care (-). The inputs are ordered according to the .ilb statement. The outputs on the right are ordered according to the .ob statement. This description is then minimized by the Espresso sum-of-products minimizer. The output in sum-of-products form is shown below:

```
.i 7
.o 13
.ilb RESET TMS STATE<2> -TCK STATE<0> STATE<1> STATE<3>
.ob NEXT_STATE<3> NEXT_STATE<2> NEXT_STATE<1> NEXT_STATE<0>
NEXT_-RESET SELECT NEXT_ENABLE NEXT_SHIFTDR UPDATEDR CLOCKDR
NEXT_SHIFTIR UPDATEIR CLOCKIR
.p 21
--11100 0000000010000
--11101 0000000000010
--0-010 0000001100000
--0-011 0000001000100
-11-1-1 0011000000000
--1-000 1010000000000
-0---10 0010000000000
-00-1-- 0011000000000
-0--01- 0010100000000
```

```
-11--0- 0010100000000
-00--0- 0010000001001
--1-1-- 0100000000000
-1--0-- 0001100000000
--0---1 1000100000000
-----11 1000000001000
-01--0- 1100100001001
-1---0- 0101000001001
---0--- 0000000001001
------0 0000110000001
----1-- 0000000001001
1------ 1111000000000
.e
```

This reduced sum-of-products form has 21 product terms. A representative pseudo-nMOS PLA layout is shown in Fig. 8.90. The registers have been arrayed across the bottom of the PLA because this allows the inputs and outputs to the PLA to be placed on the bottom of the PLA. If the registers are incorporated into the PLA itself, the inputs and outputs have to be placed on the top and bottom of the PLA to achieve a small pitch. In this implementation, the input pitch was 7.5μ, the minterm pitch was 5.25μ, and the output pitch was 4.5μ. The layout shown was 181μ wide by 270μ high. Minimum-sized n-transistors were used, although these could be increased in size to improve the speed if necessary. As designed, the state machine implemented with a PLA could operate at a worst-case cycle time of 10 *ns*. The PLA dissipates 8 mW at 10 MHz. Of this around 5 to 6 mW are due to DC dissipation in the p pull-ups. A dynamic version was estimated to dissipate about 3 mW.

A manual schematic design was also completed for the design shown in Fig. 8.89. In addition, a library file, describing the standard-cell library, was input to MISII. Each gate denotes its name, size, logic equation, and timing behavior. The library is shown below:

```
# Area is approximate virtual grid squares
# Name Area Equation
# <phase> <input load> <max load>
# <rise-block-delay> <rise-fanout-delay> <fall-block-delay>
<fall-fanout-delay>
GATE ZERO    0    O=CONST0;
GATE ONE     0    O=CONST1;
GATE XOR     420  O=A*!B+!A*B;      PIN * UNKNOWN 2 50
                                    0.52  0.4  .45  .54
GATE XNOR    420  O=A*B+!A*!B;      PIN * UNKNOWN 2 50
                                    0.5  0.4  .31  0.4
GATE OR3     300  O=A+B+C;          PIN * NONINV 1 50
                                    0.33  0.14  .81  0.12
GATE OR2     240  O=A+B;            PIN * NONINV 1 50
                                    0.36  0.13  .49  0.10
GATE NOR4    300  O=!(A+B+C+D);     PIN * INV 1 50
                                    0.31  0.5  .35  0.12
```

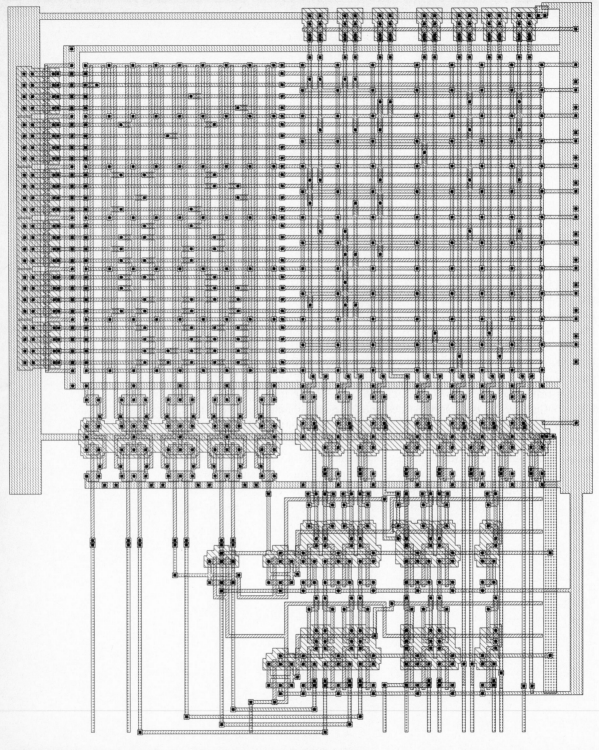

FIGURE 8.90 PLA layout for boundary-scan tap controller design

612

```
GATE NOR3        240  O=!(A+B+C);       PIN * INV 1 50
                                        0.29   0.4   .35   0.12
GATE NOR2        180  O=!(A+B);         PIN * INV 1 50
                                        0.17   0.26   .29   0.12
GATE NAND4       300  O=!(A*B*C*D);     PIN * INV 1 50
                                        0.16   0.14   .36   0.4
GATE NAND3       240  O=!(A*B*C);       PIN * INV 1 50
                                        0.14   0.14   .33   0.3
GATE NAND2       180  O=!(A*B);         PIN * INV 1 50
                                        0.18   0.14   .28   0.2
GATE INVERTER 120    O=!A;              PIN * INV 1 50
                                        0.29   0.14   .1    0.12
GATE BUFFER      180  O=A;              PIN * NONINV 1 50
                                        0.6   0.14   .2    0.12
GATE AND4        360  O=A*B*C*D;        PIN * NONINV 1 50
                                        0.73   0.14   .37   0.2
GATE AND3        300  O=A*B*C;          PIN * NONINV 1 50
                                        0.37   0.12   .64   0.1
GATE AND2        240  O=A*B;            PIN * NONINV 1 50
                                        0.47   0.12   .31   0.1
GATE AOI21       240  O=!(A+(B*C));     PIN * INV 1 50
                                        0.17   0.2   .28   0.2
GATE OAI21       240  O=!(A*(B+C));     PIN * INV 1 50
                                        0.17   0.2   .28   0.2
GATE MUX2        480  O=((A*!S)+(B*S)); PIN * NONINV 1 50
                                        0.14   0.2   .28   0.3
GATE MUX2-INV 420    O=!((A*!S)+(B*S));PIN * INV 1 50
                                        0.45   0.14   .6  0.1
```

MISII examined the logic equations, minimized the logic, and created a netlist in terms of the library gates. The netlist output of MISII is shown below:

```
.model tap-fsm-aoi.eqn
.inputs RESET TMS STATE<2> -TCK STATE<0> STATE<1> STATE<3>
.outputs NEXT_STATE<3> NEXT_STATE<2> NEXT_STATE<1> NEXT_-
STATE<0> NEXT_-RESET \
SELECT NEXT_ENABLE NEXT_SHIFTDR UPDATEDR CLOCKDR NEXT_-
SHIFTIR UPDATEIR CLOCKIR
.default_input_arrival 0.00 0.00
.default_output_required 0.00 0.00
.default_input_drive 0.14 0.12
.default_output_load 1.00
.gate INVERTER A=STATE<0> O=[324]
.gate INVERTER A=STATE<2> O=[323]
.gate NOR2 A=STATE<0> B=[323] O=[312]
.gate INVERTER A=STATE<3> O=SELECT
.gate INVERTER A=STATE<1> O=[327]
.gate OR2 A=STATE<0> B=[327] O=[329]
.gate NAND2 A=[329] B=STATE<2> O=[481]
.gate INVERTER A=RESET O=[350]
```

```
.gate INVERTER A=TMS O=[326]
.gate OAI21 A=[350] B=STATE<1> C=[326] O=[328]
.gate INVERTER A=[328] O=[449]
.gate NAND2 A=[481] B=[449] O=NEXT_STATE<2>
.gate NAND3 A=[312] B=SELECT C=NEXT_STATE<2> O=[471]
.gate NOR2 A=[323] B=[324] O=[302]
.gate NOR2 A=SELECT B=[327] O=[299]
.gate NAND2 A=[302] B=[299] O=NEXT_-RESET
.gate NAND2 A=NEXT_-RESET B=STATE<0> O=[461]
.gate NAND2 A=[461] B=TMS O=[479]
.gate NAND2 A=[326] B=STATE<0> O=[331]
.gate OAI21 A=[479] B=NEXT_STATE<2> C=[331] O=[334]
.gate OR2 A=[328] B=[334] O=NEXT_STATE<0>
.gate NAND2 A=[327] B=STATE<0> O=[335]
.gate OAI21 A=[471] B=NEXT_STATE<0> C=[335] O=[336]
.gate NOR3 A=STATE<1> B=[323] C=[326] O=[202]
.gate OAI21 A=[350] B=SELECT C=[202] O=[338]
.gate OR2 A=[336] B=[338] O=NEXT_STATE<3>
.gate AND2 A=NEXT_STATE<2> B=STATE<2> O=[272]
.gate AND2 A=NEXT_STATE<0> B=[272] O=[205]
.gate NOR2 A=STATE<0> B=STATE<3> O=[268]
.gate AOI21 A=[205] B=NEXT_STATE<3> C=[268] O=[258]
.gate NAND2 A=NEXT_-RESET B=STATE<1> O=[469]
.gate NAND2 A=NEXT_STATE<2> B=[469] O=[459]
.gate AOI21 A=RESET B=[326] C=[459] O=[255]
.gate NAND2 A=[258] B=[255] O=NEXT_STATE<1>
.gate OR2 A=STATE<2> B=[329] O=[343]
.gate INVERTER A=[343] O=NEXT_ENABLE
.gate OR2 A=STATE<3> B=[343] O=[433]
.gate INVERTER A=[433] O=NEXT_SHIFTDR
.gate INVERTER A=-TCK O=[344]
.gate NOR2 A=[323] B=[344] O=[248]
.gate NOR2 A=STATE<1> B=[324] O=[245]
.gate NAND2 A=[248] B=[245] O=[347]
.gate NOR2 A=STATE<3> B=[347] O=UPDATEDR
.gate NAND2 A=SELECT B=-TCK O=[348]
.gate OR2 A=[329] B=[348] O=CLOCKDR
.gate AND2 A=[433] B=NEXT_ENABLE O=NEXT_SHIFTIR
.gate NOR2 A=SELECT B=[347] O=UPDATEIR
.gate NAND2 A=STATE<3> B=-TCK O=[349]
.gate OR2 A=[329] B=[349] O=CLOCKIR
.end
```

This was fed to the NS Design System[43], where a program-generated schematic was created. This schematic was extracted, and netlist information and physical cell details were fed to the TimberWolf[44] placement program. The placement information was then returned to the NS VLSI design system where the circuit was automatically routed and a symbolic standard-cell layout created. After compaction to a set of CMOS-process design rules, the final mask layouts were available for backannotation, simulation, timing analysis, and size comparison. This complete integrated process (Fig. 8.91) completes automatically from changing any of the primary inputs in about five minutes. A number of "knobs" may be turned to affect the size of the layout. Starting from the top, a number of different logic-synthesis scripts and a variety of standard-cell libraries, varying from a 2-input NAND, a two-input NOR, and an INVERTER to more extensive collections of gates were tried. The number of rows that TimberWolf used was also varied.

Finally, the standard-cell height may be varied to create "performance" (large) or "area" (small) conscious layouts. Alternatively with the system described above, additional characterized symbolic standard cells may be added in about 5 minutes. The process may also be changed, thus creating a further dimension for optimization. Because all the characterization tools (simulation, timing analysis) work at the transistor level, new cells may be added with ease. Moreover with sophisticated placement programs, standard-cell layouts may be "reflowed" into unused space between larger fixed blocks. Table 8.10 summarizes the results.

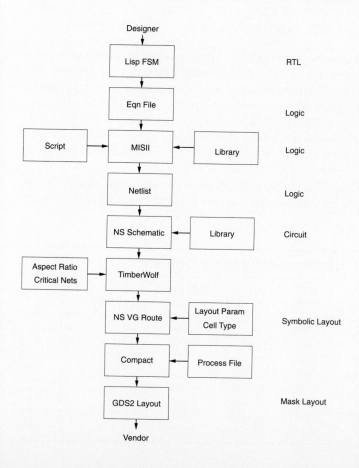

FIGURE 8.91 Control-logic design process

In Table 8.10, the *basic* library had (in addition to registers):

inverter

nand2

nor2.

The *tlw* library had the following gates (in addition to registers):

inverter

nand2,nand3,nand4,and2,and3,and4

nor2,nor3,nor4,or2,or3,or4

xor,xnor.

The *tlw-aoi* library had, in addition to the *tlw* library, the following four gates:

mux2,mux-inverting

and-or-invert $Z = !(A . (B + C))$

or-and-invert $Z = !(A + (B . C))$.

Table 8.10 shows the flexibility in aspect ratio that can be gained with standard-cell layouts. It also illustrates that metal3 provides layouts that are about half the size of the metal2 counterparts. Combined with RTL synthesis and logic optimization, the standard-cell approach provides an excellent means of capturing control logic (and other logic styles also). The final layout after compaction for one of the standard-cell layouts for a two-level-metal process is shown in Fig. 8.92. A metal3 layout is shown in Fig. 8.93 (also Plate 9).

Finally, because this state machine was very small, a gate-matrix layout was completed by hand (assuming that an automatic synthesis program was

TABLE 8.10   Standard-cell Layout Options

LOGIC	# ROWS	SIZE (mm²)	LIBRARY	STD-CELL HEIGHT	METAL LAYERS	DIM $X \times Y$ (mm)
Fig. 8.89	2	.120	tlw	50μ	2	.67 × .18
MISII	2	.118	basic	50μ	2	.62 × .19
MISII	2	.099	tlw-aoi	50μ	2	.52 × .19
MISII	3	.1	tlw-aoi	50μ	2	.4 × .25
MISII	4	.122	tlw-aoi	50μ	2	.34 × .33
MISII	2	.058	tlw-aoi	50μ	3	.49 × .12
MISII	3	.058	tlw-aoi	50μ	3	.34 × .17
MISII	3	.074	tlw-aoi	30μ	2	.4 × .19

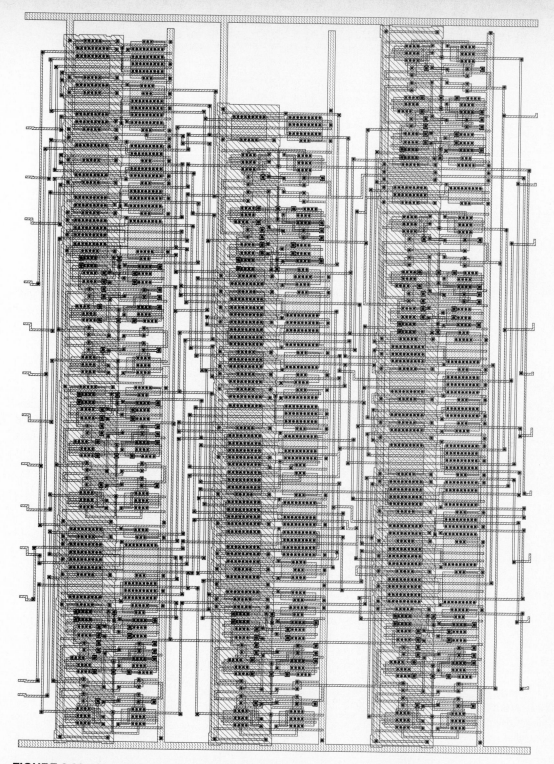

FIGURE 8.92 Metal2 standard-cell layout for boundary-scan tap controller

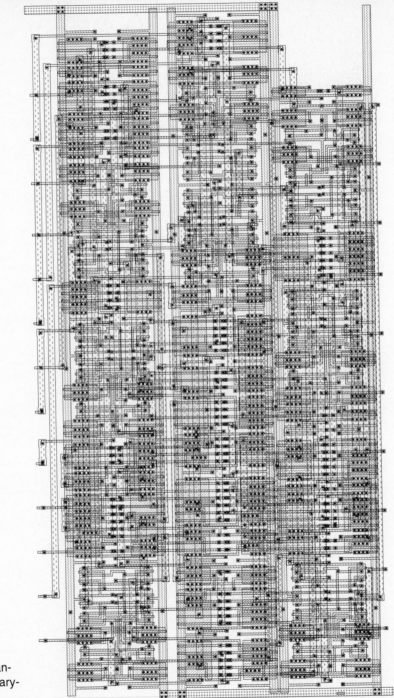

FIGURE 8.93 Metal3 standard-cell layout for boundary-scan tap controller

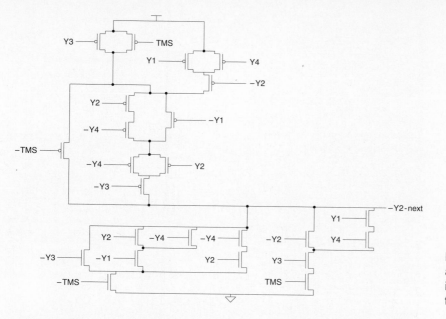

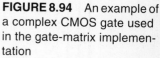

FIGURE 8.94 An example of a complex CMOS gate used in the gate-matrix implementation

available for production designs). In this design minimum-size transistors were used and large AOI gates were used to implement the more complex terms. Figure 8.94 shows the gate used to implement the Y_2-*next* term. The design could use dynamic logic and dynamic registers to reduce the size and power even further. The registers were placed to the right of the gate-matrix logic section. The layout is shown in Fig. 8.95.

Table 8.11 summarizes the area, speed, and power of the three implementations. The various implementations exhibit their strong points. The PLA is easy to design in a fully automated manner, is small, and for this application is fast enough. A dynamic version would have lower power dissipation. It is, however, fairly fixed in size. The standard-cell design may also be fully automated, as with the PLA, from a state-machine description. The speed and size may be varied over some range by logic-synthesis techniques and the avail-

Table 8.11   Area, Speed, and Power of Control Implementations

STYLE	AREA	SPEED	POWER
PLA	$.050 \text{ mm}^2$	10 ns	8 mW
Standard Cell (DLM)	$.099 \text{ mm}^2$	10 ns	6.3 mW
Standard Cell (TLM)	$.058 \text{ mm}^2$	~10 ns	~6 mW
Gate Matrix	$.032 \text{ mm}^2$	15 ns	1.5 mW

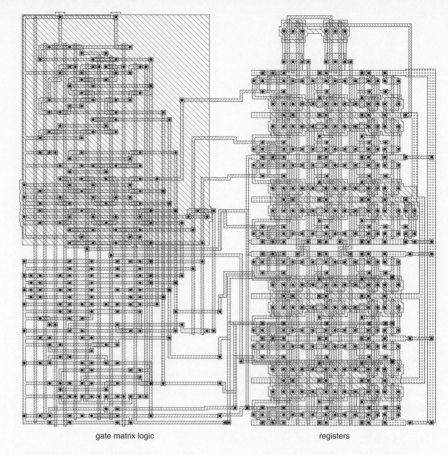

Figure 8.95 Gate-matrix layout for boundary-scan tap controller

gate matrix logic   registers

ability of a continuously variable symbolic layout library. The aspect ratio and shape of the resulting layout is primarily dictated by the placement program and in theory may be varied with a granularity down to the size of a gate (~20μ × 50μ). Finally, the gate-matrix layout represents the smallest and lowest power-dissipation case (which is largely due to the fact that this example is so simple). This comes at the cost of speed (although this is not important in this application) and time to design because an automated approach was not available. This approach is not recommended when suitable automation is not available except under extreme power or size restrictions.

8.5 Summary

This chapter has presented a range of sub-system designs in terms of data-path, memory, and control elements. Coupled with I/O structures, these form the basic building blocks from which larger systems may be hierarchically structured. How one goes about designing and implementing a given CMOS chip is largely affected by the availability of tools, the schedule, the complexity of the system, and the final cost goals of the chip. In general, the simplest and least expensive (in terms of time and money) approach that meets the target goals should be chosen. Given an ideal tool set that can draw on large libraries of predefined components, synthesis tools, and integrated VLSI backend tools, an almost continuous trade-off may be made to select a particular approach. The following rules may be applied. Where the design is of moderate complexity (a description that changes with time) and time to silicon is of paramount importance, an FPGA approach is probably suitable. If speed or complexity eliminate this approach, a gate-array is the next logical choice. To cost-reduce a gate-array or include sizable memories, a standard-cell approach is the next option. The next step to symbolic layout of regular arrays (i.e., datapaths, special memories, cellular arrays) and standard-cell control logic, is a big one. This step is usually taken on large-volume chips where the complexity, speed, and area dictate a custom approach. At this time in the evolution of VLSI, the final step of a full-custom, micron-tweaked design should almost never occur except for small, highly optimized circuits. As for capture methods, both HDLs and schematics have their strong points. It is likely that an increasing number of designers will move to HDLs, but schematics will be around for some time yet.

8.6   Exercises

1. Design an 8-bit parallel accumulator (adder and register) that is optimized for low power and has a power-down capability. Show how your circuit would retain the stored state.

2. Design a 32-bit parallel adder optimized for speed, single-cycle operation, and regularity of layout. Repeat the exercise with no layout restrictions.

3. Show how the layout of the parity generator in Fig. 8.25(a) may be designed as a linear column of XOR gates with a tree-routing channel.

4. Design an 8-bit barrel shifter (i.e., arbitrary left or right rotate) using multiplexers. Explain what the performance limitations of your design might be.

5. A four-section Finite Impulse Response (FIR) filter employs fixed coefficients and implements the function

$$Y = [2 \times X(t)] + [4 \times X(t-1)] + [8 \times X(t-2)] + [2 \times X(t-2)] + [4 \times X(t-3)],$$

where $X(t)$ is the sampled value of an 8-bit input at time t. Design a circuit to implement this function.

6. Design an FSM to control the stop lights at a four-way intersection with pedestrian crossing. Implement the control logic as a PLA, as multiple-level logic, and as a ROM microcontroller.

7. Tong and Jha[45] describe a binary divider. Design the base cell and show how you would complete a layout for the divider.

8. Complete a floorplan (showing clock and power and ground routing) and the circuit design for key cells (i.e., memory cell, row decoder, column decoder, sense amplifier) in a three-read-port, two-write-port register file. What simulations would you carry out to ensure the performance and justify the selection of power and ground bus widths?

8.7   References

1. Peter Denyer and David Renshaw, *VLSI Signal Processing: A Bit-Serial Approach,* Reading, Mass.: Addison-Wesley, 1985.
2. P. B. Denyer and D. J. Myers, "Carry-save arrays for VLSI signal processing," *VLSI 81* (John Gray, ed.), Edinburgh, Scotland: Academic Press, pp. 151–160.
3. Yasoji Suzuki, Kaichiro Odagawa, and Toshio Abe, "Clocked CMOS Calculator Circuitry," *IEEE Journal of Solid State Circuits,* vol. SC-8, no. 6, Dec. 1973, pp. 734–739.
4. Nan Zhuang and Haomin Wu, "A new design of the CMOS full adder," *IEEE JSSC,* vol. 27, no. 5, May 1992, pp. 840–844.
5. Kazuo Yano, Toshiaki Yamanaka, Takashi Nishida, Masayoshi Saito, Katsuhiro Shimohigashi, and Akihiro Shimizu, "A 3.8-ns CMOS 16*16-b multiplier using complementary pass-transistor logic," *IEEE JSSC,* vol. 25, no. 2, Apr. 1990, pp. 388–395.
6. M. Pomper, W. Biefuss, K. Horinger, and W. Kaschite, "A 32-bit execution unit in an advanced nMOS technology," *IEEE JSSC,* vol. SC-17, no. 3, Jun. 1982, pp. 533–538.
7. T. Sato, M. Sakate, H. Okada, T. Sukemara, and G. Goto, "An 8.5ns 112-b transmission gate adder with a conflict-free bypass circuit," *IEEE JSSC,* vol. 27, no. 4, Apr. 1992, pp. 657–659.
8. T. Sato, et al., *op. cit.*
9. M. Uya, K. Kaneko, and J. Yasui, "A CMOS floating point multiplier, IEEE International Solid-State Circuits Conference, *Digest of Technical Papers,* Feb. 1984, pp. 90–91.
10. J. Slansky, "Conditional-sum addition logic," *IRE Transactions on Electronic Computers,* vol. EC-9, Jun. 1960, pp. 226–231.
11. Albrecht Rothermel, Bedrich J. Hosticka, Gerhard Troster, and Juergen Arndt, "Realization of transmission-gate conditional-sum (TGCS) adders with low latency time," *IEEE JSSC,* vol. 24, no. 3, Jun. 1989, pp. 558–561.

12. T. Sato et al., *op. cit.*

13. M. Y. Tsai, "High density parity-checking circuits with pass transistors," *IBM Technical Disclosure Bulletin,* vol. 26, no. 3A, Aug. 1983, pp. 959–960.

14. W. R. Grifton and J. A. Hiltebeitel, "CMOS four-way XOR circuit," *IBM Technical Disclosure Bulletin,* vol. 25, no. 11B, Apr. 1983, pp. 6066–6067.

15. "TTL Databook," Texas Instruments, Dallas, Tex.

16. Gensuke Goto, Tomio Sato, Masao Nakajima, and Takao Sukemura, "A 54 * 54 regularly structured tree multiplier," *IEEE JSSC,* vol. 27, no. 9, Sept. 1992, pp. 1229–1236.

17. R. F. Lyon, "Two's complement pipeline multiplier," *IEEE Transactions on Communications,* vol. COM-24, Apr. 1976, pp. 418–425.

18. Peter Denyer and David Renshaw, *op. cit.*

19. Kunihiko Yamaguchi, Hiroaki Nambu, Kazuo Kanetani, Youji Idei, Noriyuki Homma, Toshiro Hiramoto, Nobuo Tamba, Kunihiko Watanbe, Masanori Odaka, Takahide Ikeda, Kenichi Ohhata, and Yoshiaki Sakurai, "A 1.5-ns access time, 78-μm^2 memory-cell size, 64-kb ECL-CMOS SRAM," *IEEE JSSC,* vol. 27, no. 2, Feb. 1992, pp. 167–174.

20. V. L. Rideout, "One-device cells for dynamic random-access memories," *IEEE Transactions on Electron Devices,* vol. ED-26, Jun. 1979, pp. 839–852.

21. Hideto Hidaka, Kazutami Arimoto, and Kazuyasu Fujishima, "A high density dual-port memory cell operation and array architecture for ULSI DRAM's," *IEEE JSSC,* vol. 27, no. 4, Apr. 1992, pp. 610–617.

22. Dong-Sun Min, Sooin Cho, Dong Soo Jun, Dong-Jae Lee, Yongsik Seok, and Daeje Chin, "Temperature-compensation circuit techniques for high-density CMOS DRAM's, *IEEE JSSC,* vol. 27, no. 4, Apr. 1992, pp. 626–631.

23. Hideto Hidaka, Kazutami Arimoto, Kazutoshi Hirayama, Masanori Hayashikoshi, Mikio Asakura, Masaki Tsukude, Tsukasa Oishi, Shinji Kawai, Katsuhiro Suma, Yasuhiro Konishi, Koji Tanaka, Wataru Wakamiya, Yoshikazu Ohno, and Kazuyasu Fujishima, "A 34-ns 16-Mb DRAM with controllable voltage down-converter," *IEEE JSSC,* vol. 27, no. 7, Jul. 1992, pp. 1020–1027.

24. Toshiaki Kirihata, Sang H. Dhong, Koji Kitamura, Toshio Sunaga, Yasuano Katayama, Roy E. Scheuerlein, Akashi Satoh, Yoshinori Sakaue, Kentaroh Tobimatsu, Koji Hosokawa, Takaki Saitoh, Takefumi Yoshikawa, Hideki Hashimoto, and Michiya Kazusawa, "A 14-ns 4-Mb CMOS DRAM with 300-mW active power," *IEEE JSSC,* vol. 27, no. 9, Sept. 1992, pp. 1222–1228.

25. Hideto Hidaka, Yoshio Matsuda, and Kazuyasu Fujishima, "A divided/shared bit-line sensing scheme for ULSI DRAM cores," *IEEE JSSC,* vol. 26, no. 4, Apr. 1991, pp. 473–478.

26. Toshio Yamada, Yoshiro Nakata, Junko Hasegawa, Noriaki Amano, Akinori Shibayama, Masaru Sasago, Naoto Matsuo, Toshiki Yabu, Susumu Matsumoto, Shozo Okada, and Michihiro Inoue, "A 64-Mb DRAM with meshed power line," *IEEE JSSC,* vol. 26, no. 11, Nov. 1991, pp. 1506–1510.

27. Walter H. Henkels, Duen-Shun Wen, Rick L. Mohler, Robert L. Franch, Thomas J. Bucelot, Christopher W. Long, John A. Bracchitta, W. J. Cote, Gary B. Bronner, Yuan Taur, and Robert H. Dennard, "A 4-Mb low-temperature DRAM," *IEEE JSSC,* vol. 26, no. 11, Nov. 1991, pp. 1519–1529.

28. Takeshi Nagai, Kenji Numata, Masaki Ogihara, Mitsuru Shimizu, Kimimasa Imai, Takahiko Hara, Munehiro Yoshida, Yoshikazu Saito, Yoshiaki Asao, Shizuo Sawada, and Syuso Fujii, "A 17-ns 4-Mb CMOS DRAM," *IEEE JSSC,* vol. 26, no. 11, Nov. 1991, pp. 1538–1543.

29. Travis N. Blalock and Richard C. Jaeger, "A high-speed sensing scheme for 1T dynamic RAM's utilizing the clamped bit-line sense amplifier," *IEEE JSSC,* vol. 27, no. 4, Apr. 1992, pp. 618–625.

30. Evert Seevinck, Petrus J. van Beers, and Hana Ontrop, "Current-mode techniques for high-speed VLSI circuits with application to current sense amplifier for CMOS SRAM's," *IEEE JSSC,* vol. 26, no. 4, Apr. 1991, pp. 525–536.

31. Travis N. Blalock and Richard C. Jaeger, "A high-speed clamped bit-line current-mode sense amplifier," *IEEE JSSC,* vol. 26, no. 4, Apr. 1991, pp. 542–548.

32. Shingo Aizaki, Toshiyuki Shimizu, Masayoshi Ohkawa, Kazuhiko Abe, Akane Aizaki, Manabu Ando, Osamu Kudoh, and Isao Sasaki, "A 15-ns 4-Mb CMOS SRAM," *IEEE JSSC,* vol. 25, no. 5, Oct. 1990, pp. 1063–1067.

33. Heribert Geib, Werner Weber, Erdi Wohlrab, and Lothar Risch, "Experimental investigation of the minimum signal for reliable operation of DRAM sense amplifiers," *IEEE JSSC,* vol. 27, no. 7, Jul. 1992, pp. 1028–1035.

34. Richard D. Jolly, "A 9-ns, 1.4-Gigabyte/s, 17-ported CMOS register file," *IEEE JSSC,* vol. 26, no. 10, Oct. 1991, pp. 1407, 1412.

35. Hirofumi Shinohara, Noriaki Matsumoto, Kumiko Fijimori, Yoshiki Tsujihashi, Hiroomi Nakao, Shuishi Kato, Yasutaka Horiba, and Akiharu Tada, "A flexible multiport RAM compiler for data path," *IEEE JSSC,* vol. 26, no. 3, Mar. 1991, pp. 343–349.

36. Alfredo R. Linz, "A low-power PLA for a signal processor," *IEEE JSSC,* vol. 26, no. 2, Feb. 1991, pp. 107–115.

37. Yoshihisha Iwata, Masaki Momodomi, Tomoharu Tanaka, Hideko Oodaira, Yasuo Itoh, Ryozo Makayama, Ryouhei Kirisawa, Seiichi Aritome, Tetsuo Endoh, Riichiro Shirota, Kazunori Ohuchi, and Fujio Masuoka, "A high-density NAND EEPROM with block-page programming for microcomputer applications," *IEEE JSSC,* vol. 25, no. 2, Apr. 1990, pp. 417–424.

38. Sateh M. S. Jalaledine and Louis G. Johnson, "Associative IC memories with relational search and nearest-match capabilities," *IEEE JSSC,* vol. 27, no. 6, Jun. 1992, pp. 892–900.

39. Sargur N. Srihari, "A special-purpose content addressable memory chip for real-time image processing," *IEEE JSSC,* vol. 27, no. 5, May 1992, pp. 737–744.

40. R. K. Brayton, G. D. Hachtel, C. T. McMullen, and A. Sangiovanni-Vincentelli, *ESPRESSO-IIC: Logic Minimization Algorithms for VLSI Synthesis,* The Netherlands: Kluwer Academic, 1984.

41. Gerard M. Blair, "PLA design for single-clock CMOS," *IEEE JSSC,* vol. 27, no. 8, Aug. 1992, pp. 1211–1213.

42. R. Brayton, R. Rudell, A. Sangiovanni-Vincentelli, and A. R. Wang, "MIS: a multiple-level logic optimization system," *IEEE Transactions on CAD of Integrated Circuits and Systems,* vol. CAD-6, no. 6, Nov. 1987, pp. 1062–1081.

43. James J. Cherry, "CAD programming in an object oriented programming environment," in VLSI CAD Tools and Applications (Wolfgang Fichtner and Martin Morf, eds.), Boston, Mass.: Kluwer Academic, 1987, Chapter 9.

44. C. Sechen and A. Sangiovanni-Vincentelli, "TimberWolf3.2: a new standard cell placement and global routing package," *Proceedings of the 23rd Design Automation Conference,* Las Vegas, Nev., 1986, pp. 432–439.

45. Qiao Tong and Niraj K. Jha, "Design of C-Testable DCVS Binary Array Dividers," *IEEE JSSC,* vol. 26, no. 2, Feb. 1991, pp. 134–141.

CMOS SYSTEM CASE STUDIES

PART 3

Part 3 comprises three case studies of CMOS chips or modules. They are included as examples to show the use of sub-systems introduced in Chapter 8 and Chapter 5. In addition, the designs were chosen to illustrate three different levels of design.

The first example describes an embeddable RISC microcontroller. This was implemented as a custom chip but could be implemented as a standard cell, gate array, or even an FPGA. The design emphasis here is from the architectural level down to the gate level.

In the second example, describing a television ghost cancellation chip, the architecture is also important but the detailed circuit and layout design have a huge impact on the commercial viability of the chip. The emphasis extends from the architectural level to the circuit level.

The final example illustrates a simple analog-to-digital converter where an individual inverter is the ultimate focus of attention.

The three case studies represent decreasing logic complexity and increasing emphasis on circuit design. As the complexity of a given CMOS system increases, the ability to individually address individual transistors, gates and sub-systems decreases. This trend is illustrated by these examples.

CMOS SYSTEM DESIGN EXAMPLES

9

9.1   Introduction

Many times VLSI design mirrors board-level system design, where standard components such as ALUs, memories, and logic gates are combined to form a specified function. This reality is reflected in CMOS standard-cell or gate-array libraries. However, the VLSI medium affords the designer the possibility of creating new components that break the barriers created by packaging. This might be a section of logic, phase-locked to a lower external clock, that operates at extremely high speed. Or it might be a special memory structure that merges logic and the structure of an algorithm to meet a speed, power, or cost objective.

In the first edition of this book, this chapter contained a number of examples that illustrated this principle. Implemented in today's technology many of these examples would form small components of much larger chips.

In this chapter, three examples are given that illustrate how the components developed in Chapter 8 and previous chapters are used in larger systems. The first is a contemporary high-speed RISC microcontroller that may be used for a variety of high-speed DSP applications. It provides an illustration of the flow of constraints from high-level decisions to the low-level implementation that results. The example also provides an example of three kinds of CMOS layout—datapath, memory, and control logic. While the implementation style used in this example is symbolic custom layout, the design could also be implemented (at an area and perhaps performance cost) as a standard-cell, sea-of-gates design or even an FPGA.

The second example might be thought of as a "classical" regularly structured CMOS VLSI design, in which one basic core cell is replicated many times to form the major portion of the chip. Many times such structures find application in high-speed digital-signal processing applications, such as video filtering and image processing.

The final example bridges the analog/digital gap by describing a simple 6-bit flash A/D converter. Each example demonstrates increasing emphasis on the detail of circuit and layout.

9.2 A Core RISC Microcontroller [†]

The first example presented is a RISC microcontroller that was designed as part of a much larger image processing chip. A block diagram of the processor is shown in Fig. 9.1. The processor had an on-chip instruction RAM and connected via a system address and data bus to the rest of the chip and special function units. As it is shown in Fig. 9.1, the processor is typical of

---

[†] This processor was designed by B. Edwards, C. Terman and N. Weste of TLW.

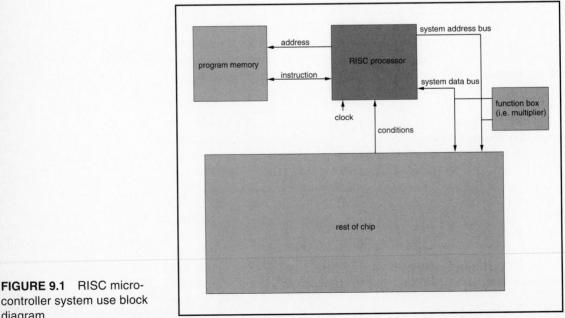

FIGURE 9.1 RISC micro-controller system use block diagram

embedded programmable cores which provide control for other dedicated processing elements.

The 16-bit processor had to run at 40MHz and affords a good example of an embedded processor that can be used for a wide variety of control and signal-processing applications. While a little more complicated than a minimal microcontroller, the concepts demonstrated in this example can be applied to a wide range of CMOS VLSI system problems.

We begin with the design of the instruction set, which in turn must be able to implement the operations required of the processor. We follow with a description of the pipeline architecture or the arrangement of memory and logic to enable the implementation of the instruction set in the required cycle time. The major logic blocks are then summarized. The layout of these blocks is then summarized. Finally, the methods of testing and verifying the processor are described.

9.2.1　Instruction Set

The instruction set defines what basic operations are possible with the processor and forms a high-level specification for the processor. The instructions in this instance are divided into two groups, namely,

- the control-transfer class.
- the ALU class.

The control-transfer class includes jump and call instructions. The ALU class includes arithmetic and logic operations. Other types of instructions might include operating system instructions (for a full microprocessor) and specialized I/O instructions (say, for a graphics accelerator).

9.2.1.1　Address Architecture

At this point there is a decision to make concerning the type of address architecture that the microcontroller is to implement. Options include a stack-address architecture, an accumulator architecture, and a multiple-register (two- or three-register) address architecture. Each dictates a particular register architecture. Figure 9.2 indicates some of the possibilities.

An accumulator architecture has a special register associated with the ALU that holds the intermediate results of computation (Fig. 9.2a). The computation for

```
c = a + b
```

would be

```
load a
plus b
store c
```

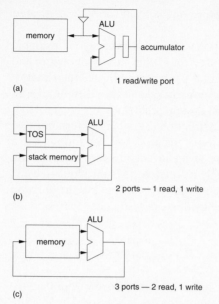

Figure 9.2 Alternative processor address architectures

In this example the register file has to be able to do one read or write at a time. At any one time only a single port is required, at the expense of taking three clock cycles to complete the add. However, a single-port register file can be implemented with a six-transistor static RAM cell, the smallest all-transistor static memory that can be implemented.

In a stack architecture (Fig. 9.2b), the register file implements a stack. The ALU uses the stack and a special register called the top-of-stack (TOS) to perform arithmetic operations. For instance, to complete the operation

```
c = a + b;
```

the following operations would be completed:

```
push a    pushes a onto stack (TOS=a)
push b    pushes b onto stack (TOS=b)
pop add   pops a and b, pushes a+b onto stack (TOS = a+b = c)
```

This requires that the register be read and written in a single cycle as the pop operation reads (a) and writes (c= a+b). The static RAM cell shown in Figs. 8.53 and 8.65 can achieve this by single-ended sensing of the bit lines and differential driving for write operations. This RAM cell is not much larger than the RAM cell that would be used in the stack architecture. Two decoders are needed, so the overall register file would be slightly larger than for the stack case. Although the add operation takes three cycles in the example above, in general as one operand may already be on the top of the stack, the add can be completed in two cycles.

A three-address architecture is what most modern RISC processors use (Fig. 9.2c). It completes the operation

```
c = a + b
```

(and most other operations) in one cycle. To do this requires a three-ported register file that can independently read two operands and write a third. This register file can be implemented using the structures shown in Fig. 8.65 (a regular RAM cell with multiple ports) or Fig. 8.66 (a register file). This register file is larger than the static RAMs for the accumulator and stack architectures due to the increase in memory-cell size and the three address-decoders that are required, but the architecture is potentially two or three times faster (for various reasons this speed increase is not always reflected in real programs).

The size (in words) of the register file has to be estimated. This is related to the maximum number of intermediate results that need to be held at one time. This might be estimated by implementing the chip function in a high-level language (C, FORTRAN, Pascal, Lisp) first and using this as a basis for estimation. In this case, 128 registers were dictated by the system architects.

At this point the designer might assess the areas of each type of register structure (if area was of importance) by completing a layout for each memory cell and decoder or, in the case of a gate-array or standard-cell design, finding the most appropriate register architecture. In the case at hand, the three-port address register architecture is chosen for the following reasons:

- Potentially provides the fastest architecture.
- Easy to program.
- Increase in area deemed not important (a custom layout is assumed).
- No complicated clocking required (minimizes design time).

An equal address space was allowed for random "external registers," yielding an 8-bit address field for reads and writes. A 1 in the MSB of the read or write address indicates a read or a write to a register on an external bus. This provides for a register-mapped I/O space for communication with external devices through the implementation of extra hardware if required.

9.2.1.2   ALU Class Instructions

With the register architecture fixed, an encoding for the instruction set may be proposed (Hex numbers are used). Because there are three addresses of 8-bits required, the following encoding for the ALU class was used, resulting in a 32-bit instruction. In the following description,

WR (write address) is one of
00-7F register file address

> `80-FF` external interface address
> RA (read port A address) is one of
> > `00-7F` register file address
> > `80-FF` external interface address
> RB (read port B address) is one of
> > `00-7F` register file address
> > `80-FF` external interface address
> OP is the op-code for the instruction.
> IT is the instruction type.

The first type of instruction is used for general arithmetic and logic operations.

IT = 1 three-address arithmetic instruction

#bits = 2	6	8	8	8
IT=1	OP	WR	RA	RB

The general operation is as follows:

```
WR = RA op RB
```

So for instance, a specific operation might include

```
WR[10] = RA[5] + RB[4]
```

In the instruction above, location 10 in the register file is replaced with the sum of locations 5 and 4.

A two-address literal instruction is provided as follows:

IT = 2 two-address with sign-extended 8-bit literal

IT=2	OP	WR	RA	LITERAL

The general operation is as follows:

```
WR = RA op LITERAL
```

A specific operation might be

```
WR[100] = RA[20] + 24
```

In the instruction above, location 100 in the register file is replaced with the sum of location 10 and the constant 24.

A single-address literal instruction is also provided. This allows a 16-bit literal to be loaded into the register file.

IT = 3 one-address with sign-extended 12-bit literal

IT=3	OP	WR	LITERAL

The general operation is as follows:

```
WR = op LITERAL (B is undefined)
```

A specific operation might be

```
WR[100] = -1 (#xFFFF)
```

that is, placing the constant FFFF in location 100.

The opcode, OP, is defined as follows

```
00      A+B                 04      A-B-1
01      A+B+1               05      A-B
02      A                   06      A-1
03      A+1                 07      A
08-0F   Undefined
10      all zeros           14      A and notB
11      A and B             15      A
12      notA and B          16      A xor B
13      B                   17      A or B
18      notA and notB       1C      notB
19      A xnor B            1D      A or notB
1A      notA                1E      A nand B
1B      notA or B           1F      all ones
20      logical left shift by SHIFTR
21-2F   logical left shift by 1 to 15 bits
30      arithmetic right shift by SHIFTR
31-3F   arithmetic right shift by 1 to 15 bits
```

SHIFTR is a special register (external address = 21) that allows a shift amount to be externally specified.

Although these assignments may seem random, they are linked to the implementation. For instance, bit 0 of the ALU opcode is the carry-in to the adder in the ALU, bit 1 forces the B bus to zero, and bit 3 inverts the B bus. Similarly, bits 3–0 are used directly by the Boolean logic to implement the functions outlined above (see Section 9.2.3.1). These assignments are used to eliminate control logic.

9.2.1.3   Control Transfer Instructions

The control transfer instructions implement jumps, call, and return. They are defined as follows:

IT=0	OP 6 bits	COND 8 bits	JA 12 bits

where

```
OP=20 Jump True
OP=22 Jump False
```

```
          OP=10 Call
          OP=08 Return
```

and

COND defines a condition code as follows:

```
00        ALU result negative
01        ALU result zero
02        Adder result had carry
03        ALU result was negative or zero
04        Boolean/Shifter result zero
05-0F     <Illegal>
1F-34     External conditions selectable by multiplexer
3F        True
40-FF     <Illegal>
```

JA specifies a 12-bit jump or call address. In the case of a CALL, the current program counter (PC) is pushed onto a stack, while a RETURN pops the PC from the stack.

9.2.2 Pipeline Architecture

The design was partitioned into six major blocks (Fig. 9.3a). The ALU_DP is responsible for performing arithmetic and logic operations. It also contains an interface to an external system-address and data bus. The register file is responsible for providing operands for the ALU_DP and storing the results of ALU operations. The PC (Program Counter) data path is responsible for calculating the next program counter value. It therefore has to deal with JUMPs, CALLs, and RETURNs. For the latter instructions, it implements an eight-deep stack. The Instruction Pipe datapath stores the instruction for a number of pipeline stages and performs comparisons to permit pass-around (see Section 9.2.2.1). The instruction RAM provides instructions for the processor and the control section provides for instruction decode and various other control operations.

The ALU_DP, register file, instruction pipe, and PC datapath were constructed using datapath techniques although they could be implemented as standard cells or sea-of-gates structures. A high-speed static RAM was used for program storage. A separate write port was provided to load the program RAM. A single control block was used to control all sections.

The operation of the microcontroller is as follows:

1. The PC presents an address to the instruction RAM, which in turn looks up an instruction to be applied to the machine. This includes the opcode for the ALU and the addresses to the register file.

2. The register file accesses the operands addressed by the instruction

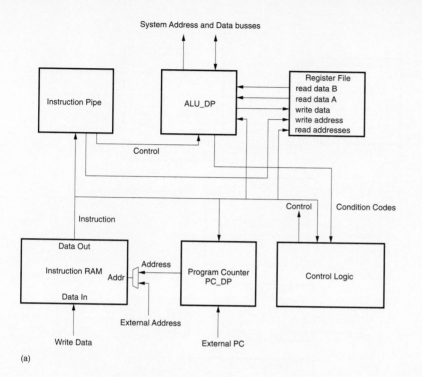

(a)

FIGURE 9.3 Processor-block diagram: (a) overview; (b) schematic

(either from the register file or external registers), and places them on the ALU input busses.

3. The ALU calculates the result including condition flags such as carry and zero.

4. The result is written to the register file or external registers.

If all of these operations were placed in one cycle, the cycle could be quite long. For instance, consider the following representative times:

Instruction RAM access	15 *ns*
Register file read	10 *ns*
ALU operation	15 *ns*
Register write	10 *ns*

which results in a total cycle time of around 50 *ns*. This exceeds the desired cycle time by more than a factor of two. Fortunately, this problem may be solved by the use of pipelining.

The microcontroller may be conveniently pipelined according to the steps outlined above, calling the stages the I, R, E, and W stages for Instruction fetch, register Read, Execute, and Write operations. One may of course choose to pipeline the machine differently; this scheme is chosen based on

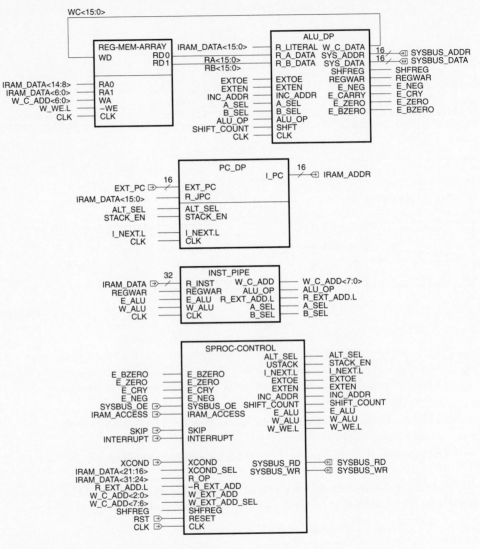

(b)

Figure 9.3 *(continued)*

the fact that the microcontroller has to run at 40 MHz and on some experience of how fast modules will run. In practice, a number of pipeline schemes might be explored, with the one that meets the speed requirement with the least design effort selected.

In the signal nomenclature, I_, R_, E_, or W_ preceding a name indicates that the data is valid in that pipeline stage. The pipeline operation can be visualized using the following pipeline diagram:

Cycle	I stage	R stage	E stage	W stage
0	Inst 1			
1	Inst 2	Inst 1		
2	Inst 3	Inst 2	Inst 1	
3	Inst 4	Inst 3	Inst 2	Inst 1
4	Inst 5	Inst 4	Inst 3	Inst 2

In cycle 0, instruction 1 is in the I stage, having been fetched from the instruction RAM. In cycle 1, it moves to the R stage and is used to address the register file to access the read operands in the instruction (RA and RB). Instruction 2 enters the I stage. Instruction 1 enters the E-stage in cycle 2, where it presents the opcode to the ALU. The read operands have also been fetched from the register file. Instruction 3 enters the I stage, and Instruction 2 moves to the R stage. Finally, in cycle 3 Instruction 1 enters the W stage, where the WR address is used to write the result of the ALU operation into the register file.

9.2.2.1   Bypassing, Result Forwarding, or Pass-around

In the above pipeline diagram it may be seen that the operands for instruction 2 and 3 have been read by the time that the result of instruction 1 is written back to the register file. This requires that the W stage write data be forwarded to the R stage or E stage if these stages require the data that is being written to the register file. For instance, if the result of instruction 1 were used in instruction 2, then the data being written to the register file in cycle 3 would also have to be passed to the E stage for instruction 2. This is done by comparing the operand addresses in the R stage and E stage with those in the W stage and controlling a set of multiplexers that feed the appropriate operand to the ALU.

Consider the following code fragment:

```
ADD A,B,C     Add A, B and place in C
SUB D,C,F     Subtract C from D and place in F
XXX
YYY
```

The following pipeline diagram represents the code sequence:

Cycle	I stage	R stage	E stage	W stage
0	ADD A,B,C			
1	SUB D,C,F	ADD A,B,C		
2	XXX	SUB D,C,F	ADD A,B,C	
3	YYY	XXX	SUB D,C,F	ADD A,B,C

In cycle 3, the result of adding A and B is being written to location C. However, C is required in the E stage to compute $D - C$. Thus the normal read path from the register file has to be bypassed to allow the current (W stage) value of C to be passed to the ALU.

9.2.2.2 Conditional Branching

The condition code (if coming from the ALU) is calculated late in the E stage. Any jump occurs in the cycle after the condition is calculated and, due to the pipelining, the instruction after a jump is executed.

Consider the following code sequence that implements a branch based on the result of an ADD instruction:

```
     ADD A,B,C        Add A, B and place in C
     JMP ZERO,FOO     Jump to location FOO if the result of
                      A+B(C)=0
     SUB E,F,G        Subtract F from E and place in G
     AND X,Y,Z        AND X and Y and place in Z
FOO: OR P,Q,R         Branch Target FOO
     ADD A,R,C
```

The following pipeline diagram represents the code sequence where the branch is not taken:

Cycle	I stage	R stage	E stage	W stage
0	SUB E,F,G	JMP ZERO	ADD C,A,B	
1	AND D,G,H	SUB E,F,G	JMP ZERO,FOO	ADD C,A,B
2		AND D,G,H	SUB E,F,G	JMP ZERO
3			AND D,G,H	SUB E,F,G

In cycle 0 the ADD instruction is executed and the ZERO condition is calculated. The JMP instruction is executed in cycle 1, the condition being registered to the W stage. In cycle 2 the SUB instruction is unconditionally executed, and in cycle 3 the AND instruction is executed.

The following pipeline diagram represents the code sequence where the branch is taken:

Cycle	I stage	R stage	E stage	W stage
0	SUB E,F,G	JMP ZERO	ADD C,A,B	
1	OR P,Q,R	SUB E,F,G	JMP ZERO,FOO	ADD C,A,B
2	ADD A,R,C	OR P,Q,R	SUB E,F,G	JMP ZERO
3		ADD A,R,C	OR P,Q,R	SUB E,F,G

The sequence here is the same except that in cycle 1 the OR instruction moves to the I stage. In cycle 3 this instruction is executed. Note that the SUB instruction is still executed in cycle 2.

9.2.2.3 Subroutine Call and Return

Consider the following code sequence, which demonstrates a Call and Return Sequence:

```
      ADD A,B,C
      CALL FOO
      SUB E,F,G
BAZ:  XXX
      YYY
      ...
FOO:  ADD J,K,L
      RETURN
      SUB X,Y,Z
```

The execution of this sequence is shown below:

Cycle	I stage	R stage	E stage	W stage
0	SUB E,F,G	CALL FOO	ADD A,B,C	
1	ADD J,K,L	SUB E,F,G	CALL FOO	ADD A,B,C
2	RETURN	ADD J,K,L	SUB E,F,G	CALL FOO
3	SUB X,Y,Z	RETURN	ADD J,K,L	SUB E,F,G
4	XXX	SUB X,Y,Z	RETURN	ADD J,K,L
5	YYY	XXX	SUB X,Y,Z	RETURN
6		YYY	XXX	SUB X,Y,Z
7			YYY	XXX

As with the JUMP instruction, the instructions after the CALL and RETURN instructions are also executed.

9.2.2.4 I/O Architecture

The I/O architecture in this example is a condensed version of the real I/O architecture to keep the example simple. Five registers are provided in the I/O space of the processor (seven are listed below but some are read/write or otherwise utilize the same physical register). They are as follows:

- SBRAR (address = 0) Sysbus Read Address Register. When this register is written, it causes a read on the system bus. If an SBRAR is written on cycle i, then on cycle $i + 1$, the SBRAR address is output to the SYS-ADDR bus. On cycle $i + 2$, the data is returned on the SYS-DATA lines. On cycle $i + 3$, the data may be used in the ALU. Thus a code segment to read data from the system bus and increment it might be

```
WR[20] = RA[80](SBRAR)
XX
YY
WR[20] = RA[20] + 1
```

- SBWAR (address = 81 (hex)) Sysbus Write Address Register. This register holds the next address of a Sysbus write. It uses the same register as SBRAR.

- SBWDR (WR = 82) Sysbus Write Data Register. When this register is written, it causes a write on the system bus.

- SBWDR-INC (address = 83) This causes an autoincrement in the SBWAR register, which provides for a simple DMA capability.

- SBRDR (address = 84-read-only) This register contains the data from the last system bus read.

- REGWAR (address = 85) This register is used as a register-file write address in ALU class instructions when WR = C0 – DF.

- SHFREG (address = 86) This register allows a shift amount to be specified for the shifter when the ALU-OP is 20 or 30.

9.2.3   Major Logic Blocks

With the instruction set defined, and address and pipeline architectures decided, the next step is to define logic and storage elements that will realize these architectures. Just how one translates architecture into an RTL design varies. It usually requires a stepwise refinement process where a design is proposed, simulated, and modified to correspond more closely to the required behavior. Most often, previous experience will aid in determining good directions. It is a skill that improves as more designs are completed.

In this section, an RTL design will be presented in schematic form as a finished design. Because the style of processor design is fairly generic, it is hoped that this example will provide readers with a starting point for their own designs. The complete schematic for design for the controller core (sans instruction memory and I/O devices) is shown in detail in Fig. 9.3(b).

9.2.3.1   ALU_DP

The ALU_DP module is responsible for computing collecting operands, collecting the results, and interfacing with external modules. It is divided into three main sections:

- The I/O-REGS.
- The ALU proper.
- The EXT-BUS-DP.

The module is shown in Fig. 9.4.
    The inputs are:

R_A_DATA and *R_B_DATA* busses—These are the read data busses from the register file.

R_LITERAL—Literal bus from the instruction RAM.

A_SEL and *B_SEL*—A and B operand select control.

ALU_OP—The ALU op-code.

SHFT—The shift amount for the shifter.

EXTOE, EXTEN controls—Various controls to the EXT-BUS-DP module for loading registers and tristating busses.

INC_ADDR—An increment control for the autoincrementing address register (SBWDR_INC) in the external-bus module.

CLK—The clock.

The outputs are:

SYS_ADDR (system address bus)—Used to address external modules.

SYS_DATA (system data bus)—Used to transfer data to and from external modules.

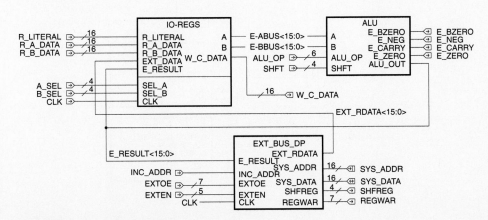

FIGURE 9.4 ALU_DP-block diagram

W_C_DATA—Write data bus to the register file.

SHFREG —Shifts amount to control block.

REGWAR—Alternates register write address to INST_PIPE.

CONDITION-CODES—Flags from ALU operation such as ZERO (*E_BZERO and E_ZERO*), NEGATIVE (*E_NEG*), and CARRY (*E_CARRY*).

First we will deal with the two modules that surround the ALU.

9.2.3.1.1 IO-REGS

The IO-REGS are responsible for providing the appropriate operands to the ALU. The A and B port of the ALU have identical structures (IO-REG) that provide operands from:

- the register file.
- the literal bus (from the instruction RAM).
- W-stage data (for pass-around) and E-stage data.
- external data from the external bus (*EXT_DATA*).

The IO-REGS circuit is shown in Fig. 9.5 and the IO-REG subblock is shown in Fig. 9.6. The modules are composed of multiplexers, registers, and

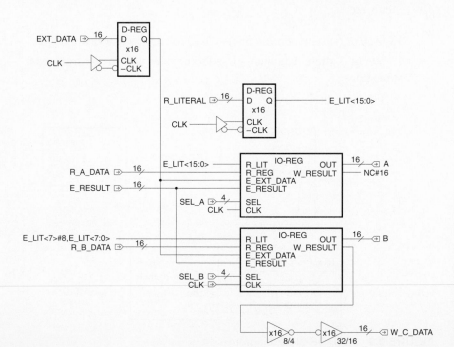

Figure 9.5 IO-REGS module schematic

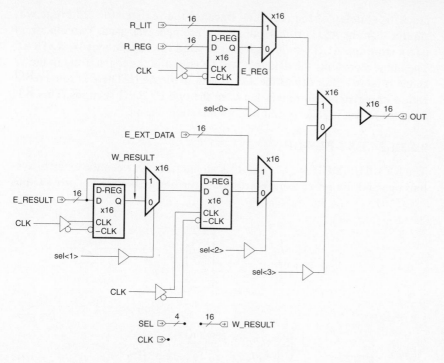

FIGURE 9.6   IO-REG module schematic

buffers. As a single clock is being used, the register structure in Fig. 9.25(a) was chosen for the following reasons:

- Static (A habit; dynamic registers could be used but if area is not that affected, static allows the clock to be stopped and the static registers are not as susceptible to noise). Static registers also allow IDDQ testing (see Chapter 7).
- Small number of transistors.

In Fig. 9.5 the literal field from the instruction RAM, *R_LITERAL<15:0>*, is registered to create the E-stage version of this data, *E_LIT<15:0>*. Note that port A receives the full 16-bit field of the literal, while the B port receives an 8-bit sign-extended version (*E_LIT<7>#7, E_LIT<7:0>*). The register-file write data is taken from the B IO-REG module. A further register stores the *EXT_DATA* bus. In the IO-REG module (Fig. 9.6), a register stores the R-stage data from the register file. The decision to place this register here rather than in the register file relates to the difference in critical path between the ALU and the read access of the register file. In this case, the relatively long delay of driving a bus from the register file to the ALU is placed in the R stage. If the register were placed in the register file, this delay would be in the E stage (because it would be added to the clock-to-Q delay of the register). Because the register-file read time was shorter than the ALU critical path, this placement was used. Two other registers store the ALU result (*E_RESULT*) and the subsequent *W_RESULT*. The latter represents the data that will be stored in the register file in the W stage. A large buffer drives the *A* or *B* data to the ALU. There is some replication in registers and muxes between the two IO-REG sections. This was done for layout regularity and bus optimization reasons.

9.2.3.1.2 EXT-BUS-DP

The EXT-BUS-DP (Fig. 9.7) provides an interface between the external system bus and the processor. A write-data register (SBWDR) is provided to

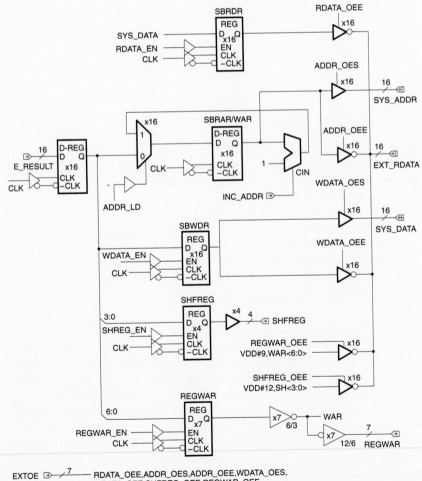

Figure 9.7 EXT-BUS-DP module schematic

which data may be written from the ALU to the *SYS_DATA* bus. An address register (SBRAR/WAR) is provided to supply addresses to the system address bus (*SYS_ADDR*). An incrementer is provided with the address register to aid in simple DMA-type operations. The *SYS_DATA* may also be registered (SBRDR) for reads from the *SYS_DATA* bus. These registers may be tristated onto the *EXT_RDATA* bus, which is an input of the IO-REGS module. The SHFREG stores a shift amount that may be used by the shifter. In the actual chip, this module had a few extra registers for other special operations. As mentioned previously, REGWAR provides a write address for the register file for ALU operations with addresses in the range C0–DF.

9.2.3.1.3   ALU

The ALU is divided into three subsections, namely,

- the adder.
- the Boolean unit.
- the shifter.

While it would be possible to merge the Boolean unit with the adder (à la the 181 ALU), the designers happened to like this partitioning because the instruction decode is simple and the modularity allows the Boolean and shifter to be dropped if those blocks are not required. Also a range of adders may be used to achieve different size and speed requirements. As shown in Fig. 9.8, each functional unit takes its inputs from the *A* and *B* buses and conditionally tristates the result onto the *ALU_OUT* bus.

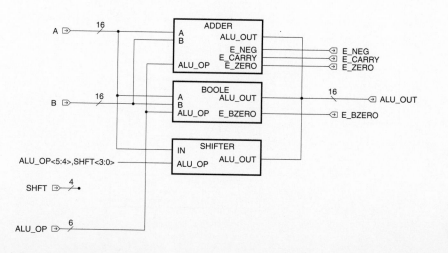

FIGURE 9.8   ALU module schematic

The adder module chosen for this design is the static Manchester carry stage shown in Fig. 8.19(a). The reasons are as follows:

- It is a static design requiring no clocking (straightforward design).
- It could be designed to have a very small pitch, which is important in keeping datapath height small and delays low.
- It had been used in previous designs successfully.
- It provides a better speed trade-off for the adder (simpler PG and SUM generators—see Figs. 8.18 and 8.34).

The adder module is designed in 4-bit sections and consists of a PG generator, a carry lookahead section, and a SUM generator. Figure 9.9 shows a schematic of a single-bit section that also represents the physical layout—i.e., the carry section is sandwiched between the PG generator on the left and the SUM logic on the right. Figure 9.10 shows the schematic for a 4-bit section. Complementary CMOS logic is used throughout the adder except for the carry lookahead gate, which is a pseudo-nMOS NAND gate. Variations of this adder might include the pseudo-nMOS carry-lookahead gate shown in Fig. 8.17. Other adders that might be used if the speed is not too stringent might include the adder shown in Fig. 8.7(b). This is potentially the smallest adder that might be used, and the speed can be altered somewhat by adjusting the size of the CARRY gate transistors. This adder, used as a carry-select adder, might also prove useful. If in doubt, the designer might complete some initial simulations of various adders at this point. If possible, you should actually design the layouts and backannotate the schematics or HDL when simulating because this will give the best indication of the final speed that might be attained. Remember to simulate at the Worst Case Speed corner of the process. Figure 9.11 shows the complete 16-bit adder complete with control circuitry. A zero and negate circuit (called BUS-OP in this

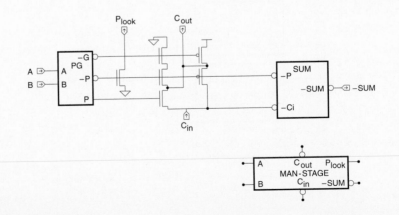

Figure 9.9 Manchester stage used in adder

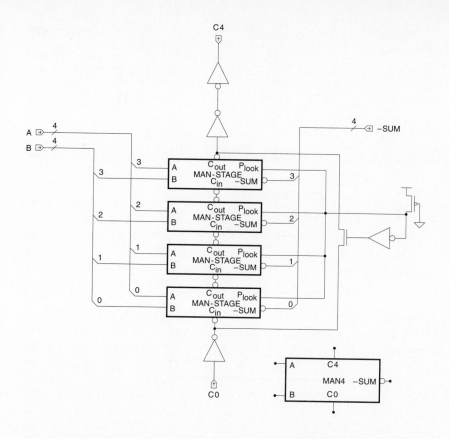

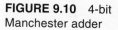

FIGURE 9.10 4-bit Manchester adder

design) is required on one input (*B*) to zero the operand or set *B* to all ones (for adding zero, −1 or just passing *A*) and conditionally inverting the operand for subtraction operations. While this could be implemented using an XOR gate and an AND gate as shown at the bottom of Fig. 9.11, the pass-gate implementation shown in Fig. 9.25(b) is faster and smaller and was used in this design. A zero detect is placed on the output of the adder (*E_ZERO*). It is placed here rather than on the *ALU_OUT* bus in Fig. 9.8 because this implementation is faster by the time it takes the tristate buffers in the adder and Boolean unit to drive that bus. As a result, the zero circuit is replicated in the Boolean unit (*E_BZERO*). The circuit used for this is a pseudo-nMOS NOR gate. Other complementary CMOS solutions could also be used (see Fig. 8.28). The ALU-OP instruction decode is also shown. This is fairly straightforward. For instance, ALU-OP=5 is A-B. Thus *INVERT* = 1, *CIN* = 1, and *ALU-OP<5:4>* = 0. This inverts the *B* operand, adds 1 to the adder via the carry in, and enables the adder tristate buffer.

The module used for the Boolean unit is a transmission-gate circuit based on the structure shown in Fig. 8.33, and is shown in Fig. 9.12. The

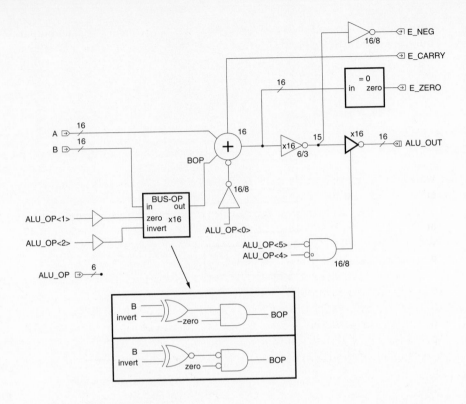

Figure 9.11 16-bit adder schematic

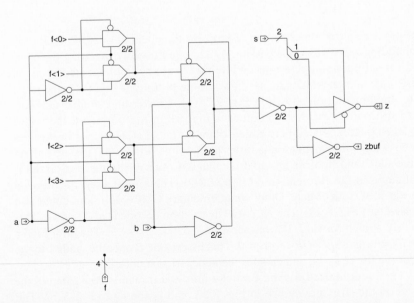

Figure 9.12 Boolean bit

complete Boolean unit is shown in Fig. 9.13 that includes a tristate output buffer, decoding logic, and a zero detect. As an example of a Boolean instruction, consider that ALU-OP=16 is A⊕B. In Fig. 9.12, $f<0>=0$, $f<1>=1$, $f<2>=1$, and $f<3>=0$. Thus the truth table is

A	B	OP
0	0	0
1	0	1
0	1	1
1	1	0

which implements the XOR operation. *ALU-OP<5:4>=01* enables the Boolean unit onto the *ALU_OUT* bus.

The shifter is a tristate-buffer multiplexer structure using a left shifter and a right shifter with direct decoding of the shift amount (see Fig. 8.47 with tristate-buffers replacing transmission gate muxes). Tristate buffers were used to achieve the desired speed.

A word about the critical path. In the ALU section, the critical path starts in a register, passes through the IO-REG muxes, through the BUS-OP circuit (conditionally negating and zeroing the operand), through the adder and into the condition code logic where the conditions are registered. The critical path as reported by the timing analyzer is shown below.

Summary of path-clock rising to clock rising delay of 22.7 *ns* due to a delay of 21.8 *ns* at node 3836 and a .9 *ns* setup time into the register:

```
CK to CK 22.7ns (setup time 0.9) data node 3836 at CK + 21.8
S>sproc>SPROC-CONTROL-1>REG-17>D-REG-MUXSTANDARD
```

The delays have the form:

```
Node name cumulative-delay (this-node-delay)
node-path
```

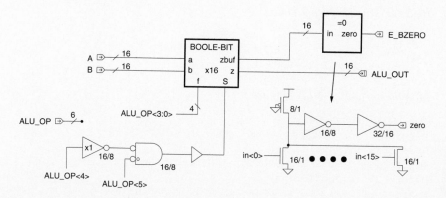

FIGURE 9.13   Boolean unit schematic

Clock buffer delay:

```
Node CK 0.0ns (0.9ns)
S>sproc>INST_PIPE-2>CLOCK-BUFFER-3>CG-INVERTER-3>NMOS-4
Node 4329 0.9ns (0.6ns)
S>sproc>INST_PIPE-2>CLOCK-BUFFER-3>CG-INVERTER-4>PMOS-4
```

Clock-to-Q delay of register:

```
Node 4338 1.5ns (1.6ns)
S>sproc>INST_PIPE-2>DP-D-REG-10
```

Instruction Pipe logic—Figure 9.18:

```
Node 4277 3.1ns (0.9ns)
S>sproc>INST_PIPE-2>CG-INVERTER-47>PMOS-4
```

IO-REGS delay (Fig. 9.6):

```
Node 1364 4.0ns (1.6ns)
S>sproc>SP-DP-1>IO-REGS-1>IO-REG-2>MUX-DRIVER-1>
      CG-INVERTER-3>NMOS-4
Node 3007 5.6ns (1.0ns)
S>sproc>SP-DP-1>IO-REGS-1>IO-REG-2>MUX-DRIVER-1>
      CG-INVERTER-4>PMOS-4
Node 3008 6.7ns (1.0ns)
S>sproc>SP-DP-1>IO-REGS-1>IO-REG-2>DP-MUX2-3#2>
      CG-TG-2>NMOS-1
Node 3082 7.6ns (0.7ns)
S>sproc>SP-DP-1>IO-REGS-1>IO-REG-2>BUS-DRV-2#2>
      CG-INVERTER-9>PMOS-4
Node 3126 8.3ns (0.6ns)
S>sproc>SP-DP-1>IO-REGS-1>IO-REG-2>BUS-DRV-2#2>
      CG-INVERTER-10>NMOS-4
```

BUS-OP gate (Figure 9.11):

```
Node 2372 9.0ns (0.9ns)
S>sproc>SP-DP-1>ALU-1>ADDER-1>BUS-OP-2#2>
      CG-INVERTER-14>PMOS-4
```

Manchester-adder delay (Figure 9.11):

```
Node 2313 9.9ns (4.0ns)
S>sproc>SP-DP-1>ALU-1>ADDER-1>MAN-16-2>MAN-4-LSB-1
Node 2381 13.9ns (0.7ns)
S>sproc>SP-DP-1>ALU-1>ADDER-1>MAN-16-2>MAN-4-1#0
Node 2379 14.6ns (0.7ns)
S>sproc>SP-DP-1>ALU-1>ADDER-1>MAN-16-2>MAN-4-1#1
Node 2380 15.3ns (4.0ns)
S>sproc>SP-DP-1>ALU-1>ADDER-1>MAN-16-2>MAN-4-1#2
```

```
Node 2306 19.3ns (1.6ns)
S>sproc>SP-DP-1>ALU-1>ADDER-1>CG-INVERTER-15#15>NMOS-4
Node 2343 20.9ns (0.7ns)
S>sproc>SP-DP-1>ALU-1>ADDER-1>CG-INVERTER-7>PMOS-4
```

LE gate in control logic (Figure 9.21):

```
Node 1436 21.7ns (0.1ns)
S>sproc>SPROC-CONTROL-1>OR-17>NOR2STANDARD>
     N-CHANNEL-MOSFET-94
Node 3836 21.8ns
```

From this, it may be seen that ~48% of the timing budget is used by the 16-bit adder, 18% is spent in the register, and 22% is spent in the input-operand switching. Because this was not the overall worst path in the design and the timing was close to the design goal of 40 MHz, this timing was deemed sufficient. If improvements were required, the adder and IO-REGS could be further scrutinized (or the design could be transferred to a smaller process).

9.2.3.2   Register File

The register is arranged as a 32×64 memory with 4:1 column multiplexing. The register-file structure shown in Fig. 8.66(a) is used as the three-port register file. Figure 9.14(a) shows the transistor sizes used in the memory cell along with the read- and write-row decoders. The ratios of the write-path transistors and the storage inverter are chosen to ensure correct writes over all process corners. Two read-access transistors are used to provide dual read ports. Two read-row decoders and a write-row decoder that are based on a complementary 5-bit AND gate are used.

A 2-bit write address performs a column select for write operations, while 2-bit read addresses employ a multiplexer decoder (Fig. 8.63) to yield a 16-bit result. The column circuit, which is a 4-bit section, is shown in Fig. 9.15(a). It includes two 4:1 multiplexers (using single n-channel transistors), a sense inverter, and a buffer, which select the data to be routed to the IO-REGS. The write circuitry consists of a register to hold the write data and buffers for the four write-data lines ($WD<3:0>$) and write-strobe logic to allow the selective writing of any of the four columns ($WAS<3:0>$).

A write operation proceeds by placing an address on the write-address lines and then deasserting the clock. This causes one of four column WRITE-STROBE (WAS) signals to be asserted, which writes data into the cells with the word line asserted. Latches were added to the WRITE-ADDRESS (WA) lines to improve the speed.

Reads are totally static. For read operations, the critical path begins in the Instruction RAM, the output of which is passed to the word-line decoder of the register file. This in turn drives the row line of the register file, accessing a register. This triggers bit-line changes, which are demultiplexed and

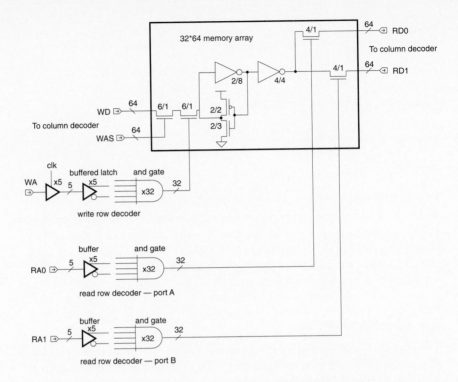

Figure 9.14 Register file partial schematic

driven to the IO-REGS module in the ALU-DP. There is a large spectrum over which the speed (and power) of the register file may be changed by sizing the row-decoder drivers, the decoder, AND gates, and the word-line drivers. In addition, moving the read registers into the register file could also be used to improve speed if necessary. Additionally, the bit lines could have been precharged to improve speed. In this case none of these improvements were required. A typical SPICE simulation for verifying the read-access time is shown in Fig. 9.15(b). This includes parasitic capacitances that are determined from a mask extraction of the layout of key cells of the register file (for instance, from a layout of the register file memory cell, the word-line and bit-line capacitances may be estimated). Figure 9.15(c) shows an address-input changing and the word-line response and the bit-line change. Figure 9.15(d) shows the bit-line response, the delay through the n-channel column multiplexer, and the final output. This shows that the delay from the address change to valid output is around 10 ns. The bit-line sense amplifier is ratioed to move the threshold voltage toward V_{SS}, which improves the sense time. For simplicity, the bit-line is not precharged, although some speed increase could be achieved by precharging and using a more exotic sensing scheme. However, this met the speed goals by almost half a clock cycle, so no further design effort was required.

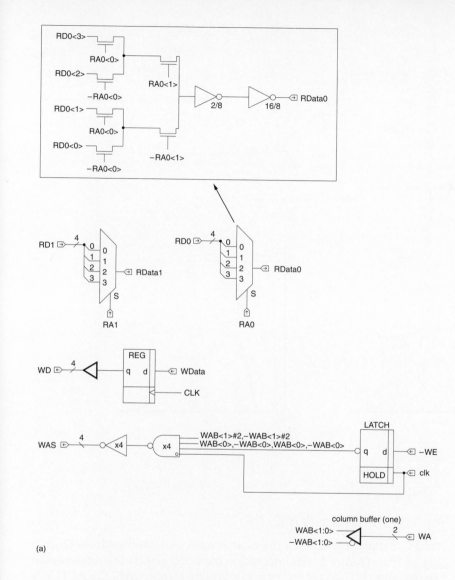

(a)

FIGURE 9.15   Register file read: (a) column decoder circuit; (b) SPICE model; (c) waveforms; (d) waveforms

A point that arises here is the importance of being able to rapidly prototype a design to assess where the speed bottlenecks are. Frequently this can be done with pencil and paper or more conveniently with a good top-to-bottom VLSI CAD system. Completing a rough "first draft" of a design can often highlight critical parts of the design ahead of time. This prevents work on areas that do not affect the performance of the overall system. Frequently, designers tend to take a myopic view of the design and can spend unnecessary time optimizing something that does not matter ("disappearing down the optimization rat-hole").

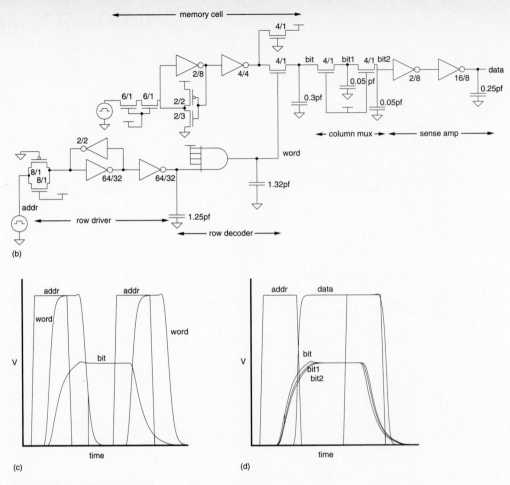

Figure 9.15 *(continued)*

9.2.3.3 PC Datapath (PC_DP)

The PC datapath computes the next program-counter address (*I_PC*) (Fig. 9.16a). During normal operation the *I_PC* selects the incremented version of the PC (*I-NOR-PC<15:0>*). When a JUMP, CALL, or RETURN operation occurs, the I_PC register is loaded with the R_JPC (from the instruction) in the case of a JUMP or a CALL, or with the stack in the case of a RETURN (*I-ALT-PC<15:0>*). The multiplexers to achieve this can be seen in the figure. The signal *I_NEXT* selects either the "normal" PC (*I-NOR-PC*) on ALU operations or conditional jumps that fail. It selects the "alternate" PC (*I-ALT-PC*) on subroutine call, return, and conditional jumps that are taken. Note that the *I_PC* signal is duplicated, feeding the *I_PC* to the incrementer. This was done to improve speed because a critical path exists from the *I_NEXT.L* signal, through the incrementer and into register NPC, that switches the *I_PC* multiplexer. The regular *I_PC* output of the

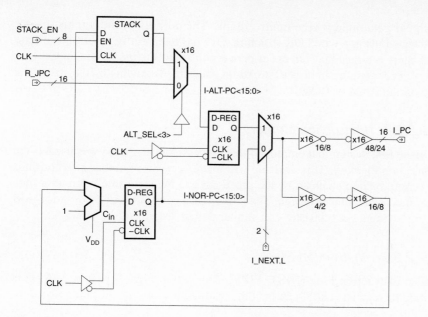

(a)

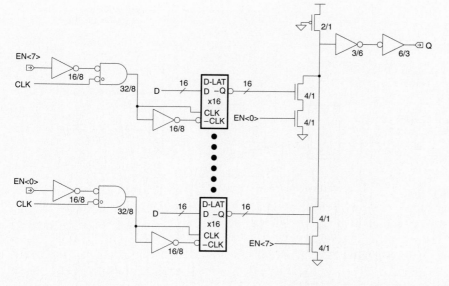

(b)

FIGURE 9.16 PC_DP

module is heavily loaded because it drives external modules. The duplicated path provides a faster *I_PC* to the incrementer, thereby improving speed.

The subroutine stack with a depth of eight is constructed from eight 16-bit latches. The stack design is shown in Fig. 9.16(b). One of eight latches may be conditionally written when the clock is low depending on an enable signal generated by the control logic. The outputs of the 8 registers are selected using a wired OR structure. One has to carefully design this circuit as there is a race that can occur between the data and the gated clock.

The module can be a 10- to 16-bit datapath, depending on the size of the program RAM or ROM.

9.2.3.4 Instruction Memory

In the first design that employed this processor, static RAM was used as the program memory. This decision was made because a number of algorithms were to be implemented by the processor and the algorithms were in a state of flux at the time of design. The RAM was a conventional, fast, static RAM employing a six-transistor cell similar to that described in Chapter 8.

9.2.3.5 Instruction Pipe

The Instruction Pipe (INST_PIPE) is shown in Figs. 9.17 and 9.18. This module registers certain parts of the instruction and generates the select con-

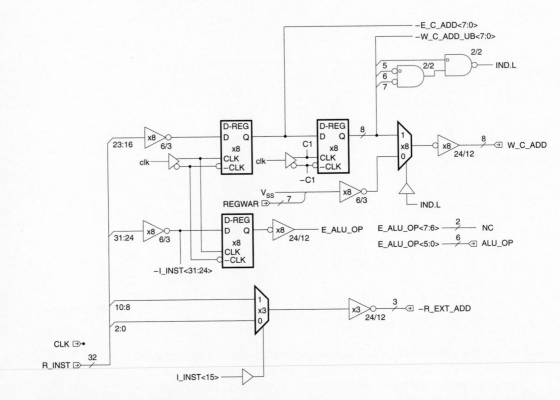

Figure 9.17 INST_PIPE
module registers

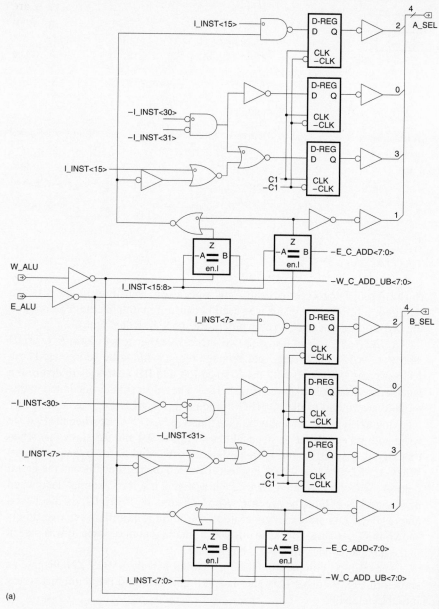

(a)

FIGURE 9.18   INST_PIPE
address comparators:
(a) schematic; (b) equality
gate

trols for the IO-REGS module that determine which operands are fed to the ALU.

The Instruction RAM has a register that stores the I-stage instruction *I_INST <31:0>* and outputs *R_INST<31:0>*. In INST_PIPE, the *R_INST<31:24>* bits are registered to form *E_ALU_OP<7:0>*, which is fed

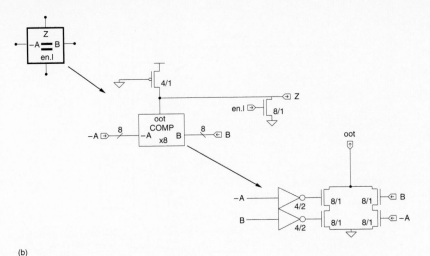

Figure 9.18 *(continued)*   (b)

to the ALU_DP to control the ALU operation. While only six bits are required, a full 8-bit register was used because it maintains the regularity of the layout (and does not take up any more space in a data path).

R_INST<23:16> is stored in two successive registers to form *E_C_ADD* and *W_C_ADD,* the E-stage and W-stage register-file write addresses. These addresses are compared with the R-stage RA and RB addresses to determine whether bypassing is required (Fig. 9.18a). The select signals generated in this module are routed to the IO-REGS module in the ALU-DP. For instance by referring to Fig. 9.18(a), it may be seen that *A_SEL<1>* is set when the E-stage write address equals the R-stage address. Figures 9.4 and 9.5 show that when this signal is asserted, the E-stage data is forwarded to the bypass register. When *A_SEL<3>* is asserted, it selects between the register-file and the literal field (*A_SEL<0>*). When *A_SEL<2>* is asserted, it selects external data; when deasserted, it selects between *E_RESULT* and *W_RESULT,* as discussed above. *B_SEL* is similarly generated. The logic required to achieve this is shown in Fig. 9.18(a). Because this was a small amount of logic, it was placed in the datapath under the metal2 bus signals.

The address comparators use an enabled pseudo-nMOS XNOR gate as shown in Fig. 9.18(b). This gate is small and fast, and it fits unobtrusively into the datapath.

9.2.3.6 Control Logic

The control-logic block is responsible for four main control functions:

- instruction decode.
- microstack address control.

- condition-code control.
- I/O control.

Figure 9.19 shows the instruction-decode logic. In addition to the *R-RET* signal the signals,

> *E-CALL,* a Call instruction
>
> *E-JFALSE,* a Jump False instruction
>
> *E-JTRUE,* a Jump True instruction
>
> *E-ALU,* an ALU instruction

are generated. The write enable for the register file (*W_WE*) is generated when there is a *W-ALU* signal while the *W_EXT_ADD_SEL<1>* signal is generated if an external register is not addressed. The *SHIFT-COUNT* signal is used by the shifter and is either derived from the op-code or an external register in the EXT_BUS_DP section.

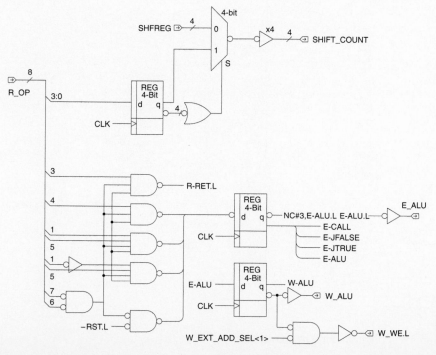

FIGURE 9.19   Control— instruction decode

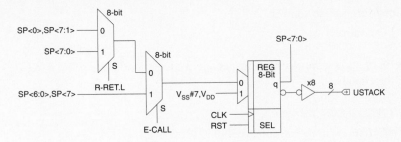

Figure 9.20 Control—stack address

The PC_DP-stack (microstack) address is generated using an 8-bit shift register (Fig. 9.20). The shift register is reset with a 1 in the LSB. When *E-CALL* is true, the register left-shifts, while when *R-RET* is true the register right-shifts. Thus this implements a pointer that points to the current return address. This could have been implemented as a 3-bit counter and a set of row decoders in the stack. However, this was deemed simpler and smaller for this size stack.

The condition-code logic is responsible for collecting the conditions, selecting the appropriate condition, and then controlling I_NEXT multiplexer in the PC_DP, as shown in Fig. 9.18. The registers are shown in Fig. 9.21, and the condition-code logic is shown in Fig. 9.22. In Fig. 9.21 the con-

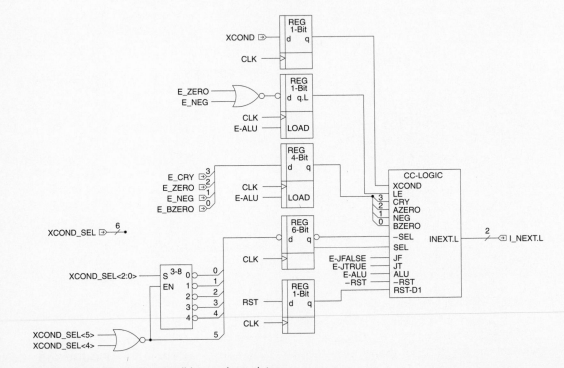

Figure 9.21 Control—condition code registers

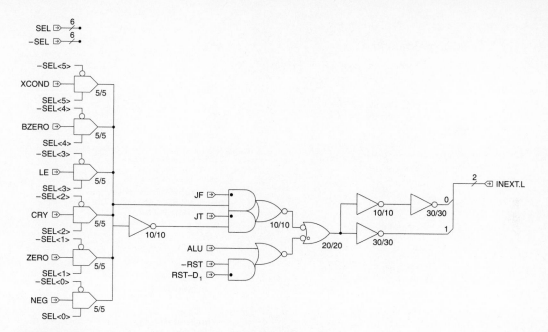

FIGURE 9.22 Control—condition code selection

ditions and the selection field (*XCOND_SEL*) are registered. Conditions that come from the ALU are

E_CRY—carry from adder

E_ZERO—adder zero

E_NEG—high bit of ALU

E_BZERO—Boolean zero.

In addition extra conditions are generated. For instance, *LE* (less-than-or-equal-to) is *E_ZERO* ORed with *E_NEG*. A condition may also be passed from an external source (*XCOND*).

Figure 9.22 shows the condition-code logic. First, a 6-input multiplexer selects the appropriate condition. This is then passed to a set of gates that control the *INEXT* signal dependent on *JF* (Jump False), *JT* (Jump True), *ALU* (an ALU instruction—no jump can be taken), and reset.

Finally, the I/O control logic is shown in Fig. 9.23. This controls the writing of registers and the tristating of busses in the EXT_BUS_DP. The *EXTOE* bus controls the enabling of external registers onto the external bus for reading by the ALU. The *EXTEN* signals control the loading of various external registers caused by writes by the ALU or the returning *SYSBUS* data.

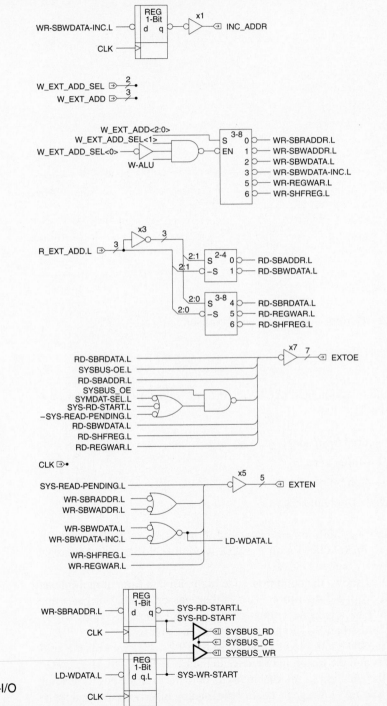

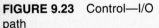

FIGURE 9.23 Control—I/O path

9.2.4 Layout

The microcontroller was constructed using three styles of layout, namely,

- datapath elements
- standard-cell layout
- memory layout

and of course, routing.

A datapath layout strategy was selected for this design, which is based on one its designers have used for a large Lisp microprocessor and which has proved useful on numerous other datapath chip layouts (Fig. 9.24a). Metal2 power busses run at the bottom and top of the cell. Sometimes these may be omitted and the power busses run vertically in metal1. Space is allowed for four metal2 busses to run through the cell (or five without metal2 power busses). The choice of four busses was originally made for

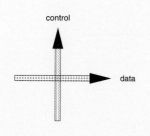

(a)

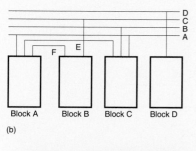

(b)

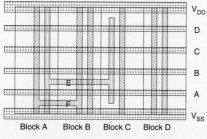

(c)

FIGURE 9.24 Datapath strategy

the microprocessor, and most datapaths have been shoehorned into this structure. A designer of course could choose more route-through busses but probably not fewer. Active circuitry is placed under the route-throughs in diffusion, polysilicon, and metal1. In a nonsilicided process, metal control lines are run vertically, while in a silicided process, the polysilicon may be used as control lines as long as the delay does not impact speed (be warned: in high-speed circuits silicide is really not adequate). More than four busses may be run through cells for short distances. For instance, adjacent connects between cells may be made in polysilicon or metal. Metal (or poly) can pass over an intervening cell if metal (or poly) transparency is provided in the cell. For instance, to achieve metal transparency, all vertical connections are made in polysilicon. This normally elongates the cell but may provide the right trade-off of increasing the length of one cell rather than adding an extra bus for every bit for the complete length of the datapath. Figures 9.24(b) and (c) show an example of 5 to 6 busses being routed. The height of the cell is determined by a combination of the metal2 pitch and the n-to-p spacing of the transistors. There is a maximum width of the horizon-

Table 9.1   Standard Cells

INVERTING FUNCTIONS	
INVERTER	
2-input NAND	2-input NOR
3-input NAND	3-input NOR
4-input NAND	

NONINVERTING FUNCTIONS	
2-input AND	2-input OR
3-input AND	3-input OR
4-input AND	BUFFER 1X, 2X, 4X, 8X drive

STORAGE ELEMENTS	
D REGISTER	D LATCH
D REG/CLEAR	D LATCH/CLEAR
D REG/SET	D LATCH/SET
D REG/MUXED	

OTHER LOGIC FUNCTIONS	
2-input MUX	Tristate-Buffer
4-input MUX	
XOR	
XNOR	
AND-OR-INVERT 221	
OR-AND-INVERT 221	

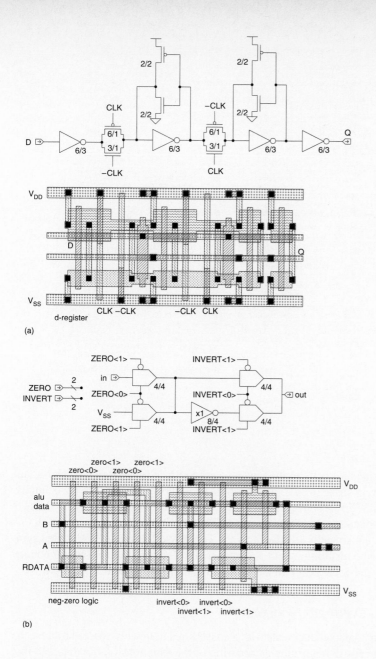

FIGURE 9.25 Datapath layouts: (a) register; (b) BUS-OP gate

tally oriented n- and p-transistor that determines the minimum height of the cell given a certain number of route-throughs. If wider transistors are required, then they must be rotated or composed of multiple smaller transistors. Figure 9.25 shows a number of examples of symbolic layouts of cells designed using this style of layout. Figure 9.25(a) shows the *D*-register cell, and Fig. 9.25(b) shows the BUS-OP cell used for conditionally negating and zeroing the *B* operand to the ALU. Figure 9.26 shows 4 bits of the Manchester adder (also Plate 10).

665

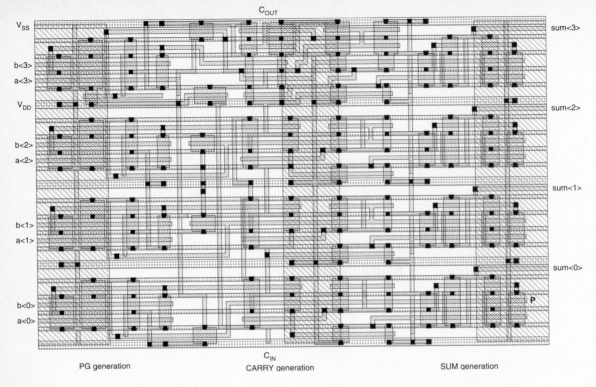

FIGURE 9.26 Manchester adder layout

The standard cells in the control section use a fairly standard two-level-metal routing strategy (see Fig. 6.29b). Metal1 power busses run horizontally, and polysilicon (silicide) runs vertically. Routing channels may be metal1/metal2 or metal1/poly. Table 9.1 gives a list of all the standard cells used.

The standard cells are placed by the TimberWolf program[1] and routed symbolically. The resulting layout is compacted to form a mask layout. Of course, any place-and-route program may be used to build the standard-cell logic.

9.2.4.1 Datapath Floorplans

When using datapath elements with a constrained number of route-throughs, the designer needs to determine an ordering of the functions on the datapath that does not require the number of feedthroughs to be exceeded. The ordering of functions on a datapath may be determined by permuting the order of the blocks and counting the number of connections between the blocks. This is done in a top-down manner from the highest level at which a single datapath is required. Common tricks employed include using tristate drivers as multiplexers (thus only requiring one common wire), replicating logic to reduce route-throughs, or using routing layers other than metal2 on adjacent or nearly adjacent modules. As noted previously, adjacent connections may be made in polysilicon or metal.

As an example of a typical floorplan, the ALU_DP module will be examined. The floorplan follows the schematic hierarchy, employing a layout block for the IO-REGS, ALU, and EXT_BUS_DP. These are placed adjacent to each other, as shown in Fig. 9.27. The register-file and literal ports enter on the left, and the system-address and data ports are accessed at the right of the ALU_DP. The control signals and clock enter at the bottom. From Fig. 9.4 it may also be seen that the *R_EXT_BUS* and *ALU_OUT* bus have to connect to all three modules. Just at this level it may be seen that IO-REGS is going to have at least six horizontal busses passing through it, so some thought has to be given to this potential problem.

Examining Fig. 9.28(a), the floorplan of the IO-REGS module may be seen. This module consists of 16 bit-slice sections vertically abutted on top of a control section. At this level *R_A_DATA, R_B_DATA, R_LITERAL,* and *W_C_DATA* enter on the left of the module. *A*, *B*, and *EXT_DATA* enter on the right. The bit-slice is shown in Fig. 9.28(b). It does not follow the schematic hierarchy in order to meet the four-bus constraint. With the arrangement shown in Fig. 9.28(b), no more than four busses are required in the datapath bit-slices.

Figure 9.29(a) shows the floorplan of the EXT_BUS_DP module. It consists of a number of 16-bit modules abutted to a common control section. Each vertical 16-bit section is associated with a register, which is shown in Fig. 9.7. Some registers (e.g., SHREG and REGWAR) have bus-throughs in their upper bits. Figure 9.29(b) shows the RDDATA register-block bit-slice, which is composed of a register and tristate buffer abutted. The controls run vertically.

Figure 9.30(a) shows a floorplan of the ALU. The ALU datapath is split according to the schematic hierarchy (Fig. 9.8), that is, an adder, a Boolean unit, and a shifter block. Figure 9.30(b) shows the floorplan of the adder. It consists of a 16-bit BUS-OP, Manchester adder, zero detect, and bus-driver horizontally abutted, with a control block at the bottom. The adder is, in turn, four 4-bit Manchester sections (Fig. 9.26). The Boolean-unit bit floorplan is shown in Fig. 9.30(c). Finally, the shifter floorplan is shown in Fig. 9.30(d). The shifter was

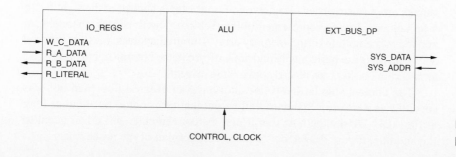

Figure 9.27 ALU_DP floorplan

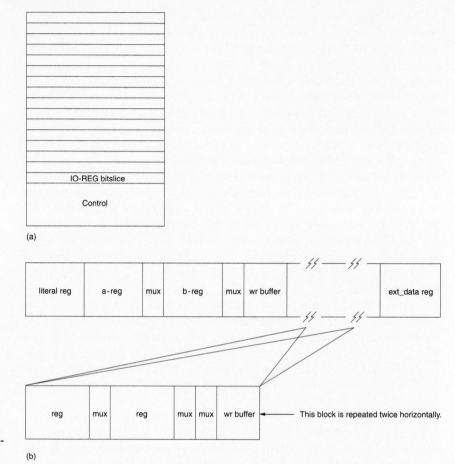

FIGURE 9.28 IO-REGS floorplan: (a) complete module; (b) 1-bit slice

designed so that one module shifted left and the same module, reflected in X and Y and abutted to a common central bus-buffer module, shifted right.

The other datapath modules are similarly constructed. As an example of a memory structure, the floorplan of the register file is shown in Fig. 9.31. The storage array (the register file shown in Fig. 9.14a) is abutted at the right by 16 4-bit column circuits, represented in Fig. 9.14(b). The three row-decoders are arrayed at the bottom of the memory array. Buffered address lines for the read and write ports run horizontally and are buffered in the bottom-right corner. Column-address buffers are placed above these drivers.

The control standard cell does not have an ordered floorplan but was specified as a two-row standard-cell layout that turns out to be about as long as the ALU_DP datapath so that it may be conveniently placed adjacent to this module. Figure 9.32 shows a possible floorplan of the processor.

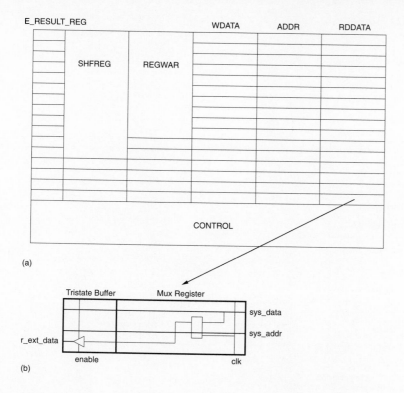

Figure 9.29   EXT_BUS_DP floorplan: (a) complete module; (b) an example bit

9.2.5   Functional Verification and Testing

A schematic for the design was first captured and the functionality of various modules checked. For instance, the adder, Boolean unit, and shifter had tests written for them. In addition for this design a C register transfer model of the processor was written as the RTL schematic was being developed. Once the overall processor was captured at the schematic and RTL levels, an assembler was written so that programs could be written using the instruction set of the machine to verify the functionality. A suite of tests were written to check each instruction class and type of operation. For instance, all arithmetic instructions were checked, a test was written to check the register file, the pass-around logic was checked, and the condition-code logic was checked. These vectors were enhanced to increase fault coverage by running them on a fault simulator.

Apart from these initial functional tests, timing simulations were run on the backannotated schematics to verify the performance of modules such as the adder. A timing analyzer was then used to report overall worst-case timing paths for the processor (and its peripherals) as a whole. This was first done with the layout incomplete, and as layouts were completed, the accu-

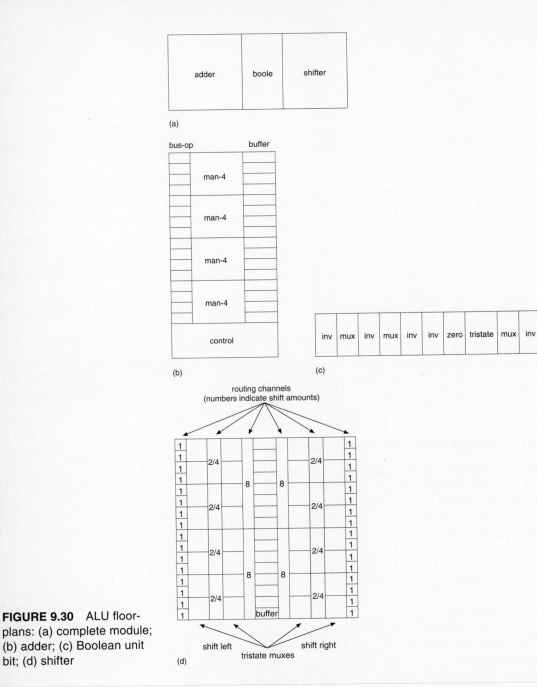

FIGURE 9.30 ALU floor-plans: (a) complete module; (b) adder; (c) Boolean unit bit; (d) shifter

rate backannotated schematics were used to achieve increasingly accurate timing analyses. Frequently, gates had to be moved between modules and between pipeline stages to achieve the desired speed.

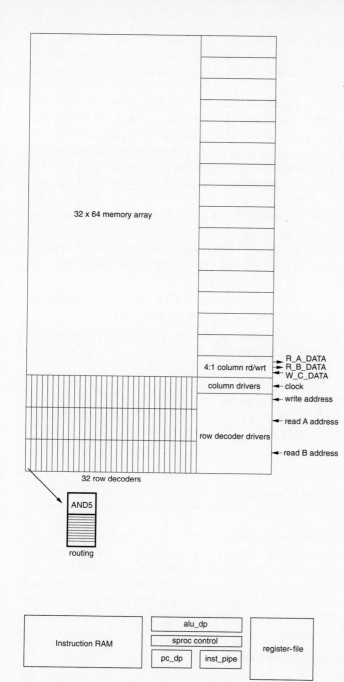

32 x 64 memory array

R_A_DATA
R_B_DATA
W_C_DATA
4:1 column rd/wrt
column drivers ← clock
← write address

row decoder drivers ← read A address

← read B address

32 row decoders

AND5

routing

Figure 9.31 Register-file floorplan

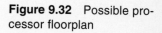

Instruction RAM

alu_dp
sproc control
pc_dp | inst_pipe

register-file

Figure 9.32 Possible processor floorplan

In addition, extensive tests were written for the C functional model. The vectors at the boundary of the symbol processor were captured and applied to the schematic version and the corresponding outputs checked for consistency.

The C RTL model was merged with other C-level models for the complete chip, and the system as a whole was tested by pumping data though the complete chip, and comparing the output values with a "golden" software model. This kind of testing verifies that the processor can be programmed to perform the required signal processing operations, but it does not give very much insight into whether all instructions work for all operands in the processor.

So in summary there were a number of levels of verification and testing:

- At the individual submodule level (functionality (C RTL simulator and transistor level simulator) and timing tested (transistor level simulator)).
- At the processor level (functionality (C RTL simulator) and timing tested (transistor level simulator and timing analyzer)).
- At the chip level (functionality and timing tested; C RTL simulator, module level timing analyzer used).

A good strategy to have is a set of regression tests that are run anytime a change is made to a module. These can be run in bottom-up mode so that bugs in low-level modules are found without having to find the bugs in time-consuming high-level simulations.

9.3   A TV Echo Canceller

The chip described in this section[2] (designed by A. Corry, B. Edwards, and N. Weste of TLW, and C. Greenberg of Philips Laboratories) is presented as an example of the kind of structure that lends itself to implementation as a regular structure. Because the chip is dominated by this regular structure, a high proportion of the engineering of the chip may be directed at the repeated structure, thereby providing effective use of the chip area.

9.3.1   Ghost Cancellation

This application is in the area of video-ghost cancellation.[3,4] Terrestrial and cable TV transmissions are subject to multiple-path propagation and transmission-line impedance discontinuities. Both of these imperfections in the communication channel lead to what is termed "ghosting," or echoes, which is familiar to most TV viewers.

Figure 9.33(a) shows a representation of the transmission path subject to ghosting. The signal at the receiver is given by

$$S_R = S_T(1 + H_G),$$

where

S_R = the received signal

S_T = the transmitted signal

H_G = the contribution of ghosts to the signal at the receiver.

If the receiver incorporates a filter structure, as shown in Fig. 9.33(b), then

$$S_C = \frac{S_R}{(1 + H_{GC})}$$

$$= S_T \frac{1 + H_G}{1 + H_{GC}},$$

where

S_C = the processed signal

H_{GC} = the response of the filter.

If $H_{GC} = H_G$, then the original signal is restored.

A typical ghosted signal is shown in the time domain in Fig. 9.34(a). It consists of the main signal and a set of ghosts that precede the main signal (pre-ghosts) and a set of ghosts that follow the main signal (post-ghosts). The filter structure shown in Fig. 9.34(b) may be used to cancel the ghosts shown in Fig. 9.34(a). A filter with characteristic H_{GC1} prior to the adder cancels the pre-ghosts, and the filter with characteristic H_{GC2} is used to cancel the post-ghosts. A delay line adds the main signal at the required point in time via a three input adder.

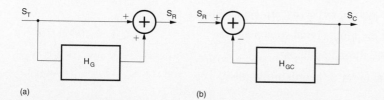

(a)   (b)

FIGURE 9.33 Ghosts: (a) transmission channel; (b) receive channel

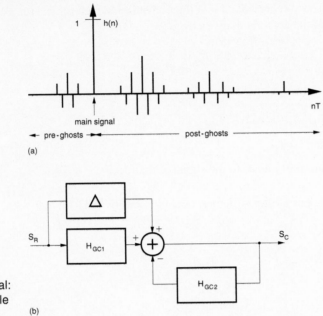

Figure 9.34 Ghosted signal: (a) time domain; (b) possible cancellation filter

The filters shown in Fig. 9.34(b) may be implemented with filters that are called Finite Impulse Response (FIR) and Infinite Impulse Response (IIR) filters. By sending a known "training signal" as part of a regular TV signal, the filter coefficients of the two filters may be determined and the ghosts canceled.[5,6]

9.3.2 FIR and IIR Filters

Mathematically, a sampled data FIR filter is represented by

$$y(t) = \sum_{i=0}^{n} h_i x(t), \qquad (9.1)$$

where

$y(t)$ = the filtered signal stream at time t

$x(t)$ = the input signal stream at time t

h_i = the filter coefficients

n = the order or length of the filter.

An IIR filter may be constructed by using an FIR filter with feedback.

There are a number of well-known forms for the sampled data FIR filter. Some are shown in Fig. 9.35. Figures 9.35(a) and (b) show two straightforward implementations of a 4-tap FIR filter. In the first implementation a delayed version of the incoming signal (X) is fed to a set of taps comprised of a multiplier and an adder. The multiplier multiplies the coefficient (H_n) by the delayed version of X. The adders are cascaded to form the final sum Y. In the second implementation, the delay is placed between adders in the taps. Each filter tap requires a register, an adder, and a multiplier. The precision of

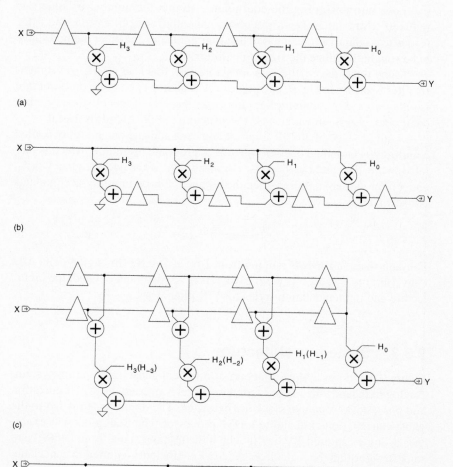

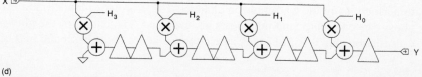

FIGURE 9.35 FIR architectures

the adder and multiplier depend on the precision of the coefficient, the length of the filter, and the desired precision or result. Implementations vary from just writing code on a microprocessor or DSP processor (for speech applications) to full hardware implementations. The multiplier can be a costly component in high throughput situations (such as video). If the filter response is fixed and the coefficients are therefore fixed, the multiplier may be simplified to contain only the product terms required. For instance, if an 8-bit coefficient is 00110011, then only four adders are required to perform this fixed multiplication. For 01111110, only two adders are required (the MSB and LSB are subtracted from the signal). This is known as Canonic Signed Digit representation.[7] In fixed FIR filters, a great deal of signal-processing expertise goes into designing the coefficients so that the number of adders is reduced. Unfortunately, in the ghost-cancellation application, the filter response is adaptive (i.e., it has to be programmable) and other methods have to be sought to reduce the size of the filters.

Many times the coefficients are symmetric, that is $H_{-n} = H_n$. A symmetric filter is shown in Fig. 9.35(c). The cost of adding two taps in this structure is only two adders, a multiplier, and two registers. Frequently, every other coefficient is zero, in which case the structure in Fig. 9.35(d) is useful.

A single stage of an FIR filter requires one add and one multiply (called a multiply-accumulate operation) at the appropriate resolution, which is dictated by the data and coefficients. For m-bit data, if the data frequency is f_D, and n filter taps are required, then the number of m-bit multiply-accumulates per second is

$$N_{mult\text{-}acc} = n \times f_D$$

For data sizes and speeds ranging from 1 bit at 100 Hz (for sigma-delta A/D converters) to 8 to 12 bits at 40 MHz (HDTV video) and beyond, the architecture and implementation styles for FIR filters vary widely.

9.3.3   System Architecture

Figure 9.36 shows a block diagram of a typical ghost cancellation system. Analog baseband video is converted to digital form, and the synchronization and phase-locked sample clock are extracted. The digital video is fed to the ghost cancellation chip and to a DSP processor. The DSP processor examines a single captured TV line in which the training signal is embedded and runs an algorithm that calculates the filter coefficients required to cancel any imperfections in the transmission channel. These are downloaded into the ghost-canceller chip, and the ghosts are canceled. The output of the ghost-cancellation chip is fed to a D/A converter and hence to the baseband port of a TV receiver.

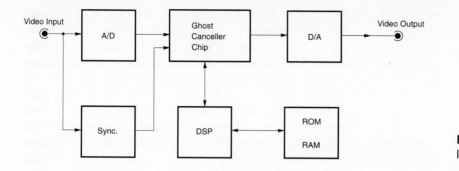

FIGURE 9.36   Ghost cancellation system

9.3.4   Chip Architecture

9.3.4.1   Filter Considerations

In this application, the following specifications are applicable:

Sampling rate	$= 4 \times \text{Fsc}$
	$= 14.32$ MHz
Ghost delays	$= -10 \ \mu s$ to $+40 \ \mu s$
	$= -150$ samples to $+580$ samples
Coefficient precision	$= 8$ bits
Sample precision	$= 8$ or 9 bits

Ideally then, a total of 730 8-bit filter taps would be required. Each tap would require an 8-bit multiply-accumulate block operating at 14.32 MHz. As an example, a 50 MHz DSP processor (16 to 32 bits), could deal with ~3.5 taps. Based on some early area estimates, it was decided that a full implementation of 730 taps was too large to provide economic die sizes in current technology. In addition, most of the filter taps would be zero, making this direct implementation very inefficient from a hardware resource point of view.

It was realized that if each tap could be positioned independently within the time domain, the filter would require one tap for each nonzero filter coefficient. The disadvantage of this approach is that a separate delay line is required for each filter tap. An intermediate approach is to group a number of successive taps into sections, with a delay line for each section. This yields reasonably effective use of the taps, since a single ghost generally requires many taps to cancel it. This architectural trade-off was verified at Philips Research Labs with a prototype system with simulated and real echo situations. As a result, the number of taps required was reduced to around 150. This was the first step toward creating a practical implementation.

The second optimization to improve chip size involves the filter-tap design because it dominates the chip area. As a starting point we might consider an 8×8 multiplier-accumulator based on the designs given in Chapter 8.

As a rough size estimate, an 8×8 Booth-recoded multiplier with an 18-bit accumulator would require 16 adders, 8 half-adders, and an 18-bit CPA adder. An alternative to a parallel multiplier is a word-serial multiplier. This requires four cycles to compute an 18-bit product for 8-bit coefficients and employs a single 18-bit adder. The former parallel implementation requires a cycle time of ~70 *ns*, while the latter requires a cycle time of 17.5 *ns*. While these decisions are justified within a few sentences, they actually required quite a bit of prototyping of various multipliers, which involved completing the layouts and simulating backannotated schematics with a transistor-level timing simulator. Simulation without the layout is usually not a good idea, because with compact structures, delays are normally due to self-loading and are hence highly affected by the actual layout.

9.3.4.2 Chip Overview

A block diagram of the chip is shown in Fig. 9.37. In essence it is the same structure as shown in Fig. 9.34(b). The main modules consist of the IIR and FIR filter sections, a 3-input adder, a main signal-variable-delay line, and various scaling and rounding logic.

   The filter sections were divided into nine sections of twenty contiguous filter taps each, resulting in a total of 180 taps. Each section was designed so it could be part of the FIR- or the IIR-filter response. In addition, each section may receive an input that has a fixed delay of 0–128 samples for a section used in the FIR filter or 0–448 samples for a section used in the IIR filter. The signal may then be delayed by a 0–63 stage programmable delay line.

   In the FIR and IIR filters, a section may be placed at an arbitrary temporal location by virtue of the programmable delay lines. Figure 9.38 shows an example where there are 40 FIR stages (Filter Blocks[8–7]) and 140 IIR stages (Filter Block[6–0]). The bold lines show the signal flow in this configuration.

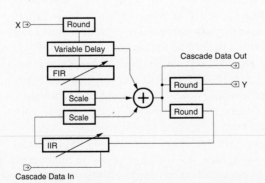

Figure 9.37 Ghost canceller chip architecture

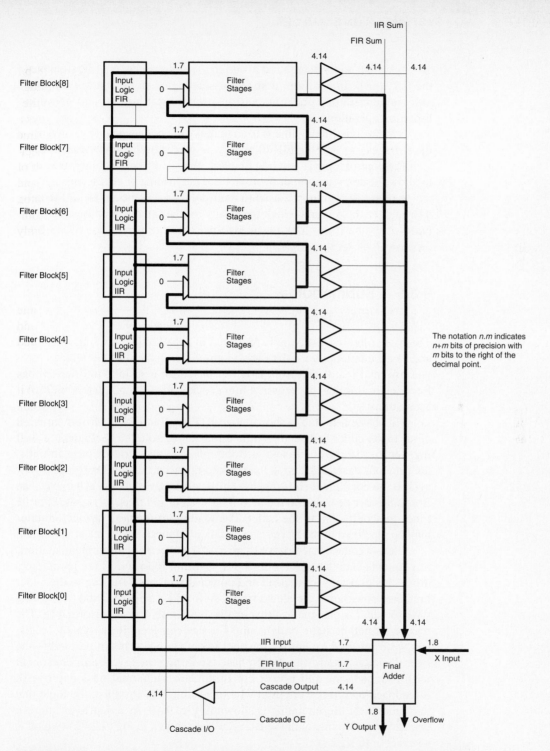

FIGURE 9.38 An example of use of the chip with 40 FIR stages and 140 IIR stages

A 3-input adder is required to add in the main signal, the FIR signal, and the IIR signal. In addition, it allows cascading of more than one chip, and associated circuitry performs various rounding, limiting, and overflow-detection operations.

A main signal-delay line is used to delay the main signal (by convention the strongest) so that the FIR filter may be placed before this signal in time.

The control logic is required to interface with an external microprocessor to allow the configuration of the filters and the loading of coefficients.

A phase-locked loop is used to multiply a $4 \times$ Fsc clock (14.32 MHz) by 4 to achieve the required clock frequency for the filter. A 2-phase clock generator uses the PLL clock or an external clock to generate a 57.28 MHz 2-phase clock for the chip.

9.3.5   Submodules

9.3.5.1   *Filter Taps*

The basic filter tap structure is reviewed in Fig. 9.39(a). An 8-bit resolution and an accumulation to 18 bits was required. A number of filter architectures were investigated, resulting in a tap based on a serial Modified-Booth Recoded multiplier that requires four clock cycles to complete a multiply-accumulate.

This basic filter tap is shown at an RTL level in Fig. 9.39(b). It consists of an 18-bit adder and accumulator, a register to hold the coefficient, a shift register to shift the coefficient, a Booth recoder, and a Booth gate. Initially, the previous cascaded sum is loaded into the accumulator. On the next four cycles, the coefficient shift register shifts left by two bits in each cycle. The Booth recoder operates on the incoming sample and passes 0, 1, –1, 2, or –2 times the coefficient to the adder. The loading of the next tap accumulator may be pipelined in the last cycle so that only four clock cycles are required.

Clearly some choices have to be made between circuit implementations of elements such as adders, registers, and multiplexers. Three parameters drove this design—size, speed and power dissipation. Turning to the adder first, size constraints dictated a ripple-style adder (Figs. 8.6 and 8.7b). This style of adder is the smallest that can be built with decent performance. The registers could be static or dynamic. For size considerations, dynamic registers were selected except for the coefficient storage. This in part led to the next decision which was the 2-phase clocking strategy. Again consistent with reduced size and power consumption, n-channel pass gates with p-feedback inverters were used where possible. For instance, the logic that implements the Booth gating is shown in Fig. 9.40. As a matter of interest, one can compare this with the BUS-OP gate shown in Fig. 9.25(b) (which is a subset of this gate).

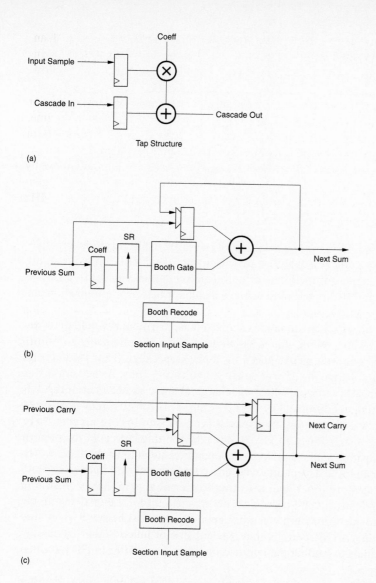

(a)

(b)

(c)

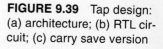

FIGURE 9.39 Tap design:
(a) architecture; (b) RTL circuit; (c) carry save version

Why was one design used in one instance and not in the other? In the
design in Fig. 9.40 area was of paramount importance, so the extra design
time spent in verifying the correct operation of this gate (it is a ratioed gate)
was well spent. In the processor example, the BUS-OP gate is one of many
different modules that had to be designed, so a "fire and forget" philoso-
phy—that is, an approach that is guaranteed to function correctly and whose
speed is verified during the normal course of transistor-level timing analy-

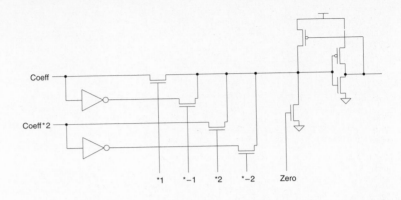

Figure 9.40 Booth gate

sis—was used. If no transistor-level tools were available, a gate-level implementation (XOR and AND) gate would be used. This microexample provides a model for the kinds of optimization that may occur in a CMOS VLSI design. It is important to achieve a balance between optimization and project completion time.

Power dissipation estimations showed that a 5-volt part would dissipate around 2–3 watts. While this is satisfactory for a ceramic package, something nearer one watt was required for a plastic package. Thus the decision was made to operate the chip at either 3.3 volts or 5 volts. As a result it was found that an 18-bit ripple adder was marginal at the slow corner of the 3.3-volt operating conditions (110°C, 3.0 volts). This forced a reevaluation of the ripple-adder strategy and was solved by employing a carry-save approach. This is shown in Fig. 9.39(c). While this increases the overall power dissipation somewhat, it allows operation at the lower voltage, which resulted in a power dissipation near one watt for a 3.3-volt operation. Both sum and carry are pipelined, resulting in a critical path that is close to a clock-to-Q delay, a worst-case adder delay, and a register setup delay. In the process in which this chip was implemented, this was below 5 *ns*. A side benefit of this architecture is that it is capable of much higher frequency operation, thus extending the application of the basic filter to HDTV video rates.

Figure 9.41 shows the final circuit diagram for a single bit of the filter tap. The complete filter tap consists of 18 of these bits with a clock driver and qualification block driving this datapath. An example of the symbolic layout of the datapath is shown in Fig. 9.42 (also Plate 11). Typical simulations that would be run on this datapath would include speed tests to ensure that the stage operated at 17.5 *ns* at all process corners. In addition, the power dissipation could be estimated to aid in overall chip planning and power distribution.

The basic floorplan of a section is shown in Fig. 9.43. A U-shaped structure, ten taps wide, is used so that the input and output from the section are

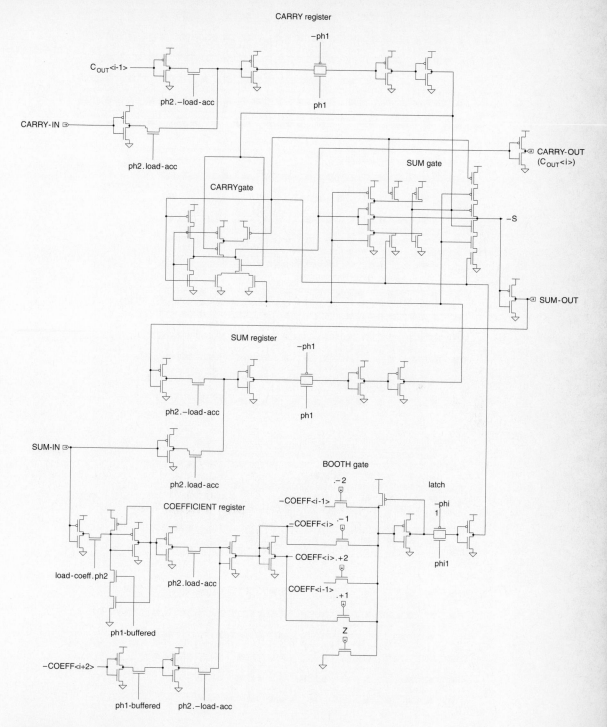

FIGURE 9.41 Tap circuit

683

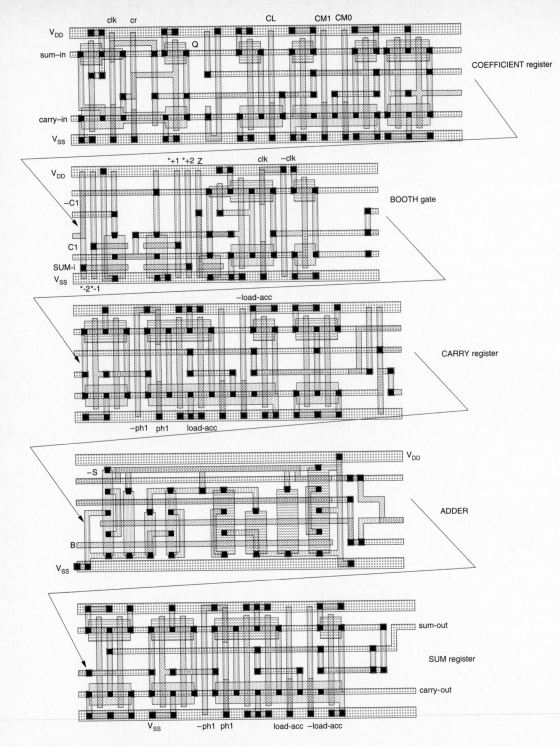

FIGURE 9.42 Representative tap layouts

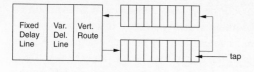

FIGURE 9.43   Section floorplan

available at a central routing trunk. Because the Booth-recoding logic is common to all taps, one recoder is used for all twenty taps in a section. An I/O section provides for driving the output of a section onto global IIR-sum and FIR-sum busses. Each section is connected to its neighbors via a cascade connection.

9.3.5.2   Delay Lines

There are at least two approaches to implementing the variable delay lines: a multiport RAM and shift register. Although the per-bit area of the RAM implementation is smaller than the shift register, the shift-register approach was chosen because it provided a more compact floorplan when considered in concert with the layout of the sections.

The programmable delay line for each section is placed at the left of each section (Fig. 9.43). To the left of the variable delay line, the fixed-delay block is situated (Fig. 9.43). This consists of the actual delay line and a distributed multiplexer. The distributed multiplexer consists of 11 8-bit busses that run vertically for the height of the nine sections. Each fixed delay takes its input from one of these busses and outputs to another. The multiplexer consists of tristate buffers placed under the vertical routing.

The shift-register bit is a simple 2-phase dynamic register and is shown in Fig. 8.70(b), and the variable delay line is shown in Fig. 8.70(a).

9.3.5.3   Phase-locked Loop- and Clock-generation

The Phase-Locked Loop is a charge-pump type PLL[8,9]; it was first introduced in Fig. 5.61 and is repeated in Fig. 9.44. A divide-by-4 counter in the feedback loop provides one $16 \times Fsc$ clock for a $4 \times Fsc$ input-clock. The phase detector measures the difference between the PLL VCO frequency (divided by 4 in this case) and the incoming reference frequency ($4 \times Fsc$—14.32 MHz). The phase

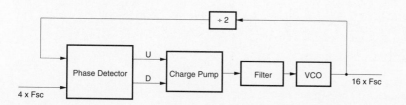

FIGURE 9.44   PLL

detector produces a sequence of UP or DOWN pulses, which are used to switch a charge pump. The charge pump either charges or discharges a capacitor with voltage or current pulses, as determined by the phase detector. A filter is used to limit the rate of change of the capacitor voltage, and the result is a slowly rising or falling voltage that depends on the frequency difference between the PLL VCO and the reference frequency. The VCO increases or decreases its frequency of operation as the control voltage is increased or decreased. Together, the components form a closed-loop feedback system whose phase and frequency response are determined by the characteristics of the charge pump, the filter, and the VCO.

The phase-detector implementation is shown in Fig. 9.45. This is a conventional phase detector implemented in static CMOS logic. If F_2 falls before F_1 falls, the signal DN is asserted. If F_1 falls before F_2 falls, UP is asserted.

The charge pump is shown in Fig. 9.46. The charge pump feeds pulses of current to a filter and capacitor, which has the effect of charging the capacitor up or down. This results in a voltage that rises or falls, and controls the VCO, either increasing or decreasing the frequency. The charge pump consists of a resistively biased constant current source (N_1, N_2, N_3) with n (N_4) and p (P_1) current mirror sources. These feed current mirror transistors N_5 and P_2. These in turn are switched to the filter via CMOS transmission gates, with complementary clocks balanced for equal delay. The current-source transistors are double the minimum length to improve the drain conductance.

Because the filter had to be monolithic (with no external components), an RC filter was constructed from MOS transistors. A CMOS transmission gate is used as a resistor, and MOS transistors are used as capacitors (Fig. 9.47a).

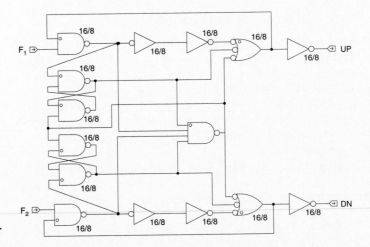

Figure 9.45   Phase detector

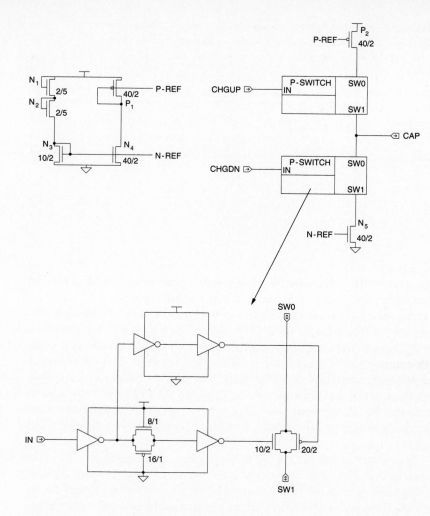

FIGURE 9.46 Charge pump

These are not ideal components but are adequate for this application. P-diffusion resistors might also be used for the resistor, while the capacitors can be poly-poly capacitors if a two-poly process is available. Alternatively, the filter can be implemented off-chip at the expense of a pad and of possible noise injection. The equivalent RC filter is shown in Fig. 9.47(b).

The VCO consists of 13 stages of a current-starved oscillator with a buffered output (Fig. 9.48). This was chosen because it had a wide range of operation and was verified to operate correctly over all process corners. The main parameter of interest for the VCO is the frequency range and the oscillator sensitivity in terms of MHz/volts. The transistors were made larger than minimum to reduce the effects of geometry biases and to swamp the routing

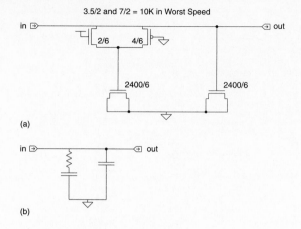

3.5/2 and 7/2 = 10K in Worst Speed

(a)

(b)

Figure 9.47 Filter

capacitance (thereby ensuring high-speed performance). The n- and p-transistors are sized to achieve equal rise and fall time.

The filter values and the VCO sensitivity vary widely over process corners. The lockup times for the VCO have a much longer time constant than the VCO itself. Thus it is time consuming (and possibly impractical) to completely verify the VCO using circuit simulation (SPICE). In order to ensure that the chosen parameters would work over the process corners, a simple analytical model was constructed that modeled the PLL. This was used to verify the VCO lock times. Various of these functional simulations were spot-checked against full circuit simulations for accuracy in the analytical model.

The output of the PLL is fed to the 2-phase clock generator, which is shown in Fig. 9.49. This consists of a conventional cross-coupled 2-phase clock generator with the final driver transistors being 4000μ (p) and 2000μ (n) wide. The clock load was 100pF per phase. A multiplexer allowed the on-chip PLL to be bypassed. This was done as a system requirement and also as a safety mechanism in case the analog PLL was inadequate (it wasn't).

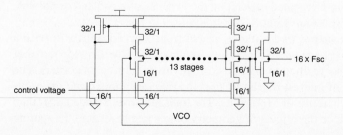

Figure 9.48 VCO

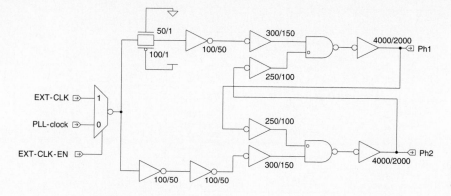

FIGURE 9.49 Clock generator

9.3.5.4 Peripheral Processing

A number of support modules surround the filters. These include

- a microprocessor interface.
- the final filter adder.
- an I/O block.

The microprocessor block is used to load the filter-block coefficients, configure the filter sections, and monitor filter overflow. The I/O block is responsible for converting the filter output to signed or unsigned form and for managing the I/O communication busses for multiple chips. The final adder combines the main pulse of the FIR with the output of the FIR and IIR filters. In addition it rounds the filter output to 8 or 9 bits and can scale the outputs or the FIR and IIR filters by factors of 2. In many ways it is this kind of "glue" and support logic that can dominate the design after the core signal processing section has been designed. One must always remember to count in the "control" portions of a design when estimating design times.

9.3.6 Power Distribution

At 5-volt operation, the chip dissipates around 2.5 watts and draws 500 mA. On average this is 55 mA per section with the peak current being much higher than this. To both achieve metal migration requirements and reduce switching noise, each section was effectively provided with a power and ground pad, shown in Fig. 9.50(a). These connections were routed in metal3. In addition, to reduce power-supply noise on-chip bypass capacitors were placed under the power lines. These consist of large gate-area n-transistors, with their gates connected to V_{DD} and source and drains connected to V_{SS} (substrate).

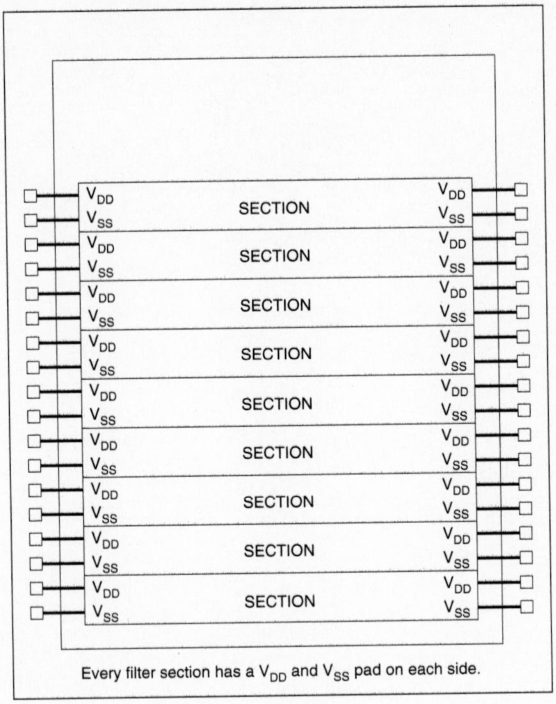

Figure 9.50 Power distribution to chip

Every filter section has a V_{DD} and V_{SS} pad on each side.

9.3.7 Chip Floorplan

The overall chip floorplan is shown in Fig. 9.51(a) (also Plate 12). The sections occupy the majority of the chip. The final datapath and I/O sections are at the top of the design. The clock driver is placed in the pad ring at the top left, while the PLL is placed in the pad ring at the top right. The PLL has its own 5-volt supply.

Figure 9.51(b) shows the bond diagram for the chip for a 144-pin plastic PGA package. This is a step of the design process in which the placement of

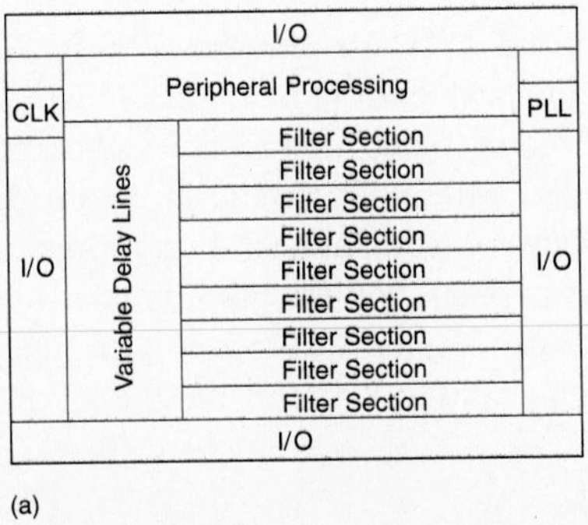

Figure 9.51 Ghost chip: (a) floorplan; (b) bond diagram

(a)

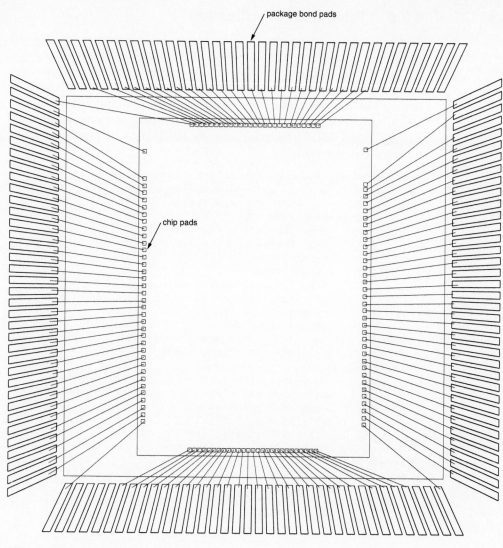

package bond pads

chip pads

FIGURE 9.51 *(continued)*

the I/O pads of the chip and the package bond leads are checked for bond-lead compliance. Usually, this means checking for correct bond angles (the angle of the bond wire with respect to the center of the package, and checking that bonds do not cross each other).

9.3.8   Testing and Verification

Similarly to the testing of the RISC microcontroller, testing of this chip was completed at various levels. Low-level SPICE simulations were completed for key modules, such as the adder bit and registers. Timing simulation was completed at the section level with backannotated schematics. Full chip simulation was completed at the gate level using a unit-delay logic simulator. Functional models were also written for a filter tap, and simulations were completed at a mixed transistor/functional level. The Lisp functional model for a tap is shown below.

```
(deffunctional-model csa-filter-stage-dp
 :inputs (("sum-in<17:0>" :capacitance .11)
          ("carry-in<17:0>" :capacitance .09)
          (Z :capacitance .26)
          (*+1 :capacitance .26)
          (*+2 :capacitance .26)
          (*-1 :capacitance .26)
          (*-2 :capacitance .26)
          (load-coeff :capacitance .08)
          (-load-acc :capacitance .18)
          (ph1 :capacitance .18)
          (ph2 :capacitance .26))
 :outputs (("sum-out<17:0>" :capacitance .08)
           ("carry-out<17:0>" :capacitance .11))
 :local-state ((*-2or*-1 :initform 'X)
               (creg-master :initform '0) (creg :initform '0)
               (sreg-master :initform '0) (sreg :initform '0)
               (recode-reg :initform '0)
               (coeff-reg :initform '0)
               (coeff-shifter :initform '0)
               (coeff-shifter-reg :initform '0))
 :model (progn
; most outputs advance on phase1
            (when (eq ph1 1)
            (setq *-2or*-1 (sim-or *-1 *-2)
              creg creg-master
              sreg sreg-master
              coeff-shifter (if (eq coeff-shifter-reg 'X) 'X
                (lsh coeff-shifter-reg 2)))
;; recoding the coefficient
            (setq recode-reg
              (cond
                ((eq Z 1)   0)
                ((eq coeff-shifter-reg 'X) 'X)
                ((eq *+1 1) coeff-shifter-reg)
                ((eq *-1 1) (logxor coeff-shifter-reg #o777777))
                ((eq *+2 1) (lsh coeff-shifter-reg 1))
                ((eq *-2 1) (logxor (lsh coeff-shifter-reg 1) #o777777))
                (t 'X))))
;; compute sum/carry output bits of the adder
```

```
      (if (or (eq recode-reg 'X)(eq sreg 'X)
              (eq creg 'X)(eq *-2or*-1 'X))
          (setq sum-out   'X
              carry-out 'X)
          (loop with sum = 0 and carry = 0
              for i below 18.
              for coeff-bit = (ldb (byte 1 i) recode-reg)
              for sum-bit = (ldb (byte 1 i) sreg)
              for carry-bit first *-2or*-1 then
                  (ldb (byte 1 (1- i)) creg)
              for adder = (+ coeff-bit sum-bit carry-bit)
              do (setq sum (dpb (logand adder 1) (byte 1 i) sum)
                    carry (dpb (lsh adder -1) (byte 1 i) carry))
              finally (setq sum-out sum
              carry-out carry)))
;; phase 2 clock asserted
          (when (eq ph2 1)
;; coefficient load cycle
              (when (eq load-coeff 1)
                  (setq coeff-reg sum-in))
;; accumulator-load cycle
              (case -load-acc
;reload coefficient
                  (0 (setq coeff-shifter-reg coeff-reg
                      creg-master carry-in
                      sreg-master sum-in))
;shift coefficient
                  (1 (setq coeff-shifter-reg coeff-shifter
                      creg-master carry-out
                      sreg-master sum-out))))
              )
 :delays
      ((ph1↑ sum-out↑ :delay 11.4 :driver-size 12/1)
       (ph1↑ sum-out↓ :delay 11.4 :driver-size 8/1)
       (ph1↑ carry-out↑ :delay 11.4 :driver-size 12/1)
       (ph1↑ carry-out↓ :delay 11.4 :driver-size 8/1))
:timing-constraints ((-load-acc ph2↑ :setup 1.0 :hold 3.0)
          (Z  ph1↑ :setup 8.4 :hold 3.0)
          (*+1  ph1↑ :setup 8.4 :hold 3.0)
          (*-1  ph1↑ :setup 8.4 :hold 3.0)
          (*+2  ph1↑ :setup 8.4 :hold 3.0)
          (*-2  ph1↑ :setup 8.4 :hold 3.0)
          (sum-in  ph2↑ :setup -3.0 :hold 3.0)
          (carry-in  ph2↑ :setup -3.0 :hold 3.0)
          (load-coeff ph↑ :setup 3.25)) ;write pulse setup
)
```

At the start of the model, inputs and outputs are defined by the keywords
: inputs and :outputs. Each input is denoted by a name and a load
capacitance, while each output has a name, a capacitive load, and a drive
strength. The :local-state keyword denotes the internal registers in the
model and their initialization values. The functionality is specified within the

: `model` section. Finally, the :`delay` section specifies important signal-to-signal delays, while the :`timing-constraints` section specifies :`setup` and :`hold` times for the registers in the design. The timing values are derived from SPICE or timing simulations. The model above is useful for both simulation and timing analysis. In the latter case, the modularity of the tap is well specified by the I/O and timing specifications that are included. Extensive timing analysis was completed, using backannotated schematics with parasitics extracted from compacted mask layouts.

The critical path in this design consisted of the clock generator and distributed skew in the clock lines (remember that the design also had to operate at 3.3 volts).

9.3.9 Summary

This section has presented an example of a CMOS VLSI design that employs a high degree of regularity. Accordingly, the design style changes from the processor design, with much more emphasis being placed on circuit and layout design for the key replicated cells. A key talent in CMOS-system design is knowing when and where to optimize. For example, had the ghost-cancellation chip been implemented as a standard-cell or gate array, it would have been from 4 to 10 times larger, resulting in a chip cost that would have made the implementation economically infeasible.

9.4 A 6-bit Flash A/D[†]

9.4.1 Introduction

This final example has been included as an example of a simple analog circuit that is almost digital (hence its inclusion in a digital CMOS text). The circuit is a 6-bit A/D converter implemented as a "flash" converter. This provides for a very fast converter at the expense of area. Although limited to about 8 bits of resolution, as CMOS circuits are becoming smaller over time, this circuit architecture is also becoming smaller, and faster. This style of converter or variants is of particular use at video-sample rates.

Figure 9.52 shows the basic architecture. An analog input is presented to a sample-and-hold circuit, which feeds one input of 2^N comparators, where N is the desired digital precision. A clock input samples the input and strobes the sample-and-hold. The other input to the comparators is connected to a resistor string connecting a reference voltage ($+V_{REF}$ and $-V_{REF}$). For a given input voltage, after the sampling process and the comparators have

---

[†]Designed by N. Weste.

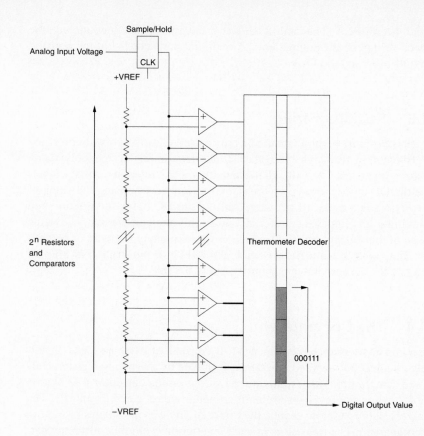

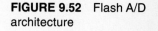

FIGURE 9.52   Flash A/D architecture

switched, the output of the comparators is a string of 1s where the highest 1 represents the highest comparator that switched. This is represented in the diagram by a gray bar. A decoder may be used to convert this "thermometer code" to an N-bit number.

9.4.2   Basic Architecture

A CMOS implementation of a 6-bit flash A/D converter is shown in Fig. 9.53. It consists of a polysilicon resistor string, 64 sampling comparators, 64 registers, and a thermometer decoder that consists of 64 3-input NOR gates

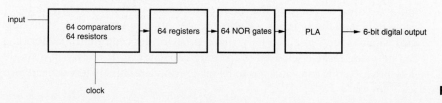

FIGURE 9.53   Flash circuit

and a 64-term PLA. The analog parts that have to be designed include the resistor string and the comparators. The digital parts consist of the register, the NOR gate, and the PLA.

9.4.3   Resistor String

The resistors may be implemented as diffusion resistors, polysilicon resistors, or metal in very high speed flash converters. In this design, polysilicon resistors were used. The value of the resistor and the reference voltage determine the DC current drawn by the reference ladder. Choosing the value of the resistors is a trade-off between limiting the DC power dissipation and achieving a low impedance reference to supply the comparator. The value chosen in this design was 20 Ω, which is approximately 1 square of polysilicon. This yields a nominal resistance of 1280 Ω for the string. For a 1-volt reference the current drawn is approximately 0.8 mA.

9.4.4   The Comparator

The comparator is shown in Fig. 9.54. It consists of two cascaded, capacitively coupled, auto-zeroed inverters[10] followed by a dynamic register. During the sample time, the input value is stored on the capacitor C_1 via pass gate P_1, while the inverters are auto-zeroed via pass gates P_3 and P_4. The auto-zero step is used to reduce the effect of any offset voltages present in the comparator. The transistors in each inverter have slightly different characteristics (such as threshold voltage and beta). This in turn leads to slightly different transfer characteristics. If the inverters were not auto-zeroed, the comparators would switch at slightly different voltages. The auto-zero step reduces this offset to below the precision of the converter.

When *sample* is false, the reference input is connected to the capacitor, thereby transferring charge to or from the capacitor. This causes a voltage change at the input of the first comparator, which causes the output of the comparator to rise or fall by an amount proportional to the gain of the inverter (and the capacitive divider formed by C_1 and the input capacitance of the inverter). This signal is further amplified by the second comparator. This is then passed to a register.

As stated, the gain of the comparator depends on the gain of the inverter and the ratio of $C_1(C_2)$ to the input capacitance of the inverter. The input capacitance of the inverter is approximately two minimum-size transistor gate capacitances. In a two-polysilicon process, the coupling capacitor (C_1) can be made large compared to the gate capacitance. However in a single poly process, this capacitor has to be made from a relatively low capacitance poly-metal-metal2 sandwich (poly and metal2 connected to one plate, metal to the other). In this design this capacitor was made .05pF. The combined gain of the capacitor and an inverter was around 10 (the capacitor value used roughly halves the gain). Thus two comparators result in a gain of about 100. Figure 9.54(b) shows the frequency response of two capacitively coupled comparators. This would indicate a sensitivity of around 10mV. In practice the converter demonstrated 6-bit resolution with an input reference of 600mV, which agrees with these simulated results.

The transistor size and n/p β ratio of the inverter transistors affect the DC-transfer characteristic (gain, linearity, and dynamic range) and the AC performance (bandwidth and phase response). A design often involves mak-

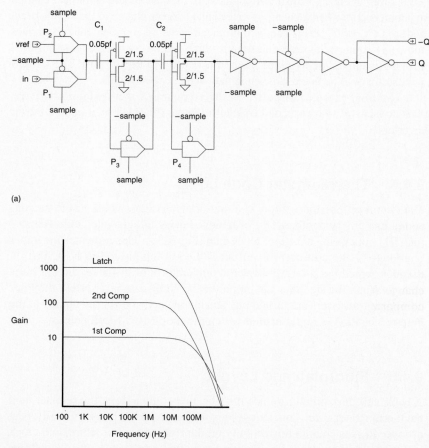

(a)

(b)

FIGURE 9.54 Comparator: (a) circuit; (b) frequency response

ing a compromise between conflicting requirements. For instance, the inverter gain is improved by using transistors with slightly longer than minimum length at the expense of input-loading capacitance. Increasing the width of the p with respect to the n moves the quiescent point ($V_{in} = V_{out}$) of the DC-transfer characteristic toward midrail but increases the input capacitance, which reduces the gain. The main point to note here is that with this extremely simple analog circuit, a relatively large effort goes into the design of one inverter compared with that of a logic inverter.

The maximum sample rate of the converter is dependent on the bandwidth of the comparators (and reference resistor string) and the maximum clock frequency of the digital circuitry. A SPICE-frequency analysis (Fig. 9.54b) revealed that the comparator used here (for the process used) had a 3dB bandwidth of 10 MHz. Measured devices were successfully operated at 10 MHz with supply voltages of both 3 and 5 volts.

Complementary transmission gates are used as analog switches. Charge injection occurs when a changing gate signal couples charge to the source/drain node via C_{gd} or C_{gs}. This can be minimized by using complementary switches and delay-balanced clocks. The sample clock and its complement are buffered with an equalized delay clock generator (shown in Fig. 5.52b). The CMOS switches have to be made large enough to achieve speed goals.

The final piece of circuitry in the comparator is a dynamic register. This consists of two cascaded tristate inverters followed by a pair of buffer inverters.

9.4.5   Thermometer Code Logic

The output of the comparators is a thermometer code, which has to be converted to a binary number. This is achieved using a logic gate, which checks for a 011 code using a 3-input NOR gate (Fig. 9.55). This indicates the upper boundary of the comparator output. This is then passed to a PLA-style decoder, which has a term for every comparator. For instance, the 8th bit is shown. When the signal at the output of the NOR gate is asserted, the PLA NOR gate transistors are turned on, causing the code 000111 to appear at the output. The PLA is implemented with a pseudo-nMOS NOR gates.

9.4.6   Floorplan and Layout

A basic cell horizontally joins a resistor, two comparators, the register, and the thermometer gate. This structure is then arrayed 64 times vertically. The decoder PLA is abutted horizontally on the right of this structure, and clock and I/O buffers are placed on the top and bottom of the structure. This floor-

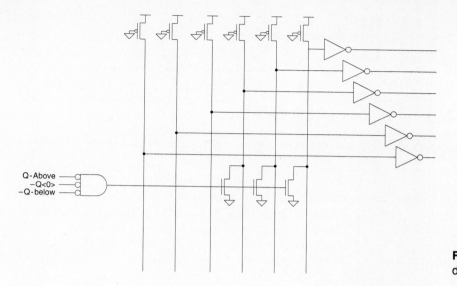

FIGURE 9.55 Thermometer decode logic

plan is shown in Fig. 9.56. A generator was written to automatically generate the layout for a given resolution. A chip micrograph is shown in Plate 13.

A portion of a 2-bit section of the converter layout is shown in Fig. 9.57. The resistor string may be seen at the bottom, with the resistors placed vertically. The input/reference switches are above the resistors. The comparator capacitor is the long structure in the center of the layout. The capacitor is

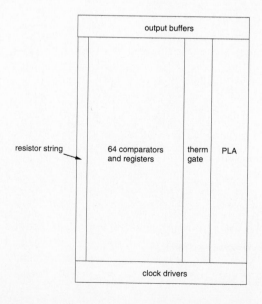

FIGURE 9.56 Floorplan of A/D

clock and−clock run vertically

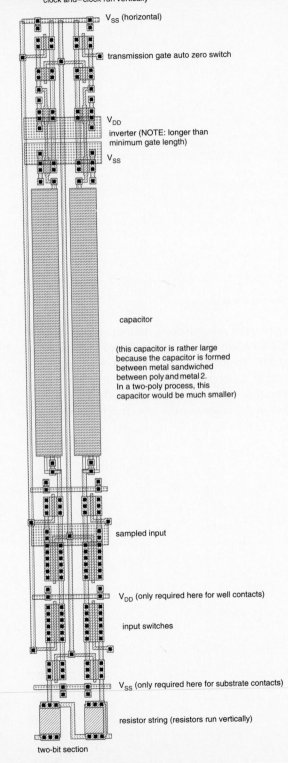

V_SS (horizontal)

transmission gate auto zero switch

V_DD

inverter (NOTE: longer than minimum gate length)

V_SS

capacitor

(this capacitor is rather large because the capacitor is formed between metal sandwiched between poly and metal 2. In a two-poly process, this capacitor would be much smaller)

sampled input

V_DD (only required here for well contacts)

input switches

V_SS (only required here for substrate contacts)

resistor string (resistors run vertically)

FIGURE 9.57 Partial cell layout for A/D

two-bit section

700

about 100μ long. This would be much smaller in a double polysilicon process. Above the capacitor is the comparator inverter and auto-zero switch. Another comparator, register, NOR gate, and PLA gate are placed above this section in the converter.

9.4.7 Summary

A modest implementation of a flash A/D converter has been presented in order to illustrate some of the issues that are addressed as we leave the digital world. There is a wealth of literature on A/D conversion, and the reader is encouraged to investigate these sources if this subject is of interest. For a more efficient 8-bit A/D converter employing a similar technique see Dingwall and Zazzu[11] and Tsukada et al.[12]

9.5   Summary

This chapter has presented three case examples of CMOS designs. The first described a contemporary RISC microcontroller with a mix of datapath, memory, and control logic. This design could be implemented in a wide range of CMOS logic styles and design methods. The second described a high-performance signal-processing circuit that is representative of video-rate architectures. To achieve commercial viability this design required custom layout and innovative architecture and circuit design. The final design featured a straightforward flash A/D converter representing the interface between the analog world and the CMOS digital domain. In this design the focus extends to basically one inverter which is extensively simulated. From these examples it may be seen that the more complex a system becomes, the less time is available to spend on low-level details; notwithstanding, it is possible to create denser, faster designs in a given technology if the appropriate amount of design effort is invested. In these days of short product cycles, time to market is almost always the dominant concern. This leads to the requirement for short design times. This in turn is achieved by the use of highly automated design systems, the use of libraries, and the reuse of other components of interest.

9.6   Exercises

1. In order to achieve a short cycle time, the RISC microcontroller in Section 9.2 used four pipeline stages. Redesign the processor to run in a single (albeit longer) cycle. How does each module change? What simplifications may be made?

2. If the critical path of the processor is that shown in Section 9.2.3.1.3, what changes to the architecture would you make to improve this speed?

3. Sketch out the logic design for a multiplication-function block that can be attached to the processor in the EXT_BUS_DP module. (Could you implement multiplication on the processor as described?)

4. Examine alternative architectures to implement the FIR/IIR filters in the ghost cancellation chip (i.e., 8×8 multiplier running 4 taps).

9.7 References

1. C. Sechen and A. Sangiovanni-Vincentelli, "TimberWolf3.2: a new standard cell placement and global routing package," *Proceedings of the 23rd Design Automation Conference,* 1986, Las Vegas, Nev., pp. 432–439.

2. Bruce Edwards, Alan Corry, Neil Weste, and Craig Greenberg, "A Single Chip Ghost Canceller," *Proceedings of the IEEE 1992, Custom Integrated Circuits Conference, May 1992,* pp. 26.5.1–26.5.4.

3. Walter Ciciora, Gary Sgrignoli, and William Thomas, "A tutorial on ghost cancelling in television systems," *IEEE Transactions on Consumer Electronics,* vol. CE-25, Feb. 1979, pp. 9–44.

4. Stephen Herman, "The development of commercial echo cancellers for television," *National Association of Broadcasters, 1991 Broadcast Engineering Conference Proceedings,* pp. 102–106.

5. David Koo, "Developing a new class of high energy ghost cancellation reference signals," *International Conference on Consumer Electronics, Digest of Technical Papers,* Jun. 1992, Boston, Mass., pp. 76–77.

6. Craig B. Greenberg, "Ghost cancellation system for high energy GCE," *International Conference on Consumer Electronics Digest of Technical Papers,* June 1992, Boston, Mass. pp. 78–79.

7. H. Samueli, "An improved search algorithm for the optimization of the FIR filter coefficients represented by a canonic signed digit code," *IEEE Transactions on Circuits and Systems,* vol. CAS-34, Sept. 1987, pp. 1192–1202.

8. F. A. Gardner, "Charge-pump phase-locked loops," *IEEE Transactions in Communications,* vol. COM-28, Nov. 1980, pp. 1849–1858.

9. Deog-Kyoon Jeong, Gaetano Borriello, David A. Hodges, and Randy H. Katz, "Design of PLL-based clock generation circuits," *IEEE Journal of Solid State Circuits,* vol. SC-22, no. 2, Apr. 1987, pp. 255–261.

10. Andrew G. F. Dingwall, "Monolithic expandable 6 bit 20 MHz CMOS/SOS A/D converter," *IEEE JSSC,* vol. SC-14, no. 6, Dec. 1979, pp. 926–932.

11. Andrew G. F. Dingwall and Victor Zazzu, "An 8-MHz CMOS subranging 8-bit A/D converter, *IEEE JSSC,* vol. SC-20, no. 6, Dec. 1985, pp. 1138–1143.

12. Toshiro Tsukada, Yuuischi Nakatani, Eiki Imaizumi, Yoshitomi Toba, and Seiichi Ueda, "CMOS 8b 25MHz flash ADC," *Proceedings IEEE International Solid State Circuits Conference,* New York, N.Y., Feb. 1985, pp. 34–35.

INDEX